Electron-Emissive Materials, Vacuum Microelectronics and Flat-Panel Displays

MATERIALS RESEARCH SOCIETY
SYMPOSIUM PROCEEDINGS VOLUME 621

Electron-Emissive Materials, Vacuum Microelectronics and Flat-Panel Displays

Symposium held April 25–27, 2000, San Francisco, California, U.S.A.

EDITORS:

Kevin L. Jensen
Naval Research Laboratory
Washington, D.C., U.S.A.

Robert J. Nemanich
North Carolina State University
Raleigh, North Carolina, U.S.A.

Paul Holloway
University of Florida
Gainesville, Florida, U.S.A.

Troy Trottier
LumiLeds
San Jose, California, U.S.A.

William Mackie
Linfield Research Institute
McMinnville, Oregon, U.S.A.

Dorota Temple
MCNC
Research Triangle Park, North Carolina, U.S.A.

Junji Itoh
Electrotechnical Laboratory
Umezono, Tsukuba, Japan

Materials Research Society
Warrendale, Pennsylvania

CAMBRIDGE UNIVERSITY PRESS
Cambridge, New York, Melbourne, Madrid, Cape Town,
Singapore, São Paulo, Delhi, Mexico City

Cambridge University Press
32 Avenue of the Americas, New York NY 10013-2473, USA

Published in the United States of America by Cambridge University Press, New York

www.cambridge.org
Information on this title: www.cambridge.org/9781107413061

Materials Research Society
506 Keystone Drive, Warrendale, PA 15086
http://www.mrs.org

First published 2001
First paperback edition 2013

Single article reprints from this publication are available through
University Microfilms Inc., 300 North Zeeb Road, Ann Arbor, MI 48106

CODEN: MRSPDH

ISBN 978-1-107-41306-1 Paperback

This work was supported in part by the Office of Naval Research under Grant Number ONR:
N00014-00-1-0251. The United States Government has a royalty-free license throughout the
world in all copyrightable material contained herein.

CONTENTS

SYMPOSIUM R: ELECTRON EMISSIVE MATERIALS AND VACUUM MICROELECTRONICS

FIELD EMISSION AND DISPLAY APPLICATIONS

FUNDAMENTAL PROCESSES

*Invited Paper

FIELD EMITTER ARRAYS

POSTER SESSION:
FEAs, NANOTUBES, AND
OTHER ELECTRON SOURCES

*Invited Paper

DIAMOND AND WIDE BANDGAP

CARBON NANOTUBES AND NANOSTRUCTURES

SYMPOSIUM Q: FLAT-PANEL DISPLAY MATERIALS

FIELD EMISSION AND DISPLAY APPLICATIONS

ELECTROLUMINESCENCE

POSTER SESSION:
FPD

TFT I: DEPOSITION
AND CRYSTALLIZATION

TFT II: LASER CRYSTALLIZATION

*Invited Paper

TFT III: DEVICES

PREFACE

Electron emissive materials are poised to improve device performance when physical size, weight, power consumption, beam current, and/or high pulse repetition frequencies are an important issue for competitive performance of a particular technology. Cold cathodes, (e.g., field emitter arrays (FEAs), diamond and other wide bandgap materials, carbon nanotubes (CNTs), and other exotic examples) offer potential improvements over conventional thermionic sources in one or more of these properties: instant ON/OFF performance, high brightness, high current density, large transconductance to capacitance ratio, or low voltage operation characteristics. Many extraordinary and exciting technologies are developing, some well known, and some new to the electron sources scene. Flat panel displays are undoubtedly the largest commercial endeavor to which cold cathode source technology is directed. Advanced microwave power tubes (used, e.g, in radar and communications) are one of the most technically challenging applications pursued. Other important applications with considerable impact include electron sources for micro-propulsion systems (e.g., Hall Thrusters) and tethers for satellites. Small and stable sources are needed for multi-beam lithography, electron microscopes, and other high-tech tools and applications. Symposium R, "Electron Emissive Materials and Vacuum Microelectronics," and Symposium Q, "Flat-Panel Display Materials," both held on April 25–27 at the 2000 MRS Spring Meeting in San Francisco, California, benefitted from a large and enthusiastic attendance. The papers contained herein are a fair representation of a dynamic and still advancing technology captured in profile at these two complementary symposia.

Symposium R began with two joint sessions with Symposium Q on Field Emission and Display Applications (R1/Q1) and Field Emission Display/Cathodoluminescence (R2/Q2). Subsequently, a session on Fundamental Processes (R3) contained theory and characterization presentations. Field Emitter Arrays (R4) constituted an entire section, and shared the poster session with FEAs, Nanotubes, and Other Electron Sources (R5). Finally, sessions were devoted to Diamond and Wide Bandgap (R6) materials as well as Carbon Nanotubes and Nanostructures (R7). In all, the entire symposium was a very good mix of all facets of the research and development enterprise, from novel fabrication techniques, to theory/modeling and characterization, and on to some excellent presentations on applications and related issues.

The invited speakers for Symposium R give a good indication of the flavor. John Robertson spoke on "Field Emission From Carbon Systems." Ron Musket spoke on "Applications of Ion Track Lithography in Vacuum Microelectronics" as a method of making exceptionally small geometry field emitter arrays. The "Theory and Modeling of Field-Induced Electron Emission" presentation by Richard Forbes revisited fundamental issues. Jong Duk Lee discussed "Mo and Co Silicide FEAs" and was followed by Dev Palmer who discussed "Stabilization of Emission Current in Silicon Field Emitter Arrays." Two back-to-back presentations on field emitters concerned their application and protection: David Whaley spoke on "Application of Field Emitter Arrays to Microwave Power Amplifiers," and L. Parameswaran's inclusion to the conference proceedings concerns "Field Emitter Array Cold Cathode Arc Protection Methods." The final invited talk was given by Otto Zhou, who spoke on the "Fabrication and Electron Field Emission Properties of Carbon Nanotubes." Coupled with the regular submissions as sampled in this proceedings volume, a strong indication of the state of electron emissive materials and their potential to make possible a variety of commercially and militarily critical technologies is contained in this volume.

Symposium Q, "Flat-Panel Display Materials," started with two joint sessions with Symposium R. J. M. Kim gave an invited talk on the display application for CNTs in "Carbon Nanotube FEDs for Large Area and Full Color Applications." David Tallant discussed "Mechanisms Affecting Emission In Rare-Earth-Activated Phosphors" in his invited talk. The next session was co-organized with Symposium S on Electrically Active Polymers. Organic Light Emitter Devices were discussed in Session Q3/S4 which featured invited talks by

Jim Sturm, Princeton, on large area polymer patterning and by Isayoshi Fujikawa of Toyota Central R&D Labs discussing novel hole transport layer materials for Organic Light Emitting Devices (OLEDs). Inorganic Electroluminescent (EL) materials were discussed in Q4, featuring a talk by Shosaku Tanaka of Tottori University presenting the latest results on blue EL thin films. Pat Green of Planar Systems discussed the evolution of flat-panel displays for the market place, particularly EL displays. A poster session (Q5) featured 11 papers, one of which won honorable mention for the best poster paper of the evening session. Thin film transistor (TFT) array materials and processing were featured in Sessions Q6-Q9. Greg Parsons, North Carolina State University, discussed low-temperature insulators for TFTs. Seung-Ki Joo of Seoul National University discussed lateral crystallization for TFTs at low processing temperatures. S.D. Brotherton of Philips Labs reported on the use of laser annealing to produce polycrystalline TFT films, while Patrick Smith of FlexICs demonstrated TFTs on plastic substrates. The symposium was finished by a 15-paper in-room poster session featuring materials and processing for TFTs.

This year, Symposium R was able to award a limited number of the R/Q proceedings to graduate students who attended Symposium R. Our ability to do so is directly the result of the generosity of Dr. John Pazik (Office of Naval Research) and Dr. Purobi M. Phillips (Altair Technologies), whose critical support is correlated with the overall success of the symposium, and who therefore deserve considerable thanks.

Kevin L. Jensen
Robert J. Nemanich
Paul Holloway
Troy Trottier
William Mackie
Dorota Temple
Junji Itoh

ACKNOWLEDGMENTS

Symposium R wishes to thank the following sponsors for their generous support:

Office of Naval Research
Altair Technologies

Symposium R also wishes to thank the invited speakers who, in addition to their fine presentations, gave invaluable assistance as Session Chairs and Co-Chairs. They are

John Robertson
Ron Musket
Richard Forbes
Jon Duk Lee
W. Dev Palmer
David Whaley
Lalitha Parameswaran
Otto Zhou

Symposium Q similarly thanks the other co-organizers James Im of Columbia University, and Sey-Shing Sun of Planar Systems. Sessions Chairs included:

Sey-Shing Sun
Patrick Smith
Ki-Bum Kim
S.D. Brotherton
Ryoichi Ishihara
James Im
Alexander Limanov
Silke Christiansen

All of the authors in the present volume deserve thanks for the time and effort spent in rendering their presentations and manuscripts of high quality. Finally, we would like to thank the staff of the Materials Research Society for their uncommon indulgence of our needs and demands, as well as assistance beyond what was expected with the preparation of the present volume.

MATERIALS RESEARCH SOCIETY SYMPOSIUM PROCEEDINGS

MATERIALS RESEARCH SOCIETY SYMPOSIUM PROCEEDINGS

Volume 605— Materials Science of Microelectromechanical Systems (MEMS) Devices II, M.P. deBoer, A.H. Heuer, S.J. Jacobs, E. Peeters, 2000, ISBN: 1-55899-513-7

Volume 606— Chemical Processing of Dielectrics, Insulators and Electronic Ceramics, A.C. Jones, J. Veteran, D. Mullin, R. Cooper, S. Kaushal, 2000, ISBN: 1-55899-514-5

Volume 607— Infrared Applications of Semiconductors III, M.O. Manasreh, B.J.H. Stadler, I. Ferguson, Y-H. Zhang, 2000, ISBN: 1-55899-515-3

Volume 608— Scientific Basis for Nuclear Waste Management XXIII, R.W. Smith, D.W. Shoesmith, 2000, ISBN: 1-55899-516-1

Volume 609— Amorphous and Heterogeneous Silicon Thin Films—2000, R.W. Collins, H.M. Branz, S. Guha, H. Okamoto, M. Stutzmann, 2000, ISBN: 1-55899-517-X

Volume 610— Si Front-End Processing—Physics and Technology of Dopant-Defect Interactions II, A. Agarwal, L. Pelaz, H-H. Vuong, P. Packan, M. Kase, 2000, ISBN: 1-55899-518-8

Volume 611— Gate Stack and Silicide Issues in Silicon Processing, L. Clevenger, S.A. Campbell, B. Herner, J. Kittl, P.R. Besser, 2000, ISBN: 1-55899-519-6

Volume 612— Materials, Technology and Reliability for Advanced Interconnects and Low-k Dielectrics, K. Maex, Y-C. Joo, G.S. Oehrlein, S. Ogawa, J.T. Wetzel, 2000, ISBN: 1-55899-520-X

Volume 613— Chemical-Mechanical Polishing 2000—Fundamentals and Materials Issues, R.K. Singh, R. Bajaj, M. Meuris, M. Moinpour, 2000, ISBN: 1-55899-521-8

Volume 614— Magnetic Materials, Structures and Processing for Information Storage, B.J. Daniels, M.A. Seigler, T.P. Nolan, S.X. Wang, C.B. Murray, 2000, ISBN: 1-55899-522-6

Volume 615— Polycrystalline Metal and Magnetic Thin Films—2000, L. Gignac, O. Thomas, J. MacLaren, B. Clemens, 2000, ISBN: 1-55899-523-4

Volume 616— New Methods, Mechanisms and Models of Vapor Deposition, H.N.G. Wadley, G.H. Gilmer, W.G. Barker, 2000, ISBN: 1-55899-524-2

Volume 617— Laser-Solid Interactions for Materials Processing, D. Kumar, D.P. Norton, C.B. Lee, K. Ebihara, X. Xi, 2000, ISBN: 1-55899-525-0

Volume 618— Morphological and Compositional Evolution of Heteroepitaxial Semiconductor Thin Films, J.M. Millunchick, A-L. Barabasi, E.D. Jones, N. Modine, 2000, ISBN: 1-55899-526-9

Volume 619— Recent Developments in Oxide and Metal Epitaxy—Theory and Experiment, M. Yeadon, S. Chiang, R.F.C. Farrow, J.W. Evans, O. Auciello, 2000, ISBN: 1-55899-527-7

Volume 620— Morphology and Dynamics of Crystal Surfaces in Complex Molecular Systems, J. DeYoreo, W. Casey, A. Malkin, E. Vlieg, M. Ward, 2000, ISBN: 1-55899-528-5

Volume 621— Electron-Emissive Materials, Vacuum Microelectronics and Flat-Panel Displays, K.L. Jensen, W. Mackie, D. Temple, J. Itoh, R. Nemanich, T. Trottier, P. Holloway, 2000, ISBN: 1-55899-529-3

Volume 622— Wide-Bandgap Electronic Devices, R.J. Shul, F. Ren, M. Murakami, W. Pletschen, 2000, ISBN: 1-55899-530-7

Volume 623— Materials Science of Novel Oxide-Based Electronics, D.S. Ginley, D.M. Newns, H. Kawazoe, A.B. Kozyrev, J.D. Perkins, 2000, ISBN: 1-55899-531-5

Volume 624— Materials Development for Direct Write Technologies, D.B. Chrisey, D.R. Gamota, H. Helvajian, D.P. Taylor, 2000, ISBN: 1-55899-532-3

Volume 625— Solid Freeform and Additive Fabrication—2000, S.C. Danforth, D. Dimos, F.B. Prinz, 2000, ISBN: 1-55899-533-1

Volume 626— Thermoelectric Materials 2000—The Next Generation Materials for Small-Scale Refrigeration and Power Generation Applications, T.M. Tritt, G.S. Nolas, G. Mahan, M.G. Kanatzidis, D. Mandrus, 2000, ISBN: 1-55899-534-X

Volume 627— The Granular State, S. Sen, M. Hunt, 2000, ISBN: 1-55899-535-8

Volume 628— Organic/Inorganic Hybrid Materials—2000, R.M. Laine, C. Sanchez, E. Giannelis, C.J. Brinker, 2000, ISBN: 1-55899-536-6

Volume 629— Interfaces, Adhesion and Processing in Polymer Systems, S.H. Anastasiadis, A. Karim, G.S. Ferguson, 2000, ISBN: 1-55899-537-4

Volume 630— When Materials Matter—Analyzing, Predicting and Preventing Disasters, M. Ausloos, A.J. Hurd, M.P. Marder, 2000, ISBN: 1-55899-538-2

Prior Materials Research Society Symposium Proceedings available by contacting Materials Research Society

SYMPOSIUM R:
ELECTRON EMISSIVE MATERIALS
AND VACUUM MICROELECTRONICS

Field Emission and
Display Applications

Mat. Res. Soc. Symp. Proc. Vol 621 © 2000 Materials Research Society

FIELD EMISSION FROM CARBON SYSTEMS

John Robertson

Engineering Dept, Cambridge University, Cambridge CB2 1PZ, UK. jr @eng.cam.ac.uk

ABSTRACT

Electron field emission from diamond, diamond-like carbon, carbon nanotubes and nano-structured carbon is compared. It is found that in all practical cases that emission occurs from regions of positive electron affinity with a barrier of ~5 eV and with considerable field enhancement. The field enhancement in nanotubes arises from their geometry. In diamond, the field enhancement occurs by depletion of grain boundary states. In diamond-like carbon we propose that it occurs by the presence of sp2-rich channels formed by the soft conditioning process.

INTRODUCTION

Various forms of carbon such as carbon nanotubes, diamond and diamond-like carbon (DLC) can show electron field emission at low applied fields of order 2-20 V/μm. This opens up a number of opportunities for electronic devices. Each device requires a different set of material parameters and so this favours the use of a different type of carbon.

The electronic device of primary interest in vacuum microelectronics is the microwave power amplifier [1,2]. The high frequency limit of semiconductor devices is set by their dimensions and the relatively small limiting velocity of electrons in semiconductors, eg 10^5 m.s^{-1} for Si. Vacuum microelectronic devices can in principle operate up to THz, as the limiting electron velocity in space is 3.10^8 m.s^{-1}. These devices require a large, stable emission current density, but place no limits on the substrate or emission site density. It is likely that carbon nanotubes are best here, as they can carry the largest current.

Field emission displays (FEDs) are flat panel displays in which the image is formed from a large array of pixels each addressed by field emission sources [3,4]. The use of thin film emitters rather than Spindt tips allows the use of wider gate holes. This makes lithography easier and lower cost for large display areas. A significant constraint is that glass substrates are used for low cost, so deposition and processing temperatures must be kept under about 500°C or lower. The cathodes are ~2 μm in diameter, so it is necessary that there is one emission site per source, equivalent to an emission site density (ESD) of 10^6 cm^{-2}. FEDs require emission current densities of about 1 mA.cm^{-2}.

Carbon field emitters could be used in vacuum power switches [5,6]. Solid state power switches have considerable difficulty in simultaneously withstanding the large breakdown voltages and having small forward voltage drops, as this places opposite constraints on the doping density of the semiconductor. A vacuum emission device can separate these two requirements – the vacuum withstands the off voltage, while the forward voltage drop can be low for a good field emitter.

Carbon nanotubes can make field emission electron guns with extremely high brightness and narrow energy distributions for use in scanning transmission electron microscopes [7]. The sharp energy distribution helps to minimise the chromatic aberration.

Finally, the decreasing size of Si devices is reaching the limits of UV lithography. Various high resolution methods such as electron beam writing are impractical, because it is necessary to write the images in two dimensions to transfer the image information in a reasonable time. This problem can be resolved by using parallel electron beam writing, for example from a controllable array of field emission cathodes. The electron beams should be sharply focused for high resolution, which is easier for flat cathodes rather than tips. This favours carbon cathodes over Spindt tips [8].

The requirements for each application are summarised in Table 1; more dots signify a greater need.

	Current density	Low Field	Site density	Current stability	Energy width
Microwave amp	**			**	
FE Display		*	**	**	*
Power switch	**	**			
FE gun	***			**	**
Lithography		*	**	*	*

The emission properties of the various types of carbon are summarised in Table 2. Again, the more dots the better.

	Current density	Low Field	Site density	Current stability	Deposition Temperature
Diamond				*	
Nano-diamond	*	**	**	**	
Diamond-like carbon		*			**
Nano-structured Carbon	*	**	**	**	**
Nanotubes	***	**	**	*	(*)

DIAMOND

We now consider the emission properties of each of the various forms of carbon in more detail. The original motivation for the study of field emission from diamond was its negative electron affinity [9-11]. Diamond is a semiconductor with a band gap of 5.5 eV. When its surface is terminated by hydrogen it has a negative electron affinity (NEA), so that its conduction band edge lies above the vacuum level. This means that any electrons in its conduction band could pass into the vacuum with no energy barrier. However, field emission requires electrons to travel round a complete circuit. This is a problem, as there is still a large potential barrier at the back contact [15](Fig. 1), and diamond has a very high resistivity. To date, the best electron emission from diamond phases is from microcrystalline and nano-crystalline diamond [16-20]. In this case, the emission is found to vary inversely with the grain size.

Emission should be possible from n-type diamond as this places electrons near the conduction band. However, the most soluble donor is nitrogen, but this gives a deep donor level at 1.7 eV below the conduction band, so the film remains highly resistive [21]. Phosphorus has a more shallow donor level ~0.5 eV below the conduction band [22]. This system could finally test the ideas of NEA field emission. However, to date, even phosphorus doped films still have a high resistivity, which will limit emission. In addition, phosphorus doping has only been carried out on homo-epitaxial films on single crystal diamond, so that making electrical contact to the back of the doped film is difficult and has required back-side etching of the diamond substrate [23].

The origin of the field emission can be determined by measuring the electron energy distribution (EED) of the emitted electrons in conjunction with for example a photoemission measurement (Fig. 2). The mean energy can show if the electrons originate from the conduction band, valence band, gap states or the Fermi level, while the EED width is proportional to the size of the local field. Bandis and Pate [24] found that emission from boron doped single crystalline diamond originates from the valence band. They later found that emission from some N-doped diamond originated from states in the lower gap [25], but it turned out that the material has a large tungsten impurity. Recently, Groning et al [20] have shown that emission from nano-crystalline diamond originates from the Fermi level, which in that case lies about 1 eV above the valence band edge. Thus, to date, no field emission has truly utilised the NEA property of diamond.

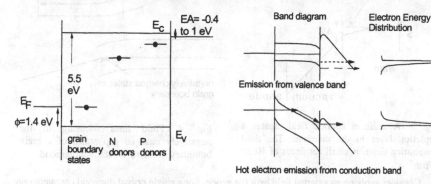

Fig. 1. Schematic band diagram for diamond [12,14], showing large barrier for electrons at the back contact, and the energy levels of grain boundary states and N and P donors.

Fig. 2. Schematic electron energy distributions for (a) emission from the valence band or Fermi level, and (b) hot electron emission from the conduction band.

The spatial location of emission can be deduced by using a scanning tunnelling microscope arrangement. Karabutov et al [26] has correlated the more emitting regions with the topological features of micro-crystalline diamond. They found that emission tends to come from the troughs rather than the peaks of the surface. This is consistent with emission from the grain boundaries, rather than the grains. Similar results were found by Rakhimov et al [27]. The grain boundaries of diamond are more conducting, sp² bonded and have states lying about 1.4 eV above the valence band edge. Thus, the results suggest that electrons cross the film along the grain boundaries and are emitted from them. This is consistent with early reports of preferential emission from the positive affinity, sp² regions of CVD diamond films [28,29]. It is also consistent with the observed EED.

On the other hand, some groups such as Groning et al [20] find that emission comes mainly from the grain peaks, not dips. While diamond is usually very resistive, CVD diamond taken from a hydrogen plasma can have a large, hydrogen-induced p-type conductivity [30]. The origin of this conductivity is not yet understood, but it could allow conduction from the grain boundary to the peaks and emission from the grain peaks.

Emission from grain boundaries is consistent with a field enhancement model [19]. The grain boundary is sp^2 bonded, and it gives rise to π states on either side of the Fermi level, which lies about 1.4 eV above the valence band edge. If the grain boundary is very narrow, these states are simply localized π states in the diamond band gap. The π states are neutral when half filled. When filled, they are negatively charged and when empty they are positively charged. The emission barrier is 4-5 eV, the energy from Fermi level to the vacuum level - the work function.

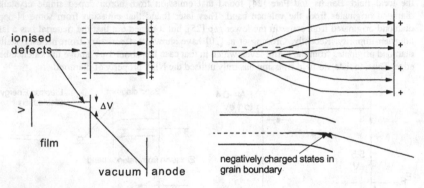

Fig. 3. A uniform applied field creates a depletion layer in the diamond. The field penetration depth is small for fields of 10-20 V/μm.

Fig. 4. Field lines focusing to the negatively charged gap states of a grain boundary in micro-crystalline diamond.

Consider applying an external field by a flat anode. For a single crystal diamond containing no defects, the field will be lower in the diamond by the dielectric constant. If the diamond possessed a uniform distribution of these π defect states, the field would induce negative charges on the states, and form a negatively charged depletion layer at the diamond surface (Fig. 2). This results in field penetration, but the depth is quite small (~1 nm) for the fields of interest (10 V/μm) as the defect density is quite large. Hence this field penetration does not result in a significant band bending or lowering of the surface potential barrier (<0.1 V). If instead, the defects are concentrated in grain boundaries, the field lines focus into these regions (Fig. 3). This produces strong local fields, which extends along the grain boundary. The grain boundary acts like an *internal tip*. This mechanism can provide an adequate narrowing of the 5 V barrier, and this allows the easy emission by tunneling.

DIAMOND-LIKE CARBON

Diamond-like carbon (DLC) is amorphous carbon (a-C) or hydrogenated amorphous carbon (a-C:H) containing a substantial amount of sp^3 bonding [31]. It consists of very smooth films, usually grown at room temperature, which is a great advantage for displays. Two main types of DLC have been studied, the a-C:H grown by plasma enhanced chemical vapour deposition (PECVD), and the highly sp³ bonded tetrahedral amorphous carbon (ta-C) grown by filtered cathodic vacuum arc (FCVA) or by pulsed laser deposition (PLD).

DLC is a reasonable emitter, when doped with nitrogen [32-38]. However, the emission site density from DLC tends to be low, which limits its use in field emission displays [39]. Many types such as undoped a-C:H are not very useful for emission because their resistivity is too high (10^{14} Ω.cm). The hydrogen can also lead to rather unstable emission, so many groups have focused on hydrogen-free a-C. Emission from ta-C has been found to depend strongly on its sp3 content, the incorporation of nitrogen, and on any plasma surface treatment [33-35]. The emission when tested in the parallel plate configuration can occur at threshold fields as low as 5 V/μm.

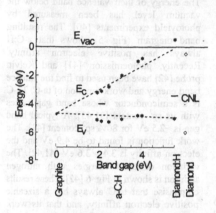

Fig 5. Schematic band diagram for a-C:H [14], from photoemission data [42]. The valence band edge remains at roughly 5 eV below the vacuum level, for all band gaps.

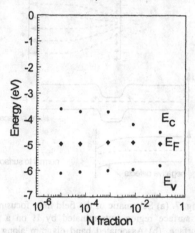

Fig. 6. Fermi level, valence and conduction band edge energies from Kelvin probe and photemission measurements [41,42]

Emission, particularly from the more resistive types of DLC, requires a forming or conditioning process in which the voltage is applied in cycles of increasing size [34]. For highly resistive a-C:H which has a large threshold field, it is clear that the conditioning process creates surface damage, which is then the cause of the emission. In ta-C's with lower threshold field, there is a form non-destructive conditioning. STM experiments suggest that this acts to increase the local conductivity at the emitting spots by a local conversion of sp³ to sp² bonding [38]. The need for conditioning indicates that the film conductivity is still too high for the most efficient emission.

The emission mechanism in DLCs has been difficult to understand. The macroscopic emission current densities roughly obey the Fowler-Nordheim equation.

$$J = af \frac{(\beta F)^2}{\phi} . \exp(-\frac{b\phi^{3/2}}{\beta F})$$ (1)

where ϕ (eV) is the barrier height, F is the applied field, β is a dimensionless geometric field enhancement factor, f is the fraction of area over which emission occurs and a and b are constants. The slope of the Fowler-Nordheim plot of 0.01 to 0.4 eV derived [12] from the Fowler-Nordheim plots by takingβ=1 is clearly not the real barrier, as such small barriers would allow temperature-dependent thermionic emission or Schottky emission which is not observed.

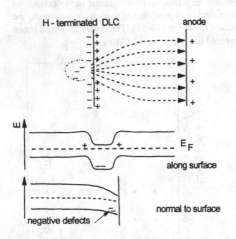

Fig. 7. (a) Schematic of the field lines focusing to a surface region unterminated by H on a ta-C surface. (b) Associated band diagram along and normal to surface [14].

It is necessary to construct a band diagram for ta-C to understand the emission mechanism [12-15]. A-C:H films have a band gap which varies with their sp³ content. The energy of their valence band below the vacuum level has been measured by photoyield experiments [40]. The resulting band diagram (Fig. 5) shows that a-C:H always has positive electron affinity. Recently, photoemission [41] and Kelvin probe [42] have been used to find the valence band energy and work function of ta-C. Ta-C is a semiconductor whose band gap varies with its sp³ content [31,43]. A typical band gap is ~2.5 eV for 80% sp³ content [43]. The work function is found to be 4.9 eV and the electron affinity is 3.4 – 3.6 eV [41,42]. The variation of band energies with nitrogen addition is shown in Fig. 6 [42]. These results emphasise that ta-C always has a sizeable positive electron affinity, and that its work function is relatively independent of sp³ content and nitrogen addition at 4.8 to 4.9 eV. This value is quite similar to that of graphite 4.9 eV [20].

The EED of a-C with a large nitrogen incorporation has been measured, and emission is found to occur from ~5 eV below the vacuum level [44]. The EED width is consistent with the presence of a high local internal field. This indicates the existence of a strong field enhancement mechanism.

It has been suggested that emission from DLC is due to hot electrons created by strong band bending at the back contact [32], as shown schematically in Fig. 2(b). However, photoemission measurements of the valence band energy of ta-C films of increasing thickness show no evidence of any band-bending at the back contact [41]. The EED also shows no evidence of hot electrons as a high energy tail. Thus, this mechanism does not seem to be relevant to DLC.

The band diagram and the absence of hot electron effects mean that there is a large, ~5 eV emission barrier at the front surface, and that there are large local fields. The main problem in any emission model is that DLC films are very smooth and generally homogeneous, so the origin of a large field enhancement is not clear [14]. The first proposal [14] was based on the observation that the EA and work function of diamond could be shifted by 1.5 eV by varying the surface coverage of hydrogen. A local surface region without hydrogen termination would have a more positive EA than the surrounding H-terminated surface. Ta-C has a large density of amphoteric defects. The common Fermi level leads to charge transfer to create negatively charged defects behind the un-terminated region. This is a depletion region. The application of a field from the external anode causes the field lines to focus towards this negatively charged zone [14](Fig. 7). This creates a large local field. The field size is limited by the potential step ΔV caused by the differences in termination, and the defect density, N, in the depletion layer,

$$F = (\frac{2N.e. \, \Delta V}{\varepsilon})^{1/2} \qquad (2)$$

For $\Delta V=1.5$ V, $N=10^{25}$ m^{-3}, $\varepsilon_0=8.9.10^{-12}$ F/m, and $\varepsilon_r=5$, then $F = 4.10^8$ V/m outside the surface, in the tunnel barrier. This field is a similar size to that seen by the EED [42]. The length scale of these large fields is of order 1 nm. However, the problem with this mechanism is that although it creates a large local field, it only lowers the emission barrier by 1.5 eV, not by the full 5 eV. Hence another mechanism is needed.

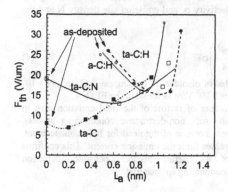

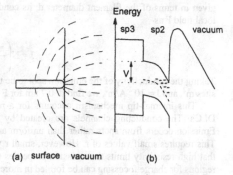

Fig. 8. Threshold field for 1 μA/cm^2 current density versus annealing for various forms of DLC films. Plotted against apparent sp^2 cluster size as derived from Raman [47]

Fig. 9. Schematic of the emission mechanism for ta-C. An internal, more sp^2 bonded channel is created, which becomes negatively charge on the application of an external field, causing barrier lowering.

The barrier lowering requires other types of heterogeneity. Heterogeneity can be deliberately introduced into DLC by deposition at higher temperatures, or by thermal annealing. Annealing creates graphitic sp^2 bonded clusters within the sp^3 matrix [45]. The emission threshold field F_{th} is found to first decrease and then increase as the annealing temperature is raised. The size of the clusters can be monitored by Raman spectroscopy. The visible Raman spectrum is dominated by the scattering of sp^2 sites. They give rise to two modes, the G peak around 1600 cm^{-1} and the so-called D (disorder) peak around 1360 cm^{-1}. The D peak is derived only from the vibration of graphitic clusters, and the intensity ratio I(D)/I(G) can be used to derive their size. For small cluster sizes, as here, I(D)/I(G) is proportional to the number of rings in the cluster [46].

We have measured the threshold field of various types of DLC as a function of annealing, A-C:H is quite convenient for this test, because although it is a poor emitter, it anneals at accessible temperatures. The emission threshold field F_{th} is found to first decrease and then increase as the annealing temperature is raised. We converted this into a dependence on cluster size, L_a, as shown in Fig. 8. An optimum cluster size is found, about 0.5 to 1 nm. This is quite small.

The emission model for DLC shown in Fig. 9 is similar to the internal tip of grain boundaries in micro-crystalline diamond. The emission site density is quite small. We assume that the conditioning process creates sp^2-rich channels extending from the film surface. The field lines focus onto their gap states, creating a negative depletion charge, a large local field and barrier lowering. A similar model has been proposed by Forbes [47].

Why is the critical dimension so small? The field enhancement factor β is given by the ratio of height to radius of a tip,

$$\beta = h/r \tag{3}$$

r is a trade-off between the need for small r values to give large enhancement factors of 200- 400, and the emitting region being able to conduct the emission current. The emission current density J is given in terms of the filament diameter d, its conductivity σ, and emission site density N and the local field F as

$$J = N\left(\frac{\pi d^2}{4}\right)\sigma F \tag{4}$$

Taking the conductivity of the emission spot to be that of disorder sp^2 carbon, say $1\ \Omega^{-1}m^{-1}$, $N \sim 10^6$ sites/m^2, and $J = 10^{-2}$ A/m^2, we find d~0.7 nm for F = 5.10^8 V/m as found in the EED.

This internal-tip mechanism accounts for a number of factors of the field emission seen in DLCs. The conducting channels are created by the soft, non-destructive conditioning process. Emission occurs from spots rather than uniform areas because of the need for field enhancement. This requires small values of r. However, small r values limit the emission current. This confirms that high resistivity limits the emission properties of DLC. Better emission is possible if depletion regions for charge focussing can be formed in more conducting sp^2 material.

NANOTUBES

Carbon nanotubes show many of the best field emission qualities, such as a large current density, large emission site density and a narrow electron energy distribution [48-55,7]. Emission has been studied from isolated nanotubes – open, closed and multi-walled; from mats of spagetti-like tubes, and from mats of oriented nanotubes. The work function of nanotubes is 4.9 to 5 eV, the same as graphite [55]. Single wall nanotubes may have a smaller work function.

It is natural to imagine that the easy field emission from nanotubes arises because they have the ultimate tip structure with a very large field enhancement factor or height/radius ratio (3). This is only partly true, and other factors also seem to be relevant. Firstly, is the effect only geometric, or does it depend on the electronic structure, for example if the nanotube is open or closed at the end? Open tubes appear to give higher current, according to Saito et al [50]. Open tubes have broken bonds at the end, and the states associated with these lie at the Fermi level.

Fransen et al [7] observed that nanotubes can give the sharpest EED of any emitting system known to date, ~0.12 V. In the Stratton model [56], the current density at any energy J(E) is given as the product of the supply function (the density of states N(E)), and the escape probability P(E).

$$J(E) = N(E).P(E) \qquad (5)$$

For a metal, N(E) is constant, below the Fermi level, and the EED width of a metal is proportional to the local field, so the EED should be wide. Thus the EED of a metallic nanotube should be wide. A narrow EED suggests that it is for a semiconducting nanotube, in which emission is occuring from a sharp density of states feature at the Fermi level – for example broken bonds of an open nanotube. A similar behaviour may occur for graphitic sheets [57].

The third interesting point is that the emission from spagetti nanotubes, which are lying down, seems to be as good as from nanotubes aligned or oriented normal to the surface [49,48]. It is not clear why spagetti would have a high enhancement factor. Indeed, Shaw et al [55] find that tubes oriented normal, at 45° or lying down have similar emission, with lying down tubes having the *best*!

Extensive work on emission from nanotubes has been carried out by Dean [52,53]. He shows that emission occurs at normal temperatures not from bare nanotube tips, but with a molecule adsorbed on the tip. The best emission occurs with the adsorbed molecule, and it reduces if the molecule desorbs at higher temperature. However, the effect of molecules is not so good news, as this introduces an instability into the emission process.

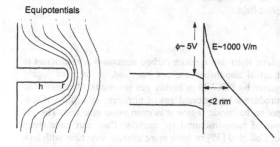

Fig. 10. Barrier lowering by field enhancement requires not just a large field, but the vacuum level to be lowered to the emission level, and within a tunnelling distance of ~ 2 nm.

Much of the work has been carried out with nanotubes growth by laser ablation or the graphite arc, purified by acids, and then stuck onto a substrate. Recent work has been directed to the chemical vapour deposition (CVD) of nanotube mats for field emission and other applications. This involves the coating of the substrate with a Ni or Co catalyst. At present, the structures grown by CVD tend to have quite large diameter (50-200 nm) and are fibers rather than nanotubes (1-2 nm in diameter). High deposition temperatures are desirable for atomic perfection of the graphene sheets, but lower temperatures could be possible for field emission applications. Ren et al [58] claim to have grown aligned nanofiber mats at below the softening temperature of glass, which is very useful for display applications.

The geometric field enhancement model gives a very simple picture, Fig. 10. The equipotentials for an applied field in the absence of any tubes lie parallel to the substrate. If a tube is metallic, the equipotentials are now pushed out in front of the tube, and give the large field in front of the tube tip. If the tube length is h and its radius is r, then the tip field F_{loc} is related to the applied field F by

$$V = Fh = F_{loc} \, r \qquad (6)$$

That is, the voltage dropped along the length of the tube in its absence is now dropped over a distance r in front of the tube. In fact, V is the barrier lowering by field enhancement mechanism. If r = 1 nm, F = 1 V/μm, and V = 5V, then this means that tubes 5 μm long are needed to lower the barrier sufficiently for emission. This clearly does not occur, or otherwise tubes lying down would not emit. Another source of voltage is needed.

A final point about emission from nanotubes is that a close array of nanotubes can screen the field enhancement of its neighbour as shown in Fig. 11. Thus, the spacing of nanotubes should roughly equal their height to maintain the field enhancement. The close spacing of nanotubes has recently been shown to limit the emission site density in nanotube mats [59]. The local geometry gives rise to a range of field enhancement factors, which leads to an exponential variation of emission site density on applied field.

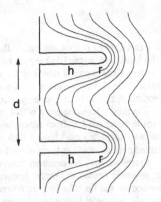

Fig. 11. Field enhancement at tips is screened by nearby tips.

NANOSTRUCTURED CARBON

For some applications such as displays, sharp but random carbon structures are sufficient to obtain good emission properties and actual nanotubes are not required. A form of highly structured, mainly sp^2 carbon can be grown by PECVD at higher gas pressures. One way is to use the cathodic vacuum arc but to introduce a background gas of nitrogen, or helium. If it is introduced directly into the cathode spot or the anode region it is even more effective. The gas leads to the formation of various types of nanostructured sp^2 carbon. They can have sharp external features such as the corraline of Coll et al [39] or have more internal structure, with leek ordering [60-63]. The advantage of nanostructured carbons over nanotubes is that they can be grown at room temperature. These materials show a high emission site density (up to 10^6 cm^{-2} at higher test fields), fairly high current density and low emission fields (1-5 V/μm) [60,62].

SUMMARY

Numerous types of carbon show field emission at relatively low threshold fields of 1-20 V/um. None of the better emitters uses the negative electron affinity of diamond, and all seem to involve emission from about 5 eV below the vacuum level. There seems to be a general emission mechanism of geometric field enhancement at work. This even seems to operate in the apparently homogeneous diamond-like carbon films, where we believe that conditioning creates the inhomogeneities. Low conductivity can be a serious limit on emission from a-C:H and diamond films. Low conductivity may also be why a-C:H and ta-C have rather low emission site densities. Carbon nanotubes have many of the best field emission features, of large current density, low emission field and sharp energy spectrum. Although nanotubes appear to be the classical case of geometric field enhancement, this does not account for why emission can occur for tubes lying down, which have quite moderate enhancements.

ACKNOWLEDGEMENTS

The author wishes to thank A Ilie for many of the experiments on ta-C, and to W I Milne, A Hart, B.S. Satyanarayana and G Amaratunga for discussion of their results. This work is partly funded by EU project CARBEN.

REFERENCES

1. P R Schwoebel, I Brodie, J Vac Sci Technol B **13** 1391 (1995)
2. M W Geis, et al, IEEE Trans ED Let **12** 456 (1991)
3. J E Jaskie, MRS Bulletin **21** (March 1996) p59
4. W B Choi, D S Chung, J H Kang, H Y Kim, Y W Jin, I T Han, Y H Lee, J E Jung, N S Lee, G S Park, J M Kim, App Phys Lett **75** 3129 (1999)
5. T Sakai et al, Abstracts, Diamond Films (2000)
6. R Rosen, W Simendiger, C Debbault, H Shimoda, L Fleming, B Stoner, O Zhou, App Phys Lett **76** 1668 (2000)
7. M J Fransen, T L van Rooy, P Kruit, App Surface Sci **146** 312 (1999)
8. V I Merkulov, et al, Extended Abstracts of 11th International Vacuum Microelectronics Conference, (1998).
9. F J Himpsel, J S Knapp, J A VanVechten, D E Eastman, Phys Rev B **20** 624 (1979)
10. B B Pate, P M Stefan, C Binns, P J Jupiter, M L Shek, I Lindau, W E Spicer, J Vac Sci Technol **19** 249 (1981)
11. J.B. Cui, J Ristein, L Ley, Phys. Rev. Lett. **81**, 429 (1998).
12. J. Robertson, Mat Res Soc Symp Proc **417** 217 (1997)
13. J Robertson, Mat Res Soc Symp Proc **498** 197 (1998)
14. J. Robertson, J. Vac. Sci. Technol. B **17**, 659 (1999).
15. M W Geis, J C Twichell, T M Lyszczarz, J Vac Sci Technol B **14** 2060 (1996).
16. W Zhu, et al, J Appl Phys **78** 2707 (1995); Science **282** 1471 (1998)
17. A A Talin, L S Pan, K F McCarty, T E Felter, H J Doerr, R F Bunshah, App Phys Lett **69** 3842 (1996)
18. F Lacher, C Wild, D Behr, P Koidl, Diamond Related Mats **6** 1111 (1997)
19. A R Krauss, D M Gruen, D Zhou, T G McCauley, L C Qin, T Corrigan, O Auciello, R P H Chang, Mat Res Soc Symp Proc **495** 299 (1998)
20. O. Gröning, O.M Küttel, P Gröning, L Schlapbach, J. Vac. Sci. Technol. B **17** 1064 (1999).
21. K Okano, S Koizumi, S R P Silva, G A J Amaratunga, Nature **381** 140 (1996)
22. S Koizumi, H Ozaki, M Kamo, Y Sato, T Inuzuka, App Phys Lett **71** 1065 (1997)
23. T Sugino, C Kimura, K Kuriyama, S Koizumi, M Kamo, Phys Stat Solidi A **174** 145 (1999)
24. C. Bandis, and B. Pate, Appl. Phys. Lett. **69**, 366 (1996).
25. K Okano, T Yamada, A Sawabe, S Koizumi, R Matsuda, C Bandis, W Chang, B B Pate, App Surface Sci **146** 274 (1999)
26. A.V. Karabutov, V.D. Frolov, S.M. Pimenov, and V.I. Konov, Diamond Relat. Mater. **8**, 763 (1999).
27. A T Rakhimov, N V Suetin, E S Soldatov, M A Timofeyev, A S Trifonov, V V Khanin, A Silzars, J Vac Sci Technol B **18** 76 (2000)
28. C Wang, A Garcia, D Ingram, M Lake, M E Kordesch, Electron Lett **27**, 1459 (1991)
29. N.S. Xu, Y. Tzeng, and R.V. Latham, J. Phys. D **26**, 1776 (1993).

30. H J Looi, et al, Carbon **37** 801 (1999)
31. J Robertson, Phys Rev B **53** 16302 (1996); Prog Solid State Chem **21** 199 (1991)
32. G.A.J. Amaratunga and S.R.P. Silva, App. Phys. Lett. **68**, 2529 (1996)
33. B.S. Satyanarayana, A. Hart, W.I. Milne, and J. Robertson, App. Phys. Lett. **71**, 1430 (1997)
34. A Hart, B.S. Satyanarayana, J. Robertson, and W.I. Milne, App. Phys. Lett. **74**, 1594 (1999)
35. L.K. Cheah, X Shi, E Liu, B K Tay, J. Appl. Phys. **85**, 6816 (1999).
36. R D Forrest, A P Burden, A R P Silva, L K Cheah, X Shi, App Phys Lett **73** 3784 (1998)
37. N Missert, T A Friedmann, J P Sullivan, R G Copeland, App Phys Lett **70** 1995 (1997)
38. T W Mercer, N J DiNardo, J B Rothman, M P Siegal, T A Friedmann, L J Martinez-Miranda, App Phys Lett **72** 2244 (1998)
39. B.F. Coll, J.E. Jaskie, J.L. Markham, E.P. Menu, A.A. Talin, and P. von Allmen, Mat. Res. Soc. Proc. **498**, 185 (1998); A Talin et al, in 'Amorphous Carbon, State of Art', ed S R P Silva et al (World Scientific, Singapore 1999)
40. J Ristein, J Schafer, L Ley, Diamond Related Mats **4** 508 (1995)
41. N Rupersinghe, G A J Amaratunga, Diamond Related Mats (2000)
42. A Ilie, A Hart, A J Flewitt, J Robertson, W I Milne, submitted to J App Phys (2000)
43. M Chhowalla, J Robertson, C W Chen, S R P Silva, G A J Amaratunga, J App Phys **81** 139 (1997)
44. O. Gröning, O.M. Küttel, P. Gröning, L. Schlapbach, Appl. Phys. Lett. **71**, 2253 (1997).
45. A Ilie, T Yagi, A C Ferrari, J Robertson, Appl. Phys. Lett **76** 2627 (2000).
46. A C Ferrari, J Robertson, Phys Rev B **61** 14095 (2000)
47. R G Forbes, this symposium (2000)
48. W A deHeer, A Chatelain, D Ugarte, Science **270** 1179 (1995)
49. J M Bonard, J P Salvetat, W A deHeer, L Forro, A Chaletain, App Phys Lett **73** 918 (1998)
50. Y Saito et al, App Surface Sci **146** 345 (1999)
51. W Zhu, C Bower, O Zhou, G Kochanski, S Jin, App Phys Lett **75** 873 (1999)
52. K A Dean, B R Chalamala, App Phys Lett **75** 3017 (1999)
53. K A Dean, P vonAllmen, B R Chamamala, J Vac Sci Technol B **17** 1959 (1999)
54. O Groning, O M Kuttel, C Emmenegger, P Groning, L Schlapbach, J Vac Sci Technol B **18** 665 (2000).
55. Y Chen, D T Shaw, L Guo, App Phys Lett **76** 2469 (2000)
56. R Stratton, Phys Rev **135** A794 (1964)
57. A N Obratztsov, I Y Pavlovsky, A P Volkov, J Vac Sci Technol B **17** 674 (1999)
58. Z F Ren, et al, Science **282** 1105 (1998)
59. L Nilsson, O Groning, C Emmenegger, O Kuttel, E Schaller, L Schlapbach, H Kind, J M Bonard, K Kern, App Phys Lett **76** 2071 (2000)
60. G.A.J. Amaratunga, M Baxendale, N Rupersinghe, I Alexandrou, M Chhowalla, T Butler, A Munindradasa, C J Kiely, T Sakai, New Diamond Frontier Carbon Technol **9**, 31 (1999)
61. G A J Amaratunga, et al, Nature **383** 321 (1996)
62. B S Satyanarayana, J Robertson, W I Milne, J App Phys **87** 3126 (2000)
63. A C Ferrari, B S Satyanayana, J Robertson, W I Milne, E Barborini, P Piseri, P Milani, Europhys Lett **46** 245 (1999)

Mat. Res. Soc. Symp. Proc. Vol. 621. © 2000 Materials Research Society

Applications of Ion Track Lithography in Vacuum Microelectronics

Ronald G. Musket[1]
Materials Science and Technology Division
Lawrence Livermore National Laboratory
Livermore, CA 94550, USA
[1]Present address: 3877 Meadow Wood Drive,
El Dorado Hills, CA 95762, USA

EXTENDED ABSTRACT

When a high-velocity (i.e., typically > 0.1 MeV/amu) ion passes through a material it can change the properties of the material within a cylindrical zone centered on the essentially straight trajectory of the ion. The electronic bonding, phase, and density are among the properties modified in the zone, which is called a latent nuclear, or ion, track. Because the diameters of latent ion tracks are typically less than 20 nm, selective chemical etching is generally employed to improve the detection and assessment of the tracks. Historically, etched nuclear tracks have been used mainly for nuclear particle identification, geochronology, measurement of extremely low-dose radiation levels, and creation of membrane filters.

An introduction to the basic concepts involved in the creation and etching of ion tracks to make masks for lithographic processes has been detailed [1]. Ion track lithography provides two main advantages over other lithographic techniques. Ions can be used to generate structures with the largest aspect ratios (i.e., length/width) and the smallest lateral widths. The main limitation of ion track lithography is that the tracks are stochastically distributed over the exposed surface

For fabrication of arrays of field emitters the random location of the tracks is not a limitation, because all the field emitters within a given area (e.g., sub-pixel of a video display) will be turned on at the same time. However, the track density should be low enough to provided essentially only isolated tracks after etching the track diameters out to the desired dimensions. As an example, only ~5% of 40-nm diameter tracks will be overlapping at a density of $10^9/cm^2$. Such isolated tracks in positive resists (e.g., polycarbonate films) can be etched to form molds for deposition of structures and to form masks for transferring the track pattern to levels below the tracked resist. Specific examples of these two approaches are described here.

The first is the creation of arrays of free-standing nano-emitters. Approximately 600-nm thick polycarbonate was spun onto a nickel-covered silicon wafer. Latent ion tracks with a density of ~$10^8/cm^2$ were created with 13.6 MeV Xe^{+4} ions incident perpendicular to the surface of the polycarbonate. The tracks were etched near room temperature with 6M KOH to a diameter of ~80 nm, and then nickel was electroplated into the molds created by the etched tracks to nearly fill the cylindrical holes. Removal of the polycarbonate mold by immersion in the KOH solution resulted in the field emitter array shown in Fig. 1. The I-V characteristics of such arrays were found to follow the Fowler-Nordheim predictions, demonstrating that the currents were due to field emission.

The hole diameter in the polycarbonate has been shown to be a linear function of the etching time for diameters greater than about 20 nm [2]. The length of the nano-structures is limited by the thickness of the polycarbonate film.

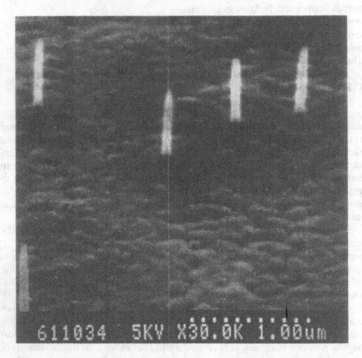

Figure 1. *Scanning electron micrograph of an array of ~80-nm diameter, 600-nm tall Ni field emitters with a density of ~10^8/cm^2.*

The role of ion track lithography in a novel process for producing cathode arrays with individual emitter structures having gates with < 300 nm diameters has been described [2]. Since that work was completed, Candescent Technologies Corporation (CTC) have shown that a similar, but refined, ion track lithography process can lead to ~120-nm diameter individually gated emitter structures [3]. These structures can be switched between fully off and fully on with only an 11 volt change on the column or gate bias (with the row-to-column bias fixed at 18 volts) [3]. More recently, at the CTC web site (candescent.com), a working 13.2" diagonal SVGA color display was shown. Fabrication of this display utilized ion track lithography for creation of the nano-scale gated, field emitter structures of each sub-pixel.

CONCLUSIONS

These results demonstrate that ion track lithography can be the key enabling technology leading to cost-effective processes for creating high-aspect-ratio nano-structures and nano-scale arrays of gated and ungated field emitters suitable for a variety of applications.

ACKNOWLEDGMENTS

The processes for creation of the arrays of field emitters discussed or referenced above were originally developed at Lawrence Livermore National Laboratory (LLNL) during a close collaboration among LLNL and CTC personnel. The main contributors to the development of the initial processes are either the authors or support staff of reference 2. Work performed under the auspices of the U. S. Department of Energy by the Lawrence Livermore National Laboratory under contract W-7405-ENG-48, including CRADA TC-774-94 with Candescent Technologies Corporation.

REFERENCES

1. R. Spohr, *Ion Tracks and Microtechnology: Principles and Applications* (Vieweg, Braunschweig, Germany, 1990).

2. A. F. Bernhardt, R. J. Contolini, A. F. Jankowski, V. Liberman, J. D. Morse, R. G. Musket, R. Barton, J. Macaulay, and C. Spindt, J. Vac. Sci. Technol. B **18**, 1212 (2000).

3. T. S. Fahlen, presented at the 12th International Vacuum Microelectronics Conference, Darmstadt, Germany, 1999 (unpublished).

Mat. Res. Soc. Symp. Proc. Vol. 621 © 2000 Materials Research Society

SURFACE TREATMENT ON SILICON FIELD-EMISSION CATHODES

M.Hajra, N.N.Chubun, A.G. Chakhovskoi and C.E.Hunt,
Electrical and Computer Engineering Department, University of California, Davis, CA. 95616,
Phone (530)-752-2735, Fax (530)-754-9217, e-mail: mshajra@ece.ucdavis.edu

ABSTRACT

Arrays and single-tip p-type silicon micro-emitters have been formed using a subtractive tip fabrication technique. Following fabrication, several different surface treatments have been attempted for comparison. We utilized ion and electron bombardment at elevated pressures (with interrupted pumping), and also hydrogen seasoning during field emission operation. The objectives of these treatments include stabilization of the emission, lowering the effective workfunction, and reducing low-frequency noise. The tips were evaluated using I-V measurements in the diode configuration. A flat Si anode, spaced nominally 6 μm and 150 μm from the cathode, was used. For the purpose of treatment, the field emission characteristics are measured in a high vacuum chamber at a pressure range between 10^{-5} and 10^{-8} Torr. The results suggest that the emitters benefit from seasoning or conditioning, for optimal performance, low noise, minimum work function and maximum reproducibility and reliability over the lifetime of the cathode.

INTRODUCTION

Silicon field emitters have demonstrated their viability as electron sources for various vacuum microelectronics applications. In recent years, the attention is drawn towards post-fabrication seasoning processes aiming to improve the performance of the field emitters. In the current work, we attempt various methods of surface treatment of single-crystal silicon emitter arrays fabricated by the well-developed subtractive manufacturing process [1]. The arrays of 50x50 tip emitters were fabricated from p-type (4-6 Ωcm) Si (100) substrate by the above process. Firstly, thermal grown oxide of 2000 Å thick and a 1000 Å thick chromium layer on the Si were patterned into a 1.8 μm-diameter disk. Using the chromium and the SiO$_2$ disk as a mask, the outline of the emitter tip was formed by means of ion reactive etching with SF$_6$ as shown in figure 1 (a). The tips were then sharpened using the method of oxidation sharpening as we have previously described elsewhere [2]. Tip caps were subsequently removed by wet etching of the silicon dioxide. The final silicon emission tip is shown in Fig.1 (b). The typical tip curvature radius is estimated using microscopy, to be on the order of 15 nm.

We investigate three methods of seasoning during the operation of the field emitters – conditioning of the emission surface using low-energy electron-stimulated desorption, surface treatment by residual gas ions, and surface cleaning using hydrogen-enhanced residual gas atmosphere.

Figure 1. a) SEM picture showing silicon tip before oxidation sharpening. Oxide cap is on top of the tip. b) After oxidation sharpening

RESULTS AND DISCUSSION

We tested the electrical and emission properties of the cathodes in a diode configuration as shown in figure 2. A flat Si anode, spaced nominally 150 μm or 6 μm from the cathode by using a quartz spacer of a proper thickness, was used. Field emission properties of the cathodes were measured in a vacuum chamber under a residual gas pressure of 10^{-8} Torr. Characterization was performed without bakeout of the vacuum system. The Hewlett-Packard 4142B modular DC source/monitor was used to acquire the emission data. A positive potential (up to 100 V) was applied to anode, and the cathode had a negative bias. The field emission properties of the cathodes were measured after the tips were conditioned for 3 days.

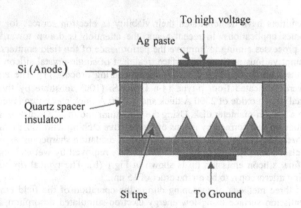

Fig. 2. Test structure. The Cathode is mounted on to a TO₃ Transistor package

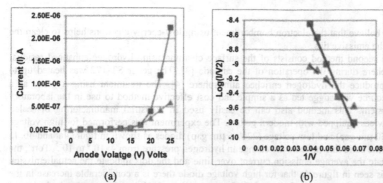

(a)   (b)

Fig. 3. *I-V characteristics (a) and the F-N characteristics (b) before (lower curve) and after (upper curve) the surface conditioning by low-energy electron stimulated desorption under low current loading.*

First method of conditioning consists of cleaning of the surface by low-energy electron stimulated desorption under low-current loading. The first treatment method was performed in a slightly different configuration compared to one shown in Fig.2: instead of the flat silicon anode we have used the same tip structure as the cathode to be treated. The reverse bias voltage was in the region from 16 to 22 V, which allowed us to obtain the emission from the anode tip array to the cathode tip array. We have observed the initial I/V characteristics and the short-term current behavior of the "fresh" silicon cathodes (Fig. 3a, lower curve), then we reversed the polarity of the electric field by applying negative voltage to a silicon anode and using the silicon field emitter as an electron collector without changing the geometry of the cathode-anode "sandwich". In this configuration we ran the diode for a period up to 16 hours. The current density during the treatment was maintained on the level of 10^{-4} A/cm^2. The electron energy was in the range between 16 to 22 eV. The Coulomb dosage applied to the cathode surface was estimated to be on the level of $6*10^{-2}$ C/cm^2. No changes in the emission from the anode were observed during the entire treatment. After the treatment, an increase of the emission current by a factor of 5 (Fig. 3a, upper curve), along with stabilization of the I/V characteristics were observed.

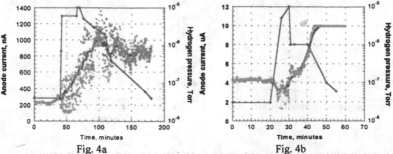

Fig. 4a   Fig. 4b

Fig. 4. *The effect of hydrogen treatment on silicon field-emission array cathodes under (a) high (1500 V) and (b) low (22V) anode voltage.*

We believe that the electron bombardment using low-energy electrons helps to clean the surface of the emissive tips.

The second method consists of the surface cleaning using hydrogen-enhanced residual gas atmosphere during the operation of the cathode [3]. The getter ST-172 was heated up to 700°C to produce the hydrogen enriched atmosphere [4]. The treatment using a getter as a hydrogen source was suggested as a simple and cost effective method to use in the laboratory vacuum system. This method also can be easily used in industrial manufacturing of sealed vacuum devices equipped with modern getters. The experiment was performed for high voltage (1500V, 150 μm gap) and low voltage (22V, 6 μm gap) diodes as shown in figure 4a. and 4b. In the figures the solid line indicate the change in hydrogen pressure from 4×10^{-8} to 10^{-5} Torr , the curve indicate the average emission current over time and the dots indicate the actual emission current. It is seen in figure 4a that for high voltage diode there is a considerable increase in the emission current for about a factor of 5 after approximately 1 hr of treatment in hydrogen atmosphere. The same effect is observed for the low voltage diode during the time range of 20 min to 30 min. After the surface was cleaned using hydrogen-enhanced residual gas atmosphere, the tips were tested for a day at a pressure of 4×10^{-8}Torr. It was observed that the emission current remained approximately at the value immediately following the surface treatment. We suggest the following mechanism for the FEA performance improvement: that the electrons emitted from the cathode help to break the hydrogen molecules and/or ionize the hydrogen atoms from the hydrogen enhanced atmosphere. These ions and atoms react with the surface contamination like oxide, carbon residue etc. of the tips and clean the surface lowering the work function.

The third method, accelerated "natural" surface conditioning was performed by increasing the pressure of residual gases. This was done by interrupting the pumping in the vacuum chamber during prolonged operation of silicon emitters from several hours to several days. The I-V characteristic seen in figure 5a corresponds to the two curves measured before (lower curve) and after (upper curve) the experiment. It is seen from the two graphs that there is a considerable increase in the emission current.

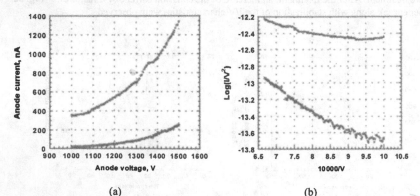

(a)   (b)

Fig 5. *I-V characteristic and the F-N characteristic before (lower curve) and after (upper curve) the "natural" surface conditioning was performed by increasing the pressure of residual gases.*

We believe that the short-term exposure to the enhanced residual gases atmosphere helps to remove an initial film of contaminants and adsorbents by ion sputtering. The two main disadvantages of this method are the following: first it requires a prolong treatment to achieve a desirable value for the emission properties, and second this method is not very advantageous to achieve the highest emission properties as obtained in the above two processes.

It should be noted that all the above methods resulted in comparable improvement in reproducibility of I-V characteristics taken after the treatment.

CONCLUSION

Three methods of surface treatment were applied to silicon field emitter arrays fabricated by subtractive process. All three methods – low energy electron-stimulated desorption, hydrogen seasoning, and residual gas ion conditioning, were performed in a simple small-volume stainless vacuum system. The methods clearly show improvement of emission characteristics, increase of emission current and stabilization of cathode performance. The hydrogen treatment seems to be suitable for improving the properties of the gated field-emission cathodes.

ACKNOWLEDGEMENTS

The Authors acknowledge the helpful input of Dr. Si-Ty Lam, and his colleagues, of Hewlett Packard Labs. This work was funded through the DARPA-MTO MEMS program under contract agreement F30602-98-3-0232, administered by the US Air Force Material Command.

REFERENCES

1. J. T. Trujillo and C. E. Hunt, *low-voltage silicon gated field-emission cathodes for vacuum microelectronics and e-beam applications*, JVST B, 11 (2), 1993, pp. 454-458,
2. R.B Marcus, T.S Ravi, and T.Gmitter, K.Chin, D.Liu, W.J Orvis and D.R.Ciarlo, C.E.Hunt and J.Trujillo, *Formation of Silicon tips with <1nm radius*, Appl.Phys. Lett. 56(3), 15 January 1990, pp 236-238.
3. R.Meyer, presented at the 9th IVMC Conference, Kyongju, Korea, 1997 (unpublished).
4. *Advanced Porous Getters. St 172 Series.* Technical publication G870530. SAES getters S.p.a. Milano, 1987.

Mat. Res. Soc. Symp. Proc. Vol. 621. © 2000 Materials Research Society

Fabrication and Characterization of Singly-Addressable Arrays of Polysilicon Field-Emission Cathodes.

N.N. Chubun, A.G. Chakhovskoi, C.E. Hunt and M.Hajra
Electrical and Computer Engineering Department, University of California,
Davis, CA. 95616, U.S.A.

ABSTRACT

Polysilicon is a promising candidate material for field-emission microelectronics devices. It can be competitive for large-size, cost-sensitive applications such as flat-panel displays and micro electro-mechanical systems. Singly-addressable arrays of field-emission cells were fabricated in a matrix configuration using a subtractive process on Polysilicon-On-Insulator substrates. Matrix rows were fabricated as insulated polycrystalline silicon strips with sharp emission tips; and matrix columns were deposited as gold thin film electrodes with round gate openings. Ion implantation has been used to provide the required conductivity of the poly-Si layer. To reduce radius of curvature of the polysilicon tips, a sharpening oxidation process was used. The final device had polysilicon emission tips with end radii smaller than 15 nm, surrounded by gate apertures of 0.4 μm in diameter. Field emission properties of the cathodes were measured at a pressure of about 10^{-8} Torr, to emulate vacuum conditions available in sealed vacuum microelectronics devices. It was found that an emission current of 1 nA appears at a gate voltage of 25 V and can be increased up to 1μA at 70 V. Over this range of current, no "semiconductor" deviation from the Fowler-Nordheim equation was observed. I-V characteristics measured in cells of a 10x10 matrix, with a cell spacing of 50 μm demonstrated good uniformity and reproducibility.

INTRODUCTION

Low-voltage field-emission cathodes are a promising type of electron source for vacuum microelectronics devices. During the last ten years, a broad collection of new emissive materials such as diamond and diamond-like carbon, compound semiconductors, noble metals, etc., have been used for field-emission cathode fabrication. Single crystal silicon, because of its advantages in VLSI technology still remains one of the most frequently used materials for many applications including flat-panel display, multi-beam lithography, microscopy, and data-storage devices [1,2]. However as the size of the single-crystal Si substrates increases, the factors related to manufacturing cost and technological complexity impose further limitation on fabrication of the cathodes for flat-panel displays and micro electro-mechanical systems. Polycrystalline silicon may be considered as a good alternative to single crystalline silicon because it can be cost-effectively deposited over larger size dielectric substrates and still be compatible with traditional semiconductor manufacturing methods.
Recently, it was found by E. Boswell and colleagues in diode emission tests, that the polycrystalline silicon field emitters fabricated using wet etching method, exhibit an emission behavior similar to single crystal silicon emitters [3]. It was observed that the oxidation sharpening had little effect on the field-emission characteristics due to the presence of sharp emitting silicon tips in a polycrystalline material. In a structure fabricated by H. Uh and others, both single and polysilicon gated field-emission arrays were fabricated on oxidized silicon

substrates using RIE etching and sharpening oxidation. Stable emission currents of 0.1 μA/tip were measured at 82 V for polycrystalline tips with a gate aperture of 1.2 μm and at 80 V for single-crystal tips with gate aperture 1.6 μm [4]. This result is similar to the data obtained previously from excellent, uniform, very-low turn-on voltage single crystal silicon field emission arrays described by M.Ding et al [5].

In this paper we present a new version of a low-cost and dependable method of fabrication of singly-addressable arrays for multi-beam vacuum microelectronics device applications.

EXPERIMENTAL DETAILS

The singly-addressable field-emission cathodes matrix consist of polysilicon "stripes" in which the emission tips are formed, over coated with oxide and having metal stripes, perpendicular to the polysilicon lines, which is used to form the gates.

The fabrication process requires a silicon-oxide-silicon structure consisting of a silicon substrate with at least 1 μm of oxide and 3 μm of polycrystalline silicon. Initial poly-silicon deposition techniques, performed at standard 620 C°, resulted in a surface with significant roughness and large silicon grain size. This influenced the minimum feature size obtainable (we use Karl Suss MA-6 vacuum contact lithography) and had a negative effect on the shape and geometry of the silicon tips. To reduce the grain size and to provide a more smooth coating, we lowered the deposition temperature to the lower value of 590 C°.

The initial resistivity of the polycrystalline silicon layer exceeded 100 kΩ-cm, which was excessively high for our application. To reduce the resistivity, the layer of polysilicon was doped by phosphor ion implantation using a dose of 3×10^{14} cm^{-2} at of 80 keV. No subsequent annealing was performed. The polysilicon was oxidized with total time 2 hours at 1100°C to obtain a 0.15 μm thick SiO_2 layer. The oxidation time was sufficient for annealing and impurity activation. After oxidation, the measured bulk resistivity of the polysilicon layer was close to 3 Ω-cm and the sheet resistance was 10 kΩ/ . This resistivity inherent to the polysilicon enabled us to obtain series resistors of 1 MΩ for each emission tip. The resistors proved to stabilize the field emission current from the individual tips, as expected.

Following the ion implantation step, the oxide layer was coated with 0.1 μm chromium to provide an etch mask. The process flow of the singly addressable array fabrication includes two chrome-oxide-polysilicon structure-etching steps. The first step produces a simple system of insulated polysilicon cathode stripes, which do not require high-resolution lithography. A variety of wet etching processes of chromium, silicon dioxide and polysilicon layers were tested for obtaining these stripes. Good results were obtained, including tilted sidewalls of the polysilicon stripes in which the cathodes are later etched. The cross-section shape provided effective step-coverage of the cathode stripes afterward with the gate metal layer. Although dry etching of polysilicon was also investigated, the sidewalls were excessively vertical and it was found that wet etching resulted in better uniformity across a 4-inch wafer.

The second masking and etching step forms the tip portion of the cathode electrodes, leaving a 1 μm diameter chrome/oxide tip cap where the tip is to be formed. Because of the grain structure of polysilicon it is extremely important doing tip etching to use isotropic reactant. Plasma etching in a SF_6/O_2 gas mixture as well as pure SF_6 was attempted. It is found that pure SF_6 provides more smooth and uniform etching in comparison with mixtures using 10% and 50% O_2.

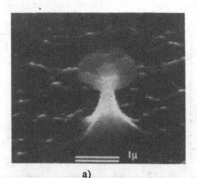 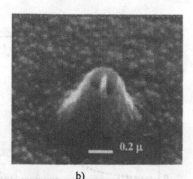

a)   b)

Figure 1. a) Polycrystalline silicon tip before oxidation sharpening. Oxide cap is on top of the tip. b) A completed volcano-type field-emission cell, having 0.3 μ gate opening and approximately 15 nm tip curvature radii.

The tip etch is continue until the diameter of the narrowest part of the tip decreases down to 0.2 μm as it is shown on Fig.1. At this point oxidation sharpening is carried out. A combination of wet and dry oxygen is used because tip sharpening is more effective in dry oxygen at low temperature; however the oxidation time would be excessively long if only dry oxidation is used. This oxidation also increased the thickness of oxide on the surface of cathode stripes and reduced the gate-cathode current leakage.

The gold film used as a gate electrode was deposited by electron-beam evaporation using the method we have previously described elsewhere [6]. As a result, the gate metal is coated on the sidewall of the tip and is barely masked by the oxide cap. This method allows us to form the gate aperture, having a diameter of 0.3-0.4 μm using conventional optical lithography technology with resolution normally producing larger feature sizes. Tip caps were subsequently removed by wet etching of the silicon dioxide. The final "volcano-type" emission cell is shown in Fig.1b. The typical tip curvature radius is estimated using microscopy, to be on the order of 15 nm.

DISCUSSION.

We tested the electrical and emission properties of the cathodes in a matrix configuration. No electrical cross-leakage between the cathode stripes and the gate electrodes in air was observed (up to 10 V DC and 1 nA sensitivity). Field emission properties of the cathodes were measured in a vacuum chamber under a residual gas pressure of 10^{-8} Torr. Characterization was performed without bake out of the vacuum system. No tip conditioning or field forming steps was performed. The Hewlett-Packard 4142B modular DC source/monitor was used to acquire the emission data. The gate electrode was grounded, positive potential (up to 100 V) applied to stainless steal foil anode, and the cathode had a negative bias. Anode-cathode spacing was approximately 1 mm.

Figure 2a shows current-voltage characteristic, obtained from a single-cell cathode with a gate opening of 0.4 μm. It was found that emission current of 1 nA is first registered at a gate voltage 25 V, increasing to 10 nA at 30 V. The Fowler-Nordheim (FN) plot for this region is a straight line, as it shown on Fig. 2b. To achieve 1 μA we needed to increase the voltage between the tip

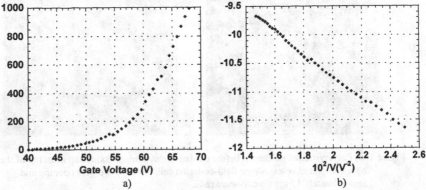

Figure 2. a) Anode current vs. gate voltage plot and b) the Fowler-Nordheim plot for the anode current up to 1000 nA.

and the gate to 68 V. For the tested range of current no "semiconductor" deviation from Fowler-Nordheim equation was observed. We consider that combination of sufficient doping level within silicon grains and high sheet resistivity on the grain borders are responsible for this behavior. The minor shifts in the plot on Fig. 2b could be explained by fluctuations of emission current, which are usually significant for single tip field emitters. The dependence of the anode current varying with gate and anode voltage as a parameter (e.g. "triode characteristics") is shown in Fig. 3 Lifetime of an individual cathode cell was up to 200 hours. Several cathodes were damaged under attempts to extract and maintain maximum possible levels of emission current. It was observed that tip damage by emission current occurred under loading of 1.5 μA per tip. These attempts often resulted in the tip destructing, as shown in Fig. 4.

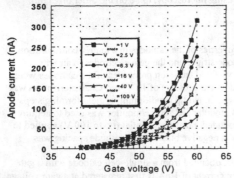

Figure 3. Anode current vs. gate voltage at different anode potentials relative to gate potential (e.g. "triode characteristics").

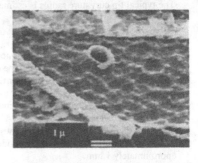

Figure 4. SEM picture showing an exploded cell, caused by excessive steady-state cathode current

High electrical and heat conductivity of the gold gate film and MΩ resistivity of cathode strip minimized damage of the structure, which is important for high-density matrix applications. I-V characteristics of separated cells in the 10x10 matrix, with the distance between adjacent emitters of 50 micron were tested demonstrating good uniformity and reproducibility. The emission current does not appear to be significantly affected by the random crystalline orientation of the emitting tip crystallite of each cathode.

CONCLUSION

Polycrystalline silicon on an insulated substrate is suitable for field-emission cathode fabrication instead of SOI wafers as it allows us to avoid the cost and size limitation.
We developed subtractive technology of singly-addressable cathode matrix fabrication, which provides a gate opening as small as 0.3 micron using contact lithography with a resolution of 1 micron. Loading resistors of any required resistance could be integrated with the cathode because of high intrinsic resistivity of polycrystalline silicon layer. An anisotropy of resistance of polycrystalline layer in lateral and orthogonal directions improve linearity of Fowler-Nordheim characteristics of the field-emission cathode.
The emission current up to 1 μA was extracted from a single tip at a gate voltage not exceeding 70 V.

ACKNOWLEDGEMENT

The Authors acknowledge the helpful input of Dr. Si-Ty Lam, and his colleagues, of Hewlett Packard Labs. This work was funded through the DARPA-MTO MEMS program under contract agreement F30602-98-3-0232, administered by the US Air Force Material Command.

REFERENCES

1. J. T. Trujillo and C. E. Hunt, low-voltage silicon gated field-emission cathodes for vacuum microelectronics and e-beam applications, JVST B, 11 (2), 1993, pp. 454-458,
2. J.H. Lee, S.W. Kang, S.G. Kim, Y.H. Song, K.I. Cho, H.J. Yoo A New Fabrication method of Silicon Field Emitter Array with Local Oxidation of Polysilicon and Chemical-Mechanical-Polishing. Proceedings of the 9th International Vacuum microelectronics Conference, July 7-12, 1996, St. Petersburg, Russia, pp.415-418.
3. E.C. Boswell, S.E. Huq, M. Huang, P.D. Prewett, P.R. Wilshaw, Polycrystalline silicon field emitters. JVST B 14(3) May/Jun 1996, pp. 1910-1913.
4. H.S. Uh, S.J. Kwon, J.D. Lee Process design and emission properties of gated n[+] polycrystalline silicon field emitter arrays for flat panel display applications. JVST B 15(2) Mar/Apr 1997, pp. 472-476.
5. M. Ding, H. Kim, and A.I. Akinwande, High Uniformity and Low Turn-on Voltage Si FEAs Fabricated with CMP. Proceedings of the 12th International Vacuum microelectronics Conference, July 6-9, 1999, Darmstadt, Germany, pp.370-371.
6. J.T. Trujillo, A.G. Chakhovskoi and C. E. Hunt, Low voltage silicon field emitters with gold gates, Proceedings of the 8th International Vacuum Microelectronics Conference, Portland, OR, July 30-Aug. 3, 1995, pp. 42-46.

Fundamental Processes

Mat. Res. Soc. Symp. Proc. Vol. 621 © 2000 Materials Research Society

Theory and Modelling of Field-Induced Electron Emission

Richard G. Forbes

University of Surrey, School of Electronic Engineering, Guildford, Surrey GU2 7XH, UK

ABSTRACT

This paper addresses issues in the theory of field-induced electron emission. First, it summarises our present understanding of the theory of Fowler-Nordheim (FN) plots, and shows the relationship between a recent precise (in standard FN theory) approach to the interpretation of the FN-plot intercept and older approximate approaches. Second, it comments on the interpretation of FN plots taken from semiconductor field emitters. Third, it summarises the main points of a recent hypothesis about the mechanism of field-induced emission from carbon-based films and other electrically nanostructured heterogeneous (ENH) materials. Weaknesses in previous hypotheses are noted. It is hypothesised that thin films of all ENH materials, when deposited on a conducting substrate, will emit electrons in appropriate circumstances. Such films emit electrons at low macroscopic fields because they contain conducting nanostructure *inside* them: this structure generates sufficient geometrical field enhancement near the film/vacuum interface that more-or-less normal Fowler-Nordheim emission can occur. In connection with experiments on amorphous carbon films carried out by a group in Fribourg, it is shown that nanostructure of the size measured by scanning probe techniques should be able to generate field enhancement of the size measured in field electron spectroscopy experiments. This result provides a quantitative corroboration of other work suggesting that emission from amorphous carbon films is primarily due to geometrical field enhancement by nanostructures inside the film. Some counter-arguments to the internal-field-enhancement hypothesis are considered and disposed of. Some advantages of ENH materials as broad-area field emission electron sources are noted; these include control of material design.

INTRODUCTION

Development of the theory of field-induced electron emission requires progress on five fronts. (1) Consolidation of the standard theory of Fowler-Nordheim (FN) emission, and of a general theory of FN-plot interpretation. (2) Consolidation and development of theory that addresses the various weaknesses of the standard theory (see later). (3) Development of an improved theory of field emission from semiconductors. (4) Application of existing theory to problems of modelling and interpretation of experimental data. (5) Scientific exploration of unsolved problems in deciding the mechanism of field-induced emission from complex materials.

This paper aims to deal with the first issue above and also with one issue of type (5), namely the mechanism of field-induced emission from carbon-based films. The latter is the subject of a review [1]: the present paper will outline its main arguments.

THE THEORY OF FOWLER-NORDHEIM PLOTS

The theory here applies to emission from a single free-electron metallic band, but can be modified to apply to other materials. Fowler-Nordheim theory was introduced in 1928 [2], but Murphy and Good's 1956 discussion [3] is now standard. Surprisingly, there has not until recently [4] been a precise theory for the interpretation of FN plots. An aim of this paper is to show more clearly how this new theory relates to older approaches [5-7] based on numerical approximations.

I recently suggested [4] writing the emission current magnitude I in the *generalised* form

$$I = \lambda A a \phi^{-1} F^2 \exp[-\mu b \phi^{3/2}/F], \qquad (1a)$$

where F and ϕ are the local field and work-function, λ and μ are *generalised correction factors*, A is the notional area of emission, and a and b are the usual universal constants whose 1999 values are:

$$a = 1.541\,434 \times 10^{-6} \text{ A eV V}^{-2}; \quad b = 6.830\,890 \times 10^{9} \text{ eV}^{-3/2} \text{ V m}^{-1}. \qquad (1b)$$

In so-called Fowler-Nordheim coordinates, this becomes

$$\ln\{I/F^2\} = \ln\{\lambda A a \phi^{-1}\} - \mu b \phi^{3/2}/F, \qquad (2)$$

where $\ln\{B\}$ denotes the natural logarithm of 'B expressed in well-specified units (preferably SI units)'. Eq. (2) defines a *theoretical curve* in a Fowler-Nordheim plot.

At given field value F_1, the *tangent* to this curve can be written

$$\ln\{I/F^2\} = \ln\{r_1 A a \phi^{-1}\} - s_1 b \phi^{3/2}/F, \qquad (3)$$

where r_1 and s_1 are the values, *taken at field F_1*, of the functions r and s given by

$$r = \lambda \exp[-b\phi^{3/2} \, d\mu/dF]; \quad s = \mu - F \, d\mu/dF. \qquad (4)$$

These functions r and s are called the *generalised intercept correction function* and the *generalised slope correction function*. It has been assumed here that λ and A may be taken as independent of F, which is not always the case. In standard FN theory, under typical emission conditions, s_1 is typically of order unity but r_1 is typically of order 100 [4].

In practice, values of the local field F are not easily measurable; so we write F in terms of the measured *macroscopic field* F_M by $F = \gamma F_M$, where γ is a *field enhancement factor* assumed independent of F_M. Eq. (3) can then be written

$$\ln\{I/F_M^2\} \;=\; \ln\{r_1 Aa\phi^{-1}\gamma^2\} - s_1 b\phi^{3/2}/\gamma F_M \;\equiv\; \ln\{R_M\} + S_M/F_M, \tag{5}$$

where R_M and S_M are defined by eq. (5). In particular,

$$S_M \;=\; -s_1 b\phi^{3/2}/\gamma. \tag{6}$$

The theoretical quantities $\ln\{R_M\}$ and S_M are the predicted intercept and slope of the tangent to the macrosopic-field version of eq.(2), taken at a macrosopic field corresponding to local field F_1.

The function s is normally a slowly-varying function of field in the range where measurable electron emission occurs. So, if an *empirical* FN plot is 'nearly a straight line', and we take F_1 as some 'average local field' corresponding to the middle of the range of F_M values over which measurements are made, then the formulae above for $\ln\{R_M\}$ and S_M provide theoretical predictions for the intercept and slope of the empirical FN plot. Conversely, the empirical intercept and slope provide estimates of R_M and S_M, so we can use the corresponding value of s_1 (approximately unity in standard FN theory) to interpret the empirical S_M value via eq. (6).

For intercept interpretation, it is better to use the combination $R_M S_M^2$ and the formula:

$$A \;=\; R_M S_M^2/ab^2\Gamma, \tag{7}$$

where $ab^2 = 7.192\,494 \times 10^{13}$ A m^{-2} eV^{-2}, and the *emission area extraction function* $\Gamma = rs^2\phi^2$.

Values of Γ calculated (as a function of work-function and current density) using standard FN theory are given in Ref. [4]. An estimate of A and the uncertainty in A can be obtained by considering the ranges of uncertainty in the values of work-function and current-density applicable to the experiments in question. A more precise result can be obtained by iteration.

Similar arguments can in principle be applied to voltage-based FN plots, by writing $F = \beta V$ (where β is the voltage-to-local-field conversion factor), and replacing γ, F_M, R_M, S_M in the above equations by β, V, R_V, S_V, respectively.

To relate this new approach towards intercept interpretation to older approaches, we rewrite eqs. (4) in terms of the *Nordheim parameter* $y_N = c\,F^{1/2}/\phi$, where the universal constant c has the value $3.794\,686 \times 10^{-5}$ eV V$^{-1/2}$ m$^{1/2}$. On defining an *auxiliary function* u [8] by

$$u \;=\; -\tfrac{1}{2}y_N^{-1}\,d\mu/dy_N, \tag{8}$$

we obtain $d\mu/dF = -c^2\phi^{-2}u$, $Fd\mu/dF = -uy_N^2$, and hence can write

$$r \;=\; \lambda\,\exp[bc^2\phi^{-1/2}u], \tag{9}$$

$$s \;=\; \mu + uy_N^2, \tag{10}$$

where $bc^2 = 9.836\ 238\ \text{eV}^{1/2}$.

The above equations are all true in generalised FN theory. In standard FN theory, we replace λ by t_N^{-2}, and μ by v_N, where v_N and t_N are the well-known *Nordheim functions* (evaluated for barrier height equal to ϕ), and let u_N, r_N and s_N denote the quantities defined by eqns (8) to (10) after this replacement has taken place. Since v_N is a function only of y_N, it follows from eqns (8) and (10) that u_N and s_N are functions only of y_N, and are easily evaluated [9,10].

So, in standard FN theory, eq. (10) can be re-arranged as

$$v_N = s_N - u_N v_N^2. \tag{11}$$

It turns out that s_N and u_N are both slowly varying functions of y_N, so we can take some 'average' value y_a, and the corresponding values s_a and u_a, and write an *approximate* expansion of v_N (in the vicinity of y_a) in the form:

$$v_N = s_a - u_a y_N^2 = s_a - u_a (c^2 \phi^{-2} F) \tag{12}$$

There are an infinite number of (s_a, u_a) pairs that can be derived *self-consistently* in this way.

The old approach [5-7] to the determination of (what I call) the intercept correction factor relied on finding an expansion of the Nordheim function v_N in the form of eq. (12). On substitution into the standard FN equation, the second term in this expansion took the field-independent form $\exp[(u_a bc^2)\phi^{-1/2})]$, and this was transferred into the pre-exponential to become the exponential part of the correction factor I denote by r_N.

In the old approach, the expansion in form (12) was in some cases carried out empirically. Since we now have a precise theory based on the auxiliary function u_N, we can check the self-consistency of the expansions conventionally quoted. I do this by finding the y_a value that gives the quoted value of s_a, and evaluating the corresponding precise value of u_a, as shown below:

Originators	y_a	s_a	u_a (empirical)	u_a (precise)
Charbonnier and Martin [5]	0.4966	0.956	1.062	1.064
Dobretsov and Gomoyunova [6]	0.4413	0.965	0.739	1.105
Spindt et al. [7]	0.5306	0.950	1.000	1.041

From this we conclude that the Charbonnier and Martin approximation is self-consistent, that the Spindt et al. approximation is not self-consistent but should be satisfactory for practical purposes, and that the Dobretsov and Gomoyunova approximation is the least satisfactory of the three.

In practice, approximations of this kind are no longer necessary since precise values of u_N (and of r_N and Γ_N) are easily evaluated on laptop computers using spreadsheets (a suitable Excel spreadsheet is available from the author) [10]. The new approach suggested in Ref. [4], in effect, considers what is the possible range in u_N corresponding to the ranges of current density and local work-function thought to be present in the experiments, and makes this part of an input to the possible range in the quantity Γ appearing in eq. (7). The resulting uncertainty in Γ can then be reduced by an iterative process based on the actual experimental data.

In the cases where it has been tested, in the context of standard FN theory, the new approach yields notional emission-area values greater than those yielded by the Spindt et al. approximation by a factor of about 1.5. A factor of this size would often be less than uncertainties associated with the validity of the standard theory. The new approach has the merit that it can be applied to detailed theories that are more advanced than standard FN theory.

SOME GENERAL COMMENTS ON FIELD ELECTRON EMISSION THEORY

More generally, we may now conclude that: standard FN theory is precisely understood; that we now have a generalised theory of FN-plot interpretation, that can be made precise in the context of a detailed theory of field electron emission; and that the relationship between this generalised theory and the older approaches to FN plot interpretation (made in the context of standard FN theory) is well understood.

However, standard FN theory is based on a number of physical assumptions, several of which are known to be poor approximations to reality (see Ref. [4]). In particular: (1) The area of emission may not be constant but may be a function of applied voltage. (2) Particularly with sharp emitters and non-metallic emitters, it may not be satisfactory to assume that the actual potential barrier can be represented as a Schottky barrier. (3) Electron transport in the actual emitter band-structure may be significantly different from that in a free-electron metal band. (4) There may be wave-function matching effects in the surface layers of the emitter, or at the inner edge of the barrier, that enhance or inhibit electron transmission through the barrier. The consequences of effects (3) and (4) may be that the correction factor λ is significantly different from unity. Correction-factor values will be material and crystallography dependent, and Modinos (private communication) argues that λ might, in special circumstances, be anything from 0.01 to 100. Caution should therefore be exercised in interpreting area estimates derived from eq. (7), even for emission from metals.

In the case of semiconductors, emission can in principle take place from the conduction band, the valence band, and/or a band of surface states. (And, in the case of amorphous or highly defective semiconductors, possibly from a band of localised states lying in the mobility gap.) Further, emission from the conduction band may be degenerate or non-degenerate. The book by Modinos [11] gives a good account of semiconductor field emission theory (except, perhaps, in the case of emission from a non-degenerate conduction band - see comments below).

If a semiconductor FN plot is straight, and if the emission is tunnelling emission coming from a single band, then FN theory may be adapted to interpret at least the slope of the FN plot. In this case we write eq. (6) in the form:

$$S_M = -s_1 b H^{3/2}/\gamma \tag{13}$$

where H is the 'height (before image-force-type reduction) of the relevant barrier'. There is a long-standing unresolved controversy [6] about the nature of the correlation-and-exchange effects seen by an electron leaving a semiconductor, but usually it should be adequate to approximate s_1 as unity. The interpretation of H then depends on which band it is assumed that emission is coming from: thus H may represent: electron affinity χ (non-degenerate conduction band); work-function ϕ (degenerate conduction band); work-function ϕ_{ss} for surface states (not always the same as ϕ) (surface-state band); or the quantity $\chi + E_g$, where E_g is the band gap (valence band).

Complications can occur in this interpretation if significant field penetration exists, or if for any reason (such as heating of the electron distribution as a result of resistive voltage drop in the emitter, or transport-limited effects) the electron distribution in the emitting band ceases to be in thermodynamic equilibrium at the bulk emitter temperature. Space prevents discussion here.

The interpretation of the FN plot intercept is more problematical in the case of semiconductors than it is for metals, because the assumption (used in the simplest theories of semiconductor field electron emission) that a semiconductor band can be treated as 'free-electron-like' is nearly always somewhat doubtful. There is also, in the present author's view, a particular difficulty with Stratton's theory [12] of field emission current density from a non-degenerate conduction band: it is far from obvious to me that the integration of transmission probability over the band structure has been carried out in a satisfactory approximation, even for an assumed free-electron band. This matter will be discussed in detail elsewhere. Since nearly all subsequent authors have used or adapted Stratton's approach, it is possible that systematic error exists in the literature. Depressing as it may seem, the best advice is to be very cautious about attempts to interpret FN plot intercepts from semiconductors until the theoretical situation is clearer.

FIELD-INDUCED EMISSION FROM CARBON-BASED FILMS

Background

The potential of carbon-based films as broad-area electron sources has been well demonstrated [13-17]. An essential characteristic of the films is that they emit electrons at very low macroscopic applied fields, typically in the range 1 to 10 V/μm. (Cold field emission normally occurs at local fields in the V/nm range.) The nature of the emission mechanism(s) has been unclear. This section discusses the author's hypothesis [1] concerning emission mechanism. It aims to cover the main questions about the *primary* cause of electron emission, but limited space means that it is not intended as a detailed review and does not include full discussion of secondary effects that may help emission. Some significant issues have, with regret, been omitted from discussion.

The central questions are: "How do electrons escape across the film/vacuum interface ?" and "What factor controls the emission current ?" Until recently, there have been four main emission hypotheses, none of which has been fully satisfactory. The first three concern the film front surface. Some of the early papers [18,19] assumed that low or negative electron affinity reduced the surface barrier sufficiently for electron emission to occur. This hypothesis was given (false) encouragement by the very low values of barrier height H (0.1 eV or less) derivable from eq. (13) if S_M is found from a carbon-film FN plot, s_1 is taken as 1 and γ is arbitrarily taken as 1. However, if electron emission is a thermodynamic-equilibrium front-surface-controlled process, the parameter that determines the emission should not be the electron affinity alone, even with non-degenerate conduction-band emission (because in this case there is a thermal-activation term as part of λ). The total barrier height (before image-force-type reduction) to be overcome, by thermal activation and/or tunnelling, is the work-function; and flat-band work-function values for carbon are typically 3 eV or (usually) more. It is difficult to think up a convincing front-surface control mechanism for the combination of a flat-band work-function of 3 eV or more and a local field of 10 V/μm or less.

A second hypothesis, proposed by Robertson [20,21], was that emission was associated with 'patch field' effects, generally similar to those known in the context of thermionic emission [22] and early work in field electron emission [23]. Patch-field effects seem to occur with nanocrystalline diamond films [24], but the minimum barrier to be surmounted is still about 3 eV, and it is again difficult to think up any convincing potential configuration that would allow easy electron escape for an applied local field of 10 V/μm or less.

A third hypothesis was to assume that there are field-enhancing structures at the film surface, with γ-values of between 100 and 1000. Plausible structures can be created when the film has been 'activated' by some form of vacuum discharge. But if care is taken to avoid such effects, then - with amorphous carbon films - scanning tunnelling microscope investigations find no such structures but indicate that such films have surfaces that are flat on the scale of about ± 2 nm [25].

Such experiments led Amaratunga and Silva [26] to conclude that, for their amorphous carbon films, emission control must be at the back interface (between the substrate and the carbon film).

This fourth hypothesis, that emission control is at the back surface, had previously been proposed by Geiss and co-workers [27], in connection with emission from bulk diamond. There is evidence to support the hypothesis in this case, where: the diamond is relatively defect free; the front surface may genuinely have low or negative electron affinity; and there is a large voltage drop across diamond, that allows electrons to pick up kinetic energy from the electric field and retain it. However, it seems unsustainable for the amorphous carbon films used by Amaratunga and Silva.

First, a back-surface control mechanism requires that the applied field can penetrate to the back surface. Even for the highest macroscopic fields (50 V/μm) used in their experiments, it takes only a surface charge density of 0.003 electrons/nm^2 (or one electron for every 360 nm^2 of surface) to totally screen the macroscopic field from the film interior. Provided that there are sufficient available states at the film front surface, at energies below the *substrate* fermi level, it is impossible to imagine that electrons will not become present in such densities at the front surface, either in the conduction band or in surface states. Since Ref. [26] assumes that the electron affinity for their films is 2 eV, there seems no possibility that such electrons would be able to escape sufficiently rapidly either by thermal activation or by tunnelling (the time-constant associated with thermionic emission is the shorter, but is of order 10^{12} years).

There also seems no possibility that electrons injected into the conduction band at the back surface of the film will normally pick up and retain enough energy to escape over the electron affinity barrier as a 'hot electron'. The scattering length in an amorphous carbon film must certainly be less than 10 nm, so even at an assumed internal field strength of 50 V/μm the maximum energy to be gained from the field before scattering is only 0.5 eV, which is insufficient to ensure escape across the electron affinity barrier.

[The situation may in fact be worse than just described, since much research confirms that in such films conduction normally occurs by hopping via localised states in the energy gap below the mobility edge associated with the conduction band. In such circumstances the surface barrier would be greater than the electron affinity.]

Thus none of the older hypotheses is able to provide a satisfactory explanation of the mechanism of electron emission. There is also the crucial evidence provided by the field electron spectroscopy (FES) experiments of the Fribourg group (see, for example, Refs. [28-30]). It is well established [31,32] that, by measuring both field emission energy distributions and current/voltage characteristics from a given emission site, it is possible to determine both the local work-function and the local field at this site. The Fribourg group, especially Küttel and O. Gröning, have applied this technique to a variety of carbon films, and have always deduced local work-functions that are 'normal' (i.e. about 4 eV, or more) and local fields that are 'normal' (i.e. in the V/nm range). The macroscopic fields in the Fribourg experiments are, however, much the same as those measured by other workers with carbon films, namely in the vicinity of 10 V/μm. The experiments are thus 'measuring' field enhancement factors in excess of 100. The conclusion must be that emission is taking place by a more-or-less 'normal' field emission process, and that field enhancement is being created by a mechanism to be determined.

The Fribourg group observed (by means of scanning probe microscopy experiments) that small conductive clusters were present in their films, and associated these with good field-emitting properties [30]. As will be seen below, the present author concluded [1,33] that the observed field enhancement is directly associated with these small conductive clusters, and that sufficient (or nearly sufficient) field enhancement can be predicted using a simple formula.

In parallel with this work, both the Cambridge research groups had reached very similar conclusions. Amaratunga et al. [34] reported a new form of thin-film amorphous carbon, in which emission was assumed to be associated with the presence of 'nanoparticle inclusions and nanotube fragments', that were assumed to generate field enhancement. Ilie et al. [35] and Robertson [36] report a careful series of experiments that involved the formation of small graphitic clusters by annealing amorphous carbon (a-c:H or ta-C-N) films, finding the lowest field emission threshold for a cluster size of 1-2 nm; it is assumed [36] that these clusters generate field enhancement.

The structure of all these carbon films in fact has similarities with the composite field emitters described by Latham and colleagues [37], and used by Printable Field Emitters Ltd as the basis for their field emission displays [38].

ENH materials and the 'internal field enhancement' hypothesis

Against this background, Ref. [1] proposes that carbon-based films emit electrons because they are members of a class of materials there termed *electrically nanostructured heterogeneous materials (ENH materials)*. Electron emission in the presence of low applied macroscopic fields should be a general property of thin films of such materials when deposited on a conducting substrate, given appropriate conditions.

ENH materials may have a structure consisting of the following:
  (1) A dielectric matrix (usually a 'leaky' dielectric).
  (2) Small conducting or semiconducting inclusions within the dielectric.
  (3) Conducting channels formed between the substrate and inclusions.
  (4) An emitting channel formed between an inclusion and the vacuum.

Structures where the inclusions are pin-like are included, as are structures where inclusions actually protrude into the vacuum.

The hypothesis was made [1,33] that "Field induced electron emission from thin films of ENH materials occurs because there is conducting nanostructure *inside* these films, that generates sufficient field enhancement to cause more-or-less normal field electron emission (FN emission) to take place near the film/vacuum interface". I call this the *'internal field enhancement'* hypothesis.

This hypothesis does not exclude the possibility that other mechanisms, not discussed here, contribute to enhancing the emission current, via secondary effects. But it does imply that geometrical field enhancement is the primary effect.

Obviously, the background described above makes this a plausible hypothesis in the case of the amorphous carbon films. But a strength of the hypothesis is that a field enhancement factor of sufficient (or nearly sufficient) size can be predicted from the measured properties of the Fribourg films. The relevant data [30] are as follows:

Film thickness (h):	300 nm
Diameter (2ρ) of conducting clusters:	20-40 nm
Macroscopic field:	20-25 V/μm
Local field (from FES):	6500 V/μm
Apparent value of γ	≈ 300

It is hypothesised that two sources of field enhancement exist. First, the conducting cluster acts as a 'hemisphere on a post' or a 'floating sphere at substrate potential'. From the usual formula, this produces a field enhancement factor γ_1 given to a good approximation by

$$\gamma_1 \approx h/\rho \approx 30. \tag{14}$$

With a 20 nm diameter cluster, it also seems plausible that should be smaller-scale field-enhancing structure on top of the cluster. It is easily plausible that this should create a further field enhancement γ_2 of around 5, and not too implausible that γ_2 should reach 10. The total predictable field enhancement $\gamma (= \gamma_1 \gamma_2)$ seems to lie in the range 150 to 300.

It is arguable that the estimated upper limit (300) on γ is at the limit of plausibility, especially since the lower limit of cluster size has been taken, but there are other uncertainties in the situation and I would argue that the agreement between the predicted and the empirically derived values of γ is 'good enough in present circumstances', and justifies the formulation of the 'internal field enhancement' hypothesis.

I further assume that internal field enhancement of this kind is responsible for the electron emission from all the carbon-based films whose structure consists of good conducting structures within a moderately conducting matrix [for want of a better term I refer to these as 'partially graphite like' (PGL) materials]. I do not, for the present, hypothesise that this mechanism will explain *all* the behaviour of films that consist of (poorly conducting) nanocrystalline diamonds with moderately conducting material between them [I refer to these as 'partially diamond-like' (PDL) materials]. An obvious test of this hypothesis for PGL films is conduct on them field electron spectroscopy experiments of the Fribourg type.

Perhaps the best reason for finding the internal-field-enhancement hypothesis comfortable is not the numerical agreement demonstrated above, but the following qualitative reasoning. For materials similar to those just discussed, the hypothesis provides a ready and robust qualitative explanation for the range of threshold macroscopic fields quoted in the literature. If we make the assumption that the onset of field emission for different partially graphite-like films occurs at a more-or-less constant value of *local* field, then the reported range of threshold macroscopic fields corresponds to a range of field enhancement factors (with some influence of different emission areas). It is easy to suppose that this range of γ-values could be associated with differences in the formation of film nanostructure during deposition.

For example, one could speculate that the conditions under which these carbon films are laid down are, in a certain sense, the energetic equivalent of the formation of steel under blast-furnace conditions. For steels, the formation of fine nanostructures during cooling is well attested by atom-probe field-ion microscopy [39]. It seems not implausible that broadly similar 'carbon metallurgical' processes occur during film deposition, and that the nature of the nanostructures would be critically and sensitively dependent on the parameters of formation and deposition. So a range of enhancement factors could be expected. And it would not be surprising if researchers were able to 'tweak' the deposition chemistry and the deposition conditions empirically in order to achieve internal nanostructures that give rise to higher-than-average enhancement factors and lower-than-average threshold fields. Whether this particular speculation is true or not, it is easy to imagine that a range of nanostructures could exist in the films.

It seems significantly more difficult to explain this variability by other hypotheses put forward to explain film emission. Even a good qualitative explanation is difficult to construct.

We also need to reconsider the 'Latham-style' composite field emitters. For graphite flakes in resin, Bajic and Latham [40] specifically suggested that emitting channel formation was due to field enhancement. However, field enhancement (*before* the emitting channel turns on) has not been a major part of the emission mechanism proposed for the other composite emitters. Rather the escape has been attributed mainly to a 'hot electron mechanism' [38].

In this case, I hypothesise that, once a conducting path to the top of a 'near-vacuum' inclusion is established the emission mechanism involves *both*: (1) geometrical field enhancement by the inclusion, that thins some barrier at the front edge of the inclusion; and (2) resulting electron penetration through a thin dielectric film, via an emitting channel. The latter may conceivably take place by a variety of mechanisms, including some involving a degree of hot electron emission, but to the present author it seems important that there should be 'normal-size' fields (in the V/nm range) at *both* ends of the emitting channel. As with the carbon-based films, a test of this hypothesis would be to take FES measurements under conditions of low peak shift, and to derive local field values from the energy distributions, as in the Fribourg measurements.

It is possible that the mechanisms of conduction within the dielectric and through/around the inclusions will be different in detail for these Latham-style emitters, as compared with the carbon-based films. Such differences in detail do not detract from the validity of the overall internal-field-enhancement hypothesis.

Comments on the internal field-enhancement hypothesis

Lateral potential differences. We next review possible counter-arguments. The first is that, with a 'leaky' dielectric, differences in electrostatic potential cannot exist across the film surface.

In the 1970's the present author conducted field electron spectroscopy experiments: differential voltage drops, as between closely adjacent parts of a field emitter surface, were *observed* to occur when emission current is flowing [41,42]. These observations were on a high resistivity semiconductor (GaP). But, later, Mousa [43] saw related effects on carbon-fibre field emitters. There is no doubt that lateral potential variations can exist.

There is also no difficulty in finding mechanisms able to sustain such effects, once initiated. The Modinos 1974 calculations on surface-state behaviour [44] show that, if there is slow transport into a surface-state band but fast transport out via tunnelling emission into vacuum, then the surface-state fermi level decouples from the bulk fermi level and moves to lower energy. This implies that the surface-state band becomes less negatively charged (or positively charged); the ionized surface then acts as an electrostatic attractor to electrons in the bulk, especially those in conducting channels leading to the surface.

Further, with a nanostructured material, where the total size of a conducting entity at the surface may be small, a similar effect might occur for the 'bulk' of this entity. If electron transport into the entity were slow but there could be fast transport out via tunnelling emission into vacuum, then the entity might become less negatively charged and act as an electrostatic attractor for electrons in the bulk, especially those in relevant conducting channels. Obviously, it will be necessary to make sure, with a more detailed theory, that such an argument is self-consistent.

The variation of threshold field with film thickness. A second requirement is that the internal-field-enhancement hypothesis should be able to explain observed trends in the variation of macroscopic threshold field with thickness. Typical results [45] are that the threshold field initially falls as the film thickness increases, passes through a minimum, and then increases again.

Qualitatively, the observed initial fall-off is to be expected because, in the simplest approximation, we have

$$F_{M,on} = F_{on}/\gamma = (\rho/h) F_{on}. \tag{15}$$

where h is here the *effective electrical thickness* of the film and ρ is the radius of the surface conductive feature responsible for emission, and F_{on} and $F_{M,on}$ are the threshold values of local field and macroscopic field, respectively. Data from the a-C:H films grown at Surrey [45], re-plotted to show $F_{M,on}$ as a function of $1/h$, give a good straight-line relationship for thicknesses up to the 69 nm data point [1]; above this $F_{M,on}$ increases again. Theoretically, on purely electrostatic grounds, we expect $F_{M,on}$ to level out towards a constant value as thickness increases. However, this effect cannot readily explain the observed minimum and subsequent increase in threshold macroscopic field with film thickness (unless there is a correlated change in the surface density of emitting features).

A possible cause may be the accumulation of *negative* space charge within the dielectric, at the back, towards the substrate/film interface. If this space-charge is distributed relatively uniformly *laterally* across the surface, as might be expected, then this will have the effect of creating a *virtual electrical base-plane* within the dielectric film. This virtual base-plane cannot be particularly well-defined: nevertheless, the quantity h that appears in the field enhancement formula above is, in principle, measured from a suitably defined virtual base-plane.

Effects observable in high resistivity films. It is also convenient to note an effect found with higher-resistivity partially diamond-like films, but which might also occur with partially graphite-like films. With PDL films, both Göhl et al. [46] and Gröning et al. [29] have found that voltage drop can occur across the film and that emission current is limited by the conductivity of the film. The current is controlled by a Poole or Poole-Frenkel hopping-type conduction process, rather than by emission through a barrier. A question arises as to how this situation is to be made compatible with the theory just presented. Space prevents full discussion, but comments may be helpful.

At low emission currents a film/vacuum-barrier controlled Fowler-Nordheim-like regime might be expected, that goes over into a film-conductivity controlled regime as the applied field is increased. When the film conductivity becomes the limiting factor, this means that the current is lower than it would be if the front barrier were the limiting factor. This implies that, in thehigher field regime, the electron transport out of the emitting surface region by tunnelling into the vacuum is intrinsically faster than electron transport into surface region by 'hopping transport'. The effect, as with the surface states in the Modinos calculation [44], should be to cause the surface region to become locally less negatively (or positively) charged. This in turn should have the dual effect of: (a) increasing (locally) the field across the film, thereby increasing the transport rate-constant into the surface region; and (b) reducing (locally) the field that controls the rate of tunnelling emission into vacuum. So an equilibrium or quasi-equilibrium situation should be reached.

If the rate of electron transport across the film is determined by processes taking place across its whole width, then this rate is probably not much influenced by a local charge variation at the surface. But emission into vacuum is likely to be much more a *local* effect (laterally), so local charge changes could significantly affect the local rate-constant for emission into vacuum. So the most probable outcome is that the nature of the surface barrier (and possibly the surface electron energy distribution) will adjust so that the current through the local barrier is equal to the current through the film. In which case the emission process will (probably) be tunnelling or energetically assisted tunnelling, but the emission current will be controlled by the bulk film conductivity.

Several different mechanisms may be possible. The suggestion of Gröning et al. [29], that enhanced field penetration into a conical emitter would effectively decrease the local field in the barrier through which tunnelling emission occurs, is one of the options.

ENH materials as electron emitters

In the design of broad-area field-electron sources there are some key requirements. These apply to all sources, including Spindt arrays [47], and include:
(1) The need to be able to generate sharp features that field emit.
(2) The need to generate these controllably, with sufficient density and uniformity.
(3) Emission currents shall be stable and not unduly noisy.
(4) Emitters shall be robust against poor vacuum conditions.
(5) Emitting sites shall be stable against burn out and other forms of degradation.

In the case of the Spindt arrays, 'ballast resistance' (or other forms of current control) is included in the substrate in order to meet requirements (3) and (5) [47].

I would argue that ENH materials can be good candidates for broad-area electron sources, for the following main reasons. First, by suitable design of the manufacturing process, one can control the size and shape of the inclusions, and hence their field-enhancing abilities. Second, again by suitable design of the manufacturing process, one can control the dispersion and surface density of the inclusions, and hence the density of electron emission sites. Third, provided that the matrix itself (a) has a relatively inert surface under industrial vacuum conditions, and (b) is relatively transparent to the flow of electrons through a thin matrix layer, then vacuum robustness can be achieved because field emission into vacuum can take place without the actual field-enhancing features being exposed to 'poisoning' or corrosion by vacuum system contaminants. Fourth, if the matrix has suitable electronic characteristics, (and, to some extent, if the matrix/inclusion interfaces also have appropriate electronic characteristics) then one may be able to design for facilitating the creation of conducting and emitting channels, and for operating these channels in a mode close to the onset of saturation. For channels close to saturation, emission currents should less noisy, burn-out may be less likely, and (possibly) the effects of physical inhomogeneity as between emission sites may be less influential on emission current variability as between sites.

Quite apart from these technological considerations, there seems some exciting new physics and materials science waiting to be learnt from these materials. Clearly, we have only just begun to unravel a very complex and subtle scientific story.

REFERENCES

1. R. G. Forbes, *Solid State Electronics* (submitted for publication).
2. R. H. Fowler and L. W. Nordheim, *Proc. R. Soc. Lond.* **A119**, 173 (1928).
3. E. L. Murphy and R. H. Good, *Phys. Rev.* **102**, 1464 (1956).
4. R. G. Forbes, *J. Vac. Sci. Technol.* **B17**, 526 (1999).
5. F. M. Charbonnier and E. E. Martin, *J. Appl. Phys.* **33**, 1897 (1962).
6. L. N. Dobretsov and M. V. Gomoyunova, *Emission Electronics* (Izdatel'stvo "Nauka", Moscow, 1966) (trans. by: Israel Programme for Scientific Translations, Jerusalem, 1971).
7. C. A. Spindt, I. Brodie, L. Humphrey and E. R. Westerberg, *J. Appl. Phys.* **47**, 6248 (1976).
8. In the defintion in Ref. [4], a minus sign has been omitted: *u* needs to be positive.
9. R. F. Burgess, H. Kroemer and J. M. Houston, *Phys. Rev.* **90**, 515 (1953).
10. R. G. Forbes, *J. Vac. Sci. Technol.* **B17**, 534 (1999).
11. A. Modinos, *Field, Thermionic, and Secondary Electron Emission Spectroscopy* (Plenum, New York, 1984).
12 R. Stratton, *Phys. Rev.* **125**, 67 (1962).
13. L. S. Pan, *Mat. Res. Soc. Symp. Proc.* **436**, 407 (1996).
14. M. Shah, *Physics World* **10** (6), 39 (1997).

15. B. R. Chalamala and B. E. Gnade, *IEE Spectrum* (April 1998) 42.
16. V. V. Zhirnov and J. J. Hren, *MRS Bulletin* (September 1998) 42.
17. S. R. P. Silva, J. Robertson, W. I. Milne and G. A. J. Amaratunga (eds.), *Amorphous Carbon: State of the the Art* (World Scientific, Singapore, 1998).
18. M. W. Geiss, N. N. Efremow, J. D. Woodhouse, M. D. McAleese, M. Marchywka, D. G. Socker and J. F. Hochedez, *IEEE Electron Dev. Lett.* **12**, 456 (1991).
19. C. Wang, A. Garcia, D. C. Ingram, M. Lake and M. E. Kordesch, *Electronics Lett.* **27**, 1459 (1991).
20. J. Robertson, *J. Vac. Sci. Technol.* **B17**, 659 (1999).
21. J. Robertson, *Carbon* **37**, 759 (1999).
22. C. Herring and M. H. Nichols, *Rev. Mod. Phys.* **7**, 95 (1935).
23. L. K. Hansen, *J. Appl. Phys.* **37**, 4498 (1966).
24. J. B. Cui, J. Ristein and L. Ley, *Phys. Rev.* **B 60**, 16135 (1999).
25. S. R. P. Silva, private communication.
26. G. A. J. Amaratunga and S. R. P. Silva, *Appl. Phys. Lett.* **68**, 2529 (1996).
27. M. W. Geiss, J. C. Twichell, N., N. Efremow, K. Krohn and T. M. Lyszczarz, *Appl. Phys. Lett.* **68**, 2294 (1996).
28. O. M. Küttel, O. Gröning, L. Nilsson, L. Diederich and L. Schlapbach, *Electron Field Emission from Carbon Films*, in Ref. [17].
29. O. Gröning, O. M. Küttel, P. Gröning and L. Schlapbach, *J. Vac. Sci. Tech.* **B17**, 1 (1999).
30. O. Gröning, O. M. Küttel, P. Gröning and L. Schlapbach, *Appl. Phys. Lett.* **71**, 2253 (1997).
31. R. D. Young and H. E. Clark, *Phys. Rev. Lett.* **17**, 351 (1966).
32. T. V. Vorburger, D. Penn and E. W. Plummer, *Surface Sci.* **48**, 417 (1975).
33. R. G. Forbes, *1st European Field Emission Workshop*, Toledo, November 1999 (unpublished abstracts).
34. G. A. J. Amaratunga, M. Baxendale, N. Rupesinghe, I. Alexandrou, M. Chhowalla, T. Butler, A. Munindradasa, C. J. Kiley, L. Zhang and T. Sakai., *New Diamond and Frontier Carbon Tech.* **9**, 31 (1999).
35. A. Ilie, A.C. Ferrari, T. Yagi and J. Robertson, *Diamond 99*, Prague, September 1999 (unpublished abstracts).
36. J. Robertson, *these proceedings*.
37. S. Bajic, M. S. Mousa and R. V. Latham, *Colloque de Physique* **50 (C8)**, 79 (1989).
38. R. A. Tuck, W. Taylor, P. G. A. Jones and R. V. Latham, *Proc. 4th International Devices Workshop, Nagoya*, 1997, p. 723.
39. M. K. Miller, A. Cerezo, M. G. Heatherington and G. D. W. Smith, *Atom Probe Field Ion Microscopy* (Clarendon, Oxford, 1996).
40. S. Bajic and R. V. Latham, *J. Phys. D: Appl. Phys.* **21**, 200 (1988).
41. E. Braun, R. G. Forbes, R. V. Latham, J. M. Pelmore and D. E. Sykes, *22nd International.Field Emission Symp., Atlanta, August 1975* (unpublished abstracts, p. 33).
42. R. G. Forbes, *Technical Digest, 9th International Vacuum Microelectronics Conference* (ISBN 5-86072-081-5) (Bonch-Bruevich Univ. of Telecommunications, St Petersburg, 1996) p. 58.

43. M. S. Mousa, *Appl. Surf. Sci.* **94/95,** 129 (1996).
44. A. Modinos, *Surface Sci.* **42,** 205 (1974).
45. R. D. Forrest, A. P. Burden, S. R. P. Silva, L. K. Cheah and X. Shi, *Appl. Phys. Lett.* **73,** 3784 (1998).
46. A. Göhl, B. Günther, T. Habermann, G. Müller, M. Schreck, K. H. Thürer and B. Stritzker, *J. Vac. Sci. Technol* **B 18,** 1031 (2000).
47. I. Brodie and C. A. Spindt, *Adv. Electr. Electron Phys.* **83,** 1 (1992).

Mat. Res. Soc. Symp. Proc. Vol 621 © 2000 Materials Research Society

Simulation Of The Influence Of Interface Charge On Electron Emission

Kevin L. Jensen and Jonathan L. Shaw
Code 6840, ESTD, Naval Research Laboratory
Washington, DC 20375-5347 USA

ABSTRACT

Several materials are promising candidates for electron sources. For diamond, a tunneling interface at the back contact limits injecting charge into the conduction band, but a purely geometric model of internal field emission is inadequate to explain experimental data. The presence of a defect, modeled by a coulomb charge, within the tunneling barrier region significantly enhances transmission and, in concert with a geometrical model, may better account for observed current levels. Charge has been suggested to play a similar role in the SiO_2 covering on a single tip silicon field emitter to explain experimental data. The tunneling theory in both cases is similar. In the present work, a general method for estimating electron transport and energy distributions through potential profiles, which describe both semiconductor interfaces and field emission potential barriers when a charged particle modifies the tunneling barrier, is developed. While the model is intended for treating a metal-semiconductor interface, it is cast here in terms of in a thin SiO_2 coating over a silicon field emitter tip to enable qualitative comparisons with experimental data. Tunneling probabilities are found by numerically solving Schrödinger's Equation for a piece-wise linear potential using an Airy Function approach. A qualitative comparison to experimental energy distribution findings is possible by utilizing an analytical model of the field emitter tip from which current-voltage relations may be found.

INTRODUCTION

Microfabricated diamond structures are under investigation for high power microwave amplifiers and electronic devices [1]. Emission measurements from insulating diamond on p- and n-doped Si substrates suggest that the substrate-diamond interface plays a dominant role in the electron emission of a diamond film [2]. A WKB analysis of the interface barrier using a geometric model of interface roughness and parameters ascertained from experimental data underestimated observed current levels by more than order of magnitude for realistic parameters [3]. If transmission into the conduction band of diamond were due to geometrical field enhancement (internal field emission model) alone, then the size of the ellipsoids which approximate interface roughness would be unphysical. Other mechanisms are evidently involved, such as defects [4, 5]. A simple model is obtained by including a coulomb potential within the potential profile at the material-diamond interface [6]. Elsewhere, measurements on a single silicon field emitter tip developed by MCNC showed that changes in the local electric potential due to charges becoming trapped in the oxide accounted for changes in the emission distribution [7]. A simulation of the effects of a coulomb potential in a tunneling barrier therefore would be of use in the analysis of both emission from diamond and from semiconductor field emitter arrays (FEA's). The present work is intended to report on progress towards such a model. The work is organized as follows. First, the Airy function approach for tunneling potentials modified by image charge effects is described. Second, the extension of the Fowler Nordheim equation to account for semiconductor properties is developed. Third, a 3-D model of field emission from a silicon tip is developed. And finally, a coulomb potential inside the tunneling barrier and its effect on current density and energy distribution is qualitatively examined in light of experimental findings, and is complimentary to Ref. [6] in that it considers mechanisms other than the sequential (two-step) transport of electrons from silicon to oxide states followed by emission into vacuum.

THEORETICAL ANALYSIS

Airy Function Approach To Wave Function Matching At Boundaries

The Airy function method to solve Schrödinger's equation has been successfully applied to the problem of electron transport through resonant tunneling diodes [8] and from field emitters [9]. Its advantage is that it accurately follows wave function evolution over large regions characterized by a constant electric field $F = e|E|$, whereas a plane-wave approach must partition the potential into a large number of steps. Its disadvantage is the evaluation of Airy functions can incur an unwelcome computational cost and provoke numerical failures in implementation unless attention is paid to the exponential arguments (when F is small), whereas the plane-wave approach is trivial to calculate. We refine and utilize a recent approach to circumvent numerical issues and provide rapid evaluation [10], while retaining the matrix methodology of earlier work [9]. In a region where $V(x) = V_o + sF(x-x_n)$, where x_n is a reference point, $s = \pm 1$, $F > 0$ is the applied field, the wave function may be expressed as

$$\psi(x) = t(k)\frac{Zi(c,\omega)}{Zi(c_o,\omega_o)} + r(k)\frac{Zi(-c,\omega)}{Zi(-c_o,\omega_o)}; \quad Zi(c,\omega) = \frac{Fi(c,\omega)}{\omega^{1/4}}\exp\left(\frac{2}{3}c\omega^{3/2}\right) \quad (1)$$

where $c = 1$ or i, $\omega(x) = |v+sfx-k^2|/f^{2/3} \geq 0$, $f = 2mF/\hbar^2$, $v = 2mV_o/\hbar^2$, m is the effective electron mass, k is momentum (i.e., $E(k) = (\hbar k)^2/2m$, and Zi is related to the Airy functions via [11]

$$Zi(c,\omega) = \frac{1}{4}(1-c)(c^2+2c+3)Ai(c^2\omega) + \frac{1}{4}(1+c)(c^2+2c+3)Bi(c^2\omega) \quad (2)$$

$Fi(c,\omega)$ is defined by Eqs. (1) and (2) and its asymptotic magnitude is $1/\sqrt{\pi}$: such a formulation is useful because the troublesome portions of the asymptotic expansions of the Airy functions for large argument cancel out when ratios are taken. Sinusoidal wave functions are described by $c^2 = -1$ whereas exponential wave functions by $c^2 = 1$, as per the Airy functions for positive and negative arguments. c_o and w_o are determined at $x = x_n$. The potential is broken up into N linear regions demarcated by x_n, where $1 \leq n \leq N$, and $V(x-x_n=0^-)$ need not equal $V(x-x_n=0^+)$. ψ and $\partial_x\psi$ are matched at x_n subject to the boundary conditions $t_0 = 1$ and $r_{N+1} = 0$. This serves to determine $t_n(k)$ and $r_n(k)$ in Eq. (1). The current density J is then

$$J_{N+1} = -\frac{2f^{1/3}s|t_{N+1}|^2}{\pi|Zi(c_N,\omega_N)|^2} \quad (3)$$

where the "N" subscript indicates evaluation at the boundary ($x = x_N$). The incident (plane-wave) current density per k is $J_0 = 2ik$. The transmission coefficient is defined by $T(E(k)) \equiv J_{N+1}/J_o$. The form of Eq. (1) insures that numerically, the arguments of the implicit exponential factors rarely run into numerical problems, even when f is small. When such regions do occur (<5% of the time here), asymptotic formulae may be used. For example, for a triangular (Schottky) barrier with $s = -1$, $t_{N+1}(k) \approx 2k/[(v-k^2)^{1/2} - ik]$, and Eq. (3) becomes

$$J_{N+1}|_{f\to 0} \approx \frac{8i}{v}k^2\sqrt{v-k^2}\exp\left(-\frac{4}{3f}\left(v-k^2\right)^{3/2}\right) \quad (4)$$

$T(k) \equiv J_{N+1}/J_0$ adopts its usual form [12]: the exponential is the WKB approximation without an image charge. Integrating $(\hbar k/2\pi m)\,T(k)$ with the supply function $f_o(k)$ gives the current density.

Band Bending, Applied Field, and the Exchange-Correlation Potential

Because the density of electrons ρ in the conduction band for a semiconductor is much smaller than for a metal, an applied field causes band bending due to electrons collecting at the surface. Following Sze [13], $\rho = N_c (2/\sqrt{\pi}) F_{1/2}(\beta\mu)$, where $F_{1/2}(x)$ is the Fermi-Dirac integral, $\beta = 1/k_BT$, T is the temperature, and $\mu(T)$ is the electrochemical potential, approximated by [9, 14]

$$F^2 = \frac{2}{3}\pi^2 N_c K_s \sqrt{\frac{\beta\mu}{\pi}}\left[1 + \frac{8}{5}\left(\frac{\beta\mu}{\pi}\right)^2\right] \tag{5}$$

K_s is the dielectric constant of silicon. The smallness of μ for semiconductors indicates that Exchange-Correlation effects are more pronounced and the image charge form of the potential barrier requires modification. The Fowler Nordheim (FN) equation may be retained if we (i) assume the potential has been shifted by a term $x_o = \hbar/(2m(\mu+\phi))^{1/2}$ where ϕ is the maximum height of the barrier above μ, (ii) replace the work function with $\Phi^* = \Phi + (8/3\pi) Q k_F^3 x_i^2 + 2 F x_o$ where $Q \approx 3.6$ eV, x_i is the origin of the background positive charge and k_F is the Fermi momentum, (iii) Taylor expand $\ln\{T(E)\}$ at the maximum of $T(E) f_o(E)$ rather than at $E \approx \mu$ as done in the standard FN treatment and (iv) assume that μ, needed for $f_o(k)$, can be related to the bulk value via the relation $\rho(\mu_{Si}; x=0^-) = \rho(\mu; x=0^+)$. While simple to state, the implementation of these approximations necessitates an iterative numerical search for self-consistent values, described in greater detail in Ref. (14), because terms such as x_i and x_o depend upon barrier height V_o, which in turn depends on them. Finally, the Airy function approach was used to validate the WKB treatment inherent in assumptions (ii-iii) for the SHM model (below). Emission from surface states [12] has *not* been included – comparisons to experimental data below are therefore qualitative at best.

Model Implementation

Tip current is obtained by integrating J over the surface of a hyperbolic emitter by $I_{tip} = \int_\Omega J(F(z,\rho)) \, d\Omega$, where (z,ρ) are cylindrical coordinates and $d\Omega$ is a differential surface element. Generalization to an array (the "Statistical Hyperbolic Model" (SHM) for a gated field emitter [15, 16]) is achieved by the introduction of a statistical term such that $I_{array} = \Sigma(\Delta\Phi, \Delta a_s) I_{tip}(\Phi^*, a_s)$, where (in contrast to previous work) Σ accounts for variation in the work function due to adsorbates ($\Delta\Phi$) as well as a distribution in tip radii Δa_s). By utilizing the aforementioned approximations (i)-(iv) plus the effective work function Φ^* and Eq. (5) to relate chemical potential to applied field, the Semiconductor version of the SHM ("SSHM") results, and was used to mimic experimental data [17] with the following parameters: tip radius $(a_s) = 30$ Å, gate radius $(a_g) = 0.9$ μm, and cone half angle $(\beta_c) = 22°$.

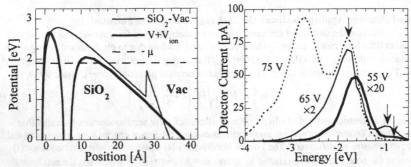

Figure 1: *(left) Heavy line is potential barrier used in analysis for $a_s = 0.1$ Å, $x_{ion} = 6$ Å. It approximates the thin line corresponding to SiO_2 potential (shown without ion potential) and changes at the SiO_2-Vacuum boundary at 28 Å (barrier lowering at surface due to electron accumulation ignored). "μ" is the chemical potential;* **Figure 2:** *(right) Exp. data from single Si emitter (E relative to E_F). Scale factors (x2) and (x20) for 65 V and 55 V respectively. Movement of peak indicated by arrows. Current: I(75 V) = 10 nA; I(65 V) = 2 nA; I(55 V) = 0.1 nA.*

Energy distributions been reported by one of us (JLS) for a single tip [18]: Figure (2) of that work is represented in Figure (2) here. The peak structure was suggested to be due to a two step (sequential) process of electrons hopping to an intermediate level then tunneling out (hence the levels require time to refill after emission): in short, the peaks are due to the presence of oxide states, and that the shift in peaks is caused by the behavior of oxide charge. The theory in the present work is complimentary to the previous work in allowing for effects of embedded charge on the Si-SiO$_2$ tunneling barrier (*e.g.,* resonant transmission), but it does not dynamically model states filling and depleting. To minimize differences with Ref. [17], we retain as much as possible the geometric and material parameters of the single-tip experiment. Accounting for some changes in parameters (*e.g.,* $a_g \approx 7500$ μm) but leaving all other parameters unchanged in the application of the SSHM, we obtain $a_s \approx 58$ Å. An oxide is presumed to exist on the emitter of sufficient width that tunneling occurs in it. The SiO$_2$ parameters used are: $m = 0.63\ m_o$ and $K_s = 2.9$ [19], as well as the Si-SiO$_2$ interface barrier = 3.25 eV [20]. Once the electron passes the Si-SiO$_2$ interface barrier, it is assumed to suffer no further impediment to emission, *e.g.,* for $F = 0.264$ eV/Å (corresponding to $V_g = 55$ V in the SSHM simulation), the drop across a $t = 2.8$ nm oxide layer will be $Ft/K_s = 2.55$ eV (Fig. (1)).

As the Airy approach to calculating $T(k)$ is a 1-D approach, a problem arises in how to include the effects of a 3-D Coulomb potential the charge. The Coulomb potential along a line adjacent to the charge will resemble $V_1(x) = e^2/4\pi\varepsilon_o(r-r_{ion}) = e^2/\{4\pi\varepsilon_o[(x-x_{ion})^2 + a_{ion}^2]^{1/2}\}$, where $x = 0$ is at the interface, x_{ion} is the perpendicular distance from the charge to the interface, and a_{ion} is the perpendicular distance from the ray (along which $T(k)$ is evaluated) to the ion center r_{ion}. The effects of the charge therefore appear as a superposition of the $T(k)$ solutions for different a_{ion} values. While not entirely satisfactory (scattering is 3-D even for incident plane waves), this method nevertheless gives a qualitative indication a coulomb charge's effect on tunneling.

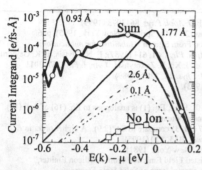 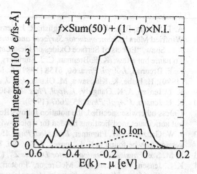

Figure 3: *(left) Individual and averaged current density evaluations for an image charge potential + coulomb potential: numbers refer to a_{ion}:* **Figure 4:** *(right) Weighted sum of "No Ion" and image charge + ion current density. The saw-tooth is an artifact of averaging the N=50 different a_{ion}: increasing N would smooth the curve to its envelope profile. Compare to "V=55 V" in Fig. (2).*

As a_{ion} increases, a transmission peak moves in relation to the Fermi level (designated μ). An indication of this effect is shown in Figure 3 for an applied field of $F = 0.26414$ eV/Å (corresponding to $V_g = 55$ V in the SSHM) where the dark curve "Sum" is the average of 50 such calculations for a_{ion} uniformly spaced between 0.1 Å to 2.6 Å (See Fig. (1)). As the coulomb potential becomes more *shallow*, the resonant peak moves to the *right*. The line "No Ion" is the solution in the absence of the coulomb potential. It is clear from the 2.6 Å line that beyond a certain radius from the ion center, the current density drops by one to two orders of magnitude. Assuming that the radius of the emission area for the tip is ≈ 26 Å (determined from the ratio of tip current to current density for no ions; a weighted emission area will be larger [16]), and take a weighted sum between the "ion" ("Sum" of Fig. 3) and the "No Ion" current, with the weight $(2.6 \text{ Å}/26 \text{ Å})^2 = 0.01 = f$, the curves of Figure 4 result. Three observations: (*i*) the majority of the current is due to the ion region (by an order of magnitude or more), (*ii*) multiple effects conspire to broaden the width of the integrand (related to the energy distribution) to on the order of 1/2 eV, comparable to the 55 V line in Fig. (2), but (*iii*) the energy distribution rises near $E \approx \mu$ ($=E_F$) – in contrast to Fig. (2), where the emission comes from oxide and interface states.

CONCLUSION

A rapid and stable Airy function approach has been developed to model transport through field emission potentials modified by the presence of a charge in the oxide. Using a coulomb potential model for the oxide charge, it was shown here that a significant enhancement of current (apart from any other additional mechanisms considered previously) can result due to the presence of such a charge. Usage of this defect model may account for higher current levels than a geometrical model of internal field emission can, and may therefore be useful in the analysis of emission data from, *e.g.*, diamond.

REFERENCES

1   J. E. Yater, A. Shih, R. Abrams, *Appl. Surf. Sci.* **146**, 341 (1999).
2   A. Göhl, V. Raiko, T. Habermannn, D. Nau, D. Theirich, G. Müller, J. Engemann, *Tech. Dig. Of 11th Int'l Conference on Vacuum Microelectronics (IVMC)* (July 12-24, 1998, Asheville, NC) p263.
3   K. L. Jensen, A. Göhl, G. Müller, *to appear in* Vol. 558 of *Proc. of Mat. Res. Soc.*, p601-606 (2000).
4   W. Zhu, G. P. Kochanski, S. Jin, *et al.*, *Appl. Phys. Lett.* **67**, 1157 (1995).

5   A. T. Sowers, B. L. Ward, S. L. English, R. J. Nemanich, *J. Appl. Phys.* **86**, 3973 (1999).
6   Winfried Mönch, *Semiconductor Surfaces and Interfaces (2nd Ed.)* (Springer, New York, 1995), p86.
7   J. L. Shaw, "Effects of Surface Oxides on Field Emission From Silicon" *(submitted for publication)*.
8   To name but a few: K. F. Brennan, C. J. Summers, *J. Appl. Phys.* **61**, 614 (1987); E. N. Glytsis, T. K. Gaylord, K. F. Brennan, *J. Appl. Phys.* **66**, 6158 (1989); C. M. Tan, J. Xu, S. Zukotynski, *J. Appl. Phys.* **67**, 3011 (1990); H. Inaba, K. Kurosawa, M. Okuda, *Jpn. J. Appl. Phys.* **28**, 2201 (1989).
9   K. L. Jensen, A. K. Ganguly, *J. Appl. Phys.* **73**, 4409 (1993).
10 K. L. Jensen, *J. Appl. Phys.* **85**, 2667 (1999).
11 Unless otherwise specified, the notation follows Ref. [10]. Here, Eq. (1) is analogous to Eq. (16) of Ref. (10), but anticipates modifications needed for a piece-wise linear potential.
12 J. W. Gadzuk, E. W. Plummer, *Rev. Mod. Phys.* **45**, 487 (1973) compare to Eq. (2.9). A higher order in E version may be found in K. L. Jensen, *J. Vac. Sci. Technol.* **B13**, 520 (1995).
13 S. M. Sze, *Physics of Semiconductor Devices 2nd*, (Wiley, New York, 1981).
14 K. L. Jensen, Y. Y. Lau, D. S. McGregor, "Photon Assisted Field Emission From A Silicon Emitter," *(submitted for publication to Vacuum Microelectronics, special issue of Solid State Electronics)*.
15 A summary of the Statistical Hyperbolic Model may be found in K. Jensen, *Phys. of Plasmas* **6**, 2241 (1999).
16 An ellipsoidal model may be found in R. G. Forbes, K. L. Jensen, "Extension of Fowler-Nordheim Theory beyond the linear field + classical image charge potential," *(to appear in J. Vac. Sci. Technol. B)*.
17 D. Temple, W. D. Palmer, L. N. Yadon, J. E. Mancusi, D. Vellenga, G. E. McGuire, *J. Vac. Sci. Technol.* **A16**, 1980 (1998). a_s and β_c from Fig. 4; a_g and N_{tips} from text and Fig. 9. $I_{array}(V_g)$ from Fig. 9.
18 J. L. Shaw, "Effects of Surface Oxides on Field Emission From Silicon" *(submitted for publication)*.
19 H. J. Wen, R. Ludeke, A. Schenk, *J. Vac. Sci. Technol.* **B16**, 2296 (1998).
20 V. Filip, D. Nicolaescu, F. Okuyama, C. N. Plavitu, J. Itoh, *J. Vac. Sci. Technol.* **B17**, 520 (1999).

Mat. Res. Soc. Symp. Proc. Vol. 621 © 2000 Materials Research Society

Observation of Hot Electron Field Emission

Jonathan Shaw
Code 6844, Naval Research Laboratory
Washington DC 20375

ABSTRACT

We report energy distributions of silicon Field Emitter Arrays coated with 50A of ZnO. The distributions reflect changes in the ZnO conductivity induced by annealing in vacuum, temperature, and annealing in hydrogen. An additional coating of titanium performed in-situ produced large additional changes. Emission from the ZnO at energies near the Fermi level increased with gate voltage only after hydrogen annealing, when hot, and after Ti coating. In those same cases the emission distribution contained a tail at energies above E_F. The high-energy emission tail is due to a many-body or Auger process whereby holes injected below E_F create hot electrons. Although emission from ZnO occurred at energies up to 8eV below E_F, no high-energy tail was observed in the normal case. Thus emission appears to occur from isolated electrons in ZnO gap states in cases where the distribution lacks a high-energy tail. Conversely, emission above E_F suggests that emission occurred from a metallic state such as an accumulated conduction band.

INTRODUCTION

The field emission energy distribution is helpful in understanding field emission physics [1], a process that remains poorly understood for general surfaces (other than atomically clean metals). Unlike clean metal surfaces, field emission from both molybdenum and silicon field emitter arrays often occurs at energies several volts below E_F [2,3]. The low energy emission is related to the presence of the native oxide and the density and energy distribution of states at the interface and in the oxide [4]. A single charge near the emission site can make a big change in the emission current and energy. The density of states in the oxide appears to be increased by both emission and baking in UHV. However, the density of states can be reduced again by exposing the surface to oxidizing species and/or cooling the surface to room temperature. Thus improved performance might be achieved by coating the surface with a conductive material relatively immune to oxidation. A number of metal oxide semiconductors are typically conductive due to native defects, and a prime example is ZnO. Although ZnO is a wide band gap semiconductor (Eg=3.3eV), it is typically n-type when it is slightly oxygen deficient. Several dopants including noble metals can increase the conductivity.

EXPERIMENT

The field emitter arrays were fabricated from n-type (0.02 ohm-cm) single-crystal silicon at MCNC. The fabrication process made use of oxidation sharpening and "tip-on-post" geometry as reported previously [5]. The gate aperture diameter was about 1.5μm, the gate metal was 0.5μm thick Pt, the tip height was about 2μm, and the tip pitch was 4um. The tip apex radii of similarly prepared specimens were about 5nm. The data reported here were obtained from an array patterned with 37530 tips. Several other arrays on the same die produced similar results. A 50Å layer of ZnO was deposited by sputtering. The ZnO-coated arrays were stored in air prior to the measurements. Specimens were loaded into an ion-pumped ultra high vacuum chamber without baking via load lock. A light bulb, placed behind the molybdenum sheet supporting the FEA die, was used to heat the specimens. We were unable to measure the die temperature exactly since it appeared that the die was considerably cooler than the platform. The temperatures given are estimates, based on the platform temperature. The chamber base pressure was 10^{-10} torr. Research grade hydrogen was admitted to the chamber though a leak valve. Titanium was deposited at UHV pressure using a small e-beam evaporator. The die contained several individually gated arrays. Wire probes mounted on manipulators were used to contact the substrate and one of the gate contacts. I-V measurements made between two substrate contacts verified that the contact potential between the probe and silicon substrate, as well as the resistive loss in the substrate, was insignificant up to currents well above the emission currents used. Measurements made on heavily doped and moderately doped specimens showed similar behaviors, again showing that any potential drop within the bulk or tip shank was not significant.

A VG CLAM 100 spherical sector analyzer, operated with constant pass energy set at 10eV to provide 0.05eV resolution, facilitated the energy distribution measurements. We verified that the analyzer resolution did not significantly broaden the spectra by comparing spectra measured using higher pass energies. Commercial instruments (Keithley models 237, 2001, and 6517) generated the gate and analyzer voltages and measured currents at the substrate, gate, and Faraday cup detector. The energy distributions are measured with respect to the vacuum level at the analyzer's entrance electrode. The Fermi level of the system is taken as 4.65eV below the vacuum level at the entrance electrode based on the typical work function of a gold plated electrode and verified by a variety of field emission energy distribution measurements including metallic surfaces. The electron energy is plotted with respect to the system Fermi energy.

RESULTS

The energy distributions obtained initially were unstable. After heating the die in UHV to 300C for 15 hours and cooling, the arrays consistently produced distributions such as illustrated in figure 1(a). The corresponding I-V plot is shown in figure 2(a) marked "initial". The emission current was much lower than typically obtained from similar uncoated arrays, as might be expected assuming the 50A ZnO coating increased the tip radius from 5 to 10nm. Emission thresholds at energies starting near zero (E_F) in those distributions show that the voltage drop across the Si/ZnO interface and ZnO coating was reasonably low. However, most of the emission occurred at energies several volts below zero. The emission currents at Vg=45 and 50V are identical at energies above –2V. The additional current emitted at 50V occurred at lower energies. Roughly speaking, this pattern continued at each higher gate voltage; the emission at the high-energy part of the distribution does not increase with gate voltage. This behavior indicates that the emission current is limited by the number of electrons available for tunneling, rather than by the tunneling probability as in the case of a metal emitter. The supply of electrons is probably limited by transport from the substrate. The distributions made at Vg=60-70V split into a main peak plus a shoulder on the high-energy side of the distributions. This behavior suggests that significant band bending within the ZnO is occurring at some of the tips due to insufficient transport. Conversely, the lack of any shift in the other distributions indicates little potential drop across the ZnO at lower emission currents.

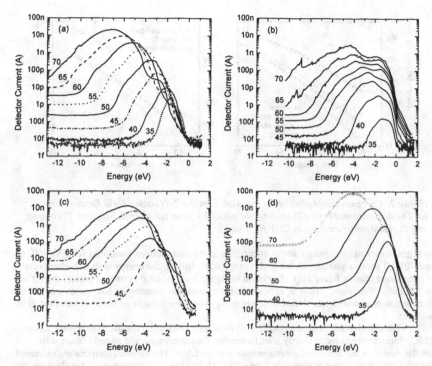

Figure 1.   Energy distributions produced by a FEA coated with 5nm ZnO. (a) Annealed in UHV, (b) after annealing in 10^{-4} torr H_2 at 250C, (c) 8 hours after H_2 anneal (after 7 hours emitting at Vg=70V), (d) after 64hours at 250C and emitting at Vg=70V.

In an attempt to reduce the ZnO (and thus increase its conductivity), the die was heated to 250C in 10^{-4} torr H_2 for one hour, followed by one hour at 300C at 10^{-9} torr to remove the hydrogen. The distributions obtained after cooling for one hour are shown in figure 1(b). The I-V curve is shown in figure 2(a). The total current was significantly lower in this condition than in the previous measurements. As in fig. 2(a), the emission at lower energies increases more rapidly with gate voltage than the emission near zero. In contrast to the previous distributions, the emission at energies near zero does increase with gate voltage, more like the behavior obtained from metals. Thus it appears that transport through the ZnO has been increased, even though the total emission current was reduced. These distributions also show significant emission at energies above E_F, with intensity decreasing exponentially with energy.

The distributions shown in figure 1(c) were measured after the array had been operating at room temperature in UHV for an additional 7 hours. These distributions are similar to the original distributions shown in figure 1(a), in that all the additional current emitted at higher gate

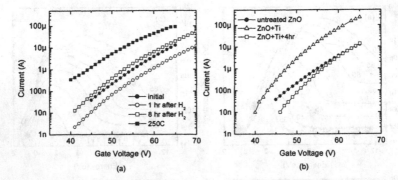

Figure 2. Current-voltage plots of the emission from the ZnO coated Field Emitter Array. (a) The data obtained before Ti coating, (b) initial I-V from baked ZnO, just after Ti coating, and Ti coating after emitting in UHV for 4 hours.

voltages occurs at increasingly lower energies, and the distributions measured with Vg=60, 65, and 70V all show a shoulder on the high energy side. In fact, the energy shift at higher currents is even larger than in figure 1(a). As shown in figure 2(a), the total current is greater than it was in the initial condition. Thus it appears that the emission probability has increased, but the distribution of electrons has shifted to lower energy (perhaps because of reduced transport to the surface).

The distributions plotted in figure 1(d) were measured while the die temperature was near 250C. Measurements made shortly after increasing the temperature and after 64 hours were similar, however the first set of measurements was unstable. The elevated temperature increased the total emission current as shown in figure 2(a). Unlike the room temperature distributions, the emission current near the Fermi energy increased as Vg increased from 35 to 40V. Above 40V the current near E_F began to saturate with the additional current emitted at lower energies. The distribution measured with Vg=70V showed a shift similar to the shift shown in fig. 1(a) and (c), however the shift was not as large. Each of the distributions shows some emission above E_F. The intensity of the emission above E_F increased with gate voltage, even in the 70V spectrum, however the increase was greatest between the 35 and 40V spectra.

At this point the die was cooled and ~2nm Ti was deposited onto the arrays. The gate leakage current increased to 1-10uA after the deposition, presumably because of deposition onto the SiO_2 dielectric surface separating the gate from the substrate. The field emission current increased initially, but degraded after operating the array in UHV several hours (as shown in figure 2(b)). However, the shape of the energy distributions did not change with time.

The energy distributions measured soon after the deposition are plotted in figure 3. These distributions are dramatically changed from the distributions measured prior to Ti coating. The distribution is sharply peaked just below E_F similar to the distributions typical of clean metal surfaces, but the peak energy occurred at -0.4eV even at the lowest gate voltage, significantly lower than might be due to thermal broadening. The distributions measured at Vg=60 and 65V are shifted to significantly lower energy. There are two or three consistent features at lower energies that shift down at higher gate voltages. The maximum intensity of the emission above

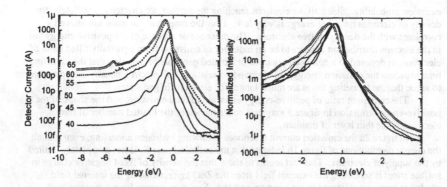

Figure 3. Energy distributions measured after in-situ deposition of 2nm of Ti. The same data is plotted in normalized format on the right.

E_F is about 0.5% of the emission peak, and the total current emitted above E_F is about 0.4% of the total current. The emission is reduced exponentially at energies above E_F. Measurable emission extends to energies greater than 4eV above E_F. The normalized distributions measured at 40,45,50, and 55V are nearly the same. The emission above zero is significantly increased in the spectra measured at 60 and 65V, where the emission peak had shifted to lower energy.

DISCUSSION

Emission at energies above E_F is known to occur during field emission from clean metal surfaces [1]. The positive energy emission occurs because the electron gas at the emitter surface is excited by hot holes created by emission from energies below E_F. The shape of the hot electron emission depends on how the hot hole energy is transferred to the surrounding electrons. The first experimental evidence of this effect showed that the emission intensity above zero energy decayed smoothly from the emission peak at zero energy [6]. Furthermore, the fraction of the total current with positive energy increased as the square of the total current. Gadzuk and Plummer provided a quantitative theory describing this behavior, based on the hot-hole scattering idea [7]. Another approach to explaining the high energy emission is that the original electron energy state is broadened by a term proportional to the tunneling time τ, $\Delta E \sim h/\tau$. Gadzuk comments: "It seems likely that both the many-body effects and the lifetimes are making nearly equal contributions...".

In contrast to the behavior of clean metals, the positive-energy portions of the distributions obtained from ZnO-Ti-coated arrays (shown in figure 3) are *independent* of the emission current. Furthermore, the emission intensity does not fall off with energy in the same way as the clean metal distributions. The spectra obtained from ZnO coated arrays showed either no emission or a much lower fraction of the emission at positive-energies. In figures 1a and 1c where no positive-energy emission was observed, the emission just below zero did not increase with gate voltage. In figures 1b, 1d, and 3, where positive-energy emission was observed, the emission near zero increased with voltage. Emission saturation with voltage implies that the emission is limited by the rate electrons reach the surface, rather than the emission probability. Since all the electrons reaching the surface are emitted immediately, the density of electrons near that energy is very low. Thus the magnitude of emission above E_F correlates with the density of free electrons. The presence or absence of the positive energy tail in the electron distribution appears to be an indicator of emission from a partially filled band of electrons, as opposed to a nearly empty band or isolated gap state. Assuming that the tunneling time broadens the emission energy, the observed correlation with electron density would appear to argue that the tunneling times are much longer for electrons in gap states.

The consistent ratio of positive-energy emission to total emission, and the shape of the positive-energy emission in figure 3 may be a consequence of the limited number of free electrons in the thin layer of titanium.

In figure 2a the emission current is reduced following hydrogen annealing, even though the energy distributions of figure 1b indicate that the emission near E_F is not completely limited by the supply of electrons. This fact seems to show that the density of fixed negative charge in surface states is screening the external field from the ZnO layer, reducing the internal field needed to transport electrons in the conduction band. Emission occurs from gap states well below E_F instead. The observed emission at energies as low as 8-10 eV below E_F simultaneously with much lower intensity emission near E_F suggests that the tunneling barrier is reduced for electrons in surface states, but not for conduction band electrons. This makes sense if we remember that the electric field can vary locally near a fixed charge. Emission from a neutral gap state will create a positive fixed charge. The local field near this positive charge can reduce the tunneling barrier in the local area, allowing electrons in other gap states to be emitted. The positive charge will also change the local potentials, including the local vacuum level. These local fields must be screened by negative charge in gap states rather than conduction band electrons.

ACKNOWLEDGEMENTS
Thanks to Gregory Exarhos (Pacific Northwest National Lab) for depositing the ZnO coatings. Thanks to DARPA, ONR, and NRL for providing funding. The Si FEAs were fabricated at MCNC under a DARPA/NRL contract.

REFERENCES
1. J. W. Gadzuk and E. W. Plummer, Rev. Mod. Phys. 45 487 (1973).
2. R. Johnston and A. J. Miller, Surf. Sci. 266 155 (1992). A. J. Miller and R. Johnston, J. Phys. Cond. Matt. 3 p.S231 (1992).
3. Purcell, Binh, and Baptist, J. Vac. Sci. Technol. B 15 1666 (1997).
4. J. L. Shaw, submitted to J. Vac. Sci. Technol. B
5. D. Temple, C. A. Ball, W. D. Palmer, L. N. Yadon, D. Vellenga, J. Mancusi, G. E. McGuire and H. F. Gray, J. Vac. Sci. Technol. B 13, p150, (1995).
6 C. Lea and R. Gomer, Phys. Rev. Lett. 25, 804 (1970).
7. J. W. Gadzuk and E. W. Plummer, Phys. Rev. Lett. 26 p92 (1971).

Mat. Res. Soc. Symp. Proc. Vol. 621 © 2000 Materials Research Society

A Novel Approach for True Work Function Determination of Electron-Emissive Materials by Combined Kelvin Probe and Photoelectric Effect Measurements

Bert Lägel, Iain D. Baikie, Konrad Dirscherl and Uwe Petermann,
Department of Applied Physics, The Robert Gordon University, Aberdeen, UK.

ABSTRACT

For the development of new electron-emissive materials knowledge of the work function (ϕ) and changes in ϕ is of particular interest. Among the various methods, the ultra-high vacuum (UHV) compatible scanning Kelvin Probe has been proven to be a superior technique to measure work function changes due to e.g. UHV cleaning processes, chemical contamination, thermal processing etc. with high accuracy (<1meV).

The Kelvin Probe measures local work function differences between a conducting sample and a reference tip in a non-contact, truly non-invasive way over a wide temperature range. However, it is an inherently *relative* technique and does not provide an *absolute* work function if the work function of the tip (ϕ_{tip}) is unknown.

Here, we present a novel approach to measure ϕ_{tip} with the Kelvin Probe via the photoelectric effect, where a Gd foil is used as the photoelectron source. This method thus provides the true work function of the sample surface with an accuracy of approx. 50meV. We demonstrate the application of the technique by *in situ* work function measurements on evaporated layers of the low work function material LaB_6 on a Re substrate and follow the changes in ϕ of LaB_6 due to the surface adsorption of residual gas molecules. Thus, the extended Kelvin Probe method provides an excellent tool to characterise and monitor the stability of low work function surfaces.

INTRODUCTION

The work function is an extremely sensitive indicator of changes in surface and interface chemical composition, adsorbate induced surface dipole and surface roughness. Further, for the characterisation of electron emissive materials, the work function is one of the most important parameters and low ϕ materials are of particular interest. However, due to the adsorption of residual gases, the work function of these materials is known to increase with time, adversely affecting their emission properties.

The Kelvin Probe [1, 2] can be used to monitor these work function changes in a non-contact, non-invasive way with high accuracy (< 1meV) [3, 4] up to temperatures of approx. 900K. It has been utilised e.g. for thin film studies [5, 6, 7], characterisation of oxides and thin films [8], semiconductor surface processing [9] and surface charge imaging [10], investigation of the adsorption kinetics of oxygen on Si(111) surfaces [11] as well as for biological applications [12].

However, the Kelvin Probe is an inherently relative technique as it measures the average work function or contact potential *difference* (CPD) between a vibrating reference electrode (the tip) and the surface under investigation. Thus, in order to obtain the absolute value of the work function of the specimen, it is necessary to know the work function of the reference electrode, ϕ_{tip}. This could be determined e.g. by a CPD measurement on a clean reference surface. The accuracy of this method however depends on tabulated ϕ values, which are often valid only for the given experimental setup and the assumption of a clean reference surface.

We have developed a relatively simple extension of the technique as an approach for a more reliable method to determine ϕ_{tip}. This is achieved via measurement of the current voltage (I-V) characteristic of the photoelectric emission from a low work function surface, such as Gd, illuminated by monochromatic light of a fixed wavelength as discussed e.g. in [13]. The work function of any conductor or semiconductor can then be obtained by subsequent Kelvin Probe CPD measurements.

We have utilised this technique to investigate low work function materials as possible targets for use in Hyperthermal Surface Ionisation (HSI), a new mass spectroscopy ionisation technique which offers high ionisation efficiency and low amounts of cracking products [14, 15]. Lanthanum Hexaboride (LaB$_6$) is particularly interesting in this respect because of its low work function and its surface re-activation capabilities after adsorption of residual gases via flash-annealing [16].

EXPERIMENTAL METHOD

The Kelvin Method

The Kelvin Probe consists of a flat circular electrode (reference electrode, tip) suspended above and parallel to a stationary electrode (the specimen, S), thus creating a simple capacitor. If an external electrical contact is made between the two electrodes their Fermi levels equalise and the resulting flow of electrons from the metal with the lower work function produces a contact potential difference, V_{CPD}, between the plates as shown in fig. 1(b):

$$eV_{CPD} = \phi_S - \phi_{tip}, \qquad (1)$$

where e is the electronic charge and ϕ_{tip} and ϕ_S are the work functions of the tip and sample, respectively. By vibrating the probe, a varying capacitance is produced which causes a current to flow back and forth between the plates.

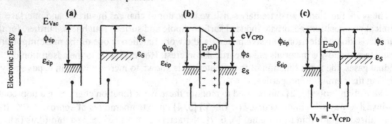

Figure 1. *Electron energy level diagrams of two different metals (a) without contact, (b) with external electrical contact, where E indicates the electrical field between the plates, and (c) with inclusion of the backing potential. ε_{tip} and ε_S refer to the Fermi levels of the tip and sample, respectively.*

Inclusion of a variable "backing potential" V_b in the external circuit permits biasing of one electrode with respect to the other: at a unique point, where $V_b = -V_{CPD}$, the electrical field between the plate vanishes, see fig. 1(c), resulting in a null output signal. The work function difference between the electrodes is thus equal and opposite to the DC potential necessary to produce a zero output signal.

The CPD measurements were performed with a UHV compatible Kelvin Probe that incorporates advanced "off-null" detection, where the balance point is determined by linear extrapolation rather than nulling with a resolution of <1mV as well as automatic control of the tip-to-sample spacing, which is an important requirement for accurate measurements [4, 17]. Measurements can be performed with a frequency of up to 1 data point per second.

Photocurrent Measurement

In the photoelectric measurement mode the capacity of the arrangement is kept constant. The photoelectrons emitted from a low work function surface (S) illuminated by monochromatic light of energy $E_{ph} = h\nu$ with $\phi_{tip} > h\nu > \phi_S$ are collected by the Kelvin Probe tip and measured as a function of the applied backing potential V_b (see figure 2). It can be seen that, at the onset of the photocurrent collected at the Kelvin Probe tip, (fig. 2b), the work function of the tip equals the photon energy E_{ph} plus a required bias potential supplied by V_b, i.e.,

$$\phi_{tip} = h\nu + V_b \text{(onset)}. \tag{2}$$

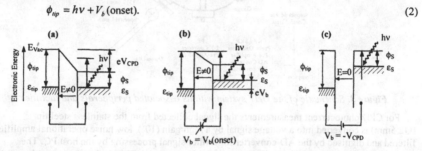

Figure 2. Energy level diagrams of the Kelvin Probe in photocurrent measurement mode with $\phi_{tip} > h\nu > \phi_S$: (a) no photoelectrons are collected at the tip, (b), the onset of the photocurrent is measured where $\phi_{tip} = h\nu + V_b$(onset) and in (c) the saturation current where $V_b = -V_{CPD}$.

With increasing bias potential, analogous to the Fowler theory of photoemisson from metals near threshold [18], the photocurrent increases with the square of V_b: $I_{ph} \propto V_b^2$. When V_b equals $-V_{CPD}$ (fig. 2c), electrons emitted from energy states below the Fermi level with virtually zero velocity are only just able to reach the collector and the photocurrent will saturate.

Experimental Setup

The experiments were carried out in a UHV chamber as shown in fig. 3 with a base pressure of $<4\times10^{-9}$Torr. A polycrystalline gadolinium foil ($25\times12\times0.1$mm^3) was used as the *photoelectron source*. The sample was cleaned by repeated resistive heating up to 1200K for several minutes. Gd has the advantage of having a relatively low work function of 3.1eV [19] and is available in form of thin foils. Further, we found that the work function of Gd remained stable within 100meV over several days. For photoemisson measurements the Gd foil was illuminated by a 100W Hg-Cd-Zn spectral lamp through a sapphire viewport which is transparent for wavelengths down to 200nm. The 312.6nm Hg emission was selected by a narrowband interference filter (FWHM = ± 20nm).

LaB$_6$ (purity 99%) films were generated via evaporation onto a polycrystalline rhenium foil (25×12×0.025mm^3, purity 99.99%). The foil was cleaned by repeated flash annealing to temperatures above 2000K. The work functions of the samples were determined by first obtaining ϕ_{tip} of the Kelvin Probe via the photoeffect I-V measurement and a subsequent Kelvin Probe CPD measurement of the respective sample. The UHV system was darkened during the measurements and all filaments were extinguished.

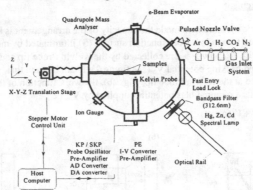

Figure 3. Schematic of the UHV system with its associated peripherals and facilities

For CPD/photocurrent measurements the signal collected from the stainless steel tip (Ø2.5mm) is converted into a voltage signal by a high gain (10^7), low noise operational amplifier, filtered and digitised by the AD-converter for further signal processing by the host PC. The computer program controls tip oscillation and backing potential. This design thus provides a fully automated measurement system.

RESULTS AND DISCUSSION

Kelvin Probe Tip Work Function Determination

Figure 4 shows typical data of the photoelectron emission from the Gd surface.

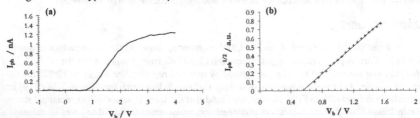

Figure 4. Typical photoemission data used to determine the work function of the Kelvin Probe: (a) the current-voltage (I-V) curve and (b) the plot $I_{ph}^{1/2}$ versus bias voltage. Note that only data from the linear portion of the curves are included in the fit.

In order to determine the onset of the photocurrent I_{ph}, the square root of I_{ph} is plotted versus the applied bias voltage V_b (fig 4b) and a least square fit of the linear part of the curve is extrapolated to zero current. ϕ_{tip} and ϕ_{Sample} are then calculated via equations 1 and 2, respectively. The results of the measurements are summarised in table I. The uncertainty of the extrapolation is approx. ± 20mV. Further, taking the measurement error due to the thermal energy distribution into account (26mV at 300K) and small changes in tip work function during the measurement (<20meV), we estimate the overall error of the work function measurement to be ± 50meV.

The work function determined for Gd (3.075 ± 0.050)eV is, within the measurement error, in agreement with that of (3.1 ± 0.150)eV given in [19]. The work function measured for the clean Re substrate was used as a calibration point for the absolute work function scale in fig. 5.

Table I. *Results of the measurements to determine tip and sample work functions.*

Measurement	E_{ph} / eV	V_b(onset) / V	V_{CPD} / V	ϕ / eV
Kelvin Probe tip	3.966	0.551± 0.016	/	4.518
Gd	/	/	-1.443	3.075
Re	/	/	0.653	5.171

Application to Work Function Measurements of LaB6 on Polycrystalline Re

In order to characterise the LaB6 evaporation process, the work function change of the Re/LaB6 system was measured at different stages of the evaporation as shown in figure 5. After an initial rapid decrease the work function changes only in small steps with further evaporation, gradually achieving approx. 3.3eV. However, if the surface is 'activated' by flashing it to 1500°C, a further work function decrease to 2.536eV is obtained. The necessity for such an activation procedure might be caused either by contamination or, indeed, a rearrangement of the LaB$_X$ structure since its work function is known to depend strongly on the surface stoichiometry of the compound [20].

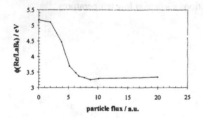

Figure 5. *Work function change during evaporation of LaB6 onto a Re foil held at 300K. The current onto the sample due to the partly ionised molecular beam was used as a measure of the particle flux.*

The work function increase due to the adsorption of residual gas molecules of the LaB6 surface after activation and, for comparison, Gd, is shown in figure 6. With respect to the LaB6 surface, we clearly observe 2 adsorption stages. The first one, marked AB, with an almost constant gradient above 10L and the second one, after approx. 80L, with a lower and rather continuous gradient leading to a work function of 3.65eV after an exposure of 200L.

This shows that LaB$_6$ is rather easily poisoned at 300K by the adsorption of the residual gas, the largest component of which is molecular hydrogen. In contrast, although having a higher initial work function, ϕ(Gd) increases by only 70meV after 200L exposure.

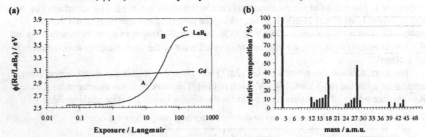

Figure 6. (a) Work function increase of LaB$_6$ and Gd due to the adsorption of residual gas molecules at 300K. (b) The relative composition of the residual gas envelope at a base pressure of 3.7×10^{-9} Torr.

CONCLUSIONS

We have demonstrated the feasibility of a novel approach to determine the work function of the reference tip of a Kelvin Probe via photoelectric current measurements on Gd using a minimum of additional instrumentation. We found that the work function of Gd remains stable over a prolonged period of time in the upper UHV pressure region of approx. 4×10^{-9} Torr and is therefore best suited for this purpose. The measurement error is estimated to be ± 50meV.

Using the new technique we have followed the evaporation of LaB$_6$ onto a polycrystalline Re substrate. After activation, the surface work function at 300K was (2.536 ± 0.050) eV, increasing to (3.65 ± 0.050) eV after an exposure of 200L to the residual gas. This indicates a very high sensitivity of the work function of LaB$_6$ to the adsorption of gas molecules which may restrict its use for applications requiring a stable low work functions surface at low temperatures.

ACKNOWLEDGEMENTS

This work was supported by EPSRC and DERA.

REFERENCES

1. Lord Kelvin, *Philos. Mag* **46**, 82 (1898)
2. W.A. Zisman. *Rev. Sci. Instrum.* **3**, 367 (1932)
3. I.D. Baikie, K.O. Vanderwerf, H. Oerbekke, J. Broeze, A. Vansilfhout, *Rev. Sci. Instrum.* **60**, 930 (1989)
4. I.D. Baikie, P.J. Estrup, *Rev. Sci. Instrum.* **69**, 3902 (1998)
5. M. Schmidt, H. Wolter, M. Nohlen, K.Wandelt, *J. Vac. Sci Technol.* **A12**, 1818 (1994)
6. E. Kopatzki, H.G. Keck, I.D Baikie, J.A. Meyer, ,R.J. Behm, *Surf. Sci.* **345**, L11 (1996)
7. I.D. Baikie, U. Petermann, B. Lägel, *Surf. Sci.* **433-435**, 770 (1999)

8. I.D. Baikie, G.H Bruggink in Materials Reliability in Microelectronics III, edited by K.P. Rodbell, W.F. Filter, H.J. Frost, P.S. Po, (*Mater. Res. Soc. Proc.* **309**, Pittsburgh, PA, 1993), pp. 35-40
9. I.D. Baikie in Chemical Perspectives of Microelectronic Materials II, edited by L.V. Interrante, K.F. Jensen, L.H. Dubois, M.E. Gross (*Mater. Res. Soc. Proc.* **204**, Pittsburgh, PA, 1991), pp. 363-368
10. B. Lägel , I.D. Baikie, U. Petermann in Defect and Impurity Engineered Semiconductors and Devices II, edited by S. Ashok, J. Chevallier, K. Sumino, B.L. Sopori, W. Goetz, (*Mater. Res. Soc. Proc.* **510**, Pittsburgh, PA, 1998), pp. 619-625
11. U. Petermann, I.D. Baikie, B. Lägel, *Thin Solid Films* **343-344**, 492 (1999)
12. I.D. Baikie, P.J.S. Smith, D.M. Porterfield, P.J. Estrup, *Rev. Sci. Instrum.* **70**, 1842 (1999)
13. L. Apker, E. Taft, J. Dickey, *Phys. Rev.* **73**, 46 (1947)
14. A. Danon, A. Amirav, *Isr. J. Chem.* **29**, 443 (1989)
15. A. Danon, A. Amirav, *Int. J. Mass Spectrom. Ion. Processes* **96**, 139 (1990)
16. J.M. Lafferty, *J. Appl. Phys.* **22**, 299 (1951)
17. I.D. Baikie, S. Mackenzie, P.J.Z. Estrup and J.A. Meyer, *Rev. Sci. Instrum.* **62**, 1326 (1991)
18. R.H. Fowler, *Phys. Rev.* **33**, 45 (1931)
19. D.E. Eastman, *Phys Rev.* **B2**, 1 (1970)
20. E.K. Storms, B.A. Mueller, *J. Appl. Phys.* **50**, 3691 (1979)

Field Emitter Arrays

Mat. Res. Soc. Symp. Proc. Vol. 621 © 2000 Materials Research Society

Mo and Co Silicide FEAs

Jong Duk Lee, Byung Chang Shim, Sung Hun Jin and Byung-Gook Park
Inter-University Semiconductor Research Center(ISRC) and School of Electrical Engineering,
Seoul National University, San 56-1, Shinlim-dong, Kwanak-gu, Seoul 151-742, Korea

ABSTRACT

For enhancement and stabilization of electron emission, Mo and Co silicides were formed from Mo mono-layer and Ti/Co bi-layers on single crystal silicon FEAs, respectively. Using the slope of Fowler-Nordheim curve and tip radius measured from SEM, the effective work function of Mo and Co silicide FEAs are calculated to be 3.13 eV and 2.56 eV, respectively. Compared with silicon field emitters, Mo and Co silicide exhibited 10 and 34 times higher maximum emission current, 10 V and 46 V higher failure voltage, and 6.1 and 4.8 times lower current fluctuation, respectively. Moreover, the emission currents of the silicide FEAs depending on vacuum level are almost same in the range of $10^{-9} \sim 10^{-6}$ torr. This result shows that silicide is robust in terms of anode current degradation due to the absorption of air molecules.

INTRODUCTION

Metal silicides including Ti [1, 2], Co [2, 3], Nb [4], Mo [5, 6], Pd [7, 8], Cr [9] and Pt [10] were previously reported in various literatures to improve electron emission characteristics of Si FEAs. The advantages of silicide emitters have been mentioned as lower effective work function, higher electrical conductivity, and better chemical and thermal stability than those of silicon emitters.

However, due to different emitter structure design and silicide formation process, the emitter materials including silicide phase and emitter structure including tip radius are not same. Thus, the results presented in the literatures are not always consistent. Therefore, it is still important to study effective work function, and emission current characteristics for the silicide emitters, especially Mo and Co silicides.

To obtain the effective work functions, slopes of Fowler-Nordheim (F-N) curves are drawn and tip radii are measured by field emission scanning electron microscopy (FESEM). The emission current fluctuation and stability at a fixed gate bias and the emission current change depending on vacuum pressure will be described.

EXPERIMENTAL

The 625 tip single crystal silicon(c-Si) FEAs with a gate opening of 1.4 μm were fabricated by dry etching and sharpening oxidation [11]. Some of the c-Si FEAs were splitted for metal deposition and followed by annealing for the silicide emitters. The Mo silicide was formed by coating of 25-nm thick Mo mono-layer on Si-tips and subsequent annealing by rapid thermal process (RTP) at 1000 °C in inert gas (N₂) ambient. Coating with Co mono-layer (25 nm), Co/Ti (18 nm/6 nm) and Ti/Co (15 nm/12 nm) bi-layers on Si-tips, and subsequent annealing by RTP at 800 °C, 900 °C and 800 °C in inert gas (N₂) ambient produced Co silicides. After formation of the silicides, the unreacted metals and unwanted by-products were removed by wet etching.

As a novel process to form Co silicide by electrical stress, after Co (250 Å) is coated on c-Si FEAs, the FEAs are loaded into an UHV chamber with a base pressure of 1.2×10^{-9} torr and then pulse voltages are applied to the FEAs. The pulses were fixed at duty of 0.7 and frequency of 10Hz. Only pulse amplitude was changed with 2V or 4V step to increase local heating on the tips.

RESULTS AND DISCUSSION

In order to examine the constituent and the dominant bindings of the metal-Si compounds, X-ray diffraction (XRD) measurement was performed. It was confirmed that Mo mono-layer and Co mono-layer, Co/Ti and Ti/Co bi-layers were completely transformed into $MoSi_2$ and $CoSi_2$ phases, respectively, as shown in Fig. 1.

SEM is used to analyze the Co silicide samples to check the surface morphology, thickness uniformity of Co silicide layers and $CoSi_2$/Si interfaces. As shown in Fig. 2, the sample formed from Ti/Co layer shows a very flat surface, uniform thickness of about 39 nm and smooth interface, because Ti prevents oxygen adsorption on the Co film during silicide formation [12]. However, the sample formed from Co mono-layer shows a rough surface and agglomeration due to adsorption of oxygen on the Co film. The quality of the silicide formed from Co/Ti layers in terms of surface roughness and thickness uniformity is in-between the two competitive samples, as shown in Fig. 2.

Figure 3 shows SEM photographs of c-Si emitter, Mo silicide emitter and Co-silicide emitters formed on c-Si. The gate metal surface of Mo silicide emitter is somewhat roughened, compared with c-Si emitter. The measured radii of silicon, Mo silicide, Co/Ti and Ti/Co tips were 35 Å, 50 Å, 115 Å and 80 Å with 10 % error, respectively. In the case of Co silicide tip formed by electrical stress, the tip radius is about 125 Å. The Mo silicide tip is much sharper than the Co silicide tips. The Ti/Co tip is sharpest among the Co silicide emitters, although all the silicide tips are not as sharp as the initial c-Si emitter.

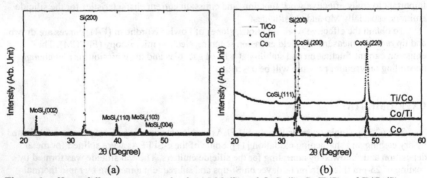

Figure 1. X-ray diffraction spectra for the (a) Mo/Si and (b)Co/Si, Co/Ti/Si and Ti/Co/Si annealed in N_2 ambient using RTA, after removal of the unreacted metals. Metal peaks were not shown to confirm successfully grown Mo and Co silicides.

Figure 2. SEM cross sectional views of Co-silicides formed by coating (a)Co(250 Å), (b)Co(180 Å)/Ti(60 Å), and (c)Ti(150 Å)/Co(120 Å) layers and annealing 800 ℃, 900 ℃, and 800 ℃, respectively.

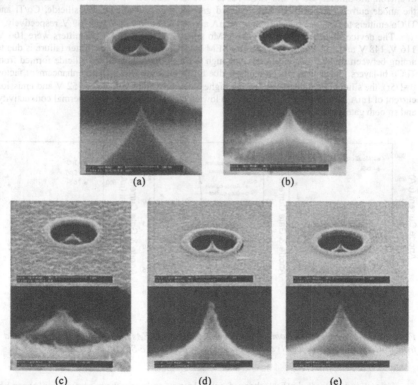

Figure 3. SEM micrographs for the fabricated (a)c-Si emitter, (b)Mo silicide emitter and Co silicide emitters formed from (c) Co, (d)Co/Ti, and (e)Ti/Co. The tip apexes of Mo and Co silicide tips became blunt compared with that of c-Si tip.

Field emission properties of the fabricated emitters were characterized in an ultra high vacuum chamber at a base pressure of 6.6×10^{-9} torr. The anode plate was 3 mm above the gate top and biased to +300 V. Figure 4 shows the I-V characteristics of c-Si and the silicide emitters with 625 tips. Fig. 4(a) shows that the anode currents at the gate voltage 100 V for the Mo silicide and the c-Si emitters are 176 μA and 21 μA, respectively. The Mo silicide emitters have lower turn-on voltage about 8V than silicon emitters. As shown Fig. 4 (b), Co silicide emitters formed from a Co mono-layer using RTA show about 46 V higher turn-on voltage than c-Si emitters and have high gate current which is about 38.5 % of the anode current at the gate voltage of 150 V, due to tip blunting and rough surface, as shown in Fig. 3(c). In the case of Co silicide FEAs formed by electrical stress, the turn-on voltage is 8 V lower than that of c-Si emitters and the gate current is about 1.4 % of the anode current at the gate voltage of 100 V. In Fig. 4(c), the anode current form c-Si, Co/Ti and Ti/Co emitters are compared. The gate currents of silicon, Mo silicide, Co/Ti and Ti/Co FEAs were less than 1.0 %, 1.0 %, 11.3 % and 2.9 % of the anode current, respectively. The required gate voltages of silicon, Mo silicide, Co/Ti and Ti/Co emitters to obtain anode current of 30 μA are 104 V, 86 V, 118 V and 109 V, respectively.

The device failure voltages of silicon, Mo silicide, Co/Ti and Ti/Co emitters were 106 V, 116 V, 148 V and 152 V, respectively. Our SEM study indicates that the emitter failure is due to arcing between tip and gate electrode. Although the turn-on voltage of Co silicide formed from Ti/Co bi-layers higher than silicon emitters due to tip blunting (lower field enhancement factor, $\beta = 1/5r$), the silicide FEAs demonstrate the highest device failure voltage of 152 V and emission current of 1mA, as shown in Fig. 4(c), due to low electrical resistivity, high thermal conductivity, and smooth gate surface morphology [13].

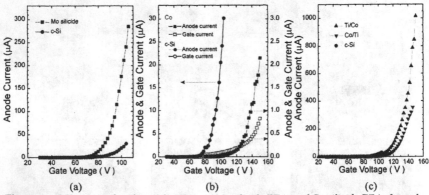

Figure 4. *I-V curves taken from c-Si FEAs, (a)Mo silicide FEAs and Co silicide FEAs formed from (b)Co mono-layer, (c)Co/Ti and Ti/Co bi-layers based on c-Si with 625 tips*

Using Fowler-Nordheim(F-N) theory for the electron emission, the work function(ϕ) can be written as the following equation [13, 14].

$$\phi = \left(\frac{b\beta}{0.95B}\right)^{2/3} = \left(\frac{S}{0.95B \times \log e \times 5r}\right)^{2/3}$$

Where B is 6.87×10^7, β is the field enhancement factor, r is the tip radius and S is the slope of F-N curve.

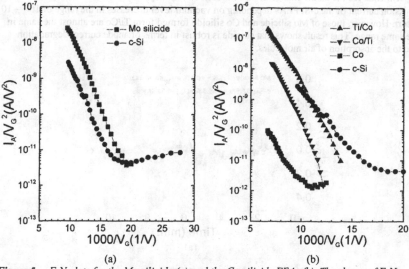

(a)   (b)

Figure 5. *F-N plots for the Mo silicide (a) and the Co silicide FEAs (b). The slopes of F-N curve of c-Si, Mo silicide, Co, Co/Ti and Ti/Co emitters are 0.4058 A/V, 0.3925 A/V, 0.4195 A/V, 0.6015 A/V, 0.6319 A/V and 0.4725 A/V, respectively.*

From the slope of Fowler-Nordheim plots in Fig. 5 and the tip radii measured from SEM photographs in Fig. 3, the effective work functions of c-Si, Mo silicide and Co/Ti and Ti/Co emitters about 4.06 eV, 3.13 eV, 2.46 eV and 2.56 eV, respectively, which are similar to the previously reported results [2, 5]. Co silicide FEAs formed by Co mono-layer using RTA have such a rough and non-uniform tip surface. So that it is not easy to measure the accurate tip radius. We tried to investigate the effective work function from the Co silicide FEAs formed by electrical stress. The slope of F-N curve is 0.358 A/V and tip radius is 125 Å in the Co silicide FEAs by electrical stress. The effective work function of Co silicide FEAs formed by electrical stress is calculated to 2.44 eV.

The emission current variations of the Mo silicide FEAs normalized with average anode current at a fixed gate bias during 12 minutes was ranging from –7.1 % to +8.1 %. In case of Co silicide FEAs formed form Co, Co/Ti and Ti/Co, the variations were ranging from –13.2 % ~

+15.8%, -16.9 % ~ +17.2 % and –8.9 % to +10.3 %, respectively. Whereas, that of silicon FEAs was ranging from –37.1 % to +55.3 %, as shown in Fig. 6. The notable reduction of emission current variation in the silicide FEAs supports that the silicide emitters are less influenced by gas adsorption and desorption and/or destruction of sharp emission site than that in silicon emitters.

For the practical applications of field emitter arrays, it is essential to investigate the vacuum dependence of emission characteristics, because field emission property depends sensitively on the work function change of the emitter surface by gas absorption [13]. Figure 7 show experimental results on the change of the emission current depending on the vacuum level. The emission current of the c-Si FEAs depending on vacuum level decreases in the range of 10^{-9} ~ 10^{-6} torr. However, those of Mo silicide and Co silicide formed form Ti/Co are almost the same in the same range. This result shows that silicide is robust in terms of anode current degradation due to the absorption of air molecules.

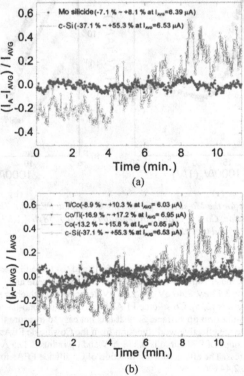

Figure 6. *The emission current variations of the c-Si and the silicide FEAs. The variation is defined as $(I_A-I_{AVG})/I_{AVG}$ at constant applied voltages, where I_{AVG} is the average anode current during the measurement.*

In order to see inertness of the silicide emitters to air molecules, the emitters were operated at high vacuum level of 10^{-9} torr(period I) and increased air pressure up to 10^{-6} torr(period II) and then the vacuum level increased to 10^{-9} torr again(period III). During period II in Fig. 8, the emission current of the silicide FEAs gradually decreased and then remained constant. On the other hand, that of c-Si FEAs inconsistently changed. This is thought to be the formation and deformation of native oxide, and/or absorption and desorption of unwanted gas. The anode current of Mo silicide FEAs was completely recovered to the original value when the pressure was reduced to 10^{-9} torr, whereas those of Si FEAs and Co silicide FEAs formed form Ti/Co were not recovered by just reducing pressure. After baking at 250 °C, anode currents of Si and Co silicide FEAs were increased. However, the anode currents of Si and Co silicide FEAs were not completely recovered to original values probably due to incompletely desorption of air molecules at the baking temperature and/or tip surface modification during device operation, as shown at period III in Fig. 8.

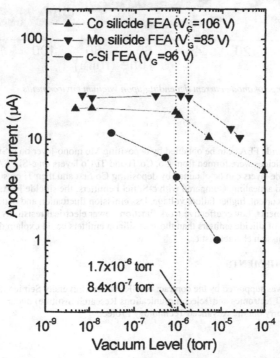

Figure 7. Anode currents of Si, Mo silicide and Co silicide formed from Ti/Co FEAs with 625 tips depending on the vacuum level.

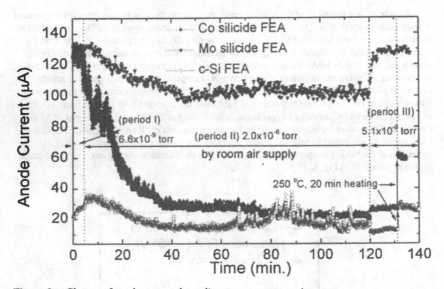

Figure 8. *Change of anode current depending upon vacuum environments.*

CONCLUSIONS

The Mo silicide FEAs can be obtained by depositing Mo mono-layer on c-Si tip and its annealing. Co silicides were formed from Co, Co/Ti and Ti/Co layers on c-Si FEAs. Uniform and smooth Co silicide layers can be obtained by depositing Co first and then Ti on silicon tips, followed by rapid annealing. Compared with c-Si field emitters, the silicide FEAs exhibited higher emission current, higher failure voltage, less emission fluctuation and better immunity to vacuum environments. Lower effective work function, lower electrical resistivity and better surface inertness of silicide emitters than those of silicon emitters could explain the reasons for the improved emission characteristics.

ACKNOWLEDGEMENTS

This work was supported by the contract of ISRC(Inter-University Semiconductor Research Center) to ETRI(Electronics and Telecommunications Research Institute) under Grant No. ISRC-99-X −5512 to carry out the Advanced Technology Project.

REFERENCES

1. J. D. Lee, H. S. Uh, B. C. Shim, E. S. Cho, C. W. Oh, and S. J. Kwon, Tech. Dig. of IVMC,

304 (1998).
2. E. J. Chi, J. Y. Shim and H. K. Baik, Tech. Dig. of IVMC, 188 (1996).
3. Y. J. Yoon, G. B. Kim and H. K. Baik, J. Vac. Sci. Technol. B **17**, 627 (1999).
4. J. S. Park, S. Lee, B. K. Ju, M. H. Oh, J. Jang, and D. Jeon, Tech. Dig. of SID, 576 (1999).
5. M. R. Rakhshandehroo and S. W. Pang, IEEE Trans. Electron Devices **46**, 792 (1999).
6. H. S. Uh, B. G. Park, and J. D. Lee, IEEE Electron Device Letters **19**, 167 (1998).
7. C-C. Wang, T-K. Ku, I-J. Hsieh, and H-C. Cheng, Jpn. J. Appl. Phys. **35**, 3681 (1996).
8. R. A. King, R. A. D. Mackenzie, G.D.W. Smith, and N. A. Cade, J. Vac. Sci. Technol. B **13**, 603 (1995).
9. I-J. Chung, A. Hariz, M. R. Haskard, B. K. Ju, and M. H. Oh, Tech. Dig. of IVMC, 245 (1996).
10. M. Takai, T. Iriguchi, and H. Morimoto, A. Hosono, and S. Kawabuchi, J. Vac. Sci. Technol. B **16**, 790 (1998).
11. H. S. Uh, B-G Park and J. D. Lee, Tech. Dig. of Int. Electron Devices Meeting, 713 (1997).
12. D. K. Sohn, J-S Park and J. W. Park, J. Electrochemi. Soc. **147**, 373 (1998).
13. I. Brodie and P. R. Schwoebel, Proc. of the IEEE **82**, 1006 (1994).
14. R. Gomer, Field Emission and Field Ionization. Cambridge, MA: Harvard Univ. Press, (1961).

Mat. Res. Soc. Symp. Proc. Vol 621 © 2000 Materials Research Society

Solid-State Field-Controlled Electron Emission:
an alternative to thermionic and field-emission

Vu Thien Binh, J.P. Dupin, P. Thevenard, D. Guillot and J.C. Plenet
Laboratoire d'Emission Electronique, DPM-CNRS,
University Claude Bernard Lyon 1, 69622, Villeurbanne, France.

ABSTRACT

In the solid-state field-controlled emitter (SSE), the emission barrier, which is the factor of utmost importance for surface electron emission, is tailored by a controlled extrinsic parameter like the injected space charge located near the surface. This is done by depositing an ultra-thin wide band-gap semiconductor layer on a metallic surface. It is an alternative approach to the thermionic or field emission for which the work function value is intrinsic to the material used. The emission current measurements from the SSE cold cathodes show stable emission, at low applied field ($\approx$50 V/μm) and in poor vacuum ($\approx 10^{-7}$ Torr). The new emission mechanism has been modeled, the calculations and the theoretical analysis confirm the experimental results. The fabrication of the SSE, either by a sputter deposition in vacuum or by a sol-gel technique, meets most of the demands specific to high throughput fabrication of cold cathodes with large emitting area dedicated to applications in vacuum microelectronics.

INTRODUCTION

As a paradigm shift for surface electron emission, we proposed the solid-state field-controlled emitter (SSE) [1]. The basic structure of SSE is an ultra-thin wide bandgap n-type semiconductor (UTSC) layer deposited on a metallic surface (Fig. 1). The electron emission from the SSE cathodes results from a serial two-step mechanism: the first step is the injection of electrons at the solid-state Schottky junction from the metal into the UTSC medium, followed by a second step which is the electron emission from the UTSC surface that becomes, under the control of an applied field F_{app}, a surface with a low electron affinity (LEA) or with a negative electron affinity (NEA). The injected charges induce a large band bending in the UTSC layer, with consequence a drastic lowering of the emission barrier $\Delta\Phi_E$ ($\Delta\Phi_E$ is the potential difference between the metal Fermi level and the vacuum level). $\Delta\Phi_E$ can become low enough to allow emission of electrons through it (LEA situation for $\Delta\Phi_E \leq 2$ eV) or over it (NEA situation for $\Delta\Phi_E \leq 0$). This emission concept differs from conventional electron emission mechanisms, such as thermionic and field emission, for which to obtain emission current it is necessary to apply either high temperature (>2000 K) or high field (>5000 V/μm) to overcome the work function which is fixed by the nature of the solid, the crystallography and the adsorption state of the cathode surface.

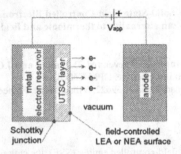

Figure 1. *Schematic draw of the SSE structure with an UTSC layer deposited on a metallic surface. The metal plays the role of an electron reservoir and the electrons are emitted from the UTSC surface.*

THE SSE CONCEPT FOR SURFACE ELECTRON EMISSION

The conventional thermionic and field emission

To obtain emission, it is necessary to remove the electrons trapped in the cathode material. This can be done by imparting excess energy to the electrons to overcome the surface barrier, either by the action of temperature, called thermionic process (TE), or by an electromagnetic radiation, called photoemission process. J_T, the TE current density in $A.cm^{-2}$, is given by the Richardson-Dushmann equation:

$$J_T \approx A_T\, T^2 \exp\left[-\frac{\phi}{kT}\right],\qquad(1)$$

where the constant $A_T \approx 120 \times (1-r)$ with r the reflection coefficient of the material, the work function ϕ is in eV. Eq. (1) means that the emission current uniformity and stability depends on the control of local temperature T, which is in the order of thousand of K, and on the state of the surface (crystallography and adsorption) which defines the value of ϕ.

A second possibility is to thin down to nanometer size the thickness of the surface barrier with the aid of an applied field F_{app}, in the range of 10^3-10^4 V/μm for $\phi \approx 4$ to 5 eV, to allow the tunneling of the electrons through it. This is known as field emission (FE) process and the current density J_F, in $A.cm^{-2}$, is given by the Fowler-Nordheim equation, that we give here in a form similar to Eq. (1):

$$J_F \approx A_F\, T_{FE}^2 \exp\left[-\frac{\phi}{BT_{FE}}\right],\qquad(2)$$

where $A_F \approx 1.55 \times 10^{-6}$, $B \approx 2.9 \times 10^6$ and $T_{FE} = F_{app} / \phi^{1/2}$ with F_{app} in $V.cm^{-1}$ and ϕ in eV. J_F is obtained at room temperature or less, but with a higher dependence on ϕ than J_T. The FE current uniformity and stability from the cold cathodes depends strongly on the local surface curvature (protrusion), which defines F_{app}, and on the adsorption from the outside environment which becomes an important factor at $T \leq 300$ K as the sticking coefficient decreases exponentially with T. Therefore, the vacuum condition is more stringent in FE than in TE in order to obtain stable current.

In both cases, the factor of utmost importance for the stability of the emission is the stability of ϕ, which is related to the state of the emitting surface and on the electronegativity of its adsorbed species. It is the main limiting situation for having a control in the reproducibility of the currents for cold cathodes, which is the present drawback for a widespread use in large area cathodes.

The SSE concept of surface electron emission

In the SSE electron emission mechanism, the barrier height $\Delta\Phi_E$ is controlled mainly by the space charge inside the UTSC layer. As the standard Richardson and Fowler-Nordheim analysis familiar to thermionic and field emission calculations from metal or semiconductor surfaces are no longer applicable, as well as the recent approaches for carbon film emission [2], a rigorous numerical simulation analysis has been developed [3]. The model consists of (i) a basic SSE structure with an UTSC layer (TiO$_2$ for example), in the range of 2 to 10 nm, deposited on a metallic surface (Pt for example) (ii) a Schottky junction between the metal and the UTSC with the conventional energy band relation [4], (iii) a triangular representation of the vacuum barrier including the image potential at the surface of the semiconductor, (iv) a numerical integration of the Poisson's equation to evaluate the equilibrium space charge distribution, Q_{SC}, inside the UTSC and (v) a calculation of the emission current density J by the resolution of the one-body Schrödinger equation using a Green's formalism based on the numerical resolution of the self-

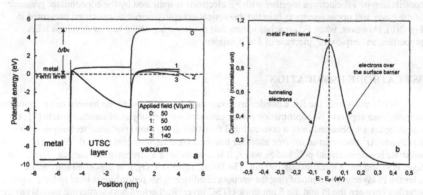

Figure 2. (a) Energy-band diagram of a SSE (5 nm TiO$_2$ UTSC layer) showing the lowering of $\Delta\Phi_E$ due to the creation of a space charge after an electron injection from the Schottky junction (diagram0 with $Q_{SC} = 1.8 \times 10^{-6}$ e/nm^2 to diagram 1 with $Q_{SC} = -2.2$ e/nm^2), followed by a modification of the barrier due to the increase of the applied field F_{app}. (b) Calculated energy distribution of the emitted electrons from a SSE showing a T-F behavior at 300 K.

consistent Lippmann-Schwinger (LS) equation [5]. The results show that at the initial stage and in the presence of F_{app} the potential variation is mainly defined by the space charge distribution due to the presence of the Schottky junction, the field penetration inside the UTSC surface and their interactions. This situation corresponds to the first equilibrium solution of the Poisson's equation and it is represented by the energy band diagram 0 in Fig. 2 with a blocking barrier at the surface of the UTSC. Through the reverse bias Schottky junction there is an electronic injection from the metal into the CB of the UTSC by a thermionic mechanism and a tunneling process. The tunneling is either resonant through the donor-states located near the CB bottom or direct when the band bending becomes sufficient. With the blocking barrier at the UTSC surface, this results in the formation of an equilibrium space charge Q_{SC} which is given by the second equilibrium solution of the Poisson's equation. This second solution is obtained when a large value of Q_{SC} (-2.2 e/nm^2 for an UTSC layer thickness of 5 nm) populated the CB of the UTSC and leads to an important band bending inside this layer with the corresponding energy band diagram 1 in Fig. 2. This means that the electronic injection starts for a threshold value of F_{app} and stops when the second equilibrium state is reached with a value of $\Delta\Phi_E$ in the range of a few tenth of eV, i.e. a situation of low electron affinity (LEA) surface. From this second equilibrium state, if F_{app} is increased the potential distribution inside the UTSC layer stays practically unmodified, only the vacuum barrier is distorted. The consequence is a lowering of $\Delta\Phi_E$ as shown by the energy band diagrams 2 and 3 in Fig. 2. Such a modification of the emission barrier means that the emission current is controlled by increasing F_{app} from the threshold value. Immediately after the threshold, the emission current is mainly obtained by a tunneling process through the barrier, i.e. FE-like behavior. As $\Delta\Phi_E$ is lowered for higher F_{app}, the contribution of electrons passing over the barrier is no more negligible, with consequence a slower increase of J vs. F_{app}. In this interval of fields the emission is a mixing of FE and TE, which means a T-F like behavior [6] but at 300 K. The contribution of TE electrons together with FE electrons is indicated by the concomitant presence of the lower and upper energy tails in the theoretical energy distribution spectrum presented in Fig. 2(b). However, when F_{app} reaches values that the lowering of the barrier develops NEA properties, an "explosive" increase of J will happen.

SSE CATHODE FABRICATION

In this study only the flat cathodes are concerned, as well as with the hairpin or tip geometry as metal base support . Nanoprotrusion planar cathodes are the subject of another article [7]. The flat cathodes are obtained from a conventional thin-film deposition sputtering technique, with a Kaufmann Ar$^+$ source in an oil-free chamber having a base vacuum less than 10^{-9} Torr. The substrate is a commercial polished Si wafer, it is used to have a reference flat surface. The metal layer is Pt and has a thickness of 100 nm. The UTSC layer is TiO$_2$, it is deposited immediately after the Pt layer without modifying the vacuum condition. This procedure is used to have a good interface between the Pt and the 5 nm thick UTSC layer. Rutherford backscattering analysis of the deposited TiO$_2$ films show a good stoechiometry. Defects resulting from the sputtered deposition of TiO$_2$ are known to induce donor states given then, to the layer, properties of a n-type wide bandgap semiconductor.

The sol-gel technology was also used for the UTSC deposition (TiO$_2$) [8]. The precursor solution was obtained by mixing titanium isopropoxide (Ti[OCH(CH$_3$)$_2$]$_4$) and propanol-2 ((CH$_3$)$_2$CHOH). Glacial acetic acid (AcOH) was then added with an AcOH to Ti molar ratio of 6.

The AcOH, which acts as a stabiliser by complexing the titanium alkoxide, is necessary to prevent precipitation of TiO2 and to maintain a stable precursor solution before coating. The water for hydrolysis comes from the esterification reaction, which occurs between excess of AcOH and alcohol. A transparent stable solution was obtained. Methanol was then used to adjust the viscosity of the sol, which controls the thickness of the deposited layer [9]. The TiO_2 film is deposited on the metal base cathode by dipping (at a constant and controlled speed of 4 cm.min^{-1} under controlled atmosphere) and then first dried at a temperature of 100°C. Finally the sample is annealed under Infra-Red lamp at a temperature of 350°C during 15 min to stabilise, densify, crystallise and harden the ultra-thin film. The thickness of such layer [10] has been measured, it was in the range of 5 nm.

SOME EXPERIMENTAL CHARACTERISTICS OF THE ELECTRON EMISSION FROM SSE's

The totality of the measurements from different SSE presented the following main characteristics.
1. There is no need for a seasoning or a preparation of the SSE after the layer depositions in order to obtain electron emission from a threshold value $_{Fapp}$. The I-V curves presented three well defined zones of emission indicated in Fig. 3(a). This is valid even if the SSE cathodes were kept in the ambient atmosphere for several days.
2. There is a complete reproducibility of the emission current after a switch off, with a duration up to days, if the SSE is kept in vacuum.
3. The electron emission is uniform over the whole SSE plane if the cathode is heated to less than 600 K to clean the surface. This was verified experimentally by observing, on a fluorescent screen, that the projection of the emission patterns of SSE deposited on the apex of a 200 nm radius tip were uniform. Using the tip apex as a support allowed a radial magnification of the emitting surface with nanometric resolution, specific to the field emission microscopy.

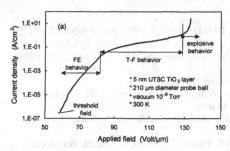
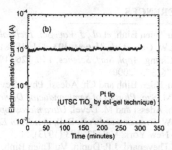

Figure 3. (a) Characteristic electron emission behavior measured from a planar SSE cathode showing the three specific emission zones. (b) Emission current stability measured from a SSE with a tip geometry cathode and an UTSC layer obtained by the sol-gel technique in 10^{-8} Torr.

4. The emission current is very stable, in particular in the high current regime. Fig. 3(b) is an example of the plot of a SSE total current for more than 3 hours of continuous emission.
5. The emission current is very sensitive to the temperature. This property is related to the strong variation of the Q_{SC} with the temperature which means a strong variation of $\Delta\Phi_E$ [3].

DISCUSSIONS AND CONCLUSIONS

The paradigm shift in the concept of SSE is the creation of the important equilibrium space charge Q_{SC} inside the UTSC in order to bring into a LEA situation the surface cathode independently of the external conditions. The needed band bending is then in the range of few eV a value that can be reached only for Q_{SC} larger than about -2 e/nm^2. Most of the specific properties of the SSE can be understood if one takes into account the covalent nature of the UTSC surface and the drastic band bending which results mainly from the electron injection from the Schottky junction. These two parameters relegate to a minor role the adsorption from the environment for the current stability. This means, from the experimental point of view, that noisy but stable emissions have been obtained for long periods in a vacuum of 10^{-7} to 10^{-6} Torr of oxygen, for example, without the destruction of the SSE cathodes. This behavior cannot be observed for conventional FE cold cathodes.

The fabrication of SSE structure needs conventional planar technology for controlled ultra-thin layer deposition like sputtering or sol-gel techniques. Such technologies permit high throughput fabrication for large area cold cathodes and opens possibilities of "intelligent emitters" such as monolithic electron sources with the integration of microlens optics [11] and regulation active matrix, either in a stand-alone structure or in array.

ACKNOWLEDGEMENTS
We acknowledge the support of Hewlett-Packard Co. (Palo-Alto) for this research.

REFERENCES

1. Vu Thien Binh, et al., *J. Vac. Sci. Technol.*, **B18(2)**, Mar/Apr 2000 ; patent #99 06 254 (1999)
2. M.W. Geis et al., *Nature*, **393**, 431 (1998) ; P.H. Cutler, N.M. Miskovsky, P.B. Lerner and M.S. Chung, *Appl. Surf. Science*, **146**, 126 (1999) ; K.A. Dean and B.R. Chalamala, *Appl. Phys. Lett.*, **76**, 375 (2000).
3. Vu Thien Binh and Ch. Adessi, Phys. Rev. Letters (in press).
4. S.M. Sze, *Physics of Semiconductor Devices*, Wiley Interscience Publ.,1981, NY, USA.
5. Ch. Adessi and M. Devel, *Ultramicroscopy* in press.
6. L.W. Swanson and A.E. Bell, in *Adv. in Electr. and Electron Physics*, L. Marton Ed., Acad. Press, New York, Vol. **32**, 193 (1973).
7. P. Thevenard, J.P. Dupin, Vu Thien Binh, S.T. Purcell, V. Semet, C. Journet, this proceeding.
8. J.C. Plenet and Vu Thien Binh, patent pending #99 14 474 (1999).
9. A.Bahtat, et al., *J. of Non-Cryst.Solids*, **202**, 16 (1996).
10. J.C. Plenet, et al., *Opt. Mat.*, **13**, 411 (2000).
11. Vu Thien Binh, et al., *Appl. Phys. Lett.*, **73**, 2048 (1998).

Mat. Res. Soc. Symp. Proc. Vol. 621. © 2000 Materials Research Society

Field-Emitter-Array Cold Cathode Arc-Protection Methods - A Theoretical Study

L. Parameswaran, R.A. Murphy
MIT Lincoln Laboratory, Lexington MA   02420-9108

ABSTRACT

Field-emitter arrays (FEAs) are desirable for use as electron emitters in microwave-tube amplifiers because they can provide such advantages as higher efficiency and faster turn-on compared to their thermionic counterparts. Calculations have shown that Spindt-type metal and semiconductor emitters operate well below intrinsic current limits due to thermal effects, even for high-current applications such as klystrodes, twystrodes, and traveling-wave amplifiers. Nevertheless, the primary barrier to FEA utilization in such applications is premature failure due to arcing. These failures appear to be produced by ionization of gas molecules and/or desorbed contaminants, which are exacerbated by a poor vacuum environment. Lifetime and stability issues have been largely resolved for less stringent applications, such as flat-panel displays, through the use of integrated passive resistors that provide current limiting. However, such an approach is not directly compatible with operation at high frequency and current density. Other more complex approaches, such as the incorporation of active control in the form of integrated transistors, have also been demonstrated, but again, only for FEAs used in displays. This paper will review some of these schemes in the context of their efficacy in improving lifetime and stability of FEA cold cathodes in high-frequency applications. A theoretical analysis will be given of the effect on high-frequency performance of incorporating arc protection structures into Spindt-type metal FEAs. Specifically, two approaches will be considered: passive protection schemes such as the use of a modified thin film resistive layer, and active schemes such as FETs and saturated current limiters.

INTRODUCTION

Field emitter arrays have been the subject of interest as electron sources for several decades, and have been considered for a variety of applications. The most successful application to date has been their use in flat panel displays [1], but more recently other more stringent applications have been investigated, such as their use as ion sources for electric propulsion systems [2], electron-beam lithography [3] and vacuum pressure sensors [4]. A few of these applications are illustrated in Figure 1, which gives estimates of the ranges of current levels, operational frequencies and ambients in which the FEAs are required to function.

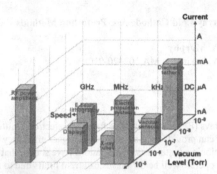

Figure 1. Applications for field emitter arrays.

Significant effort has been expended in developing FEAs as cold cathodes for high frequency inductive output amplifiers (IOAs) such as klystrodes, twystrodes, and travelling wave tubes. The gated FEA is capable of very high current densities, and can provide several advantages over its thermionic counterpart. Microfabrication techniques [5] have enabled the fabrication of extremely small devices with short gate-to-emitter lengths, resulting in low electron transit times and correspondingly high transconductances. FEAs operate at much lower voltages and since they require no heating, are capable of high pulse repetition frequencies. Because emission can be gated at the device itself, density modulation at the emitter is possible, which eliminates the need for a separate prebunching section in a travelling wave tube and results in a concomitant size reduction for the tube.

As an example of typical performance requirements desired for an FEA cathode in an IOA, in a recent NASA klystrode program undertaken to incorporate a gated FEA into a Klystrode, the FEA was required to meet the following specifications: operation at 10 GHz, peak currents in excess of 120 mA, and current densities above 100 A/cm^2. A number of groups have been able to demonstrate currents and/or current densities in excess of the required levels at lower frequencies, as well as operation in a klystrode tube with 10-GHz gate modulation, albeit at much lower current levels [6]. Extensive theoretical and modeling work, supported by experiment, has shown that it X-band operation can be achieved from a variety of microfabricated emission tip structures and materials, including tips composed of metal [7], silicon [8] and GaAs [9]. However, the primary issue in implementation continues to be premature failure of the cathodes in the tube environment, usually catastrophically, due to large current arcs from the tip to the gate or anode. This paper will present a brief review of some of the approaches that are being taken to incorporate arc protection into microfabricated field emission devices in the context of their compatibility with operation at high frequencies. After a summary of current theories on failure modes in field emitters, two categories of arc protection – passive and active – will be discussed. Passive schemes encompass such approaches as fundamental changes to the emitter structure or material (e.g. inert coatings that protect the tips from ion bombardment), as well as the incorporation of resistive elements with the tips (as implemented for FEAs used in flat panel displays). Active schemes include the integration of more complex transistor elements (JFET, MOSFET) with the tips, to provide control of emission current that is independent of the emitter extraction gate.

FAILURE MODES IN FIELD EMITTERS

Figure 2 shows an arc current failure in progress, for a typical molybdenum-tip FEA fabricated using a modified Spindt process [5]. As is often the case, the arc progresses a significant distance along the array, resulting in gate and tip melting, and subsequent device failure. Prior to the catastrophic arc, a number of non-catastrophic events were observed, presumably single- or few-tip burnouts resulting from arc currents that were not large enough to induce large-scale melting of the gate metal. The fact that the devices often remain operational in the presence of these smaller burnouts suggests that device failure may be avoided by limiting the maximum current to a level below that required for macroscopic melting of the gate and/or emitter material.

Figure 2. *Arc failure in progress, for 15000-tip array, $V_{gate-tip}$=80 V, V_{anode}=300 V, P=10^{-8} Torr.*

Traditionally, failure in FEAs has been thought to be caused by a combination of heat-induced deterioration of the emitter source, and ionization of the molecules in the surrounding ambient [10]. For an "ideal" metal emitter (composed of smooth single crystal metal), at low emission currents the Nottingham effect predominates and causes heating throughout the entire tip, whereas at higher current levels, the tip surface is cooled through the reverse Nottingham effect, while the bulk of the emitter continues to heat resistively. Typical micron-scale individual metal emitters with tip radius r_{tip} and composed of tungsten or molybdenum can accommodate maximum currents on the order of I_{max}~3000-5000r_{tip} (units are amperes and centimeters) [10], which translates to currents of milliamperes per tip, or much larger than seen under normal operating conditions. At moderate emission current levels, calculations of tip heating indicate that melting or thermal runaway is unlikely [11], especially if adequate heat sinking is provided via a conductive plane. For ideal semiconductor tips, heating is primarily ohmic in nature, and the tips are only capable of sustaining lower maximum currents on the order of I_{max}~100-200r_{tip}; but again, calculations indicate that for an ideal tip at moderate current levels, melting is unlikely [12].

Microfabricated field emission tips are not ideal but in fact are polycrystalline in nature, with both surface and bulk structural nonuniformities. Additionally, structural features that are designed to achieve higher performance (high aspect-ratio tip, high tip density) may result in reduced heat-sinking capability to the substrate, increasing the probability of thermal breakdown of the tip material and arcing. Studies of the surface of microfabricated metallic tips combined with energy distribution measurements of single and multiple-tip arrays [13,14] have shown that emission current instability is driven by the continual formation and destruction of nanoprotrusions on the surface of the tip, through ion bombardment and diffusion of the tip surface atoms. Such unstable features may also exacerbate the arcing problem, particularly at high tip-current densities.

In addition to intrinsic heating of the tip itself, failure can also be initiated by ionization and sputtering of contaminants. While most studies of FEAs have been undertaken in ultrahigh vacuum environments ($< 10^{-9}$ Torr), often FEAs are operated in much poorer vacuums, with contaminants that are sputtered or desorbed either from the emitter itself or from the surrounding chamber walls, as a result of electron beam bombardment. A sputter model developed by Brodie [15,2] to estimate lifetime in specific ambient gases indicates that the lifetime decreases rapidly with increasing gate voltage and emission current. While these estimates are pessimistic as they do not take into account competing processes such as adsorption or scattering of sputter ions, they nevertheless illustrate a problem for applications that require high tip currents, particularly in poor vacuums. One technique to alleviate this problem has been the development of "seasoning" procedures [16] to gradually burn off surface contaminants by subjecting the emitters to hydrogen plasma as they are brought to their highest emission level, and this has been effective in enabling high emission current levels at elevated pressures.

ARC PROTECTION SCHEMES

The optimal form of arc protection would have no impact on device performance in normal operating regimes, limit tip current to acceptable levels and track arcs at any frequency, and minimize added fabrication complexity. In view of this, fundamental modifications to the emitter itself to make it more robust would appear to be preferable over the incorporation of additional protective devices. Tradeoffs between optimization of operational characteristics and ease of fabrication must also be considered when selecting a protection scheme.

Passive Approaches - Robust Emitter Materials and Coatings

Recently there have been promising developments in the application of refractory metal carbides as coatings for emission tips [17]. Such coatings are thermally resistant, have good thermal conductivity for heat sinking, provide both physical and chemical passivation of the tip surface, and are resistant to ion bombardment. Materials such as HfC and ZrC have been used as emitters and as coatings on conventional molybdenum microfabricated tips. Their longevity has been impressive, with single HfC emitters operating at 5 mA for > 1000 hours, and coated ZrC/Mo emitter arrays operating at 100 µA in 5×10^{-7} Torr vacuum.

There has also been interest in implementing diamond, either as an emitter material itself or as a coating for metal or semiconductor tips. Other material possibilities include gallium nitride [18] and boron nitride [19], but for most of these materials, low emission currents and/or poor material quality have so far limited their use.

Resistive layers

Field-emission display manufacturers have for several years incorporated a resistive layer under the emitters to provide a series resistance at each tip [20]; the resistor acts as a current limiter and improves uniformity. This approach has been highly successful in improving the lifetime and reliability of FEAs used in flat panel displays, to the extent that lifetimes exceeding 10000 hrs are now routine, even in the relatively poor vacuum environment of the display packages. However, the resulting high series resistance in combination with the gate capacitance limits the frequency at which the gate-to-emitter voltage can be modulated to the MHz range, insufficient for RF applications. With proper design, the capacitance of the resistive layer can act as a bypass at X-band, while still providing enough resistance on a per-tip basis at lower frequencies to suppress arc currents. The effective impedance of the array at X-band can thus be reduced to the order of a few ohms and still retain arc protection.

Such a scheme is described in [21], where the resistive layer parameters were selected based on a desired arc current limiting level, and the impact on high frequency performance was analyzed using a simple triode model with impedance matching to the FEA. The overall effect of the added impedance in series with the emitter was expressed in terms of a frequency-dependent current modulation efficiency, and a simple SPICE model was extracted from the model, and used to investigate the operation of the system at X-band. Using experimentally determined Fowler-Nordheim parameters for a 30000-tip FEA operating at a gate-tip modulation of 50 ± 10 V at 10 GHz, a resistive layer thickness of 500 A (which presents an impedance at 10 GHz of about 3 Ω) provided a maximum in current modulation efficiency and a minimum emission current drop as compared to an unprotected device.

While the above analysis, and analogous studies done with silicon-based emitters [22], indicate that it is possible to achieve high frequency modulation with a resistive layer protection scheme, difficulties in producing a robust, defect-free thin resistive layer may prohibit the practical implementation of such a method.

Alternate Resistive Approaches

Another approach is to implement the resistor in the form of a post under the emission tip. This structure is used most notably by the Microelectronics Center of North Carolina [8] to fabricate silicon FEAs. An analysis of the high frequency performance of variants of this structure, as suggested in [22], is instructive. Figure 3 shows the results of a simple analysis of the emission and limiting currents achievable with the basic structure (with a metal tip instead of a silicon tip). Emission currents are calculated based on Fowler-Nordheim parameters extracted from a typical molybdenum cone emitter (I_{tip}=1 μA, B_{FN}=600 V, tip radius=100 A, post radius=0.2 μm) with the emitter in an impedance-matched circuit, and limiting currents are obtained assuming various bulk resistivity levels for the silicon post. All currents are normalized to the DC emission current with a perfectly conducting post. As expected, increasing the post resistivity increases the effectiveness of current limiting, but at a cost of poorer high frequency performance. A modification to the structure that incorporates capacitive bypassing is shown in Figure 4. In this case, the DC current is still limited by the resistive post, but a thin oxide layer surrounding the post shank provides bypassing at high frequencies. This does improve emission at higher frequencies, but at a cost of degraded limiting, particularly if the time scale of the failure event is in the nanosecond range where $I_{limit}/I_{emission}$ degrades to greater than 100:1. Finally, a third version, shown in Figure 5, eliminates the bypass oxide layer and replaces it with a direct contact to the post shank. In this situation, the gate-tip bias is applied to the post as well as the tip, resulting in depletion of the post and a situation analogous to pinch-off in a JFET channel. Note that the limiting current is sensitive to resistivity of the post in this case, as expected since the pinch-off voltage and saturation current in a JFET are proportional to the doping level in the channel. Also note that the post diameter in this structure is somewhat larger (r_{post}=1.4 µm) than that used in the previous variations. The electron density in the post must be sufficiently low to support the entire gate-tip voltage, which results in an increased post diameter to accommodate the depletion width in normal operation. The larger post size will significantly reduce tip densities.

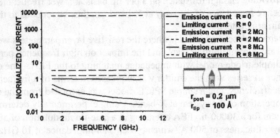

Figure 3. *Effect of resistive post on high frequency operation of FEA.*

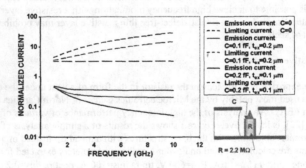

Figure 4. *Effect of resistive post with capacitive bypass on high frequency operation of FEA.*

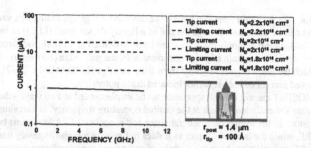

Figure 5. *Effect of resistive post with JFET-like gate on high frequency operation of FEA.*

The examples shown above indicate that it may be difficult to achieve the competing requirements of high emission currents and low limiting currents with a purely passive scheme. For that reason, we turn to active protection methods that incorporate independently modulated semiconductor devices to control emission current at all levels of operation.

Active Approaches - MOSFET Current Control

The concept of field emitters with integral regulated current control encompasses a wide range of devices and was proposed a number of years ago [23] as a natural extension of the semiconductor-based FEA. The basic approach is simple – attach the tip to the drain electrode of a transistor, and regulate emission via the induced channel and transistor gate. One of the first implementations was the incorporation of a MOSFET in series with the emission tip [24], as shown in Figure 6a. In this case, an inversion layer channel between a source electrode and the emitting tip is created to provide a path for electrons supplied by the source. The common gate not only serves to create the inversion layer, but also acts as the extraction electrode for the emitted electrons. A thick dielectric layer is used over the channel and a heavily-doped substrate is used to make the FEA extraction and FET turn-on voltages comparable. By controlling the current supplied to the tip with the FET, significant improvements in emission current stability and uniformity of silicon tips have been reported, to the extent that emission is stable to within 1% over long operational time periods [25]. In addition, stable operation in poor vacuum of as high as 10^{-5} Torr CO has been demonstrated [26].

While the single-gate FEA+FET structure is simple to construct, it does not allow for separate optimization of each element. The dual-gate structure in Figure 6b has a thin FET gate oxide providing a low MOSFET turn-on voltage, and a thick oxide surrounding the FEA, to reduce capacitance, as well as decouple the two devices [27]. With this structure, a gate-tip short resulting from an arc will not result in failure of the whole array, since the maximum current into the emitter is still controlled by the transistor gate voltage.

Typical extraction gate voltages for FEAs are several tens of volts. Since the combined structure is operated with a floating drain on the transistor, during normal operation drain potentials can rise high enough to cause impact ionization at the drain-channel junction, resulting in breakdown and loss of current control. More importantly, a failure event resulting in a direct connection between extraction gate and tip will place the entire extraction gate voltage on the drain of the transistor, which should therefore have a breakdown voltage exceeding this value. Both of these problems can be alleviated through the use of a lightly doped drain (LDD) to reduce the drain edge electric field, as is commonly done for submicron MOSFETs. In [28], two structures are compared, one in which the drain edge is concurrent with the gate edge (conventional MOSFET) and one in which the drain edge is removed from the gate edge (LDD MOSFET). The latter showed improved current control at higher induced drain potentials.

The MOSFET current control approach can be implemented in two ways - the FET can directly modulate the emission current at the desired operating frequency, or can simply be used as a variable resistor. In the former case, the transistor and associated FEA load must be capable of switching at RF, while the latter is simpler as it does not require a high-frequency transistor. Traditionally, RF operation has been the domain of compound semiconductor devices due to the higher intrinsic mobilities and lower device capacitances achievable; however, deep submicron lithography capabilities combined with silicon-on-insulator (SOI) technology to reduce junction capacitance has made it possible to fabricate silicon-based FETs that operate with unity current gain frequencies of over 1 GHz. For example, 0.25 µm fully depleted SOI n- and p-MOSFETs were recently demonstrated with f_T of 30 GHz and 13 GHz respectively [29]. The disadvantage of these devices is their very low breakdown voltage (< 5 V), which precludes them from being used directly with FEAs unless the emitter also has a very small geometry and turn-on voltage (e.g. 0.1-µm tips with $V_{turn-on}$< 13 V [30]). The solution to this is the LDMOSFET (Lateral Double-diffused MOSFET), which was designed to be used in power amplifiers. Recently, efforts to develop high frequency, high gain LDMOSFETs for use in L-band RF communications subsystems has resulted in both bulk [31] and SOI [32] devices capable of switching frequencies greater than 1 GHz and breakdown voltages in excess of 30 V. Figure 6c shows the basic structure of an SOI version of the device, with the LDD incorporated to increase breakdown voltage. The SOI version may be preferred as it offers lower on-resistance and lower parasitic capacitances over its bulk counterpart; however, it is subject to drift due to self-heating effects [33].

Normally these power MOSFETs are designed with very large channel widths, to provide output currents in the range of amps. If the structure were scaled down such that one transistor drives a pixel of 100 tips at a density of 1 µA/tip, for example, the transistor would be required to source 0.1 mA. Using measured parameters from a bulk LDMOSFET [31], a device with a gate length and width of 1 µm and 4 µm respectively can source 0.1 mA at a gate voltage of 4 V. With a somewhat conservative tip-to-tip spacing of 1.5 µm, a 100-tip pixel would require an area of about 270 µm², so that the fraction of the overall pixel occupied by the active device is kept to a minimum. Using fewer tips-per-pixel increases the area overhead fraction required by the transistor, but reduces the number of tips that are lost in the event of a gate-tip shorting failure.

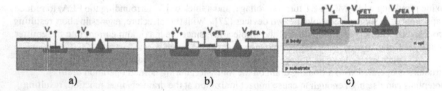

Figure 6. a) Basic MOSFET+FEA. b) Dual-gate MOSFET+FEA. c) Power SOI LDMOSFET.

Vertical Current Limiter (VECTL)

The VECTL [34] arc protection approach was developed by NEC and is illustrated in Figure 7a. This structure consists of groups of conventional molybdenum tips formed on top of silicon posts surrounded by oxide-filled trenches, and is somewhat similar to the scheme shown in Figure 5. Under normal operating conditions the series resistance of the post is low, in fact orders of magnitude lower than that used in displays. The dimensions of the trench and post have been designed such that upon the application of a high voltage directly to the tip (as happens in a gate-to-tip arc), the silicon post under the affected tip will deplete, thereby pinching off the channel and limiting the current through the tip to a designed safe value (Figure 7b). Using the VECTL scheme, NEC has demonstrated cathode lifetimes of 5000 hours in a miniaturized travelling wave tube, with output currents of 56 mA at 11.5 GHz [35]. It should be noted that the FEA was not gate-modulated, as the resistance of the post prohibited operation at X-band. Nonetheless, the approach appears to be quite efficient at suppressing arc currents, as the device did not fail under the unfavorable conditions seen in the tube.

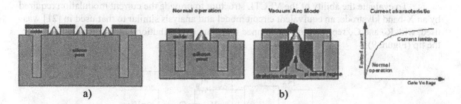

Figure 7. *a) VECTL basic structure. b) Operation under arc conditions.*

Figure 8 illustrates the operation of the current limiter under arc conditions (assumed to be a short between the extraction gate and the tip), as obtained from Pisces simulations using Silvaco tools. There are two versions of the device; the first is identical to the structure used by NEC, and the second is a proposed structure with finer geometries that may be capable of RF modulation. In both cases a direct connection between the gate and tip results in depletion of the silicon under the tip, and a saturation of the current supplied to the tip, with the saturation voltage comparable for both cases.

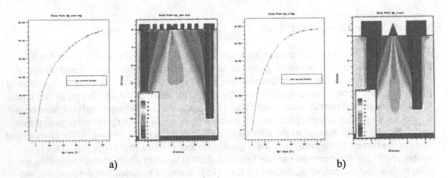

a)   b)

Figure 8. *a) Simulated gate-tip arc for NEC geometry. b) Simulated gate-tip arc for proposed reduced geometry device.*

To evaluate the ability of the VECTL structure to provide the current modulation required by an X-band klystrode, an equivalent circuit model and analysis similar to that used in [21] was used, with R_V and C_V representing the lumped impedance contribution of the silicon pillar under the tip (Figure 9).

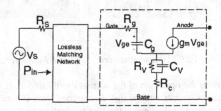

Figure 9. *Equivalent circuit model of FEA with VECTL arc protection.*

The impact on high frequency performance was evaluated by calculating a current modulation efficiency, defined as $\eta=(I_2-I_1)/I_2$, where I_2 and I_1 are the anode currents at the maximum and minimum gate-tip voltage respectively, and are functions of the input voltage swing. The RF voltage swing at the FEA input is determined by the RF voltage source, the impedance matching network, and the total input resistance of the FEA.

Two sets of geometries were compared, the existing design presented by NEC and a proposed design using reduced-geometry cathodes (Table I). Figure 10 and Table II summarize the results of the analysis, which indicate that a VECTL-based FEA can be modulated at X-band frequencies if the tip dimensions are reduced and tip densities increased. Additionally, a narrower and thinner-walled beam can be obtained with the higher achievable tip densities.

Table I. Design parameters for NEC and reduced-geometry VECTL structures.

FEA design parameter	NEC	Reduced geometry #1
Tip-to-tip spacing (μm)	1.2	0.5
Tips/column	64	64
Column width (μm)	10	4
Isolation trench width (μm)	1.2	0.5
Channel depth (μm)	10	2
Gate oxide thickness (μm)	0.5	0.35
Channel doping (cm^{-3})	2.6×10^{14}	8.0×10^{14}
Channel pinch-off voltage (V)	5	5
Limiting current per column (μA)	300	300
Channel resistance per column (kΩ)	33	19
Gate capacitance per column (fF)	11	0.84

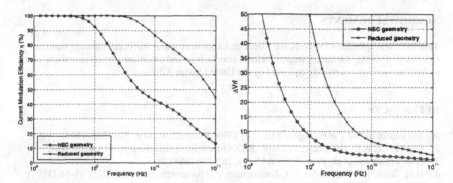

Figure 10. *a) Current modulation efficiency and b) gate modulation voltage for two tip geometries.*

Table II. RF performance comparison of NEC device and proposed reduced-geometry device.

Parameter	NEC device	Reduced-geometry #1
Number of tips per quadrant	30000	30000
Number of cells per quadrant	469	469
Annular outer diameter	900 μm	610 μm
Annular inner diameter	720 μm	526 μm
Input impedance per quadrant	18 Ω	33 Ω
Input power per quadrant	0.25 W	0.25 W
Peak current from 4 quadrants	120 mA	120 mA
Beam modulation efficiency at 10 GHz	**43%**	**87%**

CONCLUSIONS

This paper has provided a brief review and discussion of some approaches that can and are being taken to incorporate high-frequency-compatible arc protection into a microfabricated field emitter array. While the passive resistive layer approach has been highly successful for displays, and may translate to higher frequencies with suitable modifications, the availability of such active schemes as the incorporation of power RF MOSFETs and the FET-like VECTL may provide more robust current limiting. But the optimal choice providing the most flexibility in device operation may in fact be a combination of an active device and a more robust emitter material such as a carbide-coated metal or semiconductor tip.

ACKNOWLEDGMENTS

This work was sponsored by the Naval Research Laboratory under Air Force contract number F19628-95-C-0002. Opinions, interpretations, conclusions and recommendations are those of the authors and are not necessarily endorsed by the United States Air Force.

REFERENCES

1. See papers in Flat-Panel Display Materials 1998. Symposium. Mat. Res. Soc. 1998.
2. C. Marrese, A.D. Gallimore, J.E. Polk, K.D. Goodfellow, K.L. Jensen, AIAA 98-3484.
3. A.D. Feinerman et al., Proc. SPIE vol.2194, 262-73 (1994).
4. H.H. Busta, J.E. Pogemiller, B.J. Zimmerman, *J. Micromech. Microeng.* 3(2), 49-56 (1993).
5. C.A. Spindt, I. Brodie, L. Humphrey and E.R. Westerberg, *J. Appl. Phys.* **47**(12), 5248 (1976).
6. S.G. Bandy et al., presented at IEEE Int. Conf. on Plasma Science, San Diego CA, (1997).
7. R.A.Murphy et al., presented at IEEE Int. Conf. on Plasma Science, San Diego CA (1997).
8. D. Temple et al., *J. Vac. Sci. Tech. B* **16**(3), 1980-1990 (1998).
9. P. Kropfeld, et al., Proc. IVMC'98, 88-89 (1998).
10. F. Charbonnier, *J. Vac. Sci. Tech. B* **16**(2), 880-887 (1998).
11. M. Ancona, *J. Vac. Sci. Tech. B* **13**(6), 2206-2214 (1995).
12. M. Ancona, *J. Vac. Sci. Tech. B* **14**(3), 1918-1923 (1996).
13. V.T. Binh, N.Garcia, S.T. Purcell, *Adv. in Imaging and Electron Physics* **95**, 63-153 (1996).
14. S.T. Purcell, V.T. Binh, *J. Vac. Sci. Tech. B* **15**(5), 1666-1677 (1997).
15. I. Brodie, *Int. J. Electronics* **38**(4), 541-550, 1975.
16. P.R. Schwoebel, C.A. Spindt, *J. Vac. Sci. Tech. B* **12**(4), 2414-21 (1994).
17. W.A. Mackie, Tianbao Xie, P.R. Davis, *J. Vac. Sci. Tech. B* **17**(2), 613-619 (1999).
18. T. Kozawa, T. Ohwaki, Y. Taga, N. Sawaki, *Appl. Phys. Lett.* **75**(21), 3330-2 (1999).
19. T. Sugino, S. Kawasaki, K. Tanioka, J. Shirafuji, *J. Vac. Sci. Tech. B* **16**(3), 1211-14 (1998).
20. C.Py, R.Baptist, IVMC'93, 23-24 (1993).
21. L.Parameswaran et al., *J. Vac. Sci. Tech. B* **17**(2) (1999).

22. H.F. Gray, J.L. Shaw, Proc. IVMC'97, Kyongju Korea, 220-225 (1997).
23. H. Gray, U.S Patent 5359256 (1994).
24. A. Ting et al., Proc. IVMC'91, 200-201 (1991).
25. J. Itoh, T. Hirano, S. Kanemaru, *Appl. Phys. Lett.* **69**, 1578-1579 (1996).
26. S. Kanemaru, T. Hirano, K. Honda, J. Itoh, *Appl. Surf. Sci.* **146**, 198-202 (1999).
27. K. Yokoo, M. Arai, M. Mori, B. Jongsuck, S. Ono, *J. Vac. Sci. Tech. B* **13**(2), 491-3 (1995).
28. K. Koga, S. Kanemaru, T. Matsukawa, J. Itoh, *J. Vac. Sci. Tech. B* **17**(2), 588-591 (1999).
29. D.D. Rathman et al., Proc. IEEE MTT-S Digest, 577-580 (1999).
30. J.O. Choi et al., *Appl. Phys. Lett.* **74**(20), 3050-2 (1999).
31. P. Perugupalli et al., *IEEE Trans. Electr. Dev.* **45**(7), 1468 (1998).
32. S. Matsumoto et al., *IEEE Trans. Electr. Dev.* **43**(5), 746 (1996).
33. H. Neubrand et al., Proc. 6th Int. Symp. Power Semic. Dev., 123 (1994).
34. H. Takemura et al., Proc. IEDM'97, 709-12 (1997).
35. H. Makishima et al., *Appl. Surf. Sci.* **146**(1-4), 230-3 (1999).

Mat. Res. Soc. Symp. Proc. Vol. 621. © 2000 Materials Research Society

Field Emitter Cathodes and Electric Propulsion Systems

Colleen M. Marrese, James E. Polk, Juergen Mueller
Jet Propulsion Laboratory, California Institute of Technology
Pasadena, CA 91100, U.S.A.

ABSTRACT

Replacing hollow and filament cathodes with field emitter (FE) cathodes could significantly improve the scalability, power, and performance of some meso- and microscale Electric Propulsion (EP) systems. The propulsion system environments and requirements and the challenges in integrating these technologies are discussed to justify the recommended cathode configurations. Required cathode technologies include low work function coatings on Si or Mo Field Emitter Array (FEA) cathodes with arc protection and electrostatic ion filters.

INTRODUCTION

There is considerable interest now in meso- and microscale electric propulsion systems for microscale spacecraft and large inflatable spacecraft to support the robotic exploration of the solar system and characterize the near-Earth environment. Three classes of microscale spacecraft have been identified: Class I (10-20 W, 10-20 kg, 0.3-0.4 m), Class II (1 W, 1 kg, 0.1 m), and Class III (<<1 W, << 1 kg, 0.03 m).[1] Microscale spacecraft will enable the development of multiple, distributed spacecraft systems possibly in contact with a mother spacecraft to perform 3-dimensional mapping of particles and fields and global surveys of a planet with higher resolution and flexibility and lower mission risk. The fuel savings from the use of high specific impulse (thrust/propellant flow rate>1000 s) electric propulsion systems enables more difficult planetary missions with smaller launch vehicles.

Several electric thrusters being miniaturized require cathodes to provide electrons for propellant ionization and/or ion beam neutralization. Electrostatic thrusters are being developed to operate between 10s of mW and 500 W. These thrusters will require cathode current densities approximately 100 mA/cm^2 for more than 6000 hours. Filament, thermionic, and hollow cathodes require heaters and propellant feed systems which place lower limits on their power and size scalability. Filament cathodes consume 0.1 –10 W/mA.[2,3] Thermionic cathodes have demonstrated 1.5 W/mA.[4] State-of-the-art 1/8" hollow cathodes have demonstrated 50 mA with 0.2 mg/s of xenon and 8 W consumed by the keeper electrode with an anode voltage of 15 V.[5] Cesium hollow cathodes require 10 W/mA.[6] In contrast, a Mo FE cathode has demonstrated 120 mA at <1 mW with no propellant or heater. Some microscale spacecraft proposed will have less than 100 mW of power. Compatible FE cathodes may enable the use of EP systems on these spacecraft.

The primary concerns with integrating FE cathodes with EP systems are space-charge limited emission and cathode lifetime in the plasma environments generated by the propulsion systems. In an EP system the FE cathodes, with gate electrodes, will emit into a plasma anode. The space-charge current limit depends on the plasma density and temperatures, and electron beam energy and current density. The cathode lifetime will be limited because it will be subjected to constant ion bombardment. The self-generated ion population originates near the cathode when electrons emitted by the cathode ionize ambient neutrals. Slow ions and fast neutrals result from charge-exchange collisions between the fast ions accelerated by the thruster and the slow ambient neutrals, creating a second ion population bombarding the cathode. Ions generated in the discharge chamber of an ion thruster will bombard cathodes which are supplying electrons for propellant ionization.

The propulsion systems which could benefit from compatible FE cathodes are described in this paper with their demands on cathode performance. FE cathode technologies required to integrate them with EP systems are discussed.

MESO- AND MICROSCALE ELECTRIC PROPULSION SYSTEMS

The configurations of mesoscale Hall and ion thrusters, Field Emission Electric Propulsion (FEEP) systems, and Indium Liquid Metal Ion Sources (I-LMIS) are shown in Figures 1 and 2. The performance of state-of-the-art systems at low thrust levels is shown in Table 1. Each of these systems could significantly benefit from a compatible FE cathode in size, performance, and system complexity. The development of ion thrusters smaller than the configuration described Table 1 is underway. Microscale FEEPs, I-LMISs, and colloid thrusters are also under development. Table 1 shows the current and lifetime requirements of available systems and describes the cathode environments which must be tolerated. J_i is the current density of ions which will bombard the cathode.

Ion thrusters require cathodes for both propellant ionization(discharge cathode) and ion beam neutralization (neutralizer cathode). The propellant is ionized in the discharge chamber by electron bombardment. Magnets are used to impede the flow of electrons from the cathode to the anode to improve the ionization efficiency. The ion optics are composed of two multiaperture grids which are biased to focus ions which drift into the interelectrode gap and accelerate them to generate thrust. These thrusters traditionally use hollow cathodes.

Ion engines have been optimized and flight qualified to operate at power levels up to 2.3 kW. Mesoscale ion thrusters were developed to operate on cesium with considerable success; however, xenon propellant is preferred over cesium because of its toxicity. Cesium ion thrusters have demonstrated 12 μN at 4.58 W from a 1.27 cm-diameter discharge chamber up to 5.3 mN at 237 W from a 4.58 cm-diameter discharge chamber.[7] At low power levels, the thermionic cathode consumed as much as 48 % of the total power consumed by the thruster. More recently, the performance of a xenon ion thruster with a 5 cm-diameter discharge chamber was reported as 2.2 mN and 2300 s at 49% thrust efficiency and 50 W of power.[8] The efficiency calculation does not include the power and propellant consumption by the cathode.

Hall thrusters require only one cathode which supplies electrons for both ion beam neutralization and propellant ionization by electron bombardment. The axial accelerating electric field is established between the anode and cathode. Electromagnets are typically used in this system to generate a radial magnetic field near the exit plane of the thruster which creates an azimuthal drift of electrons towards the anode and a high density electron cloud near the thruster exit plane. The propellant is ionized in the electron cloud and accelerated axially by the electric field. These thrusters traditionally use hollow cathodes also.

Hall thrusters have been optimized to operate at 1.5 -3.0 kW and are currently being scaled down to mesoscale systems which will be optimized to generate 1-10 mN of thrust. Higher pressures, current densities, and magnetic field strengths are then required to reduce mean free paths in the discharge chamber and maintain the plasma discharge as the size of the thruster is reduced.[9] Xenon is the preferred propellant for these systems because of its high mass, relatively low ionization energy, and inert nature. The mesoscale X-40 has a 40 mm discharge chamber diameter. It demonstrated 7.43 mN of thrust at 100 W (150 V, 0.67 A) and 0.74 mg/s to generate 1020 s specific impulse (ion velocities of ~10,000 ms/) at 37 % efficiency (not including the cathode power and propellant).[10] At 200 W and 14.5 mN, the thruster operated at 48 % efficiency. The lifetime of this system was projected to be 850 hours, and could be improved to 2000-3000 hours by employing more sputter resistant materials. The BHT-200-X2B operates in a similar thrust regime.[11] The thrust range of this system is 4-17 mN with 100–300 W at 20-45 % efficiency (not including the cathode) and 1200-1600 s.

FEEPs and I-LMISs also accelerate ions electrostatically to generate thrust; however, unlike the ion and Hall thrusters, the propellant is not ionized by electron bombardment. The propellant is field ionized. FEEPs use cesium and I-LMISs use indium liquid metal propellant which is fed by capillary forces from the propellant reservoir through a small channel (FEEP), or wicked up the outside of a needle from the propellant reservoir (I-LMIS). The emitters are terminated with sharp edges and biased positively with respect to an extraction electrode located downstream of the emitter. The electric field applied between the electrodes deforms the surface of the liquid metal into Taylor cones[12] with a cusp (often called a jet) to maintain equilibrium between electrostatic and surface tension forces. When the electric field strength reaches 10^7 V/cm atoms of metal on the tip are ionized by field ionization or field evaporation. Ions are accelerated by the extraction electrode. These thrusters have used hollow and filament cathodes.

Indium Liquid Metal Ion Sources (LMIS) have been used for spacecraft potential control and are now being developed for propulsion applications also.[13] They are being developed to operate at 10s of μN and 4.5-6.5 kV, >6000 s specific impulse at ~15 μN/W (67 W/mN). Cesium FEEP systems have been developed to operate at ~17 mN/W (60 W/mN) for 100 μN of thrust at 8000 s specific impulse (~80,000 m/s ion velocities) with 9 W of power and a total impulse of 160 Newton-seconds (450 hours at nominal steady state operation).[14]

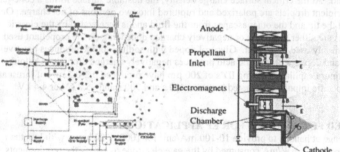

Figure 1 Ion thruster(left) and Hall thruster(right) configurations.

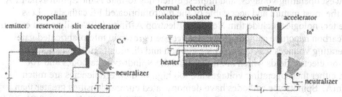

Figure 2 FEEP (left) and I-LMIS (right) configurations.

Table 1 Representative performance and cathode environment of meso- and microscale propulsion systems.

	Ion Thruster (5 cm)[8]	Hall Thruster (X-40)[10]	FEEP Thruster[14]	Colloid Thruster[15]
Thrust (mN)	2.2-4.7	5-32	0.001-1	0.2-0.5
Power (W)	50-116	80-510	0.06-60	~2
Power/Thrust	23	16	60	4.4
Specific Impulse (s)	2,300-3,100	1,160-1,933	8,000 (10 kV)	450-700
Current (mA)	230-430(disch.) 44-81(neut.)	500-1,700	0.006-6	-
Efficiency	0.49-0.61	0.31-0.55	0.98	-
Specific Mass (kg/W)	-	-	0.008	-
Propellant	Xenon	Xenon	Cesium	Glycerol
J_i ($\mu A/cm^2$)	4,000 (disch.) 0.004-0.008 (neut.)	2 -7	0.002-2	-
Pressure (Torr)	10^{-3}-10^{-6}	10^{-6}-10^{-4}	10^{-6}	-
Lifetime (hours)	6,000	950-2000	450-20,000	-

Colloid thrusters are similar to FEEPs except that charged droplets, instead of atomic ions are emitted and the propellant is not field ionized. An electric field is applied between the capillary tubes feeding the propellant and the extraction electrode to increase the surface charge density in the liquid. At the critical surface charge density, the unstable surface forms a cone-jet configuration. Incipient droplets are polarized and ruptured into two portions of net charge. One portion remains at the tip and the other escapes from the fluid and is accelerated by the electric field. The droplets are either positively or negatively charged depending on the propellant used. Glycerol is a commonly used propellant. Glycerol doped with sodium iodide produces positively charged droplets, and doped with sulfuric acid produces negatively charged droplets. Colloid thrusters with platinum capillaries having ID's of 200 μm have produced 0.2-0.5 mN of thrust at 4.4 W/mN with specific impulses estimated between 450 s and 700 s and –5.8 kV or 4.4 kV required.[15]

RECOMMENDED FE CATHODES FOR EP APPLICATIONS

The challenge at hand is to provide 10-100 mA/cm^2 for 1000-6000 hours with less than 100 V at little more than 1 mW/mA consumed by the gate electrode to satisfy the requirements of each propulsion system. The FE cathode technology which has demonstrated the highest current density, lowest operating voltages, and highest efficiency to-date is the Spindt-type FEA cathode. It is also the most mature and accessible of the microfabricated FE cathodes. Less mature FE cathode technologies include thin Negative Electron Affinity (NEA) films and carbon nanotubes. Rigid carbon nanotube cathodes have not yet been grown in microfabricated gate structures with operating voltages less than 100 V. Carbon and diamond NEA films have demonstrated turn-on electric fields which are lower than the Spindt-type cathodes and 100 mA/cm^2, however, either their operating voltages are too high or their efficiencies are much lower than 1 mW/mA. Spindt-type cathodes have demonstrated current densities greater than 2000 A/cm^2 from arrays in UHV in triode configurations with efficiencies higher than 0.01 mW/mA and lifetimes greater than 8000 hours.

At this time it is believed that to meet the current density and lifetime requirements of the EP applications, the Spindt-type cathodes should be coupled with carbide or NEA material films.

Experiments have shown that xenon will not affect the work function of Mo, Si, and C cathodes.[16] It has also been shown that the energy threshold for sputtering Mo and Si FEA cathodes with xenon ions is 49 eV and 64 eV, respectively.[16,17] The self-generated ions consist of both single and double ions if the cathode is operated above 35-37 V. At operating voltages below ~85 V, the cathode erosion process is dominated by the double ions. In a xenon environment where only the self generated ion population is considered, Mo and Si cathode operating voltages will be limited to ~37 V, approximately the double ion ionization potential of xenon, to achieve lifetimes greater than 6000 hours. In the thruster environment, the charge-exchange or discharge ion population will further limit the operating voltages. Again, this population of ions consists of both double and single ions. With operating voltages no higher than 37 V, the erosion process is again dominated by the double ions. This ion population will further limit the operating voltages to approximately 5 V (Mo) and 13 V (Si) because ions will be accelerated from the plasma potential through approximately 20 V before entering the gate electrode apertures. With optimistic FEA cathode characteristics including gate aperture radii of 0.2 μm, excellent uniformity, effective[18] tip radii of 4 nm, and packing densities of 5×10^7 tips/cm^2, modeling results show that it is impossible to attain 100 mA/cm^2 with Mo and Si FEA cathodes operating in the plasma environment generated by a Hall or ion thruster.[16]

Mo and Si FEA cathode performance can be improved by coating them with a lower work function material. If the coating decreases the sputter yield also, higher operating voltages may be tolerated while meeting the EP system lifetime requirements. Materials with these potential properties include HfC,[19] ZrC,[20] and carbon films. Mo and Si FEA cathodes have been successfully coated with carbide and carbon films.[21,22,23] These films have significantly improved the cathode performance in current and stability in UHV and in more hostile environments. With the film coatings, the cathode current and lifetime still may not meet the requirements.

Further cathode ruggedization is recommended with an electrostatic ion filter and an arc protection architecture. An ion filter should be used to retard the flow of charge-exchange or discharge ions to the cathode microtips. Without this flux of ions, the tolerable operating voltages will increase by 22-30 V up to 35-37 V and the current by several orders of magnitude while satisfying the lifetime requirements. A current limiting architecture will also limit premature cathode failure. In the hostile thruster environments the cathodes will get contaminated by thruster propellant and facility materials. Often this contamination causes arcing between the tips and the gate electrode, producing an electrical short. High resistivity substrates, field effect transistor configurations, and VErtical Current Limiting (VECTL) architectures have been recommended to limit the current through a tip to prevent arc formation. The most favorable current limiting option appears to be the VECTL architecture[24] because it is a passive configuration which is microfabricated with the cathode and permits packing densities of 5×10^7 tips/cm^2.

CONCLUSIONS

Considering all of the challenges in integrating FE cathodes and EP systems discussed, the FE cathodes required will have to incorporate the cathode technology beyond current state-of-the-art. The cathode design includes gate aperture radii ~0.2 um, packing densities of 5×10^7 tips/cm^2, effective tip radii of 4 nm, with HfC, ZrC, or C films on Si or Mo arrays to achieve 100 mA/cm for more than 6000 hours in EP systems environments. Electrostatic ion shield structures with current limiting architectures will also be necessary for the FEA cathodes to meet the EP system performance requirements.

ACKNOWLEDGEMENTS

This research described in this publication was carried out by the Jet Propulsion Laboratory, California Institute of Technology, under a contract by the National Aeronautics and Space Administration. This work was performed as part of the Advanced Propulsion Concepts task sponsored by Marshall Space Flight Center (John Cole, Project Manager, Space Transportation Research) and the Power and On-Board Propulsion (632) Program sponsored by Glenn Research Center (Joe Naninger, Thrust Area Manager). Reference herein to any specific commercial product, process, or service by trade name, trademark, manufacturer, or otherwise, does not constitute or imply its endorsement by the United States Government or the Jet Propulsion Laboratory.

REFERENCES

1 Mueller, J., AIAA-97-3058 (1997).
2 Schneider, P., *J. Chem. Phys.* 28 (4) (1958).
3 Genovese, A., Marcuccio, S., Petracchi, P., Andrenucci, M., AIAA-96-2725, 1996.
4 Tajmar, T., Ph. D. Dissertation, Vienna University of Technology (TU WEIN), Austria, 1999.
5 Domonkos, M. T., Ph. D. Dissertation, University of Michigan (1999).
6 Genovese, A., Marcuccio, S., Petracchi, P., Andrenucci, M., AIAA-96-2725 (1996).
7 Sohl, G., Fosnight, V.V., Goldner, S.J., Speiser, R. C., AIAA 67-81 (1967).
8 Gorshkov, O., Muravlev, V. A., Grigoryan, V. G., Minakov, V. I, AIAA 99-2855 (1999).
9 Khayms, V., Martinez-Sanchez, M., AIAA Progress Series-Micropropulsion, (1999) in press.
10 Belikov, M. B., Gorshkov, O. A., Rizakhanov, R.. N, Shagayda, A. A., Khartov, S. A., AIAA-99-2571 (1999).
11 Hruby, V., Monheiser, J., Pote, B., Rostler, P., Kolencik, AIAA-99-3534, (1999).
12 Taylor, G., Proceed. Royal Society, A 280, (1964).
13 Fehringer, M., Rudenauer, and Steiger, W., AIAA-97-3057 (1997).
14 Petagna, C., von Rhoden, H., Bartoli, C., Valentian, D., IEPC 88-127 (1988).
15 Perel, J., Bates.T., Mahoney, J., Moore, R. D., Yihiku, A. Y., AIAA 95-2810 (1995).
16 Marrese, C. M., Polk, J. E., Jensen, K. L., Gallimore, A. D., Spindt, C., Fink, R. L., Tolt, Z. L., Palmer, W. D., AIAA Progress Series-Micropropulsion, (1999) in press.
17 Matsunami, N., Yamamura, Y., Itikawa, Y., Itoh, N., Kazmuta, Y., Miyagawa, S., Morita, K., Shimizu, R., Tawara, H., Atomic Data and Nuclear Data Tables 31 (1984).
18 Jensen, K. L, Physics of Plasmas 6 (1999).
19 Mackie, W. A., Morrissey, J. L., Hindrichs, C. H., Davis, P. R., *J. Vac. Sci. Technol. A* 10(4) (1992).
20 Mackie, W. A., Hartman, R. L., Anderson, M. A., Davis, P. R., *J. Vac. Sci. Technol. B* 12(2) (1994).
21 Xie, T., Mackie, W. A., and Davis, P. R., *J. Vac. Sci. Technol. B* 14(3) (1996).
22 Lee, S., Lee, S., Lee, S., Jeon, D., Lee, K. R., *J. Vac. Sci. Technol. B* 15(2) (1997).
23 Jung, J. H., Ju, B. K., Lee, Y. H., Jang, J., Oh., M. H., *J. Vac. Sci. Technol. B* 17(2) (1999).
24 Takemura, H., Tomihari, Y., Furutake, N., Matsuno, F., Yoshiki, M., Takada, N., Okamoto, A., and Miyano, S., Tech. Digest IEEE-IEDM, 709 (1997).

Mat. Res. Soc. Symp. Proc. Vol. 621 © 2000 Materials Research Society

Space Based Applications For FEA Cathodes (FEAC)

B. E. Gilchrist, U. Michigan, Ann Arbor, MI 48109-2143
K. L. Jensen, Naval Research Lab, Washington, DC 20375-5000
A. D. Gallimore, U. Michigan, Ann Arbor, MI 48109-2140
J. G. Severns, Consultant

ABSTRACT

Cold cathodes such as field emitter arrays offer the potential to benefit or enable space-based applications of critical commercial, government, or military importance by providing an electron source that is low power, low cost, requires no consumables, potentially robust as well as highly reliable. Applications that would especially benefit from such cold cathodes include low power electric propulsion (EP) thruster technology, electrodyanamic tethers (ED) for propellantless propulsion in low-Earth orbit, and spacecraft negative potential charge control. In controlled environments, field emitter arrays have shown substantial capability, but have failed in harsher environments more typical of space applications. We argue that a combination of localized arc suppression coupled with a low work function, but nevertheless robust, coating such as zirconium carbide would provide the needed ruggedness to withstand energetic ions, oxygen fluxes, and adsorbates typical of a spacecraft environment. We have found that arc-protected and coated FEACs that can operate in a 1-10 microTorr pressure environment with current densities of less than 0.1 Amps/cm^2 and gate voltages between 50-100 Volts, would enable reliable, low-cost devices capable of operating in the required space environment.

1. Introduction

FEACs appear to be a technology ready for rapid development as a new class of electron charge emission space-based applications which should lower the cost and/or be an enabling technology. The use of FEACs in these applications will significantly lower power consumption and eliminate or reduce consumable requirements over competing approaches. Specifically, it will enable miniaturized fuel-efficient electric propulsion (EP) systems such as the Closed-Drift Hall thruster (CDT) that enhance performance of small satellites by reducing spacecraft propellant mass. FEACs also enable use of "propellantless" ED tether propulsion for efficient spacecraft deorbit, atmospheric drag make-up, and orbit raising (250 to 2,000 km altitude). Additionally, FEACs provide a simple method to control charging of large spacecraft solar arrays.

These space applications will require FEAC devices that: (1) emit on the order of 0.1 A/cm^2; (2) operate in the 10^{-5} to 10^{-6} Torr pressure range (survive after operational exposure to 10^{-3} Torr); (3) require bias potentials of 50-100 V or less; and, (4) have sufficient life for multi-year space operations. Recent investigations suggest that each of these requirements can be achieved. What remains is to bring together the correct set of FEAC fabrication technologies to validate and qualify them for operation in the space environment. Our investigation suggests that to demonstrate FEACs for space applications, four critical FEAC development issues must be addressed: (1) arc suppression at each tip; (2) optimizing surface coatings that lower work-function and withstand energetic ions, oxygen flux, and undesirable adsorbates; (3) minimization of bias (gate) potential; and (4) space charge management. These are discussed below.

2. FEAC Background and State-of-the-Art

FEACs are being actively developed around the world for use in vacuum devices, such as flat panel displays, coherent radiation sources, and ultra fast logic circuits because of their low turn-on voltages, high current densities, and stable operation in a pulsed mode. Typical FEACs employ molybdenum or silicon tips fabricated by ion etching techniques. Emitters can have a radius of curvature of a few tens of angstroms [1] with packing densities on the order of 10^6 tips/cm^2 [2]. Tip electric fields must be on the order of 10^9 V/m for good emission. Field emission currents of 10 μA per tip can be routinely achieved but can be 50 μA for transition metal carbides (TMCs) run in pulsed mode.

An important potential application of FEACs receiving intensive study is as replacements to thermionic cathode emitters in microwave power tube applications [3,4,5].The RF vacuum microwave application is without question the most technically challenging, both in what is required of the field emitters and in the harshness of the operating environment even in comparison to the space FEAC applications. For this application, an aggressive 5 year, multimillion dollar ARPA/NRL/NASA Vacuum Microelectronics research program was recently completed to develop a next generation emission gated vacuum rf amplifier using FEACs (which has subsequently been superseded by the Advanced Emitter Technology program at NRL). This research demonstrated that FEACs could function in a very hostile rf tube environment with performance characteristics in excess of display requirements and close to present needs. Specifically, FEAC performance in test stations (see Table II of [6]) demonstrated per-tip currents from 1 to 20 μA/tip for molybdenum and silicon arrays in pulsed mode operation, and about an order of magnitude less than that for unprotected emitters in a tight-packed ring cathode geometry under rf (10 GHz) gate modulation in a power tube environment. Recently, Mackie, et al. [7], have shown that the coating of emitters using Zirconium carbide enhances the stability of the emitted current, and further, that larger currents at lower voltages are achievable, due to a combination of the ruggedness of the coating (greater inertness to adsorbates and nanoprotrusion formation) and the carbides' lower work function.

3. Arc Protection Techniques

On a microscopic scale, three issues govern both the magnitude of the emitted current and the susceptibility to arc-initiation for the field emitter tip. They are (*i*) work function of the surface layer (*ii*) radius of curvature of the emission site, which gives rise to the field enhancement factor and (*iii*) the creation and migration of nanoprotrusions under the ultra-high electric fields which exist at the surface of the emitter. Fluctuations in the emission current are believed to be due to adsorption and desorption of adsorbates, thereby covering and uncovering lower work function layers, as well as the creation, migration, and sputtering off of nanoprotrusions. Sudden rises in current, especially if that current is directed towards the gate, can initiate processes which if unchecked can lead to arc formation and subsequent tip destruction, usually in the form of melting the tip and surrounding gate, and splattering that material over the array, possibly leading to a short.

Resistive Protection - Resistive protection schemes are known to improve FEAC reliability and are routinely employed in FEACs for display applications. Display FEACs commonly employ a resistive element ("lateral resistance") which, in addition to stabilizing the current against fluctuations, provides a voltage drop when the current increases during an arc. Because the current drawn from a field emitter is exponentially dependent upon the gate-emitter voltage difference, the voltage drop can limit the current prior to runaway and subsequent failure due to arcing. Consequently, cathode arcing is no longer considered an issue for displays (rather, the phosphors are). Lifetimes exceeding 20,000 hours are being reported for display arrays by several groups. In fact, SRI operated a FEAC emitting 5 mA (50 μA/tip) onto a phosphor screen for over 8 years until the lifetime experiment was terminated by mishap rather than FEAC failure. In addition, NEC Corporation reported reaching 5000 hours of FEAC operation in the more severe TWT power tube environment while drawing on the order of 10's A/cm^2 current levels.

VECTL Architecture – The NEC Corporation [8, 9] has recently developed a new arc-resistant FEAC using a vertical current limiter (VECTL) approach that allows FEACs in the power tube environment but without increasing the operating gate voltage. The VECTL structure is a voltage-controlled resistive element that provides arc protection but presents comparatively little resistance to the FEAC current in nominal operation. This nominal resistance is orders of magnitude lower than that of the lateral resistors used in displays. Under conditions of an arc, however, the bias across the VECTL channel substantially increases causing a constriction of the channel and limiting the current.

4. Robust Coatings

According to a popular theory [10] a major cause of nanoprotrusion formation, which affects arcing characteristics and stability of operation, is impingement by energetic positively charged ions which are accelerated toward the cathode by the same electric field that causes electron emission. It is believed that many of these impacts are sufficiently energetic to cause sputtering, removing small amounts of cathode material and leaving behind tiny pits surrounded by nanoprotrusions. If this were true then use of a cathode surface material sufficiently tough to resist sputtering should greatly improve operational stability, and this has been observed with ZrC and HfC coatings over molybdenum cathodes. Now while resistive protection holds promise for suppressing high current excursions under arc conditions, it does not address the microscopic processes which lead to arc generation. Further, the energy of the ions and their rate of impingement on the emitter (creating nanoprotrusions on the side of the tip which emit towards the gate) are related to the magnitude of the gate potential. Reduction in the size of the unit cell geometry reduces the gate potential but renders the FEAC more fragile as a consequence.

A lower work function material should enable the same current levels at lower voltage, reducing both the ion-cathode collision cross-section (not the same as the neutral particle – electron cross-section) and particle impact energy. A reduction of the work function by 1/7 eV results in a factor of 2 increase in the emitted current for a given applied gate voltage. To be advantageous, work functions should be lower than those of the base materials which are 4.41 eV for Mo and 4.05 eV electron affinity for Si. As pointed out above, the transition metal carbides (particularly zirconium carbide, ZrC, and hafnium carbide, HfC) represent a class of well-characterized and robust coatings [2]. Both show low work function characteristics (ZrC – 3.6 eV work function on Mo, HfC – 3.0 eV work function on Mo [11, Table 3] and excellent operational robustness in even relatively high-pressure environments (10^{-5} Torr). While solid carbide emitters may be fabricated, it is also possible to use the carbides to provide a thin coating over existing molybdenum or silicon emitters, thereby obtaining the low work function and robust characteristics of carbides for these arrays.

Molybdenum emitters that were cleaned and then coated with ZrC resulted in lowering the voltage required for a given current to 38% of that needed for a clean molybdenum emitter. When the emitters were exposed to atmosphere for roughly 30 minutes and after operation in a 5×10^{-5} Torr hydrogen atmosphere, only a 12% increase in gate voltage, compared to the unexposed emitters, resulted. Overall, it appears that coating FEACs with carbides may provide robust field emitters which operate at significantly reduced gate voltage levels and are capable of withstanding high fields and/or high temperature without nanoprotrusion migration or sputtering damage.

It should also be noted that the cubic form of boron nitride (BN) doped with carbon or silicon also appears to offer advantages similar to those of ZrC. The toughness and chemical inertness of cubic BN are known to be quite high [12] promising good physical and chemical cathode passivation. Preliminary electron emission results suggest that doped BN might have a low surface work function as do the carbides described above. However, the mechanism of electron emission from BN under high electric fields is not presently understood, so predictions of the usefulness of BN for field emitter cathodes is difficult. FEACs of the carbon doped species has been shown to be unaffected by increasing ambient pressure up to 10^{-4} Torr. Normal operation is resumed upon improving the vacuum to below 10^{-5} [13].

5. Space Charge Management

None of the space-based applications considered here use a physical anode to collect the emitted electron charge from the FEAC. Rather, the charge is emitted into and accommodated by a plasma acting as a virtual anode.[1] The lack of a physical, biased collecting surface of known geometry complicates characterization of the emission process. Here, we wish to establish that such operation is feasible particularly for the EP and ED tether applications.

Injection across a sheath - Classical space charge limits to current flow across a vacuum gap depend on current density, gap width, gap potentials, geometry, and on initial kinetic energy [14] with the governing equation generally referred to as the Child–Langmuir Law. Operation below the Child–Langmuir current limit is desired here for robustness of FEAC operation [15]. Here, the "gap" is an ion-rich plasma sheath transitioning from the background plasma to the FEAC emission surface. The presence of ions in the gap (sheath) improves space charge constraints as the ions act to neutralize electron charge. In a one-dimensional (planar) case with zero initial velocity for both the ions and electrons and a fixed gap width the improvement is a factor of 1.86 compared to the vacuum gap case all else being equal [14].

The one-dimensional classical (vacuum gap) Child-Langmuir Law current density limit (in MKS units) is given by [16] which we multiply by area A of the emitter

$$I_{CL}(1) = \frac{4\varepsilon_o}{9e}\sqrt{\frac{2}{m_e}}\frac{T_o^{3/2}}{D^2}\left[1+\sqrt{1+\frac{eV}{T_o}}\right]^3 A \tag{1}$$

where D is the gap spacing, V is the gap voltage, T_o is the initial energy at which the electrons are injected into the gap ($\sim$ gate voltage), e is the magnitude of the electron charge, m_e is the electron mass, and ε_o is the free space permittivity. If $V = 0$ or $V \ll T_o$ ("shorted anode") Equation 1 indicates there is a factor 8 improvement over classical 1-D Child-Langmuir current limit ($T_o = 0$).

Since Equation 1 shows that a larger Child-Langmuir current is possible with small gap spacing, we will assume here that the plasma sheath gap potential (i.e., V in Equation 1) is within a few kT_e/e of the local plasma potential and that the sheath width is on the order of several Debye length, $\lambda_D = \sqrt{\varepsilon_o k T_e/e^2 n_e}$. These assumptions therefore would tend to represent maximum possible space-charge limited current flow and would be consistent with the situation of an

---

[1] It is possible to consider a physical, grid-like anode structure above the FEAC that accelerates the emitted charge and/or spreads it out but still allows it to pass through the anode. Marrese [15] suggests use of a multi-grid FEAC structure to reject external ions from reaching the tips.

electrically isolated (floating) spacecraft or subsystem except in some extreme cases [17]. In general, the sheath dimension is set by several interdependent factors, including the sheath potential, background plasma density/temperature, emitted current density and geometry. Representative density and electron temperature values are 8×10^8 cm^{-3} and 5 eV, for just outside of the Hall thruster and 5×10^5 cm^{-3} and 0.1 eV, for electrodynamic tether applications [18]. In addition, because of the low electron temperature of ionospheric plasmas in ED tether applications, it is possible to assume $V \ll T_0$ in Equation 1.

We also note that Equation 1 can be increased by a multiplicative term if the beam of electrons can expand laterally (from the direction of beam propagation) into regions of no or less electron charge. This expansion is a natural consequence of having outward electrical fields due to spatial gradients in the charge density. Luginsland et al. [19] have determined the multiplicative term if electrons can expand in one direction from a long emitter strip of width, W, and gap, D.[2]

$$\frac{J_{CL}(2)}{J_{CL}(1)} = 1 + \frac{0.3145}{W/D} - \frac{0.0004}{(W/D)^2} \qquad (2)$$

To estimate a threshold for space charge limited current flow, we will use Equation 1 with the multiplicative factor from Equation 2 (emitter strip with 1 direction of free expansion only) as well as the factor of 1.86 for an ion-rich sheath. This results in

$$I_{CL}(across\ sheath) = 1.86 \frac{4\varepsilon_o}{9e} \sqrt{\frac{2}{m_e}} \frac{T_o^{3/2}}{D^2} \left[1 + \sqrt{1 + \frac{eV}{T_o}}\right]^3 A \left(1 + \frac{0.3145}{W/D} - \frac{0.0004}{(W/D)^2}\right) \qquad (3)$$

Using Equation 3 and the assumption of a sheath width of only $3\lambda_D$, the estimated Child-Langmuir current limit for representative electric propulsion applications can be estimated suggesting that 350 mA could be generated from a 1 cm^2 area using only 40 V bias and a W/D ratio of 2. Similarly, for the ED tether application, 2 A of current could be emitted into a low-Earth orbit plasma with approximately 110 cm^2 area and 60 V bias with a W/D ratio of 2. Detailed analysis is needed to refine and optimize the results and assumptions in the above estimates particularly for geometric factors and assumptions of sheath width.

Beam Penetration into the plasma - Besides transit of the FEAC electron beam across an ion-rich sheath, its penetration into and accommodation by the plasma must be considered. The larger the density of the electron beam (n_{ee}) relative to the background plasma density (n_e), the stronger the space charge effects will be even in the plasma. Thus, this situation will likely be most acute for ED tether applications where emitted currents are high and background plasma densities are lower. The injection of an overdense electron beam into a simulated space plasma has been studied by Okuda and Ashour-Abdalla [20] and Pritchett [21] which used full three-dimensional simulations critical for realistic results.[3] In both, it was assumed that the emission occurred from an isolated spacecraft although OAA1990 made the spacecraft very large to minimize charging. Their simulations showed that the beam of electrons would rapidly expand in the radial direction (predicted even in the sheath) until the beam density was on the order of the background plasma density. Their simulations demonstrated that beam to plasma density ratios higher than a factor of 125 for OAA1990 and greater than 1400 for P1991 could be accommodated with a significant percentage of electrons escaping. OAA1990 compared cases of

---

[2] Equation 2 of Luginsland et al. [16] has a missing "1 +" term.
[3] For convenience hereafter we refer to these papers as OAA1990 and P1991.

beam densities of 64 and 125 times the ambient density and determined that 1.3% and 8% of the electrons, respectively, were returned to the spacecraft.

We also note that OAA1990 used an electron beam injection energy of only ten times ambient thermal electrons while P1991 used several energies with 10,000 times being the lowest. Here, the most severe situation will again come from the ED tether situation where the ratio of beam to ambient plasma density is less than 600 while for the EP case the ratio is less than 10. Lower plasma densities will not necessarily make the ratio worse since space charge limits in the sheath may also force a lower beam current density.

While more detailed analysis is justified to better quantify expected performance, experimental evidence already exists for the practicality of beam emission in space [22, 23] and even emission from highly negatively charged platforms [24, 25] with ion-rich sheaths at energies similar to that analyzed in P1991. Finally, we note that the above analysis did not fully optimize or exercise all options available to improve space charge constraints including effects from additional grids and non-planar surfaces.

6. Summary

FEACs represent a unique, new capability for special applications of electric propulsion, electrodynamic tethers, and spacecraft charging. The technologies appear to be in place to develop usable devices for space. But now the right set of technologies must be brought together to demonstrate their performance. As noted in the above sections, we especially believe, a low work function, sputter resistant material would confer several benefits. First, and most importantly, these materials offer the capability to operate stably in severe environments. Second, a reduction of the work function by 1/7th of an eV results in a factor of 2 increase in the emitted current for a given applied gate voltage. Third, hard coatings may inhibit the formation of arc enhancing nanoprotrusions created by ion bombardment. Fourth, a more stable and less reactive surface would mitigate sudden release of adsorbates on the surface. Space charge management remains a real concern for analysis and experiment. Assumptions of a small sheath width allow estimates of space-charge limited current emission that appear quite practical. Additional work in this area is required, however.

References

1. Mackie, W. A., R. L. Hartman, M. A. Anderson, and P. R. Davis, *JVTSB.* **B12**, 722 (1994).
2. Mackie, W. A., et al., MRS, **Vol. 509**, W. Zhu, L. S. Pan, T. E., (1998) p173.
4. Whaley, D. R., et al., *IEEE-ICOPS,* (Raleigh, NC, June 4-5, 1998), ID01.
5. Parker, R.K., Vide - Sci. Tech. et Applications, 52 (281), 366 (1996).
6. Jensen, et al., JVSTB16, 749 (1998).
7. Mackie, Xie, and Davis, Transition metal carbide FEs for FEA devices, J. Vac. Sci. Technol. B17, 613 (1999).
8. Takemura, H., et al., *IEDM.* 38, 2221 (1991).
9. Imura, H., S. Tsuida, M. Takahasi, A. Okamoto, H. Makishima, and S. Miyano *IEDM*, p721 (1997).
10. Charbonnier F., *Applied Surface Science* **94/95**, 26 (1996).
11. Charbonnier F., *J. Vac. Sci. Technol.* B 16, no. 2, p. 880, Mar/Apr (1998).
12. Litvinov, D., C. A. Taylor II, and R. Clark, *J. Diamond and Related Materials*, 7, 360, (1998).
13. Busta, H.H. and RW Pryor, *J. Vac. Sci. Technol.* B 16, no. 3, May/June (1998).
14. Humphries, S., <u>Charged Particle Beams</u>, John Wiley and Sons, Inc., 1990.
15. Jensen, K. L., J. Appl. Physics, **82**, 845-854 (1997).
16. Luginsland, J., S. McGee, and Y. Y. Lau, IEEE Trans. Plasma Sci., **26**, p.p. 901-904, 1998.
17. Hastings, D. E. and Henry Garrett, <u>Spacecraft-Environment Interactions</u>, Cambridge University Press, 1996.
18. Marrese, C., Ph.D., University of Michigan, 1999.
19. Luginsland, J. W., et al., Two-dimensional Child-Langmuir law, Phys. Rev. Lett., 77, p. 4668, 1996.
20. Okuda, H. and M. Ashour-Abdalla, J. Geophys. Res., 95, p. 21307-11, 1990.

21. Pritchett, P. L., J. Geophys. Res. 96, p.13781-93, 1991
22. Burke, W. J., et al., *Geophys. Res. Lett.*,Vol. 25 , No. 5, p. 717-720, 1998a.
23. Burke, W. J., et al., *Geophys. Res. Lett.*,Vol. 25 , No. 5, p. 725-728, 1998.
24. Gentile, L. C., et al. *Geophys. Res. Lett.*,Vol. 25 , No. 4, p. 433-436, 1998.
25. Gilchrist, B.E., et al., *J. Geophys. Res., 95*, 2469-2475, 1990.

Poster Session:
FEAs, Nanotubes, and
Other Electron Sources

Mat. Res. Soc. Symp. Proc. Vol 621 © 2000 Materials Research Society

SILICON FIELD EMISSION ARRAYS COATED WITH CoSi2 LAYER GROWN BY REACTIVE CHEMICAL VAPOR DEPOSITION

Byung Wook Han, Hwa Sung Rhee, Byung Tae Ahn, and Nam Yang Lee*
Department of Materials Science and Engineering, KAIST, Taejon 305-701, Korea
* Orion Electric Co., Suwon, 442-749, Korea

ABSTRACT

We prepared Si emitters coated with an MOCVD $CoSi_2$ layer to improve the emission properties. The $CoSi_2$ layer was grown *in situ* by reactive chemical vapor deposition of cyclopentadienyl dicarbonyl cobalt at 650 °C. The $CoSi_2$ layer was conformally coated on the Si emitter tips and had a twinned structure at the epitaxial $CoSi_2/Si$ interface. The $CoSi_2$-coated Si emitters showed an enhanced emission due to the increase of the number of emitting site from Fowler-Nordheim plot. The fluctuation of emission current was reduced by $CoSi_2$ coating. But the long-term stability was not much improved.

INTRODUCTION

In vacuum microelectronic devices one of the ultimate goals is to develop an electron source of high-efficiency field emitter arrays. The low turn-on voltage and stable emission current are important requirements for good electron sources in the emissive flat panel displays (FPDs). Various materials such as molybdenum[1], silicon[2] and diamond[3] have been studied for field emitter tip. Especially, silicon based emitters have been intensively studied by many researchers[4] due to the extensive knowledge about the process compatibility with integrated circuits. However, silicon emitters have problems such as relatively poor electrical conductivity, poor thermal conductivity, and native thin oxide[5]. To solve the problems in Si emitter tips, therefore, surface modifications by coatings such as metal coating[6], diamondlike carbon coating[7], and metal-silicide coating[8,9] have been performed. Among them metal-silicide emitters significantly enhanced emission current and stability[10,11]. But generally, the formation of metal-silicide structure on silicon emitter requires two steps such as metal deposition and conversion to silicide by thermal annealing at high temperature. In case of cobalt silicide, a capping layer on Co metal is required to prevent the agglomeration of silicide during thermal annealing.

In this study we coated a $CoSi_2$ layer *in situ* on Si emitter tip at 650°C from cobalt metallorganic source by reactive chemical vapor deposition to simplify tip coating process[12]. CVD commonly offers advantage such as an uniform and conformal deposition over a large area. $CoSi_2$ layer was grown at lowered process temperature to 650 °C, and epitaxially grown on Si tip. We investigated the emission characteristics such as turn-on voltage, emission current, and current fluctuation of $CoSi_2$-coated field emitters and compared with those of silicon emitters.

EXPERIMENTAL PROCEDURE

A boron-doped p-type (100) Si wafer with a resistivity of 5 ~ 20 $\Omega \cdot$cm was used as a substrate. LPCVD oxide films of about 400nm were deposited and patterned as discs of 3μm in diameter for an etch-mask. The silicon was etched using SF_6 gas by inductively coupled plasma and the tips were sharpened by wet oxidation. The number of Si FEA was 1,000. Figure 1 shows the SEM micrograph of silicon emitters fabricated after sharpening oxidation. Field emission measurements were carried out at a pressure of $2\sim3\times10^{-7}$ Torr by a diode-type arrangement. Highly-polished stainless-steel anode plates were separated from the cathodes by 50μm thick insulating spacer films. The I-V curves were measured by ramping the bias voltage up and down, typically, in 20V steps.

Figure 1. SEM micrograph of silicon emitters after sharpening oxidation. The radius of Si tip is below 30nm.

Silicon emitters were coated with cobalt silicide by reactive MOCVD using cyclopentadienyl dicarbonyl cobalt, $Co(\eta^5\text{-}C_5H_5)(CO)_2$ precursor. The $CoSi_2$ was *in situ* grown at 110 mTorr with 10 sccm H_2 carrier gas at the substrate temperature ranging 650°C. The details of the $CoSi_2$ reactive CVD can be found in the literaure[12]. The growth time was 15min and the thickness of $CoSi_2$ layer was 30-35nm. The resistivity of $CoSi_2$ layer was about 17.3 μ$\Omega \cdot$cm. The microstructure of the $CoSi_2$/Si emitter interface was investigated using SEM and TEM.

RESULTS AND DISCUSSION

Figure 2 shows (a) the SEM image of Si tip coated with $CoSi_2$ grown at 650°C using $Co(\eta^5$-

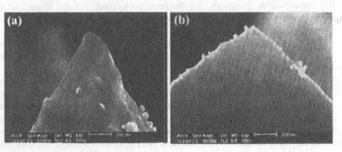

Figure 2. SEM micrographs of (a) $CoSi_2$-coated Si tip and (b) its cross section.

$C_5H_5)(CO)_2$ precursor and (b) the cross-sectional SEM image. It can be seen that the coating is smooth and uniform in the whole range of tip, which might attribute to the CVD process. The interface of $CoSi_2$/Si emitter is very compact and the thickness of the coating is estimated to be approximately 30nm.

Figure 3 shows the cross-sectional TEM micrographs of (a) Si tip with $CoSi_2$ grown at 650 °C using $Co(\eta^5-C_5H_5)(CO)_2$ precursor and (b) the selected-area diffraction pattern at the $CoSi_2$/Si interface on Si tip along the <011> zone axis. The dark $CoSi_2$ layer is conformally coated with a thickness of about 30nm along the slope of Si tip, which is similar result in figure 3(b). The diffracted spots in the SADP at the $CoSi_2$/Si interface almost coincide with each other excluding the one-directional extra spots. It is ascertained that the extra spots are originated from the twinned structures of the epitaxial $CoSi_2$. Epitaxial $CoSi_2$ on Si has two epitaxial orientations. One has the same orientation with Si substrate. The other is rotated 180° about the surface normal <111> axis (twinned structure)[13,14]. Therefore, it can be said that a $CoSi_2$ layer is epitaxially grown on Si emitter by reactive MOCVD using the $Co(\eta^5-C_5H_5)(CO)_2$ precursor.

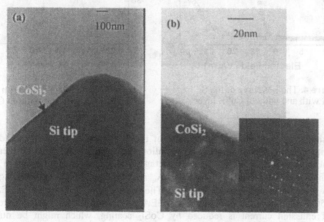

Figure 3. Cross-sectional TEM micrographs (a) $CoSi_2$ grown at 650 °C using $Co(\eta^5-C_5H_5)(CO)_2$ precursor and (b) the selected-area diffraction pattern at the $CoSi_2$/Si interface on Si tip along the <011> zone axis.

Figure 4 shows the I-V curve of the Si FEA with and without the $CoSi_2$ layer. The $CoSi_2$ layer was grown at 650 °C. The current emission of the Si FEA with the $CoSi_2$ layer is greatly enhanced compared with the Si FEA wihtout the silicide. The turn-on voltage for 1µA, which corresponds 1nA per tip, decreases from 23V/µm to 12V/µm by $CoSi_2$ layer coating on Si FEA. Therefore, the operating voltage could be reduced by about 11V/µm for the $CoSi_2$-coated Si FEA compared with the Si FEA without the silicide.

Figure 5 shows the Fowler-Nordheim plots for the Si FEA with and without the $CoSi_2$ layer at 650 °C. The slope of the $CoSi_2$-coated Si FEA is nearly same as that of the Si FEA without silicide. The results suggest that the increase of the number of emitting site might be responsible for the enhanced emission characteristics. If the blunting of Si tip and decrease of the geometrical factor by $CoSi_2$ coating are taken into account, the work function by $CoSi_2$ formation on Si FEA might be another factor of enhanced emission. Even though the diode measurement in our experiment showed the slope with and without silicide, the triode measurement exhibited work function decrease by silicide coating[15].

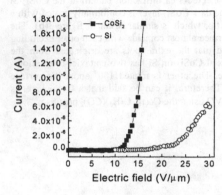

Figure 4. The I-V curve of the Si FEA with and without $CoSi_2$ layer.

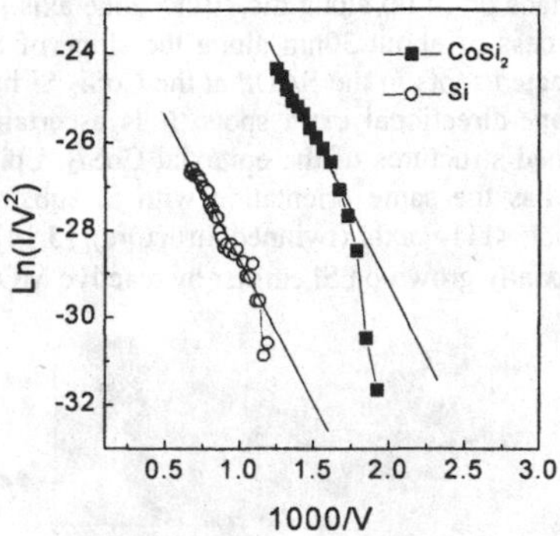

Figure 5. The Fowler-Nordheim plots of the Si FEA with and without $CoSi_2$ layer.

Figure 6 shows the anode current fluctuation of the Si FEA and the $CoSi_2$-coated Si FEA normalized with average anode current. The variations in emission current were measured at similar current level with different fixed anode voltage. The average current levels of the FEA with and without silicide were 6.28 and 1.73µA, respectively. The corresponding applied electric field were 15 and 24 V/µm, respectively. The current fluctuation of the Si FEA ranges from −34 ~ +50% and that of the $CoSi_2$-coated Si FEA ranges from −13 ~ +12%. Therefore, the fluctuation of emission current is reduced by $CoSi_2$ coating, which might be due to the chemically stable surface resulting in the decrease of the adsorption and desorption of gas species[16].

Figure 7 shows the long-term emission stability of the Si FEA and the $CoSi_2$-coated Si FEA at constant DC bias. In both case the anode current gradually decreases. Therefore, the $CoSi_2$ coating by MOCVD is not effective in the view of long-term emission stability. After DC bias test, we observed the destruction of the apex of $CoSi_2$-coated tip and the increase of slope in the F-N plot.

CONCLUSIONS

We deposited an *in situ* $CoSi_2$ layer on silicon emitters at 650°C with using cyclopentadienyl decarbonyl cobalt by reactive CVD. The $CoSi_2$ layer was conformally coated along the Si emitter and had a twinned structure of epitaxial CoSi$_2$ at the CoSi$_2$/Si interface. Therefore, we could grow epitaxial CoSi$_2$ layer on Si emitter. The emission characteristics of CoSi$_2$ coated field emitters was greatly improved. The turn on voltage for 1μA decreased from 23V/μm to 12V/μm. The enhanced emission properties were mainly due to the increase of the number of emitting site by the formation of CoSi$_2$ on silicon emitters from Fowler-Nordheim plot. The fluctuation of emission current was reduced by CoSi$_2$ coating. But the long-term stability was not much improved by CoSi$_2$ coating.

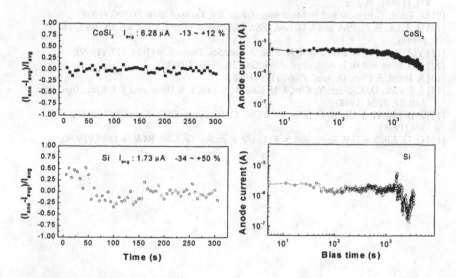

Figure 6. The anode current fluctuation of the Si FEA and the CoSi$_2$-coated Si FEA.

Figure 7. The emission stability of the Si FEA and the CoSi$_2$-coated Si FEA at constant DC bias.

ACKNOWLEDGMENTS

This work was partially supported by bk 21 program.

REFERENCES

[1] C. A. Spindt, C. E. Holland, A. Rosengreen and I. Brodie, *J. Vac. Sci. Technol*, **B11**, 468, (1993).

[2] S. Kanemaru, T. Hirano, H Tanoue and J. Itoh, *J. Vac. Sci. Technol*, **B14**, 1885, (1996).

[3] I. L. Krainsky and V. M. Asnin, *Appl. Phys. Lett.* **72,** 2574, (1998).

[4] H. S. Uh, S. J. Kwon and J. D. Lee, *J. Vac. Sci. Technol*, **B15**(2), 472, (1997).

[5] W. J. Brintz and N. E. McGrucer, *J. Vac. Sci. Technol*, **B12**, 697, (1996).

[6] D. W. Branston and D. Stephan, *IEEE Trans Electron Devices,* 38, 2329, (1991)

[7] I. H. Shin and T. D. Lee, *J. Vac. Sci. Technol*, **B 17**(2), 690, (1999).

[8] S. Y. Kang, J. H. Lee, Y. H. Song, Y. T. Kim and K. I. Cho *J. Vac. Sci. Technol*, **B 16**(2), 871, (1998).

[9] M. Takai, T. Iriguchi and H. Morimoto, *J. Vac. Sci. Technol*, **B16**(2), 790, (1998).

[10] H. S. Uh, B. G. Park and J. D. Lee, *IEEE Electron Devices Letters,* vol **19**, No 5, 167, (1998).

[11] Y. J. Yoon, G. B. Kim and H. K. Baik, *J. Vac. Sci. Technol*, **B17**(2), 627, (1999).

[12] H.S. Rhee and B. T. Ahn, *Appl. Phys. Lett.* **74**, 3176 (1999).

[13] S. Mantl, J. Phys. D: *Appl. Phys.* 31, 1 (1998).

[14] J. S. Park, D. K. Sohn, Y. Kim, J. U. Bae, B. H. Lee, J. S. Byun, and J. J. Kim, *Appl. Phys. Lett.* **73**, 2284 (1998).

[15] J. D. Lee, H. S. Uh, B. C. Shim, E. S. Cho, C. W. Oh, and S. J. Kwon *11th IVMC*, 304,(1998).

[16] Q. Li. J.F. Xu, H.B. Song, and X. F. Liu, *J. Vac. Sci. Technol*, **B14**(3), 1889, (1996).

Mat. Res. Soc. Symp. Proc. Vol. 621. © 2000 Materials Research Society

UNDER-GATE TRIODE TYPE FIELD EMISSION DISPLAYS WITH CARBON NANOTUBE EMITTERS

J. H. Kang, Y. S. Choi, W. B. Choi[1], N. S. Lee[1], Y. J. Park[1], J. H. Choi[1], H. Y. Kim, Y. J. Lee, D. S. Chung, Y. W. Jin[1], J. H. You[2], S. H. Jo[2], J. E. Jung, and J. M. Kim
The National Creative Research Initiatives Center for Electron Emission Source, Samsung Advanced Institute of Technology, P. O. Box 111, Suwon 440-600, KOREA
[1]Display Lab., Samsung Advanced Institute of Technology
[2]FED team, Samsung SDI

ABSTRACT

A new structure of triode type field emission displays based on single-walled carbon nanotube emitters is demonstrated. In this structure, gate electrodes are situated under cathode electrodes with an in-between insulating layer, so called under-gate type triode. Electron emission from the carbon nanotube emitters is modulated by changing gate voltages. A threshold voltage is approximately 70 V at the anode bias of 275 V.

INTRODUCTION

Carbon nanotubes have attracted much attention for application to the field emission sources due to their high aspect ratios, small tip radii of curvature, high chemical stability, and high mechanical strength [1]. Recently, there have been many efforts to apply carbon nanotubes to field emission sources. Wang et al. [2] grew aligned nitrogen-containing carbon nanofibers using a microwave plasma-assisted chemical vapor deposition method. They achieved a low-threshold field of 1.0 V/μm and a high emission current density of 200 mA/cm² at an applied field of 5-6 V/μm. Fan et al. [3] synthesized self-oriented carbon nanotubes on silicon substrates and analyzed their field emission characteristics. Choi et al. [4] fabricated a fully sealed field emission display(FED) panel with the 4.5inch diagonal using single-walled carbon nanotubes. Their studies, however, were restricted to diode type structures. For full gray scales and high brightness of FEDs, a triode structure is required. In this study, we present a new structure of triode type FEDs with carbon nanotube emitters, so-called under-gate triode where gate electrodes are located under cathode electrodes. Simplicity of structure and fabrication processes seems to enable the under-gate type triode structure to possess a high potential of practical applications.

EXPERIMENTAL DETAILS

Gate electrodes were fabricated by thin film processes, as described in the following. At first, aluminum with a thickness of 1500Å was deposited on the cleaned soda-lime glass by electron beam evaporation. Gate electrodes with the 400 μm line width were patterned by photolithography. SiO₂ was deposited by plasma enhanced chemical vapor deposition to insulate between the gate and cathode electrodes. In order to secure enough distance between gate and cathode electrodes, polyimide was spin-coated on the SiO$_2$ layer at 1500 rpm. After curing at 375°C, 13 μm-thick polyimide was covered with an aluminum layer by electron beam evaporation.

The aluminum cathode electrodes were patterned to form 390 μm wide lines by photolithography. The polyimide that covered gate pads was eliminated by the oxygen plasma of reactive ion etching. The SiO$_2$ layer above the gate pads was etched off by a buffered hydrofluoric acid solution. Single-walled carbon nanotubes were synthesized by a conventional arc-discharge method as described elsewhere [5]. A paste of the single-walled carbon nanotubes mixed with a slurry of metal powders and organic vehicles was screen-printed onto the cathode electrodes through the metal mesh of 20 μm in size with the same pattern of the cathode electrodes.

Following heat treatment at 350°C, surface rubbing treatment made carbon nanotube tips protrude from the surfaces. Figure 1.(a) shows an SEM image where single-walled carbon nanotubes are standing up after the rubbing treatment. The structure of our panel is shown in figure 1(b).

Green phosphor coated ITO glass was used as an anode plate. The cathode and anode plates were kept apart by 200 μm spacers. DC voltages were supplied between the anode and cathode electrodes using a high voltage power supply (Keithley model 248). Square pulses with 30 kHz were applied to the gate electrodes using the Tektronix CFG 253 function generator and the Trek P0610B high voltage amplifier. Brightness was measured by the BM7 luminance calorimeter.

RESULTS AND DISCUSSION

Figure 2 shows simulation result that the electric field strength is concentrated at the edges of the cathode electrodes in the under-gate type triode structure. Electrons seem to emit from the edges rather than the central areas of the cathode electrodes. An increase of gate voltages enhances the electric field strength around the cathode electrodes, leading to electron emission

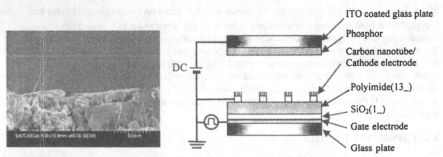

ITO coated glass plate

Phosphor

Carbon nanotube/
Cathode electrode

Polyimide(13_)

SiO$_2$(1_)

Gate electrode

Glass plate

DC

from the edges of the cathode electrodes. The device is controlled by the gate voltage modulation. For example, the electron emission turns on and off at the gate voltages of 80 V and 0 V, respectively, for a given anode voltage.

(a)   (b)

Figure 1. (a) Cross-sectional SEM image of single-walled carbon nanotubes standing-up on a cathode electrode and (b) schematic of an under-gate triode type field emission display with carbon nanotube emitters where the gate electrodes are located under the cathode electrodes with the in-between SiO₂ and polyimide insulating layers. In (b), gate voltages were supplied by DC squre pulses with the frequency of 30 kHz.

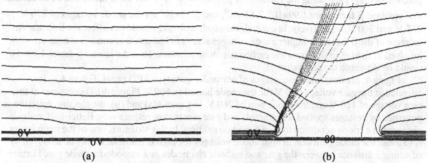

(a)   (b)

Figure 2. Simulation of electric field strength for an under-gate type triode structure when (a) 0 V and (b) 80 V are applied to the gate electrodes for a given cathode bias of 0 V.

Field emission characteristics and brightness with the gate voltages for different anode voltages are shown in figure 3. The threshold voltages are around 80 V and 70 V at the anode voltages of 250 V and 275 V, respectively. For the anode biases of 250 V and 275 V, the electron emission turns off completely below the threshold gate voltages. Emission currents occur, however, even at low gate biases for the anode voltage of 300 V. It can be explained as diode emission because the threshold voltage of our single-walled carbon nanotubes is about 1.3 V/μm in a diode condition. The brightness as high as 155 cd/m² is obtained at the gate voltage of 160 V and the anode bias of 300 V. Field emission under a strong applied electric field is usually modeled using Fowler-Nordheim(F-N) equation, i.e., I = aV²exp(-b/V), where a and b are constants[6]. An inset of figure 3 (a) shows F-N plots of the I-V curves. An inverse linearity of the F-N plots exhibits that the I-V curves are governed by a conventional field emission

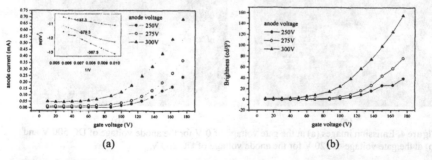

(a)   (b)

Figure 3. (a) I-V characteristics and (b) brightness of an under-gate triode type FED with gate voltages for different anode voltages. An inset in (a) exhibits Fowler-Nordheim plots of the I-V curves.

mechanism. The electric field at the surface(E) can be related to the field enhancement factor(β), and applied voltage(V) by E= βV. The β factor can be calculated either from the slope(b) of the F-N plot if the work function(ϕ) of the emitter is known or from measuring the geometry of a carbon nanotube tip. Assuming the work function of carbon nanotubes to be 5eV, same as that of graphite or C_{60}[7] and using the equation of $\beta = (6.53 \times 10^7 \phi^{3/2})/b$, the β factor can be calculated. The obtained β factors are 1.99×10^6, 2.61×10^6, and 5.32×10^6 for the anode voltages of 250 V, 275 V, and 300 V, respectively. In general, the anode voltages should not affect the β factor. An increase of the β factors at higher anode voltages seems to indicate that anode voltages are involved in field emission from the carbon nanotube emitters. Thus the lowest β factor 1.99×10^6 would be closest to the real value.

Figure 4. presents emission images of our under-gate type FED panel. Figure 4(a) is operated at the gate voltage of 0 V for the anode biased to 500 V. Figure 4(b) is obtained at the gate voltage of 120 V and the anode bias of 260 V. In figure 4(a) and (b), the electron emission is governed by voltages applied to the anode and gate electrodes, respectively. Better uniformity is observed in a triode condition than in a diode condition. In this structure, one of the main issues is to avoid the diode emission in high anode voltages for high brightness. It may be achieved by adjusting a distance between the gate and cathode electrodes or a cathode-to-anode gap. Further work to realize the high brightness in a triode condition is under progress.

CONCLUSION

We fabricated under-gate type triode FEDs using single-walled carbon nanotubes as electron emission source and characterized their field emission properties. It was observed that electron emission in the under-gate triode type panel was driven by gate voltage modulation. At higher anode voltages, the diode emission occurred even at low gate voltages. A study to overcome such a problem is under progress, including optimization of polyimide thickness and the distance between cathode and anode plates. This new structure appears to possess a high potential of practical application due to its simple structure and fabrication processes.

 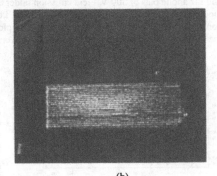

(a)   (b)

Figure 4. Emission images (a) at the gate voltage of 0 V for the anode voltage of DC 500 V and (b) at the gate voltage of 120 V for the anode voltage of DC 260 V.

ACKNOWLEDGEMENT

This work was supported by a material strategic fundamental research for creative research development sponsored by Korean Ministry of Science and Technology.

REFERENCE

1. S. J. Tans, A. Verschueren, and C. Dekker, *Nature*, **393**, 49 (1998).
2. X. Ma, E. Wang, W. Zhou, D. A. Jefferson, J. Chen, S. Deng, N. Xu, and J. Yuan, *Appl. Phys. Lett.*, **75**, 3105 (1999).
3. S. Fan, M. G. Chapline, N. R. Franklin, T. W. Tombler, A. M. Cassell, and H. Dai, *Science*, **283**, 512 (1999).
4. W. B. Choi, D. S. Chung, J. H. Kang, H. Y. Kim, Y. W. Jin, I. T. Han, Y. H. Lee, J. E. Jung, N. S. Lee, G. S. Park, and J. M. Kim, *Appl. Phys. Lett.*, **75**, 3129 (1999).
5. S. Iijima, *Nature*, **354**, 56 (1991).
6. I. Brodie and C. A. Spindt, Vacuum Microelectronics, *Advances in electronics and electron physics, vol. 83*, pp. 91-95.
7. B. W. Gadzuk et al., Acad. Sci. **B 278**, 659 (1974).

Mat. Res. Soc. Symp. Proc. Vol. 621. © 2000 Materials Research Society

A CONCENTRIC CIRCULAR TYPE IN-SITU VACUUM SEALED LATERAL FEAS EMPLOYING E-BEAM EVAPORATOR

Moo-Sup Lim, Min-Koo Han, and *Yearn-Ik Choi
School of Electrical Eng., Seoul Nat'l Univ., San 56-1, Shilimdong, Seoul, 151-742, Korea
*Dep. of molecular Sci. and Tech., Ajou Univ., Suwon 442-749, Kyung-ki do, Korea

ABSTARCT

In this paper, we propose concentric circular lateral field emitter structure which is more stable in fabricating and maintaining vacuum conditions. The new FEAs have concentric circular shape while our previous FEAs are stripe type. The anode is formed at inner circle and the cathode is outer circle respectively. We use Molybdenum or Silicon dioxide as a sealing material. After we fabricating tips, Molybdenum or Silicon dioxide is deposited with e-beam evaporator at base pressure of 3.5×10^{-6} Torr. In this process, the micro-cavity remains vacuum cavity without filling Molybdenum or oxide due to step coverage of evaporation. It should be noted that the vacuum level of the micro-cavity can be identical to the base pressure of the e-gun evaporator chamber. All of the two types of new FEAs show good reproducibility in fabricating and maintaining vacuum conditions. Field emitter structure that the anode is located at inner circle shows superior field emission characteristics. The measurements of long term stability for the anode current show that oxide is the most promising material for in-situ vacuum sealing.

INTRODUCTION

Vacuum field emitter arrays(FEAs) have attracted a considerable interest for high frequency, high temperature and flat pannel display applications[1,2]. Various field emitter structures have been reported to improve field emission characteristics. These reserches are focused on field emission characteristics such as emission current density, low turn-on voltage, and current stability. However, it is well known that FEAs can be operated in high-vacuum environment of about 1×10^{-6} Torr[3]. Therefore, most of these FEAs require an additional vacuum environments to measure its characteristics. Recently, the studys on vacuum package for FED(Field emission display) is investigated by some reserchers[4,5]. But there are few reports on sealing method for FEAs, although it is very important to apply FEAs for practical applications. We have reported in-situ vacuum sealed lateral FEAs emploiyng e-gun evaporator[6]. The structure had basically stripe shape and we used Molybdenum as sealing material. Our experiment showed that some large portion about 40% of the FEAs were failed to be sealed. It was thought that the structure had rectangular shape, and the crack of evaporated Molybdenum occurred at rectangular edges of the FEAs due to mechenical stress and poor adhesion.

In this paper, we propose concentric circular type lateral field emitter structure to improve sealing and maintaining vacuum conditions. The new FEAs have concentric circular shape to reduce mechanical stress between various thin films, while our previous FEAs are stripe type. We use silicon dioxide(oxide) and Molybdenum as sealing materials, and investigate the difference for vacuum sealing capability between two materials. We locate the anode of concentric circular type FEAs at inner circle and outer circle respectively to compare its field emission characteristics. Finally, we investigate long term stabilty of the new FEAs and our previous FEAs.

DEVICE STRUCTURE AND FABRICATION

The schematic diagram of new FEAs are shown in Figure 1. The total shape of the new FEAs is concentric circular type. We fabricate two types of FEAs. One is that the anode is located at inner circle and the other is that the anode at outer circle to compare its field emission characteristics for each FEAs.

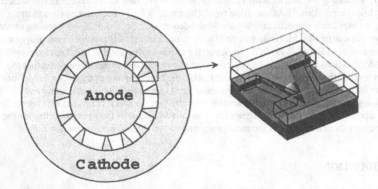

Figure 1. *Schematic diagram of concentric circular shape lateral FEAs.*

Figure 2 shows the main fabrication sequences of the new FEAs. It is almost identical our previous FEAs[5] except for the total shape of the FEAs and sealing materials. We use the starting material as the Silicon(Si) substrate which was deposited with $5000\mathring{A}$ thick nitride and $1000\mathring{A}$ thick buffer oxide. Subsequently, $1000\mathring{A}$ thick polycrystalline Silicon(poly-Si) was deposited. The $3000\mathring{A}$ thick poly-Si was doped with $POCl_3$, resulting in N^+ poly-Si layer. $1000\mathring{A}$ thick buffer oxide, $3000\mathring{A}$ thick nitride, and $1000\mathring{A}$ thick oxide were deposited.

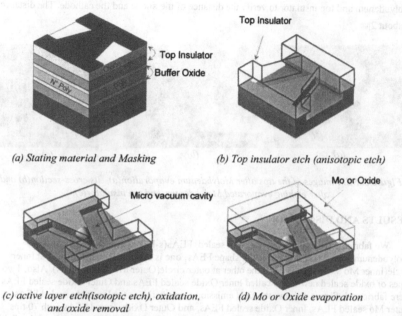

(a) Stating material and Masking   (b) Top insulator etch (anisotopic etch)

(c) active layer etch(isotopic etch), oxidation,   (d) Mo or Oxide evaporation
and oxide removal

Figure 2. *Main fabrication sequences of the new FEAs*

Then the patterning as shown in Figure 2 (a) was performed using photolithography. Then, the top insulator(1000$\mathring{A}$ thick buffer oxide, 3000$\mathring{A}$ thick nitride, and 1000$\mathring{A}$ thick oxide), poly-Si, and 1000$\mathring{A}$ thick lower buffer oxide were etched with anisotropic RIE(Reactive Ion Etch) using Cl gas. Figure 2 (b) shows the FEAs after anisotropic RIE process. Subsequently, The poly-Si layer was etched with isotropic RIE using SF_6 gas in order to make micro-cavity (Figure 2 (c)). The oxidation at 1000 $\mathcal{C}$ for 60 *minutes* in dry O_2 was performed to sharpen the poly-Si tip. The thermal oxide was removed by buffered oxide etchant (Figure 2 (c)). Molybdenum or oxide was deposited by an electron gun evaporator at a base pressure of $3.5{\times}10^{-6}$ *Torr*(Figure 2 (d)). In this process, the micro-cavity remains vacuum cavity without filling Molybdenum due to step coverage of evaporation. It should be noted that the vacuum level of the micro-cavity is identical to the base pressure of the e-gun evaporator chamber[5]. Tetraothylorthsilicate(TEOS) passivation layer was deposited and an electrical interconnections were then fabricated in proper locations by employing the mask steps.

Figure 3 (a) and (b) show SEM images of the tips after Molybdenum evaporation and its cross-section. We can find that the micro-cavity is made sucessfully though we don't know the vacuum level of it. Figure 3 (c) shows the SEM image of the tips after etching evaporated

Molybdenum and top insulator to verify the distance of the anode and the cathode. The distance is about 2μm .

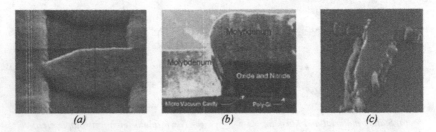

(a)   (b)   (c)

Figure 3. *SEM images of the tips after Molybdenum evaporation(a), its cross-section(b) and after etching evaporated Molybdenum and top insulator(c).*

RESULTS AND DISCUSSION

We fabricate stripe type Molybdenum sealed FEAs(s-FEAs) and two types of new Molybdenum sealed concentric circular shape FEAs; one is that the anode is located at inner circle(Inner Mo sealed FEAs), and the other at outer circle(Outer Mo sealed FEAs). Also, Two types of oxide sealed new FEAs called Inner Oxide sealed FEAs and Outer Oxide sealed FEAs were fabricated. Figure 4 shows the field emission characteristics for Inner Mo sealed FEAs, Outer Mo sealed FEAs, Inner Oxide sealed FEAs, and Outer Oxide sealed FEAs with 70 tips respectively. Field emitter structure that the anode is located at inner circle shows superior field emission characteristics as shown in Figure4.

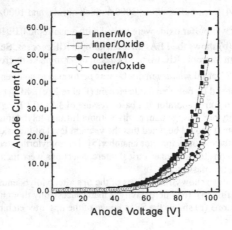

Figure 4. *Field emission characteristics of various concentric circular shape FEAs with 70 tips*

It is thought that the cohesion of emissive electrons for the case that anode is inner may be better than the opposite case, considering the conversion facters of each case are nealy identical theoretically. Also, we can assume that there are few positive ions in the micro-vacuum cavity since the anode current of the first case may be lower than the second case due to frequent collisions between electrons and ions if there are many ions in it. There are about 10-20 % difference in the anode current for each case. But, the characteristics of oxide sealed and Molybdenum sealed FEAs are almost identical.

We investigate the sealing characteristics for each case. As fabricated, three s-FEAs have the cracks in the process of the evaporation while there is only one Molybdenum sealed FEAs with the cracks. The average anode current are 0.44, 0.45, and 0.42μA/tip respectively as shown in table 1. After one week later, there are no more cracks in all case. However, the current density of the FEAs are lower than that of as fabricated FEAs. Only three s-FEAs have larger current density than 0.3μA/tip while eight FEAs in the case of oxide sealed FEAs. We can show that the oxide sealed FEAs are the most stable among the cases. Table 1 summarizes the sealing characteristics for each case.

Table 1. Comparison of sealing characteristics.

	As fabricated		one week later	
	# of FEAs with Crack	Avg. anode current (V_{ak}=90V)	# of FEAs with Crack	# of FEAs over current 0.3μA/tip
Previous FEAs	3/10	0.44μA/tip	3/10	3/10
Mo sealed FEAs	1/10	0.45μA/tip	1/10	6/10
Oxide sealed FEAs	0/10	0.42μA/tip	0/10	8/10

CONCLUSION

We fabricated the concentric circular lateral field emitter structure employing two types of sealing material and two types of anode location. The concentric circular shape FEAs are fabricated without cracks in the process of the evaporation except only one FEAs. This results show that the structure is much more stable than our previous stripe type FEAs. Furthermore, we can assume that the vacuum level of the micro cavity is rather high based on the long term stability and the fact that field emitter structure that the anode is located at inner circle shows superior field emission characteristics.

Consequently, we propose and fabricate successfully more stable in-situ vacuum lateral field emitters employing concentric circular shape in order to reduce the stresses between substrate and sealing materials.

REFERENCE

1. I. Brodie and P. R. Schwoebel, Proceeding of IEEE, Vol. 82, No. 7, pp 1006, July (1996)
2. Takao Utsumi, IEEE Trans. Electron Devices, Vol. 38, No. 10, pp. 2276(1991)
3. C. A. Spint, C. E. Holland, A. Rogengreen, and I. Brodie, IEEE Trans. Electron Devices, Vol. 38, No. 10, pp. 2355(1991)
4. J. I. Han, M. G. Kwak, and Y. K. Park, Proceeding of IVMC'97, pp706(1997)
5. S. J. Kwon, K. S. Ryu, T. H. Cho, and J. D. Lee, Proceeding of IVMC'98, pp55(1998)
6. C. M. Park, M. S. Lim, and M. K. Han, IEEE Electron Device Letters, vol. 18, no. 11, November, pp. 538 (1997)

Mat. Res. Soc. Symp. Proc. Vol. 621. © 2000 Materials Research Society

SILICON FIELD EMITTER ARRAY BY FAST ANODIZATION METHOD

Y.M. Fung, W.Y. Cheung, I.H. Wilson, J.B. Xu and S.P. Wong
Department of Electronic Engineering and Materials Science and Technology Research Center,
The Chinese University of Hong Kong, Shatin, N.T., Hong Kong, P. R. China

ABSTRACT

A new fast fabrication method entailing, two step anodization of silicon with different HF solutions was used to form a high aspect ratio silicon Field Emitter Array on n-type silicon (resistivity of $0.01\Omega cm$). A Silicon oxide mask was used to define the field emitter array. The silicon substrate was pre-anodized with low current density for 1 minute in the dark and then anodized in $HF:H_2O:Ethanol$ solution. Finally, the porous silicon was removed by isotropic solution etching. The turn-on voltage of the fabricated field emitters was approximately $27V/\mu m$ when the emission current density reaches $1\mu A/cm^2$. This compares with the turn-on field of about $35V/\mu m$ on silicon tip array fabricated by using an isotropic etching solution of HNO_3. We obtained field emitter arrays with good uniformity and reproducibility.

INTRODUCTION

The fabrication of a conventional silicon field emitter array (FEA) is by isotropic wet etching (nitric acid) or by anisotropic wet etching (potassium hydroxide). The usage of wet solution etching has several problems: (1) the etching rate is too slow and (2) the uniformity of the silicon tip size is not good because the etching rate depends on the solution flow rate onto the wafer.

Electrochemical etching by using hydrofluoric acid (HF) is widely used for the formation of porous silicon. This has been studied since the 1950s[1] and much attention has been paid to the formation mechanisms[2]. It is used for the application of fabricating thick buried silicon dioxide layer for silicon on insulator applications[3-4]. Recently, porous silicon has been applied to the fabrication of microelectromechanical systems (MEMS)[5] and silicon field emitter[6-7]. The porous silicon has been formed can be removed easily with nitric acid solution at room temperature. Using this technology for fabrication of field emitter arrays, the problem of variable chemical etching rate of silicon can been overcome.

In this work, in order to overcome the problems of conventional silicon FEA fabrication method due to the usage of wet solution etching, we propose a new fabrication method using anodization. This method can produce uniform and symmetric silicon field emitters. Anodization was used to form thick porous silicon layer which was removed by chemical etching. The microstructures of the porous silicon formed, which affects the etching rate of the porous silicon, can be controlled by varying the anodization parameters such as etchant concentration, anodization time, current density, substrate resistivity and substrate type.

EXPERIMENTAL DETAILS

The sample is etched in two different concentrations of etching solution at a constant current density of 90mA/cm^2. The schematic of the electrochemical etching cell used for the silicon tip formation is shown in Figure 1. A platinum wire is used for a counter electrode. Positive voltage is applied between the sample and the counter electrode. The constant current is supplied by a KEITHLEY 2400 SMU which was controlled by a computer and the current-voltage profile during the anodization processes could be recorded.

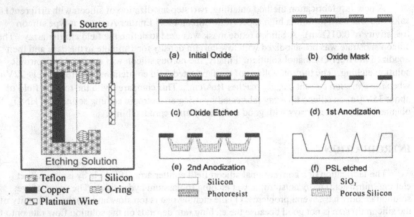

Figure 1. Anodization set-up

Figure 2. Fabrication process of silicon FEA using two step anodization method.

The process flow for the electrochemical etching is shown in Figure 2. Starting material was n-type Si (100) substrate with 0.01Ωcm resistivity. It was covered with a thermally grown, 1µm, thick silicon dioxide. The silicon dioxide was photolithographically patterned to form 20µm diameter circular discs which was used as a mask for the following anodization etching process.

Firstly, electropolishing process of the silicon surface was performed by anodization the substrate in an dilute HF:H$_2$O (4:75) solution with a current density of 90mA/cm^2 for 1 minute in the dark.

Secondly, porous silicon is formed by anodization in a HF:H$_2$O:C$_2$H$_5$OH = 20:37:16 solution with current density of 90mA/cm^2 for 1 minute in dark. Under these conditions, the anodization voltage with time is stable in these two steps which is shown in Figure 3.

Finally, the porous silicon layer formed by anodization was removed chemically by a HNO$_3$:HF:CH$_3$COOH=2:3:95 solution for 20 seconds.

Electrical testing was performed in a high vacuum chamber at a base pressure of 1x10^{-8} torr. The distance between the anode and the sample was about 20µm using a mica spacer.

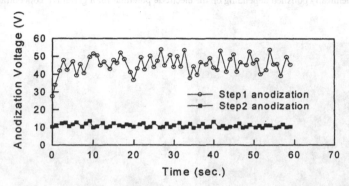

Figure 3. Voltage as a function of anodization time with current density of 90mA/cm^2

RESULT AND DISCUSSION

Electrochemical etching of silicon in a hydrofluoric acid solution has attracted much interest due to the formation of porous silicon which can easily removed by HNO$_3$. In the first anodization step, the electropolishing of silicon occurs under anodization. At this current density, fluoride ions and HF molecules directly remove silicon atoms and this reaction is responsible for the electropolishing[8]. In this step, the concentration of HF is suitable for electropolishing of n-type silicon in dark with a very good uniformity. The etching rate is about 5μm per minute. After the first anodization, silicon tip shape is formed and the thermally growth silicon dioxide is also there to protect the top of the silicon tip.

In second anodization step, the porous silicon is formed by anodization. After this, the porous silicon was removed by nitric acid. The silicon tip is formed and the etching rate is about 5μm per minute. The silicon tip is sharpened by removing the porous silicon formed.

Although, the dissolution reaction occurs readily for p-type silicon and illumination is required for n-type Si if the concentration of HF:H$_2$O is 1:1 or 1:2. In fact, at a voltage higher than the critical voltage, porous silicon will grow even if the sample is not illuminated. The etching solution concentration is suitable for the formation of porous silicon which can be easily removed by nitric acid. Unlike etching in KOH or EDP[9], anodization method is an isotropic etching method, as shown in Figure 4, the formation of the porous silicon is isotropic which is independent of the crystalline orientation. The silicon undercut formed under the spoke pattern is isotopic so that the underetch rate for various crystal plan is the same. The other advantage of the anodization process is that FEA can be fabricated onto all kind Si substrates with difference orientations. With the circular oxide mask, the tips of FEA formed are conical shape. A typical scanning electron microscope (SEM) photomicrograph of conical silicon-tip array fabricated by using a two step anodization method base diameter is about 10μm is shown in Figure 5. The porous silicon thickness increases with anodization time and the porous silicon layer is very uniform in thickness. The silicon wafer electrode is covered either by a porous silicon layer or is electrochemically polished depending on the electrode potential for a given HF concentration.

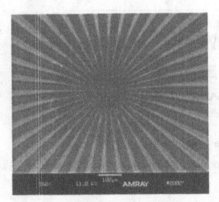

Figure 4. Isotropic etching test pattern formed by using anodization method with a silicon dioxide mask.

Figure 5. Typical SEM photomicrograph of conic silicon-tip array fabricated by using a two step anodization method. The base diameter of the tips is about 10μm.

Figure 6 shows the Fowler-Nordheim (FN) plot for FEAs fabricated by a conventional and the anodization method. The linearity of the FN plot shows that the anode current is due to electron field emission. The turn-on voltage of the fabricated field emitters by anodization was approximately 27V/μm while the emission current density reaches 1μA/cm². This compares with a turn-on field of about 35V/μm from the silicon tips fabricated by using an isotropic etching solution of nitric acid. The inverse of the slope in the FN plot allows us to determine $\beta/\phi^{3/2}$ where β is the field enhancement factor and ϕ is the work function in volts[10]. The value of $\beta/\phi^{3/2}$ are determined to be 17 and 20 for the conventional and anodized FEA samples respectively. As silicon substrates are used in both cases, they should have the same work function. Therefore the field enchantment factor was improved by the modification of surface morphology by the anodization process. Figure 7 shown atomic force microscope (AFM) micrographs of the Si surface along the bottom of the tips as the tips are too large to be investigated by the AFM. The micrographs of the sample surface were recorded after various treatments. The roughness of the two step anodized surface is much larger than the surface fabricated by nitric acid solution. The sample surface after nitric acid etching is relatively flat and the protrusions density is less. For the surface prepared by anodization, small and dense protrusions were observed. These microasperities seen after removal of the porous silicon must act as micro-cathodes on the apex of the Si tips which can improve the efficient of electron field emission from the surface[11].

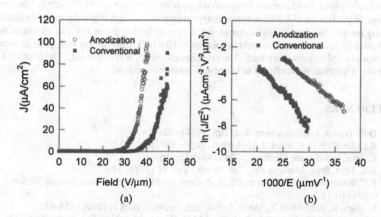

Figure 6. (a) Current-voltage characteristic for FEA fabricated using the two step anodization method and using a conventional HNO₃ etching method. (b) data plotted in Fowler-Nordheim coordinates.

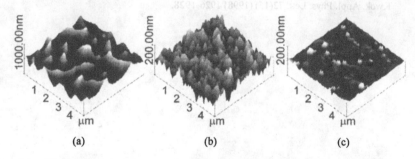

Figure 7. AFM micrographs (a) after first anodization (b) after porous silicon removal and (c) A nitric acid prepared surface.

CONCLUSIONS

Two step anodization of silicon with different HF solutions was performed to fabricate high aspect ratio silicon Field Emitter Arrays on n-type silicon (resistivity of 0.01Ωcm). The turn-on voltage of the fabricated field emitters was approximately $27V/\mu m$ while the emission current density reaches $1\mu A/cm^2$. This compares with a turn-on field of about $35V/\mu m$ on silicon tips fabricated by using an isotropic etching solution of HNO_3. Emitter arrays with good uniformity and reproducibility were fabricated. The emission efficiency was improved. The improved field emission properties are attributed to increase of surface roughness.

REFERENCES

[1] D.R. Turner, J. Electrochem. Soc. 105 (1958) 402-408.
[2] R.L. Smith, S.D. Collins, J.Appl. Phys. 71(8) (1992) R1-R22.
[3] K. Imai, Solid-State Electronics, 24 (1981) 159-164.
[4] R.C. Frye, Proc. Material Res. Soc. Symp., Vol. 33 (1984) 53-62.
[5] R.C. Anderson, R.S. Muller and C.W. Tobias, J. of Electroelectromechanical Systems, 3(1) (1994) 10-18.
[6] K. Higa, K. Nishii and T. Asano, J. Vac. Sci. Technol. B16(2) (1998) 651-652.
[7] J.R. Jessing, H.R. Kim, D.L. Parker and M.H. Weichold, J. Vac. Sci. Technol. B16(2) (1998) 777-779.
[8] B. Hamilton, Semicond. Sci. Technol. 10 (1995) 1187.
[9] T.J. Hubbard, E.K. Antonsson, J. of Electroelectromechanical Systems, 3(1) (1994) 19-28
[10] A. van der Ziel, Solid State Physical Electronics (Prentice-Hall, Engle-wood Cliffs, NJ, (1968) 164.
[11] D. Chen, S.P. Wong, W.Y. Cheung, W.Wu, E.Z. Luo, J.B. Xu, I.H. Wilson and R.W.M. Kwok, Appl. Phys. Lett. 72(15) (1998) 1926-1928.

Mat. Res. Soc. Symp. Proc. Vol 621 © 2000 Materials Research Society

Study on Depletion Mode Operation of Protruding-Tip Field Emitter Triodes

Y-C. Luo, M.Shibata, H.Okada and H.Onnagawa

Faculty of Engineering, Toyama University

3190 Gofuku, Toyama 930-8555, JAPAN

ABSTRACT

Electrical characteristics of Spindt-type Molybdenum (Mo) field emitter triode devices with varied emitter tip-height have been studied based on device modeling and experiment. Potential and electric field distributions with varied the emitter tip-height has been simulated. It is observed that the electric field of the top of the higher emitter tip was strongly affected with the anode-gate distance and the anode voltage compared to conventional field emitter triode device. Experimental results with varied different tip-height were in good agreement with that of calculated results. We present the possibility of "*depletion mode*" field emitter triode device.

INTRODUCTION

Application of field emitter triode devices has been developed for flat panel displays, high frequency devices, switching devices and so on. Because operation of the field emitter triode is based on the electron tunneling effect from solid to vacuum, electrical characteristics mainly depend on device structure and its physical dimensions. The fabrication process affects on important device parameters, such as emitter tip radius, radius of gate, half-angle of cone and tip-height. Electrical characteristics are quite sensitive to those geometrical parameters [1,2,3,4]. In this paper, we investigate potential and electric field distributions and also calculate the electron current using Fowler-Nordheim (F-N) equation varied with emitter tip-height. We compare the experimental and calculated results of the electrical characteristics varied with tip-height.

MODELS AND SIMULATION METHOD

Figure 1 shows a cross sectional view of the field emitter triode model. In this calculation, we need an assumption that all of electrons emitted from the emitter tip are collected by anode.

The calculation was carried out for various emitter tips. Parameters of field emitter triode devices, the tip heights (He), half-angle (α) of the cone, tip radius (Re) and number of the tips(N_{tip}), are listed in Table 1. Gate height (Hg) was 1.0μm and gate thickness (Wg) was 0.2μm. The gate hole radius (Rg) was 0.65 μm. The gate-anode distances (Ha) were 100 and 3μm. The gate voltage (Vg) was 60V. The anode voltage (Va) was 200V.

Finite difference method was used for simulating the electrical potential. In this model, we solve the Laplace's equation,

$$\nabla^2 V(r, z, \varphi) = 0. \tag{1}$$

The potential has cylindrical symmetry thereby reducing the problem to two dimensions, r and z.

$$\frac{1}{r}\frac{\partial}{\partial r}\left(r\frac{\partial V}{\partial r}\right) + \frac{\partial^2 V}{\partial z^2} = 0. \tag{2}$$

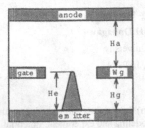

Table 1   Parameters of field emitter triode devices.

Device	He (μm)	α (deg)	Re (nm)	Ntip
A	1.6	16.0	42.0	600
B	1.2	16.0	42.0	600
C	1.0	14.0	6.0	7000
D	0.7	14.0	6.0	7000

Fig.1   The field emitter triode device model.

The following are the boundary conditions of applied voltages,

$$V(r,z) = \begin{cases} V_e, & \{\{|r| \le [R_e + Sin\,(\alpha\,)(H_e - R_e - z)/Cos\,(\alpha\,)]\} \cap \{0 \le z \le [H_e - R_e + R_e Sin\,(\alpha\,)]\} \\ & \{r^2 \le R_e^2 - (z - H_e + R_e)^2\} \cap \{[H_e - R_e + R_e Sin\,(\alpha\,)] < z \le H_e\} \\ V_g, & \{\{|r| \ge R_g) \cap [H_g \le z \le (H_g + W_g)]\} \\ V_a. & \{z \ge H_g + W_g + H_a\} \end{cases}$$

The electrical characteristics is calculated using the F-N equation which describes the relation between current density and the electric field at the tip surface [5,6,7,8,9,10] as follows

$$J = A\frac{E^2}{\phi t^2(y)}\exp\left[Bv(y)\frac{\phi^{3/2}}{E}\right],\tag{3}$$

where, J is the emission current density, E is the electric field at the tip surface, ϕ is the work function, $A=1.54\times10^{-6}$, $B=-6.831\times10^{-7}$, $y=3.7947\times10^{-4}E^{1/2}/\phi$, $t^2(y)=1.1$ and $v(y)=0.95-y^2$.
The electric field varies linearly with applied gate and anode voltages or

$$E = \beta V_g + \mathcal{W}_a.\tag{4}$$

We fit the data as [10]

$$I = N_{tip}\Phi J_{FN},\tag{5}$$

where, Φ is the emission area of the tip, the N_{tip} and the parameters β, γ and Φ are used for an empirical fitting.

DEVICES AND POTENTIAL DISTRIBUTIONS

2.1 Device A. 2.2 Potential distributions.

Fig.2   Fabricated higher tip-height device A and calculated potential distributions.

In order to study the electric field and the potential distribution around the emitter tip, the devices with different emitter tip-height were fabricated. Four types of devices where the positions of the tip-height are different, as shown in Figs.2-5, are taken into consideration in order to investigate effects of the emitter tip-height. These distributions of the potential and the electric field at the center of the tip were calculated based on the device model, as shown in Figs.2-5 and Fig.6.

3.1 Device B. 3.2 Potential distributions.

Fig.3   Fabricated conventional device B and calculated potential distributions.

4.1 Device C. 4.2 Potential distributions.

Fig.4   Fabricated conventional device C and calculated potential distributions.

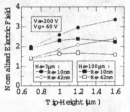

5.1 Device D.   5.2 Potential distributions.

Fig.5   Fabricated lower tip-height device D and calculated potential distributions.

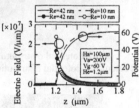

Fig.6
The calculated equi-potential curve
near the top of tip based on the
device model.

ELECTRIC FIELD DISTRIBUTIONS

Calculated potential and the electric field distributions at the center of the emitter tip are shown in Fig.7. This figure indicates that the electric field of the smaller tip radius (10nm) was stronger than that of larger tip radius (42nm).

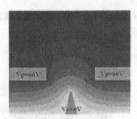

Fig.7 Distributions of potential and electric field
at the center of the tip(He=1.2μm).

Fig.8 Electric field with varied emitter
tip-height.

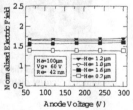

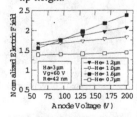

Fig.9   Electric field distributions with varied
anode voltage in the case of Ha=100μm.

Fig.10 Electric field distributions with varied
anode voltage in the case of Ha=3μm.

The calculated maximum electric field near the emitter tip-end increases with increase in the tip-height for the case of Ha=3μm, as shown in Fig.8. This figure shows that the electric field at the tip surface is strongly affected by the anode voltage for the higher tip-height device than the lower one. In the case of Ha=100μm, two type of tip radius has been calculated. As the tip radius is small, the electric field becomes stronger. However, as for the device with He=1.6μm, the electric field becomes weaker than the conventional devices.

In order to consider the variation of the electric field, we simulated electric field distributions with two types of the gate-anode distance. First, in the case of Ha=100μm, no variation in electric field are observed, as shown in Fig.9. The variation of the electric field is the negligible, because the gate-anode distances were as long as 100μm. In the case of Ha=3μm, the electric field was notably changed with varied the anode voltage for the higher emitter tip, as shown in Fig.10. No variation was obtained for the lower tip-height. In the conventional devices, the variations in electric field were smaller.

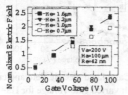

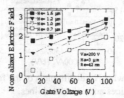

Fig.11 Electric field distributions versus gate gate voltage in the case of Ha=100μm.

Fig.12 Electric field distributions versus gate voltage in the case of Ha=3μm.

With changing of the gate voltage, the electric field distribution also simulated, as shown in Figs.11 and 12. For Ha=100μm, larger electric field variation is obtained for the higher tip-height device, as shown in Fig.11. As shown in Fig.12, the electric field due to anode potential was not negligible, because the gate-anode distance is as short as Ha=3μm.

ELECTRICAL CHARACTERISTICS

Aforementioned simulated results were compared with the experimental results. The current-voltage characteristics for the devices with different tip-height are shown in Figs.13 and 14. Higher tip-height device A and conventional device B have the same tip radius. Therefore, the same current flowed for lower voltage condition in the device B, as shown in Fig.13. The similar results were obtained in the lower tip-height device C and conventional device D.

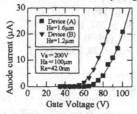

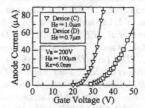

Fig.13   Current-voltage characteristics of the fabricated device A and device B.

Fig.14   Current-voltage characteristics of the device C and device D.

The F-N plots are shown in Fig.15. The current of the device B was larger than that of device A. This implies that the electric field at top of the tip of the device B is stronger than that of the device A with the same gate voltage. Calculated results(O and ●) have the same geometrical parameters and the number of the cones. This result indicates that F-N plots are very sensitive to those of geometric parameters, especially, tip-height for the same emitter structures. Fig.16 shows the calculated electric field and currents(□ and ▽) in negative gate voltage region for the two devices shown in Fig.12. In the device with He=1.6μm, it is shows that the electric field exists even in the negative gate voltage, *i.e.*, the possibility of *"depletion* mode" field emitter triode device.

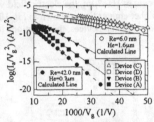

Fig.15 Fowler-Nordheim plots of the fabricated devices and calculated lines.

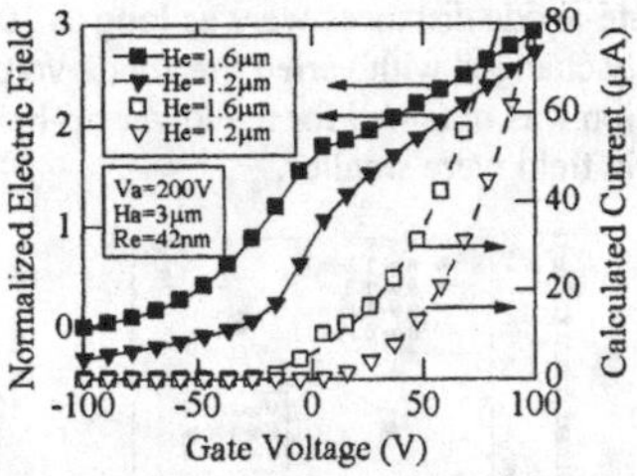

Fig.16 Calculated electric field and current with varied gate voltage for Ha=3μm.

CONCLUSION

The electrical characteristics of Spindt-type Molybdenum field emitter triode devices with various different emitter tip-height had been studied. The electric field and the potential distributions had been calculated. The electrical characteristics of the different devices in the tip-height have compared based on the experimental results. Emission characteristics have been also simulated with F-N equation. The electric field around tip-end increases with tip-height, in the case of Ha=3μm. The electrical characteristics depended obviously on tip-height and the tip radius. These electric characteristics of the fabricated devices were in good agreement with the calculated results. And we showed the possibility of the *"depletion mode"* filed emitter triode device.

REFERENCES

[1] T. Asano, *IEEE Trans. Electron Devices*, **38**, 2392 (1991).
[2] D. W. Jenkins, *IEEE Tans. Electron Devices*, **40**, 667 (1993).
[3] Z.Cui and L.Tong , *IEEE Tans. Electron Devices*, **40**, 448 (1993).
[4] N.S. Kim, Y.J. Lee and A. Hariz, *Proc. Asia Display 98*, p. 707, 1998.
[5] W. J. Orvis *et al.*, *IEEE Tans. Electron Devices*, **36**, 2651 (1989).
[6] R. M. Mobley and J. E. Boers, *IEEE Tans. Electron Devices*, **38**, 2383 (1991).
[7] R. True, *IEDM Tech. Dig.*, p.379 (1992).
[8] H. H. Busta *et al.*, *IEEE Tans. Electron Devices*, **39**, 2616 (1992).
[9] Y-C. Luo, J. Isoda, M. Shibata, H. Okada, and H. Onnagawa: *1998 Abst. joint meet. Hokuriku chapt. Phys. Soc. Jpn. and the Hokuriku-Shinetru chapt. Jpn. So. Appl. Phys.*, p.33 (1998).
[10] Y-C. Luo, J. Isoda, M. Shibata, H. Okada, and H.Onnagawa, *Trans. of the IEICE. C- II*, **J82-C-II**, pp.563-570 (1999). (in Japanese)

Mat. Res. Soc. Symp. Proc. Vol 621 © 2000 Materials Research Society

Field Emission Structure with Shottky-Barrier Electrode

Andrey P. Genelev, Alexey A. Levitsky[1], Vladimir S. Zasemkov
Krasnoyarsk State Technical Univ., Vacuum Microelectronics Lab., Krasnoyarsk, RUSSIA
[1]Krasnoyarsk State Technical Univ., Dept. of Radioelectronics, Krasnoyarsk, RUSSIA

ABSTRACT

We analyze a non-traditional version of semiconductor field emission structure, based on silicon cones arrays and destined for application in low-voltage vacuum microelectronics devices. In proposed construction, the gate electrode is used as Shottky-barrier contact on silicon tips. Due to Shottky-barrier contact we have obtained depletion layer in tip body under the gate electrode. Therefore variation of the gate electric potential provides the emitter current modulation. Experimental structures were fabricated with about 28000 silicon cones per 1 mm^2 by reactive ion etching through a silica mask. Using a quasi two-dimensional model, we have computed these emitter structures. The model takes into account non-uniformity of silicon cone cross-section. Here, we study the influence of emitter tips parameters on the structure performance. Initial results prove the possibility of cathode current electric control with the gate electrode potential. Additionally we have obtained some other electric parameters of the emitter with the Shottky-barrier contact. The results of numerical analysis and experimental study provide a guide for design of proposed field emitter structure.

INTRODUCTION

One of practice problems in vacuum microelectronics is to decrease voltage operation of field-emission devices [1,2]. Low-voltage operation can yield functional integration of semiconductor and vacuum devices. Therefore low-voltage is a goal for great variety of vacuum microelectronics applications. We have investigated a field-emission device with the gate electrode, which was formed as the Shottky barrier contact on the cathode electrode surface. Fig.1 illustrates schematically a cross-section of the emitter cone with the metal gate.

MODELING

Our goal now is to research the metal gate control. To achieve this goal we have studied the gate electric bias and carrier density influence on the depletion layer thickness in emitter body under the gate. When computing we were based on the traditional assumptions for the field-effect transistor model [3]. Figure 2 shows the model of the semiconductor emitter with the metal gate.

The basic equations for current in the emitter cone are Poisson's equation and the current continuity equation. Because of nonuniform geometry of the cathode tip we use modified forms of Poisson's equation [4,5]. Equation for electric current density $j(x,t)$ includes the dynamic component for the external circuit.

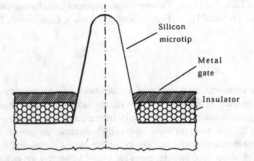

Figure 1. *Cross-section of emitter cone with metal gate.*

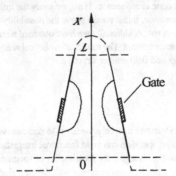

Figure 2. *Structure of the computed silicon cone with the metal gate.*

Therefore this equations for n-type semiconductor are expressed as follows:

$$\frac{\partial E_x(x,t)}{\partial x} + \frac{1}{S(x)}\frac{\partial S(x)}{\partial x} \cdot E_x(x,t) = \frac{q}{\varepsilon}[n(x,t) - N_0(x)], \qquad (1)$$

$$j(x,t) = qn(x,t)v(E_x) - qD\frac{\partial n(x,t)}{\partial x} + \varepsilon\frac{\partial E_x(x,t)}{\partial t}, \qquad (2)$$

where x - the direction coincided with the axis of an emitter cone, t - the time, $S(x)$ - cross-sectional area of the emitter tip, q - the elementary charge, ε - the dielectric permittivity, D - the diffusion coefficient, $E_x(x,t)$ - the electric field, $n(x,t)$ - the carrier concentration, $N_0(x)$ - the ionized donor density in semiconductor, $v(E_x)$ - the drift velocity of carriers, $j(x,t)=I_c/S(x)$, I_c- total current of tip.

Depletion layer modulation under the gate is described by following expression:

$$h(x,t) = \sqrt{\frac{2\varepsilon}{qN_0(x)}\left[\varphi(x,t) - \varphi_g(t) + \varphi_F\right]}, \qquad (3)$$

where $\varphi(x,t)$ - the electric potential in semiconductor, $\varphi_g(t)$ - the electric potential of the gate, φ_F - the Fermi level potential in semiconductor. Here, we put $E_x(x,t)$, $n(x,t)$, $j(x,t)$ and $v(E_x)$ as the mean values on the tip cross section.

Having been based on the system of the Poisson and the current continuity equations with corresponding initial and boundary (for $x=0$ and $x=L$) conditions, we have computed the profiles of the potential, electric field and depletion layer in semiconductor cone.

The computing process yields that the carrier concentration in semiconductor must be not more than about 10^{17} sm^{-3} for efficient gate operating.

FABRICATIONS AND EXPERIMENTAL

Arrays of emitter tips gated as above were manufactured on p and n type silicon substrates. The fabrication process consisted of three basic steps: tips forming, metal gate evaporation, and metal etching of cathode tips. Silicon emitters were formed by reactive ion etching through a silica mask. After that the thick molybdenum film was formed on the surface of cathode. Finally, the molybdenum film on tips of emitter was lift off by etching.

Figures 3(a) and 3(b) show scanning electron micrographs of the fabricated devices.

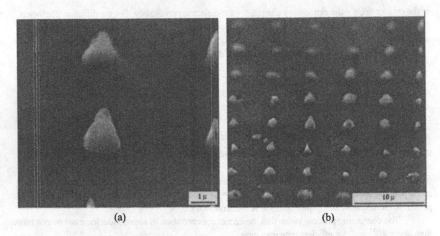

(a)

(b)

Figure 3. Scanning electron micrographs of fabricated gated emitters.

These emitters have shown stable I-V and C-V characteristics. The samples were then placed in a vacuum probe station where initial testing could be performed. Initial measurements prove the possibility of cathode current control with gate voltage.

CONCLUSION

We have studied the silicon field emitter arrays with the metal gate electrode, which was fabricated as the Shottky barrier contact on the cathode electrode surface. The results of modeling prove the possibility of low-voltage operate in this devices. Digital computation and experimental research show that it is necessary to decrease carrier density in semiconductor for effective control of cathode current with gate voltage. The ohmic heating in semiconductor tips depend critically on the doping levels. However, we can assume the Nottingham cooling at the high current levels [6].

ACKNOWLEDGMENT

The authors would like to thank A.P.Puzyr' for his support in preparing micrographs of fabricated structures.

REFERENCES

1. J.T.Trujillo, A.Chakhovskoi, C.E.Hunt, Low voltage silicon field emitters with gold gates, *Technical Digest of 8th Int. Vacuum Microelectronics Conf.*, Portland, OR, USA, July 30 - Aug.3, 1995, pp.42-46.
2. L.D.Karpov, A.P.Genelev, Y.V.Mirgorodski, A.N.Tikhonsky, Pattering and Electrical Testing of Field Emission Arrays with Novel Emitter Geometries, *Technical Digest of 9th Int. Vacuum Microelectronics Conf.*, Saint-Petersburg, Russia, July 7 - 12, 1996, pp.542-546.
3. W. Shockley, A unipolar "field-effect" transistor, *Proc. of the I.R.E.*, **v.47**, **11**, 1365 (1959).
4. A.A.Barybin, A.A.Levitsky, Modeling of semiconductor planar devices, *MRS VIMI "Engineering, Technology, Economics"*, **23**, 13 (1984) (in Russian).
5. A.A.Barybin, V.V.Perepelovsky, The Theoretical Treatment of Quasi -Two -Dimensional Equations for Computer Modeling, *Proceedings of the Int. Conf. on Mathematical methods in Science and Technology,* Vienna, June 3-6, 1995, pp.164-174.
6. M.G.Ancona, Modeling of Thermal Effects in Silicon Field Emitters, *Technical Digest of 8th Int. Vacuum Microelectronics Conf.*, Portland, OR, USA July 30 - Aug.3, 1995, pp.61-65.

Mat. Res. Soc. Symp. Proc. Vol. 621 © 2000 Materials Research Society

Micro-Structural and Field Emission Characteristics of Nitrogen-Doped And Micro-Patterned Diamond-Like Carbon Films Prepared by Pulsed Laser Deposition

Ik Ho Shin, Jae Hyoung Choi, Jeong Yong Lee and Taek Dong Lee
Department of Materials Science and Engineering, Korea Advanced Institute of Science and Technology, *373-1 Kusong-dong, Yusong-gu, Taejon 305-701, Korea*

ABSTRACT

Effect of nitrogen doping on field emission characteristics of patterned Diamond-like Carbon (DLC) films was studied. The patterned DLC films were fabricated by the method reported previously[1]. Nitrogen-doping in DLC film was carried out by introducing N_2 gas into the vacuum chamber during deposition. Higher emission current density of 0.3~0.4 mA/cm^2 was observed for the films with 6 at % N than the undoped films but the emission current density decreased with further increase of N contents. Some changes in CN bonding characteristics with increasing N contents were observed. The CN bonding characteristics which seem to affect the electron emission properties of these films were studied by Raman spectroscopy, x-ray photoemission spectroscopy (XPS) and Fourier transform infrared spectroscopy (FT-IR). The electrical resistivity and the optical band gap measurements showed consistence with the above analyses. High-resolution transmission electron microscopic (HRTEM) results also support the formation of graphite-like phase in 9 at % N-doped DLC film.

INTRODUCTION

DLC films are mechanically hard, chemically inert and of great interest as cold cathodes for field emission display. Because of low threshold field and high current density, electron field emission behaviors of DLC films have been widely investigated by many researchers.[2, 3] In the previous paper[1], improved field emission characteristics of pulse laser deposited DLC films by the micro-patterning were reported.

To understand the change in structure and electronic properties of DLC films upon nitrogen incorporation, several attempts have been made which are not found beyond disputation. Most of the researchers report an increase of conductivity by adding nitrogen in the DLC films[4, 5]. However, the role of nitrogen in DLC films in increasing conductivity is not clearly resolved. D. Mckenzie et al. and V. Veerasamy et al. found that a highly sp^3 bonded DLC films deposited by filtered cathodic vacuum arc can be doped controllably by nitrogen[6, 7]. Schwan et al. reported that nitrogen in DLC films leads to a graphitization of amorphous carbon when deposited with plasma enhanced chemical vapor deposition system[8].

In the present work, nitrogen-doped DLC films were obtained by the laser ablation system and effects of nitrogen doping on micro-structure and field emission characteristics of patterned DLC films with varying nitrogen content were studied.

To analyze the effects of nitrogen on bonding characteristics of CN bonds, Raman spectra, XPS and FT-IR analyses were employed. Also effects of nitrogen on the electrical properties and the optical band gap were studied. These effects are discussed in connection with portion of C-N bond and C=N bond.

EXPERIMENT

DLC films were deposited by a pulsed Nd-YAG ablation technique. The focused power density, 3×10^{12} W/cm^2 for wavelength of 532 nm was used. The vacuum chamber was maintained at a low 10^{-7} torr range during deposition process. Patterning method was the same as that of the previous report[1].

Field emission measurements were carried out at a pressure of $2\text{-}3 \times 10^{-7}$ torr by a diode type arrangement. The nitrogen content in the DLC films was determined by elastic recoil spectrometry using incident ion of ^{35}Cl.

For the electrical resistivity measurement, DLC films were deposited on thermally oxidized Si with 0.6 μm thick oxide layer to avoid the influence of the conductive substrates. Electrical resistivity was measured by a four-point probe system composed of KEITHLEY 236 source-measure unit and KEITHLEY 182 digital voltmeter.

HRTEM and diffraction patterns were used for analyzing microstructure of the films. The HRTEM samples were prepared by dissolving the Si substrate in HF:HNO$_3$ solution and releasing the nitrogen-doped DLC films.

RESULT AND DISCUSSION

As shown in figure 1, high emission current density of 0.3-0.4 mA/cm^2 has been obtained at the applied field of 8.5 MV/m. SEM was used for analyzing the surface of the DLC films. The patterned DLC films did not show the damage sites after the emission test. These results may be caused by great many edge-shaped areas of the patterned DLC films. Inherent each edge-shape is thought to be the direct emission sites without conditioning. The emission current density of nitrogen-doped film with 6 at % nitrogen is higher than that of the undoped film. However, further increase of nitrogen content decreased emission current density. The increase in the emission current density up to nitrogen content of 6 at % and the reduction in the emission current density at higher nitrogen content are shown.

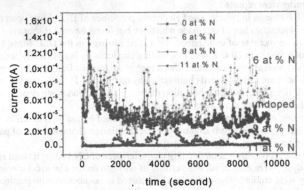

Figure 1. *The emission current vs bias annealing time for a series of the nitrogen content.*

These suggest that there is some change of CN bonding characteristics with increasing N content and the bonding characteristics may change the electron emission properties. Also the dominant CN bonding type such as C=N double bonds with the sp^2 hybridization and C-N single bonds with the sp^3 hybridization may be varied by the extent of incorporated nitrogen contents.

The I(D)/I(G) ratio and G-band width of the Raman spectra of the nitrogen-doped DLC films as a function of the nitrogen content are shown in figure 2. The I(D)/I(G) ratio increases and G-band width decreases with the increasing nitrogen content but the both showed a plateau-like region at 4 to 6 at %. The strong decrease of the G-band width and the increase of the I(D)/I(G) are observed at 6 to 9 at % nitrogen. These phenomena are consistent with the results reported by R. O. Dillon[9] suggesting the formation of sp^2 clusters.

Figure 3 shows C1s spectra of the DLC films measured by XPS. A shift of peak position of the carbon core electron excitation energy towards higher binding energy in the specimen doped with 6 at % nitrogen content was observed. This suggests that more diamond-like bond was formed[10]. Therefore, this result indicates that nitrogen is forming C-N bonds with the tetrahedral bonded carbon atoms[4] up to 6 at % of nitrogen. Further increase of nitrogen content results in shifts of the C1s peaks towards lower binding energy, e.g. more graphite-like bonding[10].

In figure 4, the deconvoluted IR spectra of the DLC films with 7.5 and 9 at % nitrogen are shown together with the whole spectra. This result shows the existence of strong C-N bonds absorption peak together with weaker C=N peak at the 6 at % N film. However, as N content increase up to 9 at %, stronger C=N bond absorption peak has been developed. It indicates that the content of C=N bond in the film with nitrogen content of 9 at % is higher than the film with 7.5 at % N. The FT-IR results are in consistent with the above mentioned Raman and XPS results.

Changes of electrical resistivity and optical band gap with varying nitrogen content are appeared in figure 5. Overall feature of the curves shows that electrical resistivity and optical band gap decrease with increasing nitrogen content. But a plateau-like region at 4 to 6 at % range and more rapid decreases with further increase of nitrogen content are observed.

It has been reported that the variation of sp^2 fraction for C and N sites as a function of N content was reported by Silva, et al.[5]. They argued that the C sp^2 fraction increased at a high

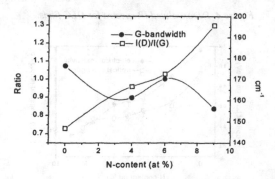

Figure 2. *The I(D)/I(G) ratio and G-bandwidth as a function of the nitrogen content*

nitrogen content and it corresponded to graphitization. Also, in the present work, the rapid decrease of electrical resistivity and optical band gap may be associated with the formation of sp^2 bond regions in the specimens over nitrogen content of 6 at %.

Figure 6(a) and 6(b) are HRTEM images and diffraction patterns showing the microstructure

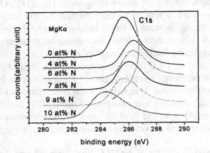

Figure 3. XPS C1s core-level electron spectra for DLC films.

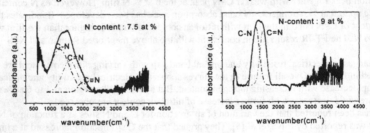

Figure 4. Deconvoluted IR spectra for DLC films.

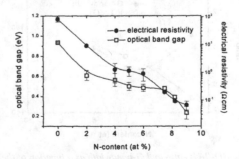

Figure 5. Electrical resistivity and optical band gap as a function of the nitrogen content.

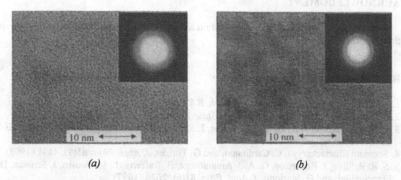

Figure 6. HRTEM images and diffraction patterns of DLC films with (a) 0 and (b) 9 at % nitrogen.

of DLC films with 0 and 9 at % nitrogen, respectively. Figure 6(a) shows that the film exhibits a typical amorphous structure. However, figure 6(b) shows the rings of graphite and several spots indicating graphite-like regions. These results are consistent with the XPS and FTIR results and the formation of sp^2 bond regions may contribute to the improved electrical conduction and to the decreased optical band gap. No appreciable changes in XPS spectra and HRTEM analyses with nitrogen content up to 6 at % suggests that nitrogen acts as a donor. This result and the formation of the graphite-like regions found at the nitrogen content about 9 at % suggest that the effect of nitrogen on micro-structure is varied with nitrogen content in the DLC films. The increase of the emission current density at the case of low nitrogen content may be associated with increased n-type carrier concentration and the decrease of the emission current density at the case of high nitrogen content may be associated with the formation of sp^2 bond regions. For more detail analyses regarding local distribution of sp^2 and sp^3 bonds in the films, DLC films are being investigated by electron energy loss spectroscopy.

CONCLUSION

The incorporation of nitrogen into the DLC films fabricated by pulsed laser deposition increased the emission current density of the films when the nitrogen content was lower than 6 at %. Further increase of nitrogen content lowered emission current density of the films abruptly. The FT-IR measurements showed that the C=N bond portion of the films increased with increase of N contents over 6 at %. This is consistent with the results of the Raman and XPS measurement. The sp^2 bond regions seem to be formed over the 6 at % N. The electrical resistivity and the optical band gap measurements supported the above mentioned analyses. The further increase of nitrogen incorporation in the film changes microstructure and emission properties abruptly.

Consequently, it is deduced that the enhancement effect on the field emission properties by the incorporation of the nitrogen is restricted by the formation of sp^2 bond regions.

ACKNOWLEDGMENT

Parts of this work was supported by 'Brain Korea 21' project. Authors would like to express their thanks.

REFERENCES

1. I. H. Shin and T. D. Lee, *J. Vac. Sci. Tech. B* **17**, 690 (1999)
2. N. Kumar, H. Schmidt and C.Xie, *Solid State Technology* **38**, 71 (1995)
3. B. S. Satyanarayana, A. Hart, W. I. Milne, J. Robertson, *Diamond and Related Materials* **7**, 656 (1998)
4. Somnath Bhattacharyya, C. Cardinaud, and G. Turban, *J. Appl. Phys.* **83(8)**, 4491 (1998)
5. S. R. P. Silva, J. Robertson, G. A. J. Amaratunga, B. Rafferty, L. M. Brown, J. Schwan, D. F. Franceschini, and G. Mariotto, *J. Appl. Phys.* **81(6)**, 2626 (1997)
6. D. Mckenzie, D. Muller, B. Pailthorpe, *Thin Solid Films* **206**, 198 (1991)
7. V. Veerasamy, J. Yuan, G. Amaratunga, W. Milne, K. Gilkes, M. Weiler, and L. Brown, *Phys. Rev. B* **48**, 17954 (1993)
8. J. Schwan, W. Dworschak, K. Jung, and H. Ehrhardt, *Diamond and Related Materials* **3**, 1034 (1994
9. R. O. Dillon, J. A. Woollam, and V. Katkanant, *Phys. Rev. B* **29**, 3482 (1984)
10. V. N. Apakina, A. L. Karuzskii, M. S. Kogan, A. V. Kvit, N. N. Melnik, Yu. A. Mityagin and V .N. Murzin, *Diamond and Related Materials* **6**, 564 (1997)

Mat. Res. Soc. Symp. Proc. Vol. 621 © 2000 Materials Research Society

Large and Stable Field-Emission Current from Heavily Si-Doped AlN Grown by MOVPE

Makoto Kasu and Naoki Kobayashi

NTT Basic Research Laboratories

3-1 Morinosato-Wakamiya, Atsugi, 243-0198, Japan

ABSTRACT

We investigated electron field emission (FE) from heavily Si-doped AlN grown by metalorganic vapor phase epitaxy. We found that, as the Si-dopant density increases, the threshold electric field decreases, which indicates that electrons are supplied to the surface effectively as a result of Si doping. We show that heavily Si-doped AlN has a maximum FE current of 347 μA (the maximum current density of 11 mA/cm^2), stable FE current (fluctuation: 3%), and a threshold electric field of 34 V/μm. We observed visible light emission (luminance: about 1200 cd/m^2) from phosphors excited by the field-emitted electrons.

INTRODUCTION

The interest in electron field emitters using wide-bandgap semiconductors such as AlN has been increasing because the electron affinity of these materials is expected to be very small [1-3]. In materials with small electron affinity, electrons can be easily extracted from the surface to the vacuum. The consequent large electron field emission (FE) will lead to very thin flat panel displays and microwave power amplifier tubes. There have been few studies on $Al_xGa_{1-x}N$ (x=0.1 in ref. 4 and $0.05 \leq x \leq 0.9$ in ref. 5) and AlN [5]. However, the reported FE current from $Al_xGa_{1-x}N$ and AlN was as low as 0.01 to 0.12 μA [4, 5]. This is probably because the $Al_xGa_{1-x}N$ and AlN samples in those studies were not intentionally doped, and, as a result, few electrons were supplied from the substrate to the surface. Therefore, we investigated the effect of Si doping on the electron field emission from AlN. We found that, as the Si-dopant density in AlN increases, the threshold electric field decreases. We show that heavily Si-doped AlN has a larger maximum FE current than that previously reported for AlN [5], a lower threshold electric field, and an extremely stable emission current. We observed visible light emission from phosphors excited by the field-emitted electrons.

RESULTS AND DISCUSSION

Heavily Si-doped 0.2-μm-thick AlN was grown at 1100°C by low-pressure (300 Torr)

MOVPE. We chose an n-type (1 x 10^{18} cm^{-3}) 6H-SiC (0001) substrate because it has high electric conductivity and because SiC and AlN have almost the same lattice constants (mismatch: 1%) and thermal expansions along the a axis. The sources were trimethylaluminum (TMA), trimethylgallium (TMG), and ammonia (NH$_3$). The Si-dopant gas was silane (SiH$_4$), the gas most widely used in GaN MOVPE growth. The Si dopant density was measured using secondary ion mass spectrography. In GaN, Si atoms form a shallow donor level. It has been reported that, in Si-doped Al$_x$Ga$_{1-x}$N ($0 \le x \le 0.33$), the resistivity increases exponentially as the Al content x increases [6]. The resistivity of the AlN with the highest Si dopant density (2.5x10^{20} cm^{-3}) was so high that the free electron density could not be measured at room temperature (RT) by Hall measurement. The full width at half maximum of (0002) reflection of the x-ray rocking curve of 0.2-μm-thick nondoped and heavily Si-doped AlN were as low as 91 and 94 arcsec, respectively. The surface roughness of the heavily Si-doped AlN surfaces was less than 50 nm. The heavily Si-doped AlN surfaces were free from cracks. For comparison, n-type Si-doped GaN and As-doped Si (111) samples were used. The GaN sample consisted of a 0.1-μm-thick Si-doped AlN buffer layer and a 1.2-μm-thick Si-doped (1 x 10^{19} cm^{-3}) GaN on an n-type 6H-SiC(0001) substrate. The highest Si dopant density in GaN at which a smooth surface could be maintained was 1 x 10^{19} cm^{-3}. The Si sample was an As-doped (n=5x10^{19}cm^{-3}, the highest value available) Si (111) wafer.

Figure 1 shows a schematic diagram of the FE measurement system. For the FE current-voltage (I-V) measurements, we used a planar sample so that the field enhancement

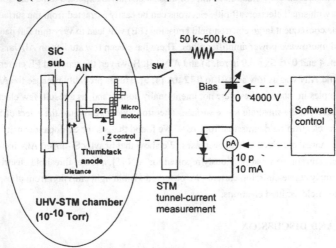

Fig. 1. Schematic diagram of the field-emission current-voltage measurement system.

factor due to the surface curvature could be neglected and the intrinsic FE effect of these materials obtained. For the FE I-V measurements, we used a W thumbtack-shaped anode probe. Its top surface was parallel to the planar sample surface. This enabled us to estimate the emission area, and consequently, the current density, accurately. The probe-sample distance is an important factor in estimating the electric field strength. In our system, the anode probe was lowered to the sample surface until the tunnel current (0.1 nA at about 10 V) was detected in the scanning tunneling microscopy (STM) mode. The probe-sample distance can be expected to be less than 1 nm when the tunnel current is detected. Next, the probe-sample distance was increased by using a piezo tube (PZT) and a micromotor. We measured I-V characteristics in the FE mode at RT. Positive bias was applied to the W probe, and negative bias was applied to the sample substrate. Series resistance was inserted to protect the pA meter and voltage source. The voltage drop in the series resistance was negligibly small with respect to the applied bias. We carefully confirmed that there was no damage on the surface before and after the field emission. Thus, the I-V data shown here are due to the field emission, not due to discharge.

Figure 2(a) shows the FE current for different Si-dopant densities (N_{Si}) in AlN [7]. As the Si-dopant density increased, the threshold voltage decreased. The threshold voltages defined at 1 nA were 90, 180, 410 V for N_{Si} = 2.5 x 10^{20}, 7.5 x 10^{19}, and 4.6 x 10^{18} cm^{-3}, respectively. For N_{Si} = 2.5 x 10^{20} cm^{-3}, the threshold voltage defined at 0.05 nA was 62 V. At the sample-probe distance of 1.8 μm, the threshold electric field was as low as 34 V/μm. We speculate that Si dopant atoms form an impurity level in the bandgap and the electrons are supplied to the surface through the impurity level. Figure 2(b) shows the FE current for N_{Si} = 2.5 x 10^{20} cm^{-3} at the sample-probe distance of about 25 μm. The maximum FE current was 347 μA. From a probe area of 3.1 x 10^{-2} cm^2, we estimated the maximum FE current density to be 11 mA/cm^2. The AlN maximum FE current density of 11 mA/cm^2 is of the order necessary for flat-panel FE displays.

Figure 3(a) shows the FE I-V characteristics of heavily Si-doped AlN (N_{Si} = 2.5 x 10^{20} cm^{-3}), Si-doped GaN, and As-doped Si. The threshold voltages defined at 0.1 nA were 68 V for AlN, 170 V for GaN, and 270 V for Si. The maximum FE currents were 19 μA for GaN and 1.9 μA for Si. From a probe area of 3.1 x 10^{-2} cm^2, we estimated the maximum FE current densities to be 0.6 and 0.06 mA/cm^2, respectively. The maximum FE current density of AlN (11 mA/cm^2) was larger than those of GaN and Si. The maximum FE current density is limited by the field evaporation of the negatively-ionized surface atoms, the sputtering of the positively-ionized residual gas, and a subsequent micro arc discharge. The reason AlN has higher maximum current density than GaN is probably that AlN has stronger bonds, and consequently, the evaporation of the surface atoms occurs less often. Figure 3(b) shows the Fowler-Nordheim (FN) plots (I/V^2 vs. 1000/V) for AlN, GaN, and Si. The negative slope corresponds to the energy barrier. This figure shows AlN has the lowest energy barrier among them. We speculate that, in heavily Si-doped

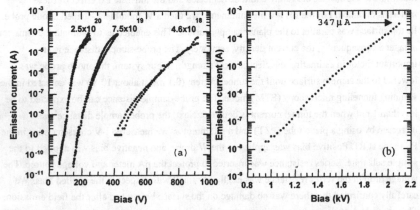

Fig. 2. Field emission current as a function of the applied bias voltage for AlN with different Si dopant densities (N_{Si}, cm^{-3}). The sample-probe distance was 1.8 μm (a). Field emission current with N_{si}= 2.5x10^{20} cm^{-3} and the sample-probe distance of about 25 μm. The maximum current (density) was 347 μ A (11 mA/cm^2) (b).

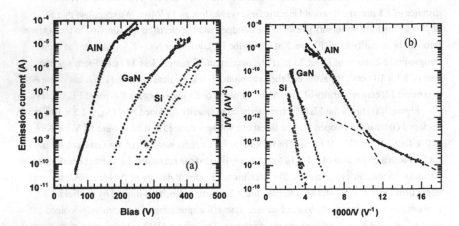

Fig. 3. Field emission current as a function of the applied bias voltage (a) and Fowler-Nordheim plots (b) for AlN (N_{Si} = 2.5 x 10^{20} cm^{-3}) and GaN (1 x 10^{19} cm^{-3}, the highest value at which a smooth surface could be maintained) and As-doped Si (n= 5x10^{19} cm^{-3}, the highest value available). The sample-probe distance was 1.8 μm.

AlN, electrons are supplied from the sample substrate to the surface by hopping conduction through the Si impurity level and then are emitted from the Si impurity level to the vacuum without transition to the conduction band.

Figure 4 shows the stability in the FE current over time for AlN and Si when the applied bias was fixed. There was no feedback system in the circuit. For AlN ($N_{Si} = 2.5 \times 10^{20}$ cm^{-3}), the fluctuation (the ratio of the root mean square to the average) was as low as 3%. To the contrary, for As-doped Si the fluctuation was 27%. The observed high stability of AlN indicates there was little change in the emission site density and a low rate of surface atom evaporation. The FE current stability for AlN is within the value necessary for flat-panel FE display application.

From the viewpoint of flat panel display application, we observed light emission from red, green, blue, and white phosphors excited by the field-emitted electrons from AlN ($N_{Si} = 2.5 \times 10^{20}$ cm^{-3}). The widely used P22-type red, green, blue or the P15-type white phosphors were coated on the top surface of the W probe. When the bias, current, distance were 3.8 kV, 9 μA, about 100 μm, respectively, the luminance of the white phosphor was estimated to be 1200 cd/m^2, which is sufficient for flat panel displays.

Fig. 4. Stability of field emission current over time for AlN ($N_{Si} = 2.5 \times 10^{20}$ cm^{-3}) and Si ($n = 5 \times 10^{19}$ cm^{-3}, the highest value available).

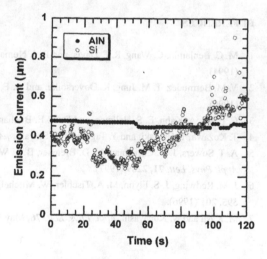

CONCLUSIONS

We investigated the electron field emission (FE) from heavily Si-doped AlN. We found that, as the Si-dopant density increases, the threshold electric filed decreases and the FE current from AlN increases. We showed that heavily Si-doped (2.5×10^{20} cm^{-3}) AlN has a larger maximum FE current density (11 mA/cm^2), a lower threshold electric field (34 V/μm) than previously reported for AlN, and an extremely stable (fluctuation: 3%) FE current. We speculate that, in heavily Si-doped AlN, electrons in the impurity level are emitted directly to the vacuum without transition to the conduction band. We observed light emission from phosphors excited by the field-emitted electrons. The luminance was estimated to be 1200 cd/m^2.

ACKNOWLEDGMENTS

We appreciate Dr. Naoshi Uesugi and Dr. Takaaki Mukai for their encouragements.

REFERENCES

1   M. C. Benjamin, C. Wang, R. F. Davis, and R. J. Nemanich, *Appl. Phys. Lett.* **64,** 3288 (1994).
2   V. M. Bermudez, T. M. Jung, K. Doversoike, and A. E. Wickenden, *J. Appl. Phys.* **79,** 110 (1996).
3   C. I. Wu, A. Kahn, E. S. Hellman, and D. N. E. Buchanan, *Appl. Phys. Lett.* **73,** 1346 (1998).
4   T. Kozawa, T. Ohwaki, and Y. Taga, and N. Sawaki, , *Appl. Phys. Lett.* **75,** 3330 (1999).
5   A. T. Sowers, J. A. Christman, M. D. Bremser, B. L. Ward, R. F. Davis, and R. J. Nemanich, *Appl. Phys. Lett.* **71,** 2289 (1997).
6   J. M. Redwing, J. S. Flynn, M. A. Tischler, W. Mitchel, and A. Saxler, *Mat. Res. Symp. Proc.* **395,** 201 (1996).
7   M. Kasu and N. Kobayashi, *Appl. Phys. Lett.* **76,** May 15 issue (2000).

Mat. Res. Soc. Symp. Proc. Vol. 621. © 2000 Materials Research Society

Electron Emission of Vertically Aligned Carbon Nanotubes Grown on a Large Area Silicon Substrate by Thermal Chemical Vapor Deposition

Cheol J. Lee[1], Jung H. Park[1], Kwon H. Son[1], Dae W. Kim[1], Tae J. Lee[1], Seung C. Lyu[1],

Seung Y. Kang[2], Jin H. Lee[2], Hyun K. Park[3], Chan J. Lee[3], and Jong H. You[3]

[1]School of Electrical Engineering, Kunsan National University, Kunsan 573-701, Korea

[2]Microelectronics Lab., Next Generation Semiconductor Dept., ETRI, Taejon 305-350, Korea

[3]Flat display Labs., R&D Center, Samsung Information Display Co., Suwon 442-391, Korea

ABSTRACT

We have grown vertically aligned carbon nanotubes on a large area of Co-Ni codeposited Si substrates by thermal chemical vapor deposition using C_2H_2 gas. The carbon nanotubes grown by the thermal chemical vapor deposition are multi-wall structure, and the wall surface of nanotubes is covered with defective graphite sheets or carbonaceous particles. The carbon nanotubes range from 50 to 120 nm in diameter and about 130 μm in length at 950 °C. Steric hindrance between nanotubes at an initial stage of the growth forces nanotubes to align vertically. The turn-on voltage was about 0.8 V/μm with a current density of 0.1 μA/cm^2 and emission current reveals the Fowler-Nordheim mode.

INTRODUCTION

Since the first observation of carbon nanotubes [1], extensive researches have been done for the synthesis using arc discharge [2-4], laser vaporization [5], pyrolysis [6], and plasma-enhanced chemical vapor deposition (CVD)[7]. Although massive production of carbon nanotubes has been realized by arc discharge and laser vaporization [5], controlling diameters, lengths, and preferable alignment of carbon nanotubes has never been easily accessible with such approaches. Synthesis of well-aligned carbon nanotubes with high quality on a large area is necessary for their applications, such as flat panel displays. Thermal CVD is a useful method to grow damageless carbon nanotubes on a substrate. We have grown carbon nanotubes on a large area substrates by thermal CVD using acetylene gas. In this work, we report the growth of well aligned carbon nanotubes on a large area of Co-Ni codeposited Si substrates, and also evaluate electron emission of vertically aligned carbon nanotubes.

EXPERIMENTAL

The p-Si substrates with a resistivity of 15 Ω cm were thermally oxidized with the layer thickness of 300 nm. Co-Ni (Co:Ni=1:1.5) metal alloys with 100 nm in thickness were thermally evaporated in a vacuum of 1.5 $\times 10^{-6}$ torr on oxidized Si(100) substrates. The samples were dipped for 100-200 sec in diluted HF solution and then loaded on a quartz boat inside the CVD quartz reactor. Argon(Ar) gas was flowed into the quartz reactor in order to prevent the oxidation of catalytic metal alloys while increasing the temperature. Samples were pretreated using NH_3 gas with a flow rate of 50-200 sccm for 10-30 min at 850-950 °C. Carbon nanotubes were grown using C_2H_2 gas with a flow rate of 20-80 sccm for 10-20 min at the same temperature. The reactor was cooled down slowly to room temperature in Ar ambient after the growth. Samples were examined by scanning electron microscope (SEM) (Hitachi, S-800, 30 kV) to measure nanotube length, diameter, alignment, and uniformity. High resolution transmission microscopy (TEM) (Philips, CM20T, 200kV) was used to determine the wall structure of individual tubes. Field emission measurement was conducted in a vacuum chamber (1x 10^{-6} Torr). A positive voltage was applied to the anode and the emission current was measured with a electrometer (Keithley 619).

DISCUSSION

Figure 1(a) shows SEM micrograph of vertically aligned carbon nanotubes which were grown at 950 °C on a large area (20 mm ×30 mm) of Co-Ni codeposited Si substrates. The carbon nanotubes are oriented vertically to the substrate and are uniformly grown with the length of about 130 μm. Figure 1(b) reveals high density of nanotubes with the diameter range from 50 to 120 nm, where nanotubes are straight. Basically, the diameter of carbon nanotubes is dependent on the size of catalytic metal particles. Top view of the vertically aligned carbon nanotubes is shown in Figure 1(c). When the density of nanotubes reaches high value, nanotubes grown in directions other than vertical direction are prohibited from growing due to the steric hindrance from the adjacent nanotubes and then change the growth direction to further grow vertically.

Figure 1. SEM micrograph of well aligned carbon nanotubes grown with gas flow rate of 40 sccm at 950 °C for 10 min on a large area (20 mm ×30 mm) of Co-Ni codeposited Si substrates. (a) A uniformly distributed morphology of vertically aligned carbon nanotubes, (b) magnified side view of (a), (c) top view of the vertically aligned carbon nanotubes.

TEM analysis was performed on the carbon nanotubes grown at 950 °C to determine the wall structures. TEM image in Figure 2 shows straight multi-wall nanotubes with a hollow inside and reveals well-ordered lattice fringes of nanotube. Outer diameter of carbon nanotube is about

60 nm and inner diameter is about 40 nm. In Figure 2, the carbon nanotubes are multi-wall structure with good crystallinity but the surface wall of nanotubes has bad crystal structure due to the defective graphite sheet structure. We have also tried various growth conditions. Carbon nanotubes were grown well independent of growth temperatures within the temperature range of 750 - 950 °C. The length of carbon nanotubes increased with increasing growth time and increasing growth temperature.

Figure 2. High-resolution TEM image of the lattice structure of multi-walled carbon nanotube. The carbon nanotubes was grown with gas flow rate of 40 sccm at 950 °C for 10 min.

Figure 3 illustrates field emission current density versus electric field from nanotubes grown on Co-Ni codeposited Si substrates. The anode was separated from nanotubes by a 200 μm spacer. The chamber pressure was maintained at 1×10^{-6} torr. The turn-on voltage was about 0.8 V/μm with a current density of 0.1 μA/cm^2, the maximum emission current density before electric breakdown was about 1.1 mA/cm^2 at 4.5 V/μm applied field. The presence of some protruded nanotubes gave rise to non-uniform emission current. The inset plot in Figure 3 shows the Fowler-Nordheim mode for the vertically aligned carbon nanotubes. In this figure, we

observed nearly straight line to indicate Fowler-Nordheim behavior even though it was slightly
deviated in the middle field region.

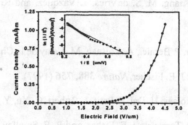

Figure 3. *Field emission current density versus electric field for well aligned carbon nanotubes
grown on Co-Ni codeposited Si substrate. (Inset) Fowler-Nordheim plot.*

Growth of vertically aligned carbon nanotubes on a large area of Si substrates using thermal
CVD will be useful for application of the field emission displays.

CONCLUSIONS

We have grown vertically aligned carbon nanotubes on a large area of Co-Ni codeposited Si
substrates using the thermal chemical vapor deposition. The carbon nanotubes are multi-wall
structure with good crystallinity but the surface of nanotube wall is consisted of defective
graphite sheet or carbonaceous particles. The turn-on voltage was about 0.8 V/μm with a current
density of 0.1mA/cm^2 and the emission current density was about 1.1 mA/cm^2 at 4.5 V/μm. The
emission current reveals the Fowler-Nordheim behavior. The current approach with the thermal
CVD is expected to be easily applicable to flat panel displays.

REFERENCES

1. S. Iijima, *Nature* **354**, 56 (1991).

2. S. Iijima and T. Ichihashi, *Nature* **363**, 603 (1993).

3. D. S. Bethune, C. H. Kiang, M. S. deVries, J. Vazquez, and R. Beyers, *Nature* **363**, 605 (1993).

4. C. Journet, W. K. Maser, P. Bernier, A. Loiseau, M. Lamy de la Chapelle, S. Lefrant, P. Deniard, R. Lee, and J. E. Fischer, *Nature* **388**, 756 (1997).

5. A. Thess, R. Lee, P. Nikolaev, H. Dai, P. Petit, J. Robert, C. Xu, Y. H. Lee, S. G. Kim, D. T. Colbert, G. Scuseria, D. Tomanek, J. E. Fisher, and R. E. Smalley, *Science*, **273**, 483 (1996).

6. M. Terrines, N. Grobert, J. Olivares, J. P. Zhang, H. Terrones, K. Kordatos, W. K. Hsu, J. P. Hare, P. D. Townsend, K. Prassides, A. K. Cheetham, H. W. Kroto, and D. R. M. Walton, *Nature* **388**, 52 (1997).

7. Z. F. Ren, Z. P. Huang, J. W. Xu, J. H. Wang, P. Bush, M. P. Siegal, P. N. Provencio, *Science* **282**, 1105 (1998).

Mat. Res. Soc. Symp. Proc. Vol 621 © 2000 Materials Research Society

Tight Binding Modeling of Properties Related to Field Emission from Nanodiamond Clusters

Denis A. Areshkin[1], Olga A. Shenderova[1], Victor V. Zhirnov[1,2], Alexander F. Pal,[3] John J. Hren[1], and Donald W. Brenner[1]
[1]North Carolina State University, Raleigh, NC
[2]Semiconductor Research Corporation, Research Triangle Park NC
[3]Troitsk Institute for Innovation and Fusion Research, Troitsk, Russia

ABSTRACT

The electronic structure of nanodiamond clusters containing between 34 and 913 carbon atoms was calculated using a tight-binding Hamiltonian. All clusters had shapes represented by an octahedron with (111) facets with the top and the bottom vertices truncated to introduce (100) surfaces. The tight-binding Hamiltonian consisted of environment-dependent matrix elements, and C-H parameters fit to reproduce energy states of the cyclic C_6 and methane. The calculations predict a density of states similar to bulk diamond for clusters with radii greater than ~2.5nm, and insignificant differences in the potential distribution between the clusters and bulk diamond for radii greater than ~1nm. Hydrogen passivated nanodiamond clusters are estimated to have an electron affinity of approximately -1.8 eV.

INTRODUCTION

Recent experimental measurements of field-emission properties of diamond nanoclusters on silicon field-emitter arrays have demonstrated desirable properties for cold-cathode applications, including low emission thresholds, high emission currents and good current stability [1]. Some of these properties, however, appear to depend on the origin and processing of the nanodiamond clusters, and hence the details of their structure, including particle size and surface conditions.

To better understand the molecular origins of field emission characteristics of nanodiamond clusters, a series of tight-binding calculations on regular-shaped structures with radii up to approximately 2.5nm have been carried out. Both pure carbon clusters and clusters with hydrogen termination were studied. These calculations, which use an environmentally dependent tight binding (EDTB) model [2], suggest that the size dependence of the density of states (DOS) is insignificant for radii greater than about 2.0-2.5 nm. In contrast, the dependence of the electron potential on cluster size is largely independent of cluster size down to radii of approximately 1 nm.

TIGHT BINDING PARAMETRIZATION

To reproduce the bulk diamond band gap and a qualitative description of the band structure, an EDTB model for carbon was used [2]. The tight-binding (TB) scaling function developed by Xu *et al.* [3] and parameterized by Davidson and Pickett (DP) [4] was used for the C-H interactions. To obtain a reasonable bulk DOS, a set of values for the C-H hopping and on-site matrix elements differing from the DP model were developed. These parameters, listed as *Present Work* in Table I, were generated by fitting the energies of the occupied electronic states in ethane to their counterparts calculated from density functional theory. The latter values were calculated using commercial software [5a].

To determine the C-H tight-binding parameters, it was necessary to first shift the on-site carbon matrix elements by a constant amount. This shift was determined by minimizing the squared differences between the EDTB and DFT values for the occupied energy levels of cyclic C_6. Plotted in Fig.1a is the DFT cyclic C_6 energy states and the EDTB states shifted by -6.95 eV. Matrix elements for the C-H interactions (given in Table I) were then determined by minimizing the least-squares difference between occupied energy levels of ethane given by the TB and DFT calculations. Because ethane has five nondegenerate occupied levels, the parameter choice is unique. The TB and DFT spectra for ethane are given in Fig. 1b.

To test the transferability of the C-H parameters, energy levels for methane and benzene molecules predicted by the TB calculations were compared to corresponding values given by DFT calculations. Plotted in Fig. 2 are the two sets of electronic energy levels for these molecules. The favorable comparison suggests that simultaneous optimization of the hopping and on-site C-H matrix elements leads to more consistent results compared to the DP scheme, which employed separate fits for on-site and two-center TB parameters. Compared to the DP on-site energy, the value of the on-site hydrogen matrix element obtained in the present work may

Table I. Carbon-carbon and carbon-hydrogen parameters for different TB models. The proposed carbon-hydrogen bond parameters compatible with the EDTB model are given as *Present Work*. The row denoted *DP* gives the Davidson-Pickett parameters for C-H bonds, and Xu *et al.* parameters for C-C bonds.

	$VC_{pp,\pi}$	$VC_{pp,\sigma}$	$VC_{sp,\sigma}$	$VC_{ss,\sigma}$	EC_s	EC_p	EH_s	$VH_{ss,\sigma}$	$VH_{sp,\sigma}$
DP	-1.55	5.5	4.7	-5.0	-2.99	3.71	-4.75	-6.52	6.81
EDTB	-1.90	5.74	4.17	-3.94	-6.04	1.02			
Present Work							2.36	-3.00	3.61

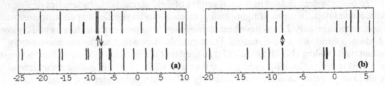

Figure 1. *Shifted TB eigenvalue spectra. The lower sections are from the TB model. The upper sections are from the DFT calculations. Energies are in eV. The length of the line segments is proportional to the level degeneracies. Arrows indicate the highest-occupied orbitals. (a) Cyclic C_6. (b) Ethane.*

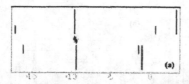 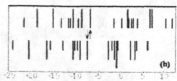

Figure 2. *Eigenvalue spectra used to check the transferability of the C-H matrix elements. The lower portions of the figures show the TB spectrum obtained in this work, the upper portions correspond to the DFT values. Energies are in eV. (a) Methane. (b) Benzene.*

seem nonphysical. However, the fact that the TB scheme is non-self-consistent implies that charge transfer should be built into the TB parameters. The DP hydrogen on-site energy is lower than both the s and p on-site energies for carbon. This choice leads to the wrong electronegativities and hence to the wrong sign for Mulliken populations. Table II lists Mulliken charge populations for benzene. Note that there was no special Mulliken population fitting in our parameterization.

RESULTS AND DISCUSSION

To study size and surface effects, two sets of carbon clusters were examined. The first set contained four carbon clusters composed of 34, 161, 435, and 913 atoms. All clusters had identical shapes represented by an octahedron with (111) facets with the top and the bottom vertices cut off to produce (100) surfaces. The second set consisted of the same four clusters with hydrogen-passivated (111) 1x1:H and (100) 1x1:2H (sym) surfaces. Optimized atomic coordinates were obtained using a many-body empirical potential [6]. The DOS for both sets of clusters are plotted in Figs. 3 and 4. Results from a DFT calculation [5b] of the DOS for the smallest cluster with hydrogen passivation are plotted in the lowest section of Fig. 4(a). These values and the Mulliken charge population for this hydrogen-passivated cluster was used to check the TB results. The energy difference between the highest occupied and lowest unoccupied molecular orbitals given by the DFT calculations and the EDTB model are reasonably close (0.4eV) to one another. However, the entire EDTB spectrum is shifted towards lower energies on the absolute scale. To quantitatively match the absolute value of the highest-occupied orbital given by the EDTB to the DFT results, the bulk and cluster EDTB spectra were shifted by -5.04 eV rather than -6.9 eV. This 1.86 eV difference suggests that the reliability with which the EDTB method yields orbital energies on the absolute energy scale is not sufficient to evaluate the electron affinity. As can be seen from Figs. 4 (c) and (d) the cluster spectra are slightly shifted towards lower energies with respect to the bulk spectrum. This shift is due to the small increase of the on-site carbon energies as the effective number of carbon neighbor atoms (as defined by the screening function) increases from 1.0 as in ethane to 4.41 for bulk diamond. Therefore the difference between hydrogen and carbon on-site energies becomes ~0.3 eV larger than it was for ethane which was used for the parameter fitting. This is equivalent

Table II. The total charge for hydrogen atoms obtained from Mulliken populations in benzene.

	This work	DP	DFT
Core charge minus Mulliken electron population	0.290	-0.340	0.301

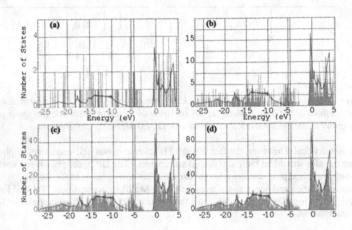

Figure 3. DOS plots for different size carbon clusters. (a), (b), (c), and (d) refer to clusters composed of 34, 161, 435, and 913 carbon atoms, respectively. Solid line shows TB bulk diamond DOS and is normalized to produce the total number of states equal to the number of states for the given cluster. Vertical line marks HOMO position. Both bulk and cluster DOS are shifted by -6.9 eV.

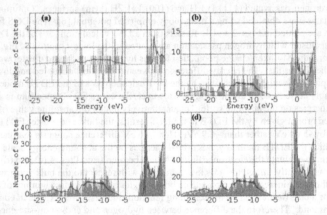

Figure 4. Hydrogen passivated cluster DOS spectra. Portions (a-d) refer to the same systems as the corresponding plots in Fig. 3. The lower part of portion (a) shows DFT/GGA DOS spectrum. Vertical line in the vicinity of the conduction band edge in portions (c) and (d) marks the lowest conduction band state which is spread over the entire cluster and is associated with the lowest bulk state. All the states below these lines are located at the cluster surface.

to a relative increase of hydrogen electronegativity, and therefore leads to the spectrum shift. However, this shift is rather small and can not account for the given ~2 eV difference between the TB and DFT values.

Mulliken charge populations are determined by the relative positions of the occupied levels and therefore are better reproduced than the DOS on the absolute scale. The average relative error between the DFT and EDTB results is less than 12%. The Coulomb potential distribution for the hydrogen passivated clusters is shown in Fig. 5. The potential was obtained from Mulliken charge populations assuming that the atoms are represented by uniformly charged spheres with radii for carbon and hydrogen atoms equal to 2.0 and 1.4 angstroms respectively. The main potential distribution feature is the steep rise at the cluster surface produced by the hydrogen monolayer. Cluster size effects are only significant for the smallest cluster. For the larger clusters the potential inside the cluster does not change appreciably with increasing cluster size.

All four clusters demonstrate a band gap size dependence; therefore size effects are apparently more pronounced for the DOS. It is possible to distinguish the surface and bulk states for the two largest clusters (radii of 1.7 nm and 2.2 nm). Analysis of the local density of states shows that the lowest unoccupied energy states are located at the surface of the clusters. The vertical lines in the vicinity of the highest unoccupied state in Figs. 4 (c) and (d) mark the transition between the unoccupied surface states and bulk conduction band (CB) states, where the latter is defined by an appreciable local density of states spread into the cluster interior. For the 2.2 nm cluster the position of the bulk CB edge is only 0.09 eV higher than the true CB edge of bulk diamond. This leads to the conclusion that the band gap size effect is insignificant itself for cluster sizes larger than about 2-2.5 nm. This conclusion, however, disagrees with the results of Ref. [7] in which an X-ray photoluminescence analysis was used to determine the CB edge position. For a 3.6 nm cluster the CB edge position was measured to be 1.2 eV above the CB edge for bulk diamond.

Because both the potential and the band structure in the 2.2 nm cluster approach bulk values it is possible to use the potential value inside the cluster to estimate the electron affinity.

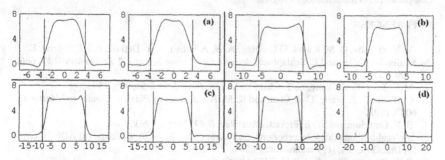

Figure 5. Coulomb potential distribution for the hydrogen passivated clusters. Sections (a-d) refer to the 34, 161, 435, and 913 carbon atom clusters, respectively. The left and right parts of each section correspond to the potential distribution along the (100) and (111) directions, respectively. Vertical lines mark the hydrogen atom positions on the cluster surface. Distance is measured in angstroms, potential is in Volts.

To do so the absolute position of the valence band edge in bulk diamond, the potential inside the cluster, and the band gap size must be added. The former can be obtained using the DFT results [4c], and equals -9.45 eV if measured relative to the vacuum level. After addition of the 5.81 eV potential we get the valence band position in a hydrogen passivated cluster. Addition of the bulk band gap results in the position of the conduction band edge with respect to the vacuum level: -9.45 + 5.81 + 5.47 ≈ 1.8 eV. Because the CB edge is above the vacuum level, hydrogen passivated clusters exhibit negative electron affinities. This conclusion is consistent with the electron affinity values for (111) 1x1:H and (100) 1x1:2H (sym) diamond surfaces. The experimentally measured electron affinity for the former was taken from Ref. [8] and equals -1.3 eV. The difference between (100) 1x1:2H (sym) and (111) 1x1:H surface affinities is -1.16 eV as calculated using *ab initio* simulation in Ref. [9]. That results in a -2.46 eV electron affinity for the (100) 1x1:2H (sym) surface. Because (100) surface constitutes about 18% of the total cluster surface, the weighted average electron affinity is -2.46 x 0.18 - 1.3 x 0.82 = -1.51 eV.

CONCLUSIONS

Size effects on the electronic properties of carbon nano-clusters have been estimated using an EDTB model. The critical size for the band structure size-effect onset is estimated to be ~2.0-2.5 nm while the critical size for the Coulomb potential distribution size effect is ~1 nm. Hydrogen dipole states result in a steep potential rise of ~6 eV at the cluster surface. Hydrogen passivated carbon nanoclusters exhibit a -1.8 eV negative electron affinity. Hydrogen passivated surface states located ~1 eV below the bulk CB edge may result in an effective band gap that is smaller than in a bulk diamond.

ACKNOWLEDGEMENTS

DAA, OAS and DWB are supported by the Office of Naval Research and the NASA-Ames Computational Nanotechnology Program.

REFERENCES

1. V. V. Zhirnov, O. M. Kuttel, O Groning, A. N. Alimova, P. Y. Detkov, P. I. Belobrov, E. Maillard-Schaller, and L. Schlapbach, *Journal Of Vacuum Science & Technology B* 17, 666, 1999.
2. M. S. Tang, C. Z. Wang, C. T. Chan, K. M. Ho, *Phys. Rev. B* 53, 979, 1996.
3. C. H. Xu, C. Z. Wang, C. T. Chan and K. M. Ho, *Journal of Physics: Condensed Matter* 4, 6047, 1992.
4. B. N. Davidson and W. E. Pickett, *Phys. Rev. B* 49, 11253, 1993.
5. Molecular Simulations, Incorporated, Software: Cerius2 (v 4.0), Builders: (a) ADF / GGA, (b) Dmol3 / GGA, (c) Faststructure / LDA.
6. D.W. Brenner, *Phys. Rev. B* 42, 9458 (1990).
7. Y. K. Chang, H. H. Hsieh, W. F. Pong, M. H. Tsai, F. Z. Chien, P. K. Tseng, L. C. Chen, T. Y. Wang, K. H. Chen, D. M. Bhusari, J. R. Yang, and S. T. Lin, *Phys. Rev. Lett.* 82, 5377, 1998.
8. J. B. Cui, J. Ristein, and L. Ley, *Phys. Rev. B* 60, 16135, 1999.
9. M. J. Rutter and J. Robertson, *Phys. Rev. B* 57, 9241, 1998.

Diamond and Wide Bandgap

Mat. Res. Soc. Symp. Proc. Vol. 621 © 2000 Materials Research Society

Anomalous Field Enhancement in Planar Semiconducting Cold Cathodes from Spontaneous Ordering in the Accumulation Region

Griff L. Bilbro[1] and **Robert J. Nemanich**[2]
[1]Department of ECE, NC State University, Raleigh, NC 27695-7911
[2]Department of Physics, NC State University, Raleigh, NC 27695-8202

ABSTRACT

Wide band gap semiconductors exhibit a low electron affinity and may prove suitable for cold cathode applications. We introduce a simple closed-form analytic approximation for the stability of electrons in the electron accumulation layer of planar Low Electron Affinity (LEA) semiconducting cathodes. This analysis extends our previous results, which used Runge-Kutta numerical integration of the linearized equations of motion for the electric potential and quasi Fermi level. The model shows conditions in which the electrons in the accumulation layer form a two dimensional array of regions of higher and lower electron density. This instability could lead to field enhancement without surface roughness and could account for observed electron emission at low applied fields.

INTRODUCTION

Wide band gap semiconductors such as diamond, AlN, or BN can exhibit negative electron affinities depending on surface termination. However, n-type doping of these materials is difficult. In contrast, $Al_{1-x}Ga_xN$ alloys (x~0.5) can be doped n-type and can exhibit a low electron affinity of ~1 eV [1]. LEA-coated cold cathodes may lead to vacuum electron devices (VEDs) with unprecedented versatility and performance [2]. Robust electron emission with high current density has been predicted for graded aluminum gallium nitride that is undoped [3] or doped [4]. LEA cathodes that require neither field enhancement nor high temperature may lead to VEDs with micrometer grid-to-cathode distances, picosecond transit times, and terahertz operating frequencies.

In this paper we provide simple closed-form expressions for our previous numerical prediction that LEA cathodes may be unstable against certain three-dimensional perturbations of the electron accumulation layer [5]. The instabilities may lead to ordered regions of higher or lower electron densities, and these regions would result in field enhancement without surface morphology.

Our effort is motivated by recent characterizations of planar cathodes which emit more electrons at lower voltages than can be readily explained without field enhancement [6] according to the usual equilibrium theory [7].

MODEL

In a Cartesian coordinate system, consider a uniform n-type semiconductor filling the region $x \leq 0$ with vacuum in the region x>0 as shown in Figure 1.

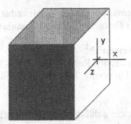

Figure 1. Coordinate system at the semiconductor/vacuum interface of a semiconductor filling the half space x < 0. The anode (not shown) is parallel to the interface at some x>0 in the vacuum.

Assuming complete ionization and Maxwell-Boltzmann statistics, the electron density for $x \leq 0$ is

$$n = N_C \exp\{ (E_{Fn} - E_C) / k_B T \}$$

where N_C is the effective density of states of the conduction band and E_{Fn} is the quasi Fermi level for electrons. E_{Fn} differs from the bulk equilibrium Fermi level E_F by

$$\phi = (E_{Fn} - E_F) / q \qquad (1)$$

measured in volts. The minimum of the conduction band E_C can be defined to be zero in the bulk, so that

$$\psi = - E_C / q$$

is the usual electric potential in volts. Then

$$n = N_D \exp\{ (\psi + \phi) / V_T \}$$

where $V_T = k_B T / q$ is the usual thermal voltage. The deviations ψ and ϕ obey Poisson's equation $\nabla^2 \psi = (q / \varepsilon_S)(N_D - n)$ which is now

$$\nabla^2 \psi = (q N_D / \varepsilon_S) [\exp\{(\psi + \phi) / V_T\} - 1]$$

in standard units where ε_S is the dielectric permittivity of the semiconductor and $q > 0$ is the fundamental charge. The deviations also conserve charge

$$-q \, \partial n / \partial t + \nabla \bullet \mathbf{J} = 0$$

where the vector current density due to drift and diffusion is $\mathbf{J} = q \mu n \mathbf{F} + q D \nabla n$, which simplifies to

$$\mathbf{J} = q \mu n \nabla \phi,$$

if the electron mobility μ and the diffusion coefficient $D = \mu V_T$ are treated as independent of electric field

$$F = -\nabla \psi.$$

The divergence of J can then be obtained from the product rule

$$\nabla \bullet J = q \mu (n \nabla^2 \phi + \nabla n \bullet \nabla \phi)$$

and, since $\partial n / \partial t = (n/V_T) \partial/\partial t (\psi + \phi)$, the continuity equation can be written as

$$\mu V_T (\nabla \bullet N_D \exp\{ (\psi + \phi) / V_T \} \nabla \phi) = \partial/\partial t \; N_D \exp\{ (\psi + \phi) / V_T \}$$

which can be simplified by performing $\partial/\partial t$ and factoring out the exponential to obtain

$$\mu V_T \{ \nabla^2 \phi + \nabla(\psi + \phi) \bullet \nabla \phi / V_T \} = \partial/\partial t (\psi + \phi)$$

in standard units. We will measure lengths in units of the extrinsic Debye length

$$L_D = [\; \varepsilon_S V_T / q N_D \;]^{1/2},$$

measure times in units of the dielectric relaxation time for electrons in the bulk

$$\tau_R = \varepsilon_S / q \mu N_D,$$

and measure voltages in units of V_T, so that Poisson's equation becomes

$$\nabla^2 \psi = \exp(\psi + \phi) - 1 \tag{2}$$

and the continuity equation becomes

$$\nabla^2 \phi + \nabla(\psi + \phi) \bullet \nabla \phi = \partial/\partial t (\psi + \phi). \tag{3}$$

PERTURBATION

We are interested in the stability of the electron accumulation layer against perturbations, so we write the electric potential and the quasi Fermi level as the sums

$$\psi = \psi_0 (x) + \delta\psi(x,y,z,t)$$

and

$$\phi = \phi_0 (x) + \delta\phi (x,y,z,t)$$

of equilibrium parts and perturbations. But the quasi Fermi level Eqn 1 coincides everywhere with the bulk E_F for any equilibrium state regardless of its stability, so $\phi_0 = 0$.

Substituting these expressions into Equations 2 and 3 yields a zeroth order equation for the usual one-dimensional equilibrium

$$\psi_0'' = \exp(\psi_0) - 1$$

and a first order system of equations for the perturbations

$$\nabla^2 \delta\psi = \exp(\psi_0) \ (\delta\phi + \delta\psi)$$

and

$$\nabla^2 \delta\phi + \psi_0' \ \delta\phi' = \partial/\partial t \ (\delta\phi + \delta\psi)$$

where prime (′) indicates ordinary differentiation with respect to x.

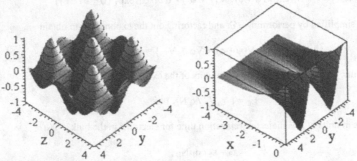

Figure 2. The (real part of the) perturbation $\delta\psi = \psi_1 \exp(ik_x x + ik_y y + Kz + st)$ of the electric potential of the equilibrium accumulation layer at t=0 for $\psi_1 = 1$. *Left:* parallel to interface: a slice in the yz plane through x=0, and *Right:* normal to interface: a slice through the xy plane through z=0.

ANALYTIC APPROXIMATION

The x-dependence of the coefficients $\exp(\psi_0)$ and ψ_0' preclude simple analysis of the first-order system. However the perturbations are confined near the surface where we replace the accumulated electron density with a constant effective density typical of the thin region near the surface

$$\rho \leftarrow \exp(\psi_0)$$

and we replace the slope ψ_0' of the zeroth order potential with a constant effective slope typical of the region

$$\sigma \leftarrow \psi_0'.$$

We estimate the region of validity for this approximation as the x for which $\rho + \sigma x > 0$:

$$-\rho /\sigma < x < 0. \tag{4}$$

In this region of x, the first order equations of motion for perturbations becomes

$$\nabla^2 \delta\psi = \rho \ (\delta\phi + \delta\psi)$$

and

$$\nabla^2 \delta\phi + \sigma \, \delta\phi' = \partial/\partial t \, (\delta\phi + \delta\psi) .$$

We consider perturbations of the form shown in Figure 2

and

$$\delta\psi = \psi_1 \exp \{ ik_x x + ik_y y + Kz + st \} \qquad (5)$$

$$\delta\phi = \phi_1 \exp \{ ik_x x + ik_y y + Kz + st \}. \qquad (6)$$

where ψ_1 and ϕ_1 are arbitrarily small constants, which are in general complex if there is a phase difference between the two perturbations.

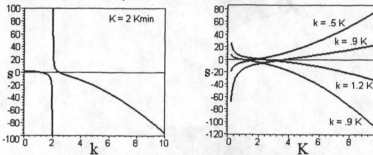

Figure 3. The approximate rate of growth in time of perturbations of wave number k parallel to the cathode surface and spatial decay rate K normal to the cathode surface for the zeroth-order equilibrium $\rho = \exp(1) \approx$ 2.7, $\sigma = \rho/2 \approx 1.4$, so that the equilibrium electric potential at the surface is $\psi_0(0) = k_B T \exp(1) \approx 68$ mV relative to the bulk. *Left:* The rate of growth s(k,K) for fixed K = 1. *Right:* The rate of growth s(k,K) for four ratios k/K which shows that instabilities may occur in two regimes: large K with low wave number k<K and small K with high wave number k>K. All curves satisfy the requirement for validity since $K_{min} = 0.5 <$ K < $15K_{min} = 7.5$ here.

We can expect the modes modeled by Eqns 5 and 6 to be weakly coupled for K is large enough to confine the perturbations to the region Eqn 4, so that perturbations with

$$K > \sigma/\rho ,$$

satisfy

$$(-k^2 + K^2) \, \delta\phi = \rho \, (\delta\phi + \delta\psi)$$

and

$$(-k^2 + K^2 + \sigma K) \, \delta\psi = s \, (\delta\phi + \delta\psi)$$

which can be solved for their rate of growth in time

$$s = -(k^2 - K^2 + \rho)(k^2 - K^2 - \sigma K)/(k^2 - K^2) . \qquad (7)$$

The one-dimensional equilibrium is stable against any excitation with s<0 since any such perturbation decays exponentially in time back toward the zeroth order equilibrium. On the other hand, modes with positive s grow in time rather than decay so that the one-dimensional zeroth-order equilibrium is unstable against any such perturbation.

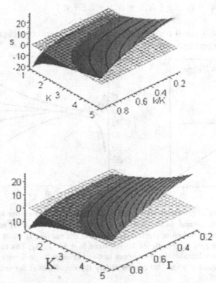

Figure 4. The approximate rate of growth in time of perturbations as a function of spatial decay rate K and the ratio k/K for modes that include the large-K-small-k instabilities shown on the right hand side of Figure 3b for the same ρ =2.7 and σ =1.4. The plane s=0 has been inserted as a grid.

RESULTS

Eqn 5 predicts positive values of s in some regions of (k, K) space. Figure 3a is a plot of s as a function of K for the case ρ = 2.7, σ =1.4, and K=1. In Figure 3a, perturbations with wave number k < 1 or with k > 1.54 decay in time but perturbations with K<k<1.53 grow if they are excited. Thermal fluctuations will excite all modes. For $Al_{0.5}Ga_{0.5}N$ doped with silicon at N_D= 10^{24} m $^{-3}$, the electron affinity is χ=1 eV, the donor ionization is small, the dielectric permittivity is ε_S = 8.85 ε_0, and the electron mobility is μ= 30 V cm^{-2} s^{-1} for high quality thin films currently reported, as previously inferred from other literature [4]. In this case L_D = 3.55 nm and τ_R= 0.16 ps at room temperature. For Figures 3 and 4 the applied electric field is F_0=1.4 $\varepsilon_S/\varepsilon_0$ in units of VT/L_D and F_0=87 V/μm in standard units. In this case, we find that the electron accumulation layer may be unstable against variations of a few nanometers wavelength in electron density along the surface.

The regions of high electron density would result in field enhancement and a lowering of the effective barrier for field emission.

REFERENCES
[1] M.C. Benjamin, M.D. Bremser, T.W. Weeks, Jr., S.W. King, R.F. Davis, and R.J. Nemanich. Applied Surface Science **104/105**, 455-460 (1996); Presented at Fifth international Conference on the Formation of Semiconductor Interfaces, Princeton, NJ, June 26-30, 1995.
[2] J. L. Shaw, H. F. Gray, K. L. Jensen, and T. M. Jung, J. Vac. Sci. Technol. **B 14**, 2072 (1996).
[3] K. L. Jensen, J. E. Yater, E. G. Zaidman, M. A. Kodis, and A. Shih, J. Vac. Sci. Technol. **B 16**, 2038 (1998).
[4] C. W. Hatfield and G. L. Bilbro, J. Vac. Sci. Technol. **B 17**, 552 (1999).
[5] G. L. Bilbro and R. J. Nemanich, Appl. Phys. Lett. **Vol. 76** No. 7, 891 (2000).
[6] A. T. Sowers, B. L. Ward, S. L. English, and R. J. Nemanich, J. Appl. Phys. **86**, 3973 (1999).
[7] R. Stratton, Proc. Phys. Soc. (London) **B46**, 746 (1955).

Mat. Res. Soc. Symp. Proc. Vol. 621. © 2000 Materials Research Society

Thermionic FEEM, PEEM and I/V Measurements of N-Doped CVD Diamond Surfaces

F.A.M.Köck, J.M. Garguilo, B. Brown, R.J. Nemanich
Department of Physics
North Carolina State University
Raleigh, NC 27695-8202
USA

Abstract

Imaging of field emission and photoemission from diamond surfaces is accomplished with a high resolution photo-electron emission microscope (PEEM). Measurements obtained as a function of sample temperature up to 1000°C display thermionic field emission images (T-FEEM). The system can also record the emission current versus applied voltage. N-doped diamond films have been produced by MPCVD with a N/C gas phase ratio of 48. The surfaces display uniform emission in PEEM at all temperatures. No FEEM images are detectable below 500°C. At ~680°C the T-FEEM and PEEM images are nearly identical in intensity and uniformity. This is to be contrasted with other carbon based cold cathodes in which the emission is observed from only a low density of highly emitting sites. The I/V measurements obtained from the N-doped films in the T-FEEM configuration show a component that depends linearly on voltage at low fields. At higher fields, an approximately exponential dependence is observed. At low temperatures employed (<700°C), the results indicate a thermionic component to the emitted current.

I. Introduction

With its unique properties diamond is a promising material for mechanical and electrical applications. Hardness, high thermal conductivity, high carrier mobilities and saturated carrier velocities, wide bandgap, radiation hardness and chemical inertness promise diamond to be superior than other materials for specific electronic devices. Recently, there is interest in the electron emission properties of diamond for use as an electron source. Part of this interest is based on the observation that diamond surfaces can exhibit a negative electron affinity (NEA). This enables electrons from the conduction band minimum to escape the diamond without an energy barrier at the surface. The NEA can be achieved by hydrogen passivation of the diamond surface. To determine whether the surface exhibits a NEA, one can employ photoexcitation of the electrons into the conduction band minimum and observe if electron emission occurs [1,2].

The technique of photo electron emission microscopy (PEEM) can be used to image the electron emission properties of a surface in a controlled UHV environment. In addition, the same apparatus can be employed to obtain field electron emission microscopy (FEEM) of the surfaces.

It is well known that substitutional boron atoms act as shallow acceptors in diamond with an energy level of 0.37eV above the top of the valence band. Another dominant impurity in natural diamond is nitrogen. Single substitutional nitrogen is a deep donor at 1.7eV below the bottom of the conduction band.

The approach of this study is to image the electron emission properties of diamond in both PEEM and FEEM modes as a function of sample temperature up to 900°C. The FEEM mode at high temperature involves thermionic field emission processes, and we have termed this imaging condition as thermionic-field electron emission microscopy (T-FEEM).

In this work we have found significant differences in the thermal emission properties of B- and N-doped diamond films. With increasing temperature N-doped diamond films exhibit strongly enhanced field emission, whereas B-doped diamond shows very low field emission that remains nearly constant over a wide temperature range.

II. Experimental Details:

N-doped diamond films were grown on 25mm diameter, $1\Omega\cdot$cm Si <100> substrates. Sample preparation included: 10min. hand polishing using 0.1μm diamond powder, ultrasonic cleaning in methanol, and drying with nitrogen gas. The CVD reactor is a 1500W ASTeX IPX3750 microwave assisted CVD system with an RF induction heated graphite susceptor. The temperature was calibrated with an optical pyrometer. The process gases were zero grade N_2, H_2 and CH_4. The gas flows were monitored and controlled with mass flow controllers. During film growth laser reflectance interferometry (LRI) was used to in situ monitor the thickness of the diamond layer.

The growth of N-doped diamond film was divided into three steps: (1) establishment of the nucleation layer, (2) growth of the N-doped film, and (3) post growth surface treatment. The conditions for the nucleation layer were 400 sccm H_2, 8 sccm CH_4, chamber pressure of 20 Torr, substrate temperature of 780°C and a microwave power of 600 W. After formation of the nucleation layer the conditions were changed to 437sccm H_2, 2.5sccm CH_4, 60 sccm N_2, pressure of 50 Torr, substrate temperature of 910°C and microwave power of 1300 W. By growing for 4 to 8 hours, films with thicknesses of 0.5μm and 1μm were fabricated. Growth was terminated by shutting off the gas flows except H_2, reducing the microwave power to 600W and reducing the substrate temperature to 780°C. The diamond film was then treated with H_2 plasma for 5 min at a pressure of 20 Torr.

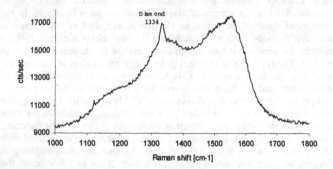

Figure 1 *Raman Spectroscopy of a nitrogen doped diamond film showing a broadened diamond peak and a spectral components attributed to sp^2 bonding. The spectrum was taken at 514.5nm at 300K.*

The PEEM and FEEM measurements were completed in an Elmitec UHV-Photo Electron Emission Microscope [3]. Measurements were obtained at a base pressure less than 3×10^{-10} Torr. The field of view can be changed from 150μm to 1.5μm with a resolution better than 15 nm. For all measurements a high voltage of 20 kV is applied between the anode and the sample surface, and the anode is a distance of 2 mm from the surface. In the PEEM measurements a mercury arc lamp was used as the UV-light source. In the FEEM measurements no UV excitation was employed and the emission was due to the high applied field.

The system has sample heating up to 1200°C to obtain T-FEEM images. The electron emission current from the sample surface can be monitored and recorded to obtain current-voltage dependence. Electrons emitted from the sample pass through a perforated anode and are imaged using electron optics. The image is intensified with a double microchannel plate and a fluorescent screen. A CCD camera is used for image capturing. The gain of the system is dependent on the voltage on the image intensifier. The images reported here have been digitally processed to remove dark regions of the intensifier.

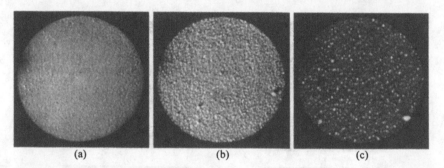

(a)　　　　　　　　　(b)　　　　　　　　　(c)

Figure 2 PEEM images obtained at room temperature with Hg-arc lamp excitation using a field of view and channel plate voltage of: (a) 150μm, 1.476kV, (b) 50μm, 1.596kV and (c) 20μm, 1.627kV. A higher channel plate voltage is required to image at higher magnification.

III. RESULTS and DISCUSSION

The influence of nitrogen on the growth and properties of diamond films has been reported in several articles [4-8]. With increasing nitrogen content in the growth chamber, increased sp^2 bonding signatures are observed and the increased FWHM of the 1332 cm^{-1} diamond peak also increases significantly. Displayed in Figure 1, the Raman scattering from the 0.5μm diamond film shows a broad spectral background extending from 1100 to1600cm^{-1}. The spectral components attributed to sp^2 bonded carbon include the 1355 – 1580 cm^{-1} doublet due to microcrystalline graphite, and the broad peak centered near 1500 attributed to sp^2 bonding in or around diamond crystals. A feature at ~1140 cm^{-1} has been attributed to disordered sp^3 bonded carbon, and the broadening of the 1332 cm^{-1} diamond peak indicates small grain size and the presence of defects or impurities [9]. It should be noted that Raman spectroscopy is about 75 times more sensitive to sp^2 bonded carbon sites than it is to crystalline diamond when 514.5nm excitation is used. It is evident that the film has a significant diamond crystal component but it also exhibits a distribution of sp^2 and the variations in bonding are on a nanometer scale.

Shown in Fig. 2 are PEEM images of different magnification of the N-doped diamond film at room temperature. The images show that the surface exhibits a uniform electron emission over the whole sample surface. The images show a fine textured grainy structure that corresponds to the morphology observed by SEM. The brighter spots in the image of the highest magnification are attributed to emission from the most raised points on the surface. These regions will have the highest field. Images resulting from PEEM measurements can be understood in terms of photoelectron emission from NEA diamond [10,11]. The photo excited electrons in the conduction band of the diamond are emitted into vacuum following the theory of photoemission. Note that the emission is uniform over the surface with no evidence of spot emission often observed in other carbon based cold cathode materials.

This uniform emission is indicative of a relatively uniform surface barrier and of a uniform carrier distribution near the surface. In the following we suggest that observation of uniform emission may be indicative of emission from the conduction band

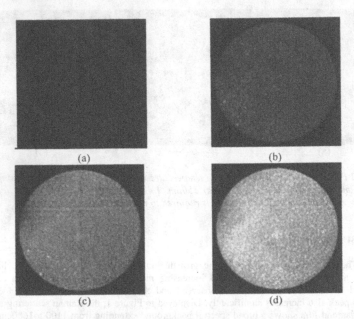

Figure 3 *T-FEEM images of N-doped diamond at 640°C (a), 680°C (b), 700°C (c) and 720°C (d). All images were obtained at a channel plate voltage of 1.55kV, and the field of view is 20μm.*

By turning off the Hg-arc lamp and increasing the sample temperature T-FEEM of the nitrogen doped diamond film can be observed. T-FEEM images were taken at various temperatures but at a constant channel plate voltage. Figure 3 (a), (b), (c), (d) shows T-FEEM images with 20µm field of view at a channel plate voltage of 1.55kV, corresponding to substrate temperatures of 640°C, 680°C, 700°C and 720°C respectively. From the T-FEEM images in Figure 3 (a), (b), (c) and (d) a very uniform electron emission can be observed that increases with increasing temperature. It is evident that the FEEM is very temperature dependent with a strong increase above 640°C. Similar measurements on B-doped, p-type diamond films do not display observable FEEM images below temperatures of 800°C. We have also repeated the measurements on a Mo plate with similar surface roughness to the diamond. Here again FEEM was observable only for temperatures greater than 900°C. It is evident that the N-doped diamond films exhibit significant emission at relatively low temperatures.

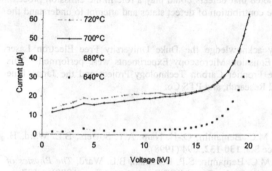

Figure 4 *The emission current vs. applied voltage obtained in the electron emission microscope for the N-doped diamond film. Note that the baseline signal increases with temperature.*

The uniformity of the emission suggests that the electrons may originate from the conduction band of the diamond. Alternatively, it is possible that the emission originates from defect states in the band gap. If this is the case then the uniformity of the emission would require that the defects are uniformly distributed throughout the film. For each sample temperature of 640°C, 680°C, 700°C and 720°C an I/V curve was recorded. Figure 4 shows a set of I/V curves obtained at various temperatures. At low fields we find a nearly constant value of the electron emission current. Increasing the anode voltage beyond 15kV results in an exponential behavior, which dominates the emission current characteristics. Electrical conductivity measurements on type IIa natural diamond, that has nitrogen as the dominant impurity, shows an exponential temperature dependence of the carrier density, and about two orders of magnitude of change that is attributed to changes in the mobility. For type IIa natural diamond S. Han *et al.* [12] report an activation energy of 1.4eV that likely reflects the excitation of electrons from nitrogen levels.

The emission characteristics at elevated temperatures suggest a thermionic contribution to the electron emission current. It appears that the emission at low fields could be attributed to thermionic emission of electrons in the conduction band of the diamond. The exponential increase of the current at higher fields indicates a tunneling process. This process may be related to electron emission at the film surface or to supply of electrons at the film substrate interface. Future research will explore these issues.

IV. CONCLUSIONS

We have grown nitrogen doped diamond films by plasma assisted CVD. Raman spectroscopy shows a broadened diamond peak and the presence of sp2 bonding in the film. PEEM investigation of the NEA nitrogen doped diamond films shows uniform electron emission from the surface at room temperature. At temperatures below 500°C FEEM does not resolve electron emission from the specimen. Increasing the sample temperature up to 720°C we find a strong increase in the electron emission suggesting that electrons from the nitrogen donor levels are excited into the CB. It is also noted that defects could play a role in the emission process. Further investigation will explore the contribution of defect states and attempt to understand the thermionic emission properties.

Acknowledgements: We gratefully acknowledge the Duke University Free Electron Laser Laboratory where all the Electron Emission Microscopy Experiments were performed. This research was supported through the Frontier Carbon Technology Project and the Japan Fine Ceramics Center, the Office of Naval Research, and PTS Co.

References

1. R.J. Nemanich, P.K. Baumann, M.C. Benjamin, O.-H. Nam, A.T. Sowers, B.L. Ward, H. Ade and R.F. Davis, Appl. Surface Sci. **130-132**, 694 (1998).
2. R.J. Nemanich, P.K. Baumann, M.C. Benjamin, S.P. Bozeman, B.L. Ward, *The Physics of Diamond,* Proc. of Int. School Physics Enrico Fermi (IOS Press, Amsterdam, 1977) p. 537.
3. H. Ade, W. Yang, S.L. English, J. Hartman, R.F. Davis and R.J. Nemanich, Surface Rev. and Lett. **5**, 1257 (1998).
4. V. Baranauskas, B.B. Li, A. Perlevitz, M.C. Tosin, S.F. Durant, J. Appl. Phys. **85**, 7455 (1999).
5. N. Jiang, A. Hatta and T. Ito, Jpn. J. Appl. Phys. **37**, 1175 (1998).
6. I.T. Han and N. Lee, J. Vac. Sci. Technol. B **16**, 2052 (1998).
7. W. Müller-Sebert, E. Wörner, F. Fuchs, C. Wild, P. Koidl, Appl. Phys. Lett. **68**, 759 (1996).
8. A.T. Sowers, B.L. Ward, S.L. English and R.J. Nemanich, J. Appl. Phys. **86**, 3973, (1999).
9. M. Park, A.T. Sowers, C. Lizzul Rinne, R. Schlesser, L. Bergman, R.J. Nemanich, Z. Sitar, J.J. Hren and J.J. Cuomo, J. Vac. Sci. Technol. B **17**, 734 (1999).
10. F.J. Himpsel, J.A. Knapp, J.A. van Vechten and D.E. Eastman, Phys. Rev. B **20**, 624 (1979)
11. J. van der Weide, Z. Zhang, P.K. Baumann, M.G. Wensell, J. Bernholc and R.J. Nemanich, Phys. Rev. B **50**, 5803 (1986).
12. S. Han, L. S. Pan, D. R. Kania, *Dynamics of Free Carriers in Diamond, Diamond, in Electronic Properties and Applications*, L. S. Pan, D. R. Kania Eds. (Kluwer Academic, New York, 1995) p. 242.

Mat. Res. Soc. Symp. Proc. Vol. 621. © 2000 Materials Research Society

CAPACITANCE AND TRANSIENT PHOTOCAPACITANCE STUDIES OF TETRAHEDRAL AMORPHOUS CARBON

KIMON C. PALINGINIS*, A. ILIE**, W.I. MILNE** and J. DAVID COHEN*
*Department of Physics, University of Oregon, Eugene, OR 97403 U.S.A.
**Engineering Department, University of Cambridge CB2 1PZ, U.K.

ABSTRACT

We have applied junction capacitance and transient photocapacitance measurements to undoped tetrahedral amorphous carbon (ta-C)/silicon carbide (SiC) heterostructures to deduce defect densities and defect distributions in ta-C. The junction capacitance measurements show two thermally activated processes. One can be related to the activation of carriers out of defects at the ta-C/SiC interface while the other one with an activation energy of 0.36eV is an intrinsic property of the ta-C. The defect density at the ta-C/SiC interface is estimated to be roughly $9 \pm 2 \times 10^9$ cm^{-2}. The transient photocapacitance measurements have allowed us to observe the broader band tail of ta-C, giving a value (Urbach energy) of 230meV.

INTRODUCTION

With optimum growth conditions, thin tetrahedral amorphous carbon (ta-C) films prepared by the filtered cathodic arc system (FCVA) method have fractions of sp^3 bonds up to 80% [1]. However, the electronic properties are greatly influenced by the residual sp^2 bonds, which, in particular lead to an optical band gap of 2 to 2.5eV [2]. Undoped ta-C is a p-type semiconductor with dark conductivities in the range of 10^{-7} to 10^{-8} Ω^{-1}cm^{-1} [3]. Films can be n-type doped by the incorporation of nitrogen or phosphorous [4] with decreases in the resistivity up to five orders of magnitude. Particular properties of ta-C, such as its significant electron emission at a low electric field threshold as well as its chemical inertness and high hardness [5] have made it an interesting material for applications such as flat-panel displays or large area coatings. However, due to difficulties in applying standard spectroscopic measurements to such thin films (< 1000Å), detailed knowledge about its electronic structure is still strongly debated. Recently we have succeeded in applying transient junction capacitance with other junction capacitance techniques to study ta-C/c-Si and ta-C:H/c-Si heterostructures [6,7]. However, because these thin ta-C films were grown on a substrate with a much smaller energy gap, this precluded being able to observe the optical transitions in the ta-C.

In this paper we report new results of transient photocapacitance measurements on a ta-C/SiC heterostructure. The high band gap of SiC (~ 3eV) provides a sufficiently wide energy window to observe optical transitions directly in the bulk of the ta-C film. In this study the sample consisted of a ta-C film of thickness less than 600Å deposited on a n-type doped silicon carbide substrate. Before we applied the photocapacitance technique, we characterized the ta-C/SiC heterostructure using admittance spectroscopy. We found that carriers in the bulk of the ta-C are thermally activated with an energy of 0.36eV. We also deduced the density of defects at the ta-C/SiC interface to be roughly $9 \pm 2 \times 10^9$ cm^{-2}, much lower as we had found for the interface in ta-C/c-Si heterostructures [6].

EXPERIMENT

A 600Å thin ta-C film was deposited on a n-type SiC substrate with a doping density of 5.4 x 10^{17} cm^{-2}. The substrate temperature during growth was roughly 300K. We evaporated aluminum contact dots with an area of 0.01cm^2 onto the sample. A nearly ohmic contact was formed on the SiC backside by evaporating titanium. This structure enabled us to apply junction capacitance measurements to the ta-C/SiC interface. To characterize the ta-C/SiC heterostructure we first applied the admittance spectroscopy technique [8] which we had already used successfully in determining some of the electronic properties in ta-C/c-Si sample configurations [6]. Recording the steady state capacitance and conductance as a function of frequency and temperature allowed us to determine the Fermi level position in the ta-C and also to estimate the density of defects at the ta-C/SiC interface. To study the distribution of localized states in the mobility gap of the ta-C film we applied the transient photocapacitance technique [9]. The basic idea of such a measurement is shown in Fig. 1. We illustrate a situation for a semiconductor with a discrete deep defect distribution within the space charge region of the p-type side of the junction. Fig. 1 (a) displays the junction in equilibrium under a certain reverse bias. To establish a non-equilibrium occupation of gap states a so called "filling pulse" is applied to the heterojunction as shown in Fig. 1 (b). Once the filling pulse is removed and the initial reverse bias restored, the initial steady-state population is recovered through the emission of trapped holes (electrons) to the valence band (conduction band) with increasing time as can be seen in Fig. 1 (c). In the dark this process is solely controlled by thermal emission of trapped carriers. However, this process can be enhanced through optical excitation, which is the basis of the photocapacitance method. To obtain a single quantity that indicates the changes that occur over a specific time scale, we multiply the transient signal $C(t,T,\omega)$ with a correlator function $A(t)$ over a fixed time window τ:

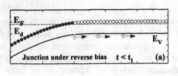

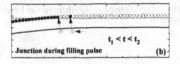

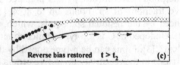

Fig.1. Basic idea of transient capacitance measurements demonstrated on a p-type semiconductor with a discrete defect level: (a) Junction under reverse bias in equilibrium showing occupied states (solid circles) below the Fermi level E_F and empty states (open circles) above E_F. (b) Situation during "filling pulse" which allows gap states to capture holes from the valence band. (c) Re-equilibration process after filling pulse is removed. Occupied states above E_F remain occupied until the holes are either thermally or optically excited to the valence band.

$$S(T,\tau,\omega) = \int_{t_1}^{t_2} A(t)C(t,T,\omega)dt \qquad (1)$$

In the photocapacitance measurement we record transient signals at a fixed temperature alternately in the dark and in the presence of sub-band-gap light [9]. The normalized photocapacitance signal P at constant temperature T is then defined as

$$P(E_{opt}, T) = \frac{S_{light}(T, \tau, \omega) - S_{dark}(T, \tau, \omega)}{\Phi(E_{opt})} \qquad (2)$$

with Φ being the photon flux at the optical energy E_{opt}. Since both S_{light} and S_{dark} contain the same contribution of thermally emitted charge, P describes purely the optically excited release of gap state carriers. Care is taken to operate at sufficiently low light levels such that strict linearity with light intensity is observed.

RESULTS AND DISCUSSION

Figure 2 shows a typical set of steady state capacitance and conductance versus temperature for five different frequencies. A 0.5V reverse bias applied to the ta-C/SiC interface. Several temperature regimes need to be discussed to understand the response of this heterostructure. At low temperatures there seems to be a frequency independent capacitance plateau; however, the magnification of this regime displayed in Fig. 3 shows clearly the existence of a thermal

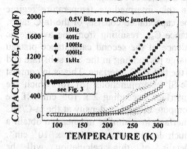

Fig. 2.Capacitance (solid symbols) and normalized conductance (open symbols) of the ta-C/SiC heterostructure under reverse 0.5V bias vs. temperature at five different frequencies.

activated process (< 140K). Since the lowest temperature employed during our measurements was 80K, we only can see the onset of a true frequency independent capacitance plateau (plateau I) towards lower temperatures as indicated by the capacitance values for 400Hz and 1kHz. We found that this lowest temperature plateau is bias dependent, varying from 695pF to 642pF and then to 615pF as the bias is changed from 0 to 0.5V and then to 1V. In general, the ac junction capacitance discloses the position of the first moment <x> of the charge responding to the applied ac voltage, $C = \varepsilon A/<x>$ [8]. In this lowest temperature region the activation of carriers in the ta-C is frozen out. Thus the value of the capacitance in this regime corresponds to the ta-C

layer of 600Å thickness acting as a static dielectric (C_{ta-C}) in series with the capacitance C_{SiC} resulting from the depletion region in the SiC substrate. We can easily calculate the depletion region width W in the SiC from the measured capacitance values C_{tot}:

$$W = \varepsilon A \left(\frac{C_{ta-C} - C_{tot}}{C_{ta-C} * C_{tot}} \right) \qquad (3)$$

where $\varepsilon = 11.7\varepsilon_0$ with ε_0 being the vacuum permittivity. Knowing the width of the space charge region in the SiC then allows us to estimate the interface potential at the SiC side of the heterojunction [10]. We obtained values of 0.2eV, 0.47eV and 0.69eV as the applied dc bias is changed from 0 to 1V.

The small frequency dependent capacitance step between plateaus I and II displayed in Fig. 3 indicates a shift in the first moment of the ac charge response in the SiC substrate closer to the

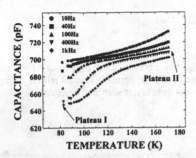

Fig. 3. Capacitance of the ta-C/SiC heterostructure vs. temperature at five different frequencies, magnifying the seemingly frequency independent capacitance plateau at low temperatures shown in Fig. 2. The capacitance step indicates a thermally activated process at low temperatures.

ta-C/SiC interface. In this temperature regime the ta-C layer is still acting merely as a static dielectric. We believe that this step results from the activation of carriers in and out of interface defect states at the ta-C/SiC heterojunction. As can be seen in Fig. 2 this small increase in the capacitance is followed by the frequency independent capacitance plateau II, which is also bias dependent. As the applied bias is changed from 0 to 1V, the capacitance value varies from 755pF to 700pF and then to 660pF (these values are the same for all frequencies but occur at different temperatures, e.g. for the 10Hz capacitance at 90K or for the 1kHz capacitance at 170K).

In this intermediate temperature regime the capacitance consists of the ta-C layer, acting as a static dielectric, in series with the capacitance C_{avg} resulting from the average ac charge response in the SiC substrate. The bias dependence of the second capacitance plateau reveals that C_{avg} is dominated by the contribution of charge responding in the SiC substrate and not by the carriers coming in and out of defect states at the ta-C/SiC interface. This is in contrast to the observation we made previously using ta-C/c-Si heterojunctions [6] for which we estimated a density of defect states at the ta-C/c-Si interface exceeding 2×10^{12} cm^{-2}. If we compute the defect density at the ta-C/SiC interface in a similar manner we obtain a much lower value: $9 \pm 2 \times 10^{9}$ cm^{-2}. Details of this calculation will be presented elsewhere.

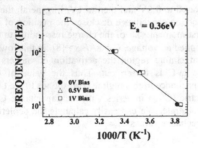

Fig. 4. Arrhenius plot of the frequency vs. inverse temperature for the middle of the large, higher temperature capacitance step. Different symbols present measurements carried out at different applied reverse biases to the ta-C/SiC junction.

Figure 2 displays that a further increase in the temperature results in a frequency dependent much large capacitance step. An Arrhenius plot of the temperature of the capacitance step as a function of frequency (determined by the corresponding peak of the conductance), shown in Fig. 4, indicates that this is a thermally activated process with an activation energy of 0.36eV. This activation energy is found to be independent of applied bias. Thus it results from the Fermi energy of the ta-C. The value obtained in this study is in good agreement to other studies [11].

In the last temperature regime, in which the ta-C conduction is activated, we can apply the transient photocapacitance technique to study the defect distribution in the energy gap of the ta-

C. An example of such a photocapacitance spectrum is shown in Fig. 5. The sample was held at 320K during the measurement and a 1kHz transient capacitance signal was recorded. In the photocapacitance measurement we monitor the change in the junction capacitance arising from the contraction of the depletion region as positive charge is lost from defect states. The photocapacitance method substracts the transient signal recorded in the dark from the transient signal recorded under light. The sign of the capacitance change tells us the dominant type of emitted carrier (electron or hole) while overall the resulting spectrum P [described by Eq. (2)] reveals the distribution of allowable optical transitions. The light intensity is kept low enough so that the electronic population of gap states is not significantly disturbed by the optical excitation. A typical spectrum obtained for the ta-C is shown in Fig. 4.

The transient photo-spectra can be interpreted in terms of a convolution between localized gap states and extended states connected by an optical matrix element for the transitions. Specifically, for transitions from the valence band to empty gap states one expects a contribution to P given by [12]

$$P(E_{opt}, T) = K(T) \int_{E_C-E_e}^{E_C} |\langle i | ex | f \rangle|^2 g(E) g_V (E - E_{opt}) dE \qquad (4)$$

where $|\langle i | ex | f \rangle|^2$ represents the optical matrix element, K is a temperature dependent constant and E_e is the thermal energy depth from which trapped carriers can be emitted during the experimental time window τ. A similar expression exists for the minority carrier transitions from filled gap states to the conduction band.

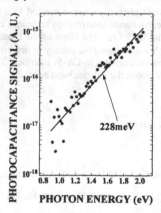

Fig. 5. Photocapacitance spectrum at 320K and 1kHz showing the sub-band-gap optical transition within the ta-C film. The ta-C/SiC junction was held under 1.4V reverse bias. A 1.2V filling pulse was applied for 200ms prior to recording the transient capacitance.

In contrast to typical photocapacitance spectra in a-Si:H [9], we could not detect a distinct band of deep defects in the ta-C. Let us first consider the simple model shown in Fig. 6 that could explain the observed spectrum for ta-C. A transition of type 1 will generally cause the capacitance to increase since it tends to increase negative charge density in the depletion region, while a transition of type 2 would have the opposite effect. However, in this p-type material the transport of holes out of the depletion region is likely to be faster than that for electrons emitted into the conduction band. Thus the transition of type 1 tends to dominate the photocapacitance signal. The sign of change in the junction capacitance under our experimental conditions confirms hole emission out of the depletion region to be the dominant effect. An exponential curve fit to the photocapacitance spectrum reveals an Urbach energy of 228 ± 20 meV. This will be the slope of the *broader* band tail in ta-C which follows from Eq. (4). The photocapacitance technique cannot itself determine the band tail revealed in Fig. 5 shown in the spectra is from the valence or conduction band. However, we tend to believe it is that of the conduction band as other studies have claimed [13].

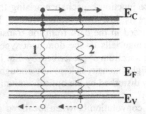

Fig. 6. Simple gap state model used to analyze the photocapacitance spectra. We only consider bandtail to conduction band tail or valence band bandtail related transitions as there seem to be no deep defects visible in the spectrum. Transition 2 appears to have little effect on the photocapacitance signal.

There might be several reasons for the absence of a distinct band in the photocapacitance spectrum due to deep defects. First one should note that our measurement is only sensitive for optical transitions above about 0.95eV in energy. This means that we are not able to see any part of the defect distribution in the energy gap of the ta-C lying closer than 0.95eV to the valence band. Since several of our earlier studies strongly suggested that the deep defect band lies close to the Fermi level, which is here determined to be around 0.36eV for this film, the photocapacitance measurement would not be able to detect it. Second, the existence of the broad band tail might overwhelm any more distinct band of defect states. That is, if the density of deep defects were relatively small and had a transition energy lying within the range of those from the band tail, one would not be able to see it in the spectrum.

CONCLUSIONS

We have applied admittance spectroscopy and the transient photocapacitance method to a ta-C/SiC heterostructure. We showed that the transient photocapacitance spectroscopy technique can be applied to study part of the defect distribution in a thin ta-C film. The slope of the broader conduction band tail was determined to be 228 ± 20meV. We also found the density of defects at the ta-C/SiC interface to be two orders of magnitude smaller than at a ta-C/c-Si interface. The Fermi energy in the ta-C film was determined to lie 0.36eV above the valence band edge.

ACKNOWLEDGEMENTS

Work at Oregon was supported by NSF Grant No. DMR 9624002.

REFERENCES

[1] M. Chhowalla, J. Robertson, C. W. Chen, S. R. P. Silva, C. A. Davis, G. A. J. Amaratunga, and W. I. Milne, J. Appl. Phys. **81**, 139 (1997)
[2] R. Lossy, D. L. Pappas, R. A. Roy, J. J. Cuomo, and V. M. Sura, Appl. Phys. Lett. **61**, 171 (1992)
[3] W. I. Milne, J. Non-Cryst. Solids **198-200**, 605 (1996)
[4] V. S. Veerasamy, J. Yuan, G. A. J. Amaratunga, W. I. Milne, K. W. R. Gilkes, M. Weiler, and L. M. Brown, Phys. Rev. B **48**, 17954 (1993)
[5] B. S. Satyanarayana, A. Hart, W. I. Milne, and J. Robertson, Appl. Phys. Lett. **71**, 1430 (1997)
[6] K. C. Palinginis, Y. Lubianiker, J. D. Cohen, A. Ilie, B. Kleinsorge, and W. I. Milne, Appl. Phys. Lett. **74**, 371 (1999)

[7] K. C. Palinginis, J. D. Cohen, A. Ilie, N. M. J. Conway, and W. I. Milne, to be published in J. Non-Cryst. Solids

[8] D.V. Lang, J. D. Cohen, and J. P. Harbison, Phys. Rev. B **25**, 5285 (1982)

[9] A. V. Gelatos, K. K. Mahavadi, J. D. Cohen, and J. P. Harbison, Appl. Phys. Lett. **53**, 403 (1988)

[10] S. M. Sze in Physics of Semiconductor Devices 1st Ed. (John Wiley & Sons, New York, 1969), p. 84 – 96

[11] V. S. Veerasamy, G. A. J. Amaratunga, C. A. Davis, W. I. Milne, P. Hewitt, and M. Weiler, Solid-St. Electron. **37**, 319 (1994)

[12] J. D. Cohen and A. V. Gelatos in Amorphous Silicon and Related Materials, edited by H. Fritzsche (World Scientific Publishing Company, 1988), p. 475 – 512

[13] N. M. J. Conway, A. Ilie, J. Robertson, W. I. Milne, and A. Tagliaferro, Appl. Phys. Lett. **73**, 2456 (1998)

Mat. Res. Soc. Symp. Proc. Vol. 621 © 2000 Materials Research Society

Field emission from heterostructured nanoseeded diamond and nanocluster carbon cathodes.

B.S.Satyanarayana[1*], K.Nishimura [2], A.Hiraki [3], W.I.Milne [1].

[1]Electronic Devices & Materials Group,Engineering Dept, University of Cambridge, Cambridge .UK
[2] KUT Academic & Industrial Collaboration Centre, Kochi University of Technology, Kochi, 782-8502, Japan.
[3] Kochi Prefectural Industrial Tech. Center, 3992-3, Nunoshida, Kochi, 781-5101, Japan.

ABSTRACT

Novel heterostructured cold cathodes made of nanoseeded diamond and cathodic arc process grown nanocluster carbon films, were studied. The nanocrystalline diamond with varying diamond concentration was first coated on to the substrate. The nanocluster carbon films were then deposited on the nanoseeded diamond coated substrates using the cathodic arc process at room temperature. The resultant heterostructured microcathodes were observed to exhibit electron emission currents of $1\mu A/cm^2$ at low fields of 1.2 - 5 V/μm. Further some of the samples seem to exhibit I-V characteristics with a negative differential resistance region at room temperature conditions. This negative differential resistance or the resonant tunneling behaviour was observed to be dependent on the nanoseeded diamond concentration.

INTRODUCTION

There is an increasing interest in field assisted electron emission from carbon based nanostructured materials. Recently low field electron emission has been reported using many nanostructured materials including nanocrystalline diamond,[1-3] nanotubes[4,5] and nanocluster carbon[6-9]. The interest stems from the diverse applications such as electron-beam lithography, electron and ion guns, sensors, electron microscopes and microprobes, low & medium power compact microwave sources and Tera hz communication devices being envisaged for these electron emitters, besides field emission displays. The need is for electron emitters capable of emitting high emission currents at low fields accompanied by a high emission site density. Further the material and the growth process should be adaptable to varying application needs. Most of the low field electron emitters reported, have been grown using high temperature processes such as hot filament CVD, microwave plasma CVD, Plasma assisted DC discharge and cathodic arc process for nanotube growth.[1-5,8] Only the coral like carbon films[6] and the nanocluster carbon films [7] grown using the cathodic arc process have been deposited at room temperature.

For a better idea, let us compare the electron emission from three different nanostructured carbon based materials. Figure-1 shows the current density against applied field plot for these different cathode materials. Figure 1 includes emission characteristics from a nanocluster carbon film grown using cathodic arc,[7] nano structured carbon films produced by supersonic carbon cluster beams generated by a pulsed plasma cluster source [9] and the nano-crystalline diamond films grown using the hot filament CVD [HFCVD].[3] It can be seen that all the samples exhibit electron emission currents of $1\mu A/cm^2$ at fields as low as 1V/μm. However the nanocluster carbon grown using cathodic arc exhibits slightly lower current than cluster

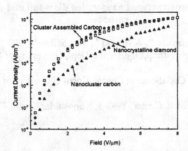

Figure -1 *Comparison of field emission from different nano/micro structured carbon films.*

assembled carbon and nanocrystalline diamond. The diamond films were grown at 825 C while both types of carbon cluster films were grown at room temperature. The cluster-assembled carbon films were observed to be relatively soft. The nanocluster carbon film grown using the cathodic arc process, were observed to be hard and adhered well to the substrate.

With the motivation to further enhance the current density and the emission site density in the case of arc grown nanocluster carbon films, the nanocluster carbon films deposited on silicon substrates precoated/seeded with nanocrystalline diamond (3-6nm) were studied. Seeding of silicon substrates with pure nanocrystalline diamond (3-6nm) particles as a pretreatment for high density diamond crystal growth [10,11] and non diamond nanostructured carbon film growth[8] have been reported earlier. We report for the first time, a study on field emission from a heterostructured cathode consisting of nanoseeded diamond and nanocluster carbon grown using the cathodic arc process.

EXPERIMENTAL CONDITIONS :

Highly conducting n++ silicon substrates were first coated or pretreated with the purified nanocrystalline diamond particles.[10] The substrates were immersed in a ultrasonic bath containing a suspension of nanocrystallline diamond grit in a colloidal solution. The nanocrystalline diamond concentration of the solution was varied from 0.005 Ct/l to 8 Ct/l. The nanocrystalline diamond seeeded substrates were subjected to a hydrogen plasma in a plasma CVD system for 2 minutes at 100 watts applied power. The nanocluster carbon films were then deposited on to all the nanoseeded diamond coated substrates simultaneously in a cathodic arc system. The films were deposited at a nitrogen partial pressure of 10^{-3} torr and helium partial pressure of 0.1 torr [7], the condition for low field electron emitting material growth. The as grown nanocluster carbon films have been shown to emit $1\mu A/cm^2$ current at 1 V/μm applied field.[7] Field emission measurements were carried out in a parallel plate configuration, where the cathode consisted of the carbon films to be tested and the anode consisted of an ITO coated glass plate.[12] The I-V measurements were made with an anode -cathode spacing of 100 μm and over a cathode area of 0.24 cm². No conditioning or forming process was required to initiate the emission in the case of the heterostructured cathodes. The Raman measurements were carried out using a Renishaw Ramanscope operating at 514 nm.

RESULT AND DISCUSSION

First the field emission characteristics of the as deposited nanocrystalline diamond layers were studied. Shown in figure 2 is the emission characteristics of the nanoseeded diamond on silicon substrates. It can be seen that the threshold field for emission is very high for the as deposited nanocrystalline diamond films on silicon substrates. The threshold field is defined the field at which an emission current density of $1\mu A/cm^2$ is emitted. The silicon substrates coated with just the nanocrystalline diamond have threshold fields in excess of 25 V/μm. Shown in figure 3 is the emission current against the applied field plots of heterostructured cathodes consisting of the nanocrystalline diamond and nanocluster carbon. It can be seen from the figure that initially when the nanocrystalline diamond concentration is very low - 0.005 Ct/l, the emission characteristics is nearly similar to the emission characteristics of the nanocluster carbon film in figure1. When the nanoseeded diamond concentration increases from 0.005 to 0.2 Ct/l, the emission current density decreases and the threshold field increases. However with further increase in nanoseeded diamond concentration emission current again improves and the threshold field decreases. Also observed in the J-E plot is the negative resistance region or a behaviour similar to resonant tunneling .

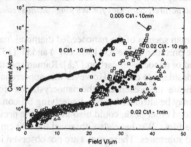

Figure 2 *Current density Vs applied field characteristics of nanocrystalline diamond seeded Silicon substrates.*

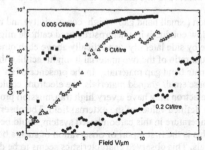

Figure 3 *Variation of emission current with applied field for varying nanoseed concentration in the case of a nanoseeded diamond and nanocluster carbon heterostructured cathodes.*

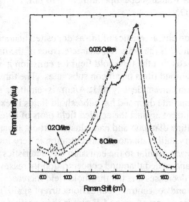

Figure 4 *Raman Spectra of nanoseeded diamond and nanocluster carbon heterostructured cathodes for varying nanoseeded diamond concentration.*

Figure 4 shows the Raman spectra of the nanoseeded diamond /nanocluster carbon heterostructured cathodes. The presence of the D (1365 cm⁻¹) and G (1575 cm⁻¹) peaks, signify the presence of nanostructured or nanocluster carbon.[7,8] Raman spectra for nanocrystalline diamond is still a subject of considerable debate. However the appearance of the small shoulder at 1440 –1450 cm⁻¹ could indicate the presence of the nanocrystalline diamond.[14,15] The small peaks around 900 cm⁻¹ may be attributed to the underlying Silicon substrate. This optical transparency to the silicon second order peak, could also indicate the presence of sp³ bonded nanocrystalline diamond in the films. We have also observed the 500 cm⁻¹ silicon first order peak in the case of nanoseeded substrates. These peaks are not observed in the case of as grown nanocluster films.[7] However the more common 1140-1150 cm⁻¹ peak associated with nanocrystalline diamond was not observed. But with increase in seeding density or size of the nanocrystalline diamond we observe a line around 1121 cm⁻¹. These are discussed in detail in another forth coming paper.

The nanocluster carbon (small band gap – high conductivity) and nanocrystalline diamond (large band gap – low conductivity) heterostructured cathode might have the two nanostructured materials side by side laterally or vertically above each other in a sandwich configuration. Under optimum ratio of the two material it might act like the quantum well structures consisting of alternate band gap material. In the presence of an high field when the barriers become finite for these small grained materials, the electrons could become quasi-bound or resonant. Thus now the electron may have a very high transmission probability, leading to very high currents at sharply defined values of the external field. The observance of this resonant tunneling effect at room temperature in this carbon based system, could be very interesting for vacuum microelectronics. This is the first report of observation of such a behaviour in carbon based nanostrucutred materials. This observed characteristics seems to be dependent on the nanoseeded diamond ratio, as the nanocluster carbon film was deposited simultaneously on all the different substrates with varying nanoseeded diamond concentration. To confirm the observed possible negative differential resistance behaviour, the measurements were repeated at three separate facilities. The observations were found to be repeatable.

CONCLUSION

Low field electron emission has been reported from a novel cathode made of heterostructured nanoseeded diamond and nanocluster carbon. Some of this films also exhibit an emission behaviour similar to resonant tunneling. This behaviour seems to be dependent on the nanoseeded diamond concentration. With further optimisation this device could be extremely useful in vacuum microelectronics. The material may also be used for cold cathodes for field emission, leading to a room temperature grown material, capable of emitting high currents at low fields, associated with high emission site density.

References

1. W. Zhu, G P Kochanski & S Jin, Science, **282**, 1471, (1998).
2. M.Q. D. M. Gruen, A. R. Krauss, O.Auciello, T. D. Corrigan and R. P. H. Chang, J Vac Sci Technol **B 17**, 705 (1999).
3. B.S.Satyanarayana, X.L.Peng, G.Adamopoulos, J.Robertson , W.I.Milne, & T.W.Clyne to be published in MRS Symp Proc. Vol 621, 2000,
4. W.A.de Heer, A.Chatelain, and D.Ugrate, Science **270**, 1179 (1995).
5. Y.Chen, S.Patel, Y.Ye, D.T.Shaw & L.Guo, App. Phys. Lett, **73**, 2119 (1998).
6. B.F.Coll, J.E. Jaskie, J.L.Markham, E.P.Menu, A.A.Talin, P.von Allmen, MRS. Sym Proc. Vol **498**, 185 (1998).
7. B S Satyanarayana, J Robertson, W I Milne, J. App. Phys. **87**, 3126, (2000)
8. O.N.Obraztsov, I.Yu.Pavlovsky and A.P.Volkov, J. Vac. Sci. Technol. **B 17** , 674, (1999).
9. A.C.Ferrari, B.S.Satyanarayana, P.Milani, E.Barborini, P.Piseri, J.Robertson and W.I.Milne. Europhys. Lett, **46**, 245 (1999)
10. H.Makita, K.Nishimura, N.Jiang, A.Hatta, T.Ito and A.Hiraki, Thin solid films, **281-282**, 279, (1996).
11. V.V.Zhirnov, O.M.Kuttel, O.Groning, A.N.Alimova, P.Y.Detkov, P.I.Belbrov,E.Maillard-Schaller and L.Schlapbach, J. Vac. Sci. Technol. **B 17** , 666, (1999).
12. B S Satyanarayana, A Hart, W I Milne & J Robertson, App Phys Lett **71**, 1430 (1997).
13. D.M.Gruen, Annu.Rev.Mater.Sci, **29**, 211, (1999).
14. A A Talin, L S Pan, K F McCarty, H J Doerr, R F Bunshah, Appl Phys Lett **69**, 3842 (1996).
15. O.N.Obraztsov, I.Yu.Pavlovsky, A.P.Volkov, V.I.Petrov, E.V.Rakova, V.V.Roddatis & S.P.Nagovitsyn, in Recent Progress in Diamond Electronics – 1998, ed., T.Ito, 169, (1998).

● **Present Address** : Sistec, KUT Academic & Industrial Collaboration Centre, Kochi University of Technology, Kochi, 782-8502, Japan.   Email : satya@ele.kochi-tech.ac.jp

Carbon Nanotubes and
Nanostructures

Mat. Res. Soc. Symp. Proc. Vol. 621. © 2000 Materials Research Society

ELECTRON EMISSION FROM MICROWAVE-PLASMA CHEMICALLY VAPOR DEPOSITED CARBON NANOTUBES

YONHUA TZENG*, CHAO LIU*, CALVIN CUTSHAW*, AND ZHENG CHEN**
* Alabama Microelectronics Science and Technology Center, Department of Electrical and Computer Engineering, Auburn University, AL 36849
** Space Power Institute, Auburn University, AL 36849

ABSTRACT

A microwave plasma CVD reactor was used for the deposition of carbon nanotubes on substrates. Hydrocarbon or oxyhydrocarbon mixtures were used as the carbon source. Hot electrons in the microwave plasma at temperatures exceeding 10,000C provided a means of dissociating the vapor or gas feedstock, heating the substrate, and allowing gas species to react in the gas phase as well as on the surface of the substrate leading to the deposition of desired carbon coatings. A high vacuum chamber was used to characterize the electron emission properties of these carbon nanotube coatings using a one-millimeter diameter tungsten rod with a hemispherical tip as the anode while the carbon nanotube coatings served as the cathode. The current-voltage characteristics of the carbon nanotube coatings were measured and used for calculating the electric field at which electron emission turned on as well as calculating the field enhancement factor of the carbon nanotubes. Field emission of electrons from carbon nanotubes starting from an electric field lower than 1 volt per micrometer has been achieved.

INTRODUCTION

Micro-engineering in the nano-meter scale using carbon atoms as the building block has resulted in amazingly excellent characteristics of cabon-based structures that finds broad applications ranging from microelectronics to cutting tools. One of these amazing carbon materials is CVD diamond that has been intensively studied in the past fifteen years or so. More recently, the science and practical applications of carbon nanotubes became popular because of its equally great commercial and technological potentials.

Carbon, when properly arranged in a 3-dimensional order, forms the hardest materials on earth, diamond. In addition to being an attractive gemstone, diamond is the champion in almost every aspect of materials properties. Its thermal conductivity is five times better than copper at room temperature making it an excellent material as the heat spreader for high power density integrated circuits. Its bandgap energy is higher than SiC and most of compound semiconductors making it desirable for fabricating high power density electronic devices or sensors and actuators for operation in high-temperature and radiation hostile environments. Its super hardness and chemical inertness make it an economic material for cutting tools or simply used as protective coatings.

By building carbon structures in the nano-meter scale, it is possible to make carbon tubes that are as small as a few nanometers in diameters and are formed by layers of graphite-like carbon structures. When these carbon nano-tubes are coated on a smooth or porous substrate with a conductive surface layer, they emit electrons without needing to be heated to such a high temperature as a hot tungsten filament cathode in a vacuum tube is usually done. It, therefore, becomes a very promising material for the application as a cold cathode for flat panel displays or other electron devices.

A number of methods for chemical vapor deposition of diamond based on different gas mixtures and energy sources for dissociating the gas mixtures have been reported [1]. These techniques include the use of high-temperature electrons in various kinds of plasmas, high-temperature surfaces of hot filaments, and high-temperature gases in combustion flames to dissociate molecular hydrogen, oxygen, halogen, hydrocarbon, and many other carbon containing gases. The substrate is usually maintained at a temperature much lower than that of electrons in plasma, the hot filaments, or the combustion flame, resulting in a super equilibrium of atomic hydrogen near the diamond-growing surface.

Most of diamond CVD processes involve in the use of one or more compressed gases. The most typical example is the use of 1% methane gas diluted by 99% hydrogen. These gases usually must be precisely controlled using electronic mass flow controllers to ensure an accurate composition in the gas feed. Diamond CVD using a microwave plasma CVD reactor and a liquid feedstock without any compressed gas was also reported [2].

Chemical vapor deposition of carbon films in the form of carbon nanotubes is not much different from that for the CVD of diamond. The key experimental parameter for the coating of carbon nanotubes instead of diamond or diamond-like carbon films is the presence of suitable catalysts for carbon nanotubes to grow in the gas phase or on the substrate. In this paper, a microwave plasma reactor that was designed and used for diamond deposition was used for depositing carbon nanotubes on metal substrates for electron emission studies. Oxyhydrocarbon vapor or hydrocarbon gas mixtures were used as the feedstock.

EXPERIMENTS

Shown in Figure 1 is the schematic diagram for the microwave plasma CVD reactor that was used for this study. Hydrocarbon or oxyhydrocarbon mixtures were fed into a vacuum chamber that was evacuated by a mechanical pump. The chamber gas pressure was controlled by a throttle vale and a manometer pressure gauge. An rod antenna was used to couple microwave power from a rectangular waveguide into a cylindrical metal cavity through a quartz window that separate the atmosphere from the vacuum chamber. The microwave power formed a plasma ball on the top of the substrate. Electrons in the plasma have very high temperatures exceeding 10,000°C. The plasma heated the substrate to a preset temperature as well as heated the gas mixtures for proper dissociation and reaction in the gas phase leading to carbon deposition on the substrate surface. Copper, nickel, and Cobalt were used as the catalysts for the carbon coatings to grow into carbon nanotubes instead of forming diamond-like carbon coatings.

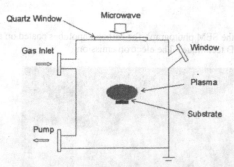

Figure 1. Schematic diagram of the microwave plasma CVD apparatus.

The carbon nanotube coated substrates were then loaded into a high vacuum chamber shown in Figure 2 that was pumped down by a turbomolecular pump and an ion pump to a vacuum of about 1×10^{-7} Torr. The carbon nanotube coated substrate served as the cathode while a tungsten rod of one-millimeter diameter with the tip grounded into a hemispherical shape was used as the anode. The distance between the anode and the cathode was adjusted using a linear vacuum feedthrough to move the tungsten rod closer or farther from the carbon nanotube coated substrate. A desktop computer controlled the output of a high-voltage power supply for applying a voltage between the anode and the cathode. The electric field is calculated by dividing the applied voltage by the gap spacing between the anode and the cathode. The electron emission current was measured by a digital ammeter and recorded by the computer for further plotting and calculation.

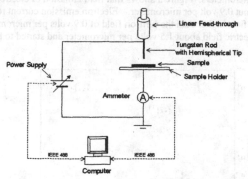

Figure 2. Experimental setup for electron emission measurement.

RESULTS

Shown in Figure 3 is the SEM photograph for carbon nanotubes coated on a nickel plate using microwave plasma CVD technique. The electron emission

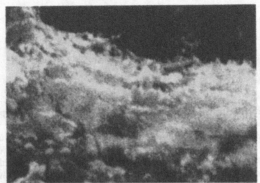

Figure 3. SEM photograph for carbon nanotubes coated on a nickel plate.

characteristics for this carbon nanotube coated nickel substrate are shown in Figure 4 and 5. The carbon nanotube coating was fabricated in 90 minutes at a temperature around 900°C in an oxyhydrocarbon plasma powered by 900 watts of microwave at 2.45GHz at a gas pressure of 36 Torr. The gap spacing between the tungsten rod and the nanotube coating for electron emission measurement was 495 micrometers. Figure 3 shows that field emission of electrons started at an electric field equal to about 0.9 volt per micrometer. Electron emission current rose exponentially at electric field higher than the turn-on field of 0.9 volts per micrometer. The slope decreased at the electric field about 1.5 volts per micrometer and started to become saturated.

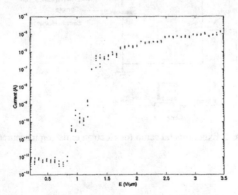

Figure 4. Field emission electron current vs. electric field.

The Fowler-Nordheim plot of the data is shown in Figure 5. A well-defined region of negative slope appears prior to the region of emission current saturation. The field enhancement factor was calculated from the negative slope using 5 eV emission barrier for graphite to be 2840.

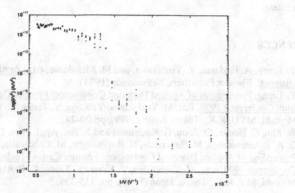

Figure 5. Fowler-Nordheim plot for field emission electron current.

DISCUSSION

The fabrication processes for CVD diamond and carbon nanotube coatings are similar. Microwave plasma CVD technique provides a means of producing very high temperature electrons for dissociating species that are difficult to be dissociated by a hot filament. From Figure 3 and 4, it is clear that the emission mechanism is of the Fowler-Nordheim type with electron field emission commencing at a low electric field because of the 2840 times of field enhancement factor [3-4]. The small diameter and very high aspect ratio of the carbon nanotube make it possible for the local electric field at the tip of the carbon nanotube to be more than two thousand times higher than that calculated by dividing the applied voltage by the distance between the anode and the carbon nanotube coating. With optimization of the coating conditions and the proper choice and applications of catalysts, high electron emission current exceeding 1 A per square centimeter at an electric field lower than 1 volt per micrometer is achievable.

CONCLUSIONS

Carbon nanotube with threshold electron emission electric field lower than 1 volt per micrometer has been coated using a microwave plasma CVD reactor similar to the one that was used for depositing CVD diamond. This is the lowest reported threshold electric field for electron emission from a cold carbon nanotube. Field emission results from carbon nanotube coatings showed clearly the Fowler-Nordheim behavior.

ACKNOWLEDGEMENTS

The authors wish to acknowledge the support from the Office of Naval Research, the NASA, the Alabama Microelectronics Science and Technology Center, and Auburn University Space Power Institute.

REFERENCES

1. Y. Tzeng, A. Feldman, Y. Yoshikawa, and M. Murakawa, eds., Applications of Diamond, Elsevier Publishers, Netherlands (1991).
2. Y. Tzeng, Proceedings of Applied Diamond Conference/ Frontier Carbon Technology Joint Conference 1999, Eds. M. Yoshikawa, Y. Koga, Y. Tzeng, C.-P. Klages, and K. Miyoshi, MYU, K.K., Tokyo, Japan (1999), pp.20-24.
3. W. Zhu, C. Bower, O. Zhou, G. Kochanski and S. Jin, Appl. Phys. Lett. **75,** 873 (1999)
4. G. A. J. Amarutunga, M. Baxendale, N. Rupesinghe, M. Chhowalla, and I. Alexandrou, Proceeding of Applied Diamond Conference/ Frontier Carbon Technology Joint Conference 1999, eds. M. Yoshikawa, Y. Koga, Y. Tzeng, C.-P. Klages, and K. Miyoshi, MY, K.K., Tokyo, Japan (1999), pp. 335-339.

Mat. Res. Soc. Symp. Proc. Vol 621 © 2000 Materials Research Society

Fabrication of High-Density Carbon-Nanotube Coatings On Microstructured Substrates

Zheng Chen
Space Power Institute, Auburn University, AL 36830, USA
e-mail: chenzhe@mail.auburn.edu
Yonhua Tzeng and Chao Liu
Department of Electrical and Computer Engineering
Auburn University, AL 36830, USA
e-mail: tzengy@eng.auburn.edu

ABSTRACT

Fabrication and characterization of carbon nanotubes deposited on microstructured Ni substrate are presented. The highly active surface-area of the microstructured Ni substrate provides high-density nucleation sites for carbon nanotubes. Coated fine Ni powder also serves as a catalyst for the nanotube growth. Hydrocarbon mixtures were used as the carbon source for the chemical vapor deposition process. Carbon nanotubes deposited on the microstructured Ni substrate were examined by SEM. An ultra high vacuum chamber was used to characterize the field emission properties of carbon nanotube coatings.

INTRODUCTION

Carbon nanotube (CNT) [1-3] has excellent properties due to its form, which is a tubular form of carbon with a diameter in the range of 1nm to several tenth micrometers. CNT exhibits extraordinary high Young's modulus and tensile strength. It is stiff as diamond. CNT also can have better electrical conductivity than other forms of carbon due to its graphite structure. In addition, CNT has large surface area and high aspect ratio. With these excellent properties, CNT has been found in many applications, such as cold field emitters, electronic device, sensors, energy storage, and mechanical reinforcements for high strength composites.

CNT has been successfully grown by several techniques (e.g. chemical vapor deposition CVD and laser ablation) in many laboratories. The size of CNT is in the range of a few to hundred nanometers in diameter and several microns long. The problem to hinder CNT for many practical applications is the slow growth rate, high cost, limited deposited size, weak interface bonding between CNT and substrate, and inability to synthesis CNT with control on the shape and size.

Intention of the present investigation is to explore an alternative method to accelerate the CNT growth rate and to improve the control on the bonding strength and growth sites. The approach implemented in this research is to carefully engineer the substrate for CNT growth. The highly active surface of the microstructured Ni substrate provides high-density nucleation sites for CNT and the chemical coated substrate serves as a catalyst for the nanotube growth. SEM and the field emission experiment were used to evaluate the quality of the CNT.

EXPERIMENTS

Ni foil with 5 mil thickness was prepared as substrate through a micro-structuring process on the surface of the foil to increase its surface area (figure 1). The substrates then were coated with

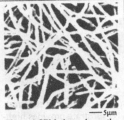

metallic catalysts (e.g. Ni based powder) to enhance the growth rate. The substrates were loaded on a substrate holder in the quartz tube that was then pumped down and subsequently purged with carrier gases such as nitrogen or argon. A resistive heating belt then heated the substrate to an appropriate temperature (i.e. 600°C). The temperature was measured by an attached thermocouple on the substrate holder. The gas pressure in the quartz tubing was controlled to be at 100 Torr by means of a throttle valve while the gas flow was controlled by electronic mass flow controllers. A schematic diagram of the set-up for coating processing is shown in figure 2.

Figure 1 SEM photo shows the microstructured substrate surface.

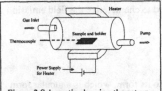

Figure 2 Schematic showing the set-up for CNT coating process.

A mixture of 20-50% acetylene diluted by nitrogen or argon was fed into the quartz tube to allow the gas mixture to flow over the surface of the substrate for carbon deposition. The total gas flow rate was in the range of 50-200 sccm. The carbon deposition time ranged from 30 minutes to 12 hours depending on the substrate temperature. At the substrate temperature of 600°C, which was selected for this paper, the deposition time was used 30 minutes and 60 minutes, respectively.

The field emission measurement was performed in a vacuum chamber, which was pumped down to about 10-8 Torr by an ion pump and a turbo-molecular pump. As schematically shown

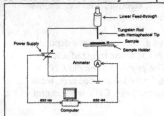

in figure 3, a voltage up to 2kV was applied to a tungsten probe, which is 1 mm in diameter and has a spherical shape at the tip, to collect emitted electrons from the grounded CNT samples. The size and shape of the probe were chosen so that the area of emission from the cathode was approximately to a 100μmx100μm area, which is the size of a display pixel in most cases. The distance between the tip of the tungsten probe and the cathode samples was controlled by a high-vacuum linear feedthrough. The emission current-voltage (I-V) characteristics were recorded by a computer.

Figure 3 Schematic showing the set-up for field emission measurement.

---

* A SPI proprietary process.

RESULTS

Two samples were prepared for this investigation. Sample A was grown at 600°C for 30

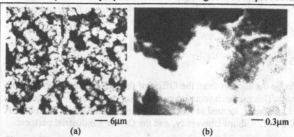

— 6μm — 0.3μm
(a) (b)

Figure 4 SEM photos show the CNTs on the microstructured substrate (sample B). (a) is low magnification photo and (b) is at high magnification. The furlike substance shown in (b) is the CNTs, which have diameter about 50nm.

minutes and sample B was grown at 600°C for 60 minutes. Figure 4 shows their surface morphologies of the sample B. The CNT was grown in an island-type on the microstructured substrate. The direction of CNT tips is perpendicular to the substrate surface and CNT covers most of the substrate surface. The bonding strength between CNTs and the substrate seems to be much better than the samples that their substrates have not been microstructured. The CNTs apparently are close-end type but have distinguished sharp tips, which naturally become effectively active field emitters. In addition, the island-type morphology of these CNTs has potential advantage for stable emission because the conductive path could be directly established between the substrate and the emitter surface rather a pile of random oriented CNTs.

Considering that the surface roughness of the CNT samples might affect the local fields, the I-V

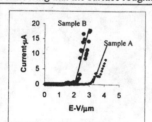

Figure 5 I-V field emission curves for sample A and B, respectively.

measurements were conducted at a large distance between the tungsten probe tip and the surface of CNT sample. Typical emission I-V curves of sample A and B are shown in figure 5 at a probe-sample distance of 414μm. The voltage used for this measurement is in the range of 20 to 1600V at a 20 V step. The turn-on fields are 2.1 V/μm for sample B and 3.2V/μm for sample A, which are determined based on the current density reaching at 10 μA/cm^2.[4]

The CNTs grew in the longer growth time showed that the CNTs have better coverage on the substrate surface and higher aspect ratio. The relatively low turn-on field is apparently attributed to the sharp tip of the CNT emitters, good bonding between the CNT and the substrate, which results in a good conductivity, and the short conductive path mentioned early.

CONCLUSIONS

In summary, we have successfully developed a new method to grow CNT, in which CNT was grown on the microstructured substrate. The growth rate, bonding strength, and size and shape are apparently superior to the CNTs grown on non-microstructured substrates. The turn-on field was measured at about 2 V/μm. This development is at a preliminary stage. More detail tests are needed to understand the effect of processing parameters on the CNT growth rate and shape and size, and the CNT characteristics (i.e. I-V emission curve, turn-on field, and current density).

ACKNOWLEDGEMENTS

The authors wish to acknowledge the support from the Office of the Alabama Microelectronics Science and Technology Center. This research was also partially supported by the Center for Space Power and Advanced Electronics, located at Auburn University, with funds from NASA Cooperative Agreement NCC3-511, Auburn University, and the Center's industrial partners

REFERENCES

[1] S. Iijima, Nature 354, 56 (1991)
[2] P.M. Ajayan, Science 256, 1212 (1994)
[3] P.G. Collins, Science 278, 100 (1997)
4 Jean-Marc Bonard and Frederic Maier, "Field emission properties of multiwalled carbon nanotubes," Ultramicroscopy, 73 (1998) pp 7-15.

Mat. Res. Soc. Symp. Proc. Vol. 621. © 2000 Materials Research Society

HOT-FILAMENT ASSISTED FABRICATION OF CARBON-NANOTUBE ELECTRON EMITTERS

YONHUA TZENG*, CHAO LIU*, AND ZHENG CHEN**
* Alabama Microelectronics Science and Technology Center, Department of Electrical and Computer Engineering, Auburn University, AL 36849
** Space Power Institute, Auburn University, AL 36849

ABSTRACT

A hot-filament CVD reactor was used for the deposition of carbon nanotubes on substrates. Hydrocarbon or oxyhydrocarbon mixtures were used as the carbon source. Hot filaments at temperatures exceeding 2000C provided a means of dissociating the vapor or gas feedstock, heating the substrate, and allowing gas species to react in the gas phase as well as on the surface of the substrate leading to the deposition of desired carbon coatings. A high vacuum chamber was used to characterize the electron emission properties of these carbon nanotube coatings using a one-millimeter diameter tungsten rod with a hemispherical tip as the anode while the carbon nanotube coatings served as the cathode. The current-voltage characteristics of the carbon nanotube coatings were measured and used for calculating the electric field at which electron emission turned on as well as calculating the field enhancement factor of the carbon nanotubes. Field emission of electrons from carbon nanotubes starting from an electric field of as low as 1-2 volts per micrometer was achieved.

INTRODUCTION

Micro-engineering in the nano-meter scale using carbon atoms as the building block has resulted in amazingly excellent characteristics of cabon-based structures that finds broad applications ranging from microelectronics to cutting tools. One of these amazing carbon materials is CVD diamond that has been intensively studied in the past fifteen years or so. More recently, the science and practical applications of carbon nanotubes became popular because of its equally great commercial and technological potentials.

Carbon, when properly arranged in a 3-dimensional order, forms the hardest materials on earth, diamond. In addition to being an attractive gemstone, diamond is the champion in almost every aspect of materials properties. Its thermal conductivity is five times better than copper at room temperature making it an excellent material as the heat spreader for high power density integrated circuits. Its bandgap energy is higher than SiC and most of compound semiconductors making it desirable for fabricating high power density electronic devices or sensors and actuators for operation in high-temperature and radiation hostile environments. Its super hardness and chemical inertness make it an economic material for cutting tools or simply used as protective coatings.

By building carbon structures in the nano-meter scale, it is possible to make carbon tubes that are as small as a few nanometers in diameters and are formed by layers of graphite-like carbon structures. When these carbon nano-tubes are coated on a smooth or porous substrate with a conductive surface layer, they emit electrons without needing to be heated to such a high temperature as a hot tungsten filament cathode in a vacuum tube is usually done. It, therefore, becomes a very promising material for the application as a cold cathode for flat panel displays or other electron devices.

A number of methods for chemical vapor deposition of diamond based on different gas mixtures and energy sources for dissociating the gas mixtures have been reported [1]. These techniques include the use of high-temperature electrons in various kinds of plasmas, high-temperature surfaces of hot filaments, and high-temperature gases in combustion flames to dissociate molecular hydrogen, oxygen, halogen, hydrocarbon, and many other carbon containing gases. The substrate is usually maintained at a temperature much lower than that of electrons in a plasma, the hot filaments, or the combustion flame, resulting in a super equilibrium of atomic hydrogen near the diamond-growing surface.

Most of diamond CVD processes involve in the use of one or more compressed gases. The most typical example is the use of 1% methane gas diluted by 99% hydrogen. These gases usually must be precisely controlled using electronic mass flow controllers to ensure an accurate composition in the gas feed. Diamond CVD using a hot filament CVD reactor and a liquid feedstock without any compressed gas was also reported [2].

Chemical vapor deposition of carbon films in the form of carbon nanotubes is not much different from that for the CVD of diamond. The key experimental parameter for the coating of carbon nanotubes instead of diamond or diamond-like carbon films is the presence of suitable catalysts for carbon nanotubes to grow in the gas phase or on the substrate. In this paper, a hot filament CVD reactor that was designed and used for diamond deposition was used for depositing carbon nanotubes on metal substrates for electron emission studies. Oxyhydrocarbon vapor or hydrocarbon gas mixtures were used as the feedstock.

EXPERIMENTAL

Shown in Figure 1 is the schematic diagram for the hot-filament CVD reactor that was used for this study. Hydrocarbon or oxyhydrocarbon mixtures were fed into a vacuum chamber that was evacuated by a mechanical pump. The chamber gas pressure was controlled by a throttle vale and a manometer pressure gauge. A tungsten filament was heated resistively by a 60Hz AC electrical current to a temperature exceeding 2000C or lower depending on feedstocks. The hot filament heated the substrate placed above or under it to a preset temperature as well as heated the gas mixtures for proper dissociation and reaction in the gas phase leading to carbon deposition on the substrate surface. Copper, nickel, and Cobalt were used as the catalysts for the carbon coatings to grow carbon nanotubes instead of forming diamond-like carbon coatings.

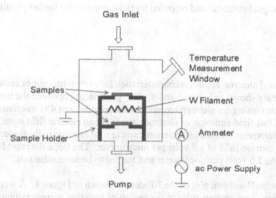

Gas Inlet

Temperature
Measurement
Window

Samples

W Filament

Sample Holder

Ammeter

ac Power Supply

Pump

Figure 1. Schematic diagram of the hot-filament CVD apparatus.

The carbon nanotube coated substrates were then loaded into a high vacuum chamber
that was pumped down by a turbomolecular pump and an ion pump to a vacuum of
about 1×10^{-7} Torr. The carbon nanotube coated substrate served as the
cathode while a tungsten rod of one-millimeter diameter with the tip grounded into a
hemispherical shape was used as the anode. The distance between the anode and the

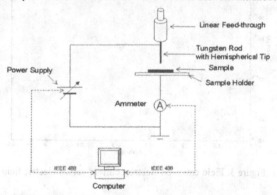

Linear Feed-through

Tungsten Rod
with Hemispherical Tip

Sample

Sample Holder

Power Supply

Ammeter

IEEE 488

IEEE 488

Computer

Figure 2.   Experimental setup for electron emission measurement.

cathode was adjusted using a linear vacuum feedthrough to move the tungsten rod closer
or farther from the carbon nanotube coated substrate. A desktop computer controlled
the output of a high-voltage power supply for applying a voltage between the anode and
the cathode. The electric field is calculated by dividing the applied voltage by the gap
spacing between the anode and the cathode. The electron emission current was

measured by a digital ammeter and recorded by the computer for further plotting and calculation.

RESULTS

The typical electron emission characteristics for carbon nanotubes coated on a nickel substrate are shown in Figure 3 and 4. The gap spacing between the tungsten rod and the nanotube coating for electron emission measurement was 495 micrometers. Figure 3 shows that field emission of electrons started at an electric field equal to about 1.5 volts per micrometer. Electron emission current rose exponentially at electric field higher than the turn-on field of 1.5 volts per micrometer. The slope decreased at higher electric field than 2.5 volts per micrometer and started to become saturated.

The Fowler-Nordheim plot of the I-E data is shown in Figure 4. A well-defined region of negative slope appears prior to the region of emission current saturation. The field enhancement factor was calculated from the negative slope using 5 eV emission barrier for graphite to be 1510.

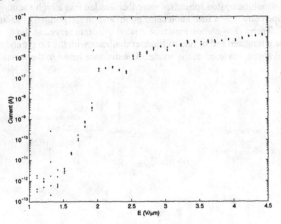

Figure 3. Field emission electron current versus electric field.

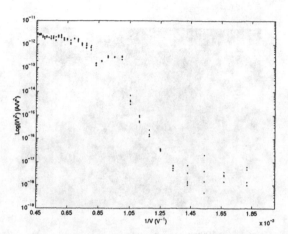

Figure 4. Fowler-Nordheim plot for field emission electron current.

The example described above was taken from a carbon nanotube coating fabricated in 80 minutes at a temperature around 900C in the environment of a mixture of methanol and ethanol vapor of 100 Torr pressure that was fed into the reactor chamber from a premixed solution of 100 gram of methanol and 6 grams of ethanol. A tungsten filament was used to dissociate the vapor and to promote the gas phase reactions for generating carbon containing species responsible for carbon nanotube deposition. The substrate was a nickel fibrous disk that was pre-coated with an appropriate concentration of nickel chloride and placed under the filament with the growth surface facing the filament. The distance between the tungsten filament and the substrate was about 1.5 cm. Shown in Figure 5 is the SEM photograph of carbon nanotubes that grew on the nickel substrate under the assistance of a hot filament.

DISCUSSION

The fabrication processes for CVD diamond and carbon nanotube coatings are similar. Hot filament CVD technique provides a means of very large-area coating of diamond films as well as carbon nanotubes. From Figure 3 and 4, it is clear that the emission mechanism is of the Fowler-Nordheim type with electron field emission commencing at a low electric field because of the 1510 times of field enhancement factor [3-4]. The small diameter and very high aspect ratio of the carbon nanotube make it possible for the local electric field at the tip of the carbon nanotube to be more than one thousand times higher than that calculated by dividing the applied voltage by the distance between the anode and the carbon nanotube coating. With optimization of the coating conditions and the proper choice and applications of catalysts, high electron emission current exceeding 1 A per square centimeter at an electric field lower than 1 volt per micrometer is our goal using this promising hot filament CVD technique.

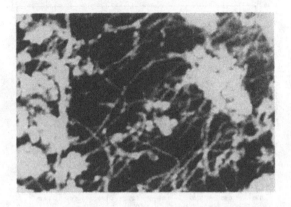

___ 100nm

Figure 5. SEM photograph of carbon nanotubes grown by the hot-filament CVD process.

CONCLUSIONS

Carbon nanotube with threshold electron emission electric field as low as 1.5 volts per micrometer has been coated using a tungsten hot-filament CVD reactor similar to the one that was used for depositing CVD diamond. The hot-filament CVD technique allows the up scaling of the coating process for depositing carbon nanotubes on a very large-area substrate. Multiple substrate coatings can also be achieved using the hot-filament CVD technique. Field emission results from carbon nanotube coatings showed clearly Fowler-Nordheim behavior.

ACKNOWLEDGEMENTS

The authors wish to acknowledge the support from the Office of Naval Research, the NASA, the Alabama Microelectronics Science and Technology Center, and Auburn University Space Power Institute.

REFERENCES

1.	Y. Tzeng, A. Feldman, Y. Yoshikawa, and M. Murakawa, eds., Applications of Diamond, Elsevier Publishers, Netherlands (1991).
2.	Y. Tzeng, in Proceedings of Applied Diamond Conference/ Frontier Carbon Technology Joint Conference 1999, Eds. M. Yoshikawa, Y. Koga, Y. Tzeng, C.-P. Klages, and K. Miyoshi, MYU, K.K., Tokyo, Japan (1999), pp.420-424.

3.   W. Zhu, C. Bower, O. Zhou, G. Kochanski and S. Jin, Appl. Phys. Lett. **75**, 873 (1999)

4.   G. A. J. Amarutunga, M. Baxendale, N. Rupesinghe, M. Chhowalla, and I. Alexandrou, <u>Proceeding of Applied Diamond Conference/ Frontier Carbon Technology Joint Conference 1999</u>, eds. M. Yoshikawa, Y. Koga, Y. Tzeng, C.-P. Klages, and K. Miyoshi, MY, K.K., Tokyo, Japan (1999), pp. 335-339.

Mat. Res. Soc. Symp. Vol 621 © 2000 Materials Research Society

In-situ Plasma Diagnosis of Chemical Species in Microwave Plasma-assisted Chemical Vapor Deposition for the Growth of Carbon Nanotubes Using $CH_4/H_2/NH_3$ Gases

Y.S.Woo[1], I.T.Han, N.S.Lee, J.E. Jung, D.Y. Jeon[1], and J.M.Kim
[1]Dept. of Materials Sci. and Eng., Korea Advanced Institute of Science and Technology, Taejon, Korea.
Display Lab., Samsung Advanced Institute of Technology, P.O. Box 111, Suwon 440-600, Korea, *jongkim@sait.samsung.co.kr

ABSTRACT

Synthesis of multi-wall carbon nanotubes (MWNTs) was attempted by microwave plasma enhanced chemical vapor deposition using $CH_4/H_2/NH_3$ gases on Ni/Cr-coated glass at low temperature. The synthesis was investigated by optical emission spectroscopy and quadrupole mass spectroscopy. It was observed that MWNTs could be grown within a very restrictive range of gas compositions. An addition of a small amount of NH_3 resulted in a decrease of C_2H_2, which can be used to estimate the amount of carbon sources in plasma for the growth of MWNTs, and an increase of CN and H_α radicals acting as etching species of carbon phases. These results show that carbon nanotubes can be grown only under an appropriate condition that the growing process surpasses the etching process. The optimum C_2H_2/H_α ratio in a gas mixture was found to be between 1 and 3 for the MWNT growth at low temperature.

INTRODUCTION

Since the first observation of multi-wall carbon nanotubes (MWNTs) by Ijima, many methods to synthesize MWNTs have been reported such as arc discharge [1,2], laser vaporization [3,4], pyrolysis [5], and plasma-enhanced chemical vapor deposition (CVD) [6-9]. For practical applications, low temperature growth of MWNTs on glass is required because glass is widely used for commercial purposes and its deformation temperature is rather low. However, low temperature deposition results in such an involvement of amorphous carbon impurities. Thus an etching process to remove such carbon impurities during the MWNT growth is critical to the synthesis of high quality MWNTs. An in-situ diagnosis of the plasma during the growth may suggest how to enhance the etching reaction. Optical emission spectroscopy (OES) has been widely used as a diagnostic tool to identify active radical species in a plasma [10-12]. It has also been well known that CH_3 radicals [13-15] and C_2H_2 molecules [15] are regarded as important precursors for the deposition of diamond films. Only a few results, however, has been reported about active chemical species for the growth of carbon nanotubes in arc discharge [16] and laser vaporization [4], which have suggested C and C_2 dimers as precursors for the carbon nanotube growth.

In this study, an NH_3 gas was added to a CH_4/H_2 gas mixture to enhance the etching reaction. And the plasma was diagnosed during the deposition process using OES and quadrupole mass spectroscopy (MS) to identify active species for the carbon nanotube growth.

EXPERIMENTAL

MWNTs were grown on Ni/Cr-coated glass substrates using microwave plasma-enhanced CVD (MPECVD). A Cr film with the thickness of 1500Å was coated on corning glass to

Table 1. Gas compositions for the growth of MWNTs and growth results with such variations.

CH$_4$ conc. (%)	CH$_4$/H$_2$/NH$_3$ (sccm)		CH$_4$ conc. (%)	CH$_4$/H$_2$/NH$_3$ (sccm)	
5%	7.5/ 141 /1.5	No MWNTs	15%	22.5/ 126 /1.5	No MWNTs
	7.5/ 139.5 /3	MWNTs		22.5/ 124.5 /3	No MWNTs
	7.5/ 136.5 /6	MWNTs		22.5/ 121.5 /6	MWNTs
	7.5/ 132 /10.5	No MWNTs		22.5/ 117 /10.5	MWNTs
	7.5/ 127.5 /15	No MWNTs		22.5/ 112.5 /15	MWNTs
				22.5/ 97.5 /30	No MWNTs
10%	15/ 135 /0	No MWNTs	20%	30/ 120 /0	No MWNTs
	15/ 132.3 /2.7	MWNTs		30/ 117.6 /2.4	No MWNTs
	15/ 128.5 /6.75	MWNTs		30/ 114 /6	No MWNTs
	15/ 121.5 /13.5	No MWNTs		30/ 108 /12	No MWNTs
	15/ 94.5 /40.5	No MWNTs		30/ 84 /36	No MWNTs

improve the adhesion between the glass substrate and a Ni layer. Subsequently, a catalytic Ni film of 330Å thick was deposited by using electron beam evaporation. The MPECVD for MWNT deposition was described elsewhere[17]. The deposition conditions are as follows: microwave power of 450W, gas pressure of 10 torr, and total gas flow rate of 150 sccm. The substrate temperatures were in the range of 500~600°C which is measured by a thermocouple and a pyrometer. For the different CH$_4$ concentrations of 5%, 10%, 15%, and 20%, the NH$_3$ concentrations were varied. The detailed gas compositions are listed in Table 1. As-grown MWNTs on substrates were investigated by scanning electron microscopy (SEM, Philips XL30SFEG). OES (Photal, MPCD-1000) was utilized to analyze the species in the plasma during the deposition. A small fraction of the gases was sampled to a differentially pumped quadrupole MS (Dycor, system 1000) through a needle valve with 2μm diameter to identify gas species during the deposition process.

RESULTS AND DISCUSSION

As shown in Table 1, the MWNTs were grown within a very restricted range of compositions of CH$_4$/H$_2$/NH$_3$ gas mixtures and their representative SEM images are give in Fig. 1. The densities of MWNTs grown at 5% and 10% CH$_4$ are higher than at 15% CH$_4$ and no MWNT was observed at the 20% CH$_4$ concentration for a range of NH$_3$ concentrations used in this study.

Fig. 2 (a) shows OES spectra detected from the plasmas of CH$_4$/H$_2$ gas mixtures with and without the NH$_3$ addition. Fig. 2(b) gives the intensities of CN and H$_\alpha$ radicals as a function of the NH$_3$ concentrations for different CH$_4$ concentrations. It is shown that the intensities of CN and H$_\alpha$ radicals are greatly increased by adding a small amount of NH$_3$. In the CH$_4$/H$_2$/NH$_3$ gas plasma, CN and H$_\alpha$ radicals are prominent, but C$_2$ radical that usually occurs in laser ablation [4] is not observed. The intensities of CN and H$_\alpha$ radicals increase linearly, and saturated when the NH$_3$ concentration reaches the same level of the CH$_4$ concentration. It may indicate that most CN and H$_\alpha$ radicals are formed by the reaction between CH$_4$ and NH$_3$ in the gas phases.

The plasma was also diagnosed using MS during the MWNT deposition process. As can be seen in Fig. 3 (a), the gas species generated by the plasma reaction are C$_2$H$_2$, HCN, and N$_2$. The intensity of the HCN peak (mass 27) increases as the NH$_3$ concentrations are higher and is saturated like CN radicals in the OES spectra of CN radicals. The behavior of HCN peak with the NH$_3$ concentrations coincides with those of CN and H$_\alpha$ radicals, as shown in Fig. 2(b). On

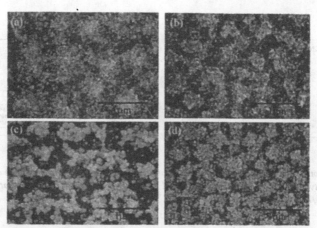

the *Fig. 1 SEM images of MWNTs grown at $CH_4/H_2/NH_3$= (a) 7.5/136.5/6scm, (b) 15/128.5/6.75sccm, (c) 22.5/112.5/15sccm, and (d) 30/114/6sccm.*

other hand, the intensity at the mass 26 which corresponds to C_2H_2 and fractured HCN (CN) decreases at higher NH_3 concentrations for given CH_4 concentrations. It seems that most mass 26 species do not come from the fracture of HCN but from C_2H_2. C_2H_2 appears to be formed by the overall reaction between CH_4 and CH_4.

Fig. 4 presents MS intensities of C_2H_2 and OES intensities of H_α as a function of the CH_4 concentrations for different NH_3 concentrations. It is shown that the C_2H_2 concentrations are increased at higher CH_4 concentrations, but H_α peaks are little affected by the change of CH_4 concentration. An addition of a small amount of NH_3 enhances the formation of atomic hydrogen and tightly bonded HCN through overall bimolecular reactions between CH_4 and NH_3, but decreases the quantity of C_2H_2. Consequently, the addition of NH_3 increases quantity of etching precursors and diminishes carbon sources in the gas phase for the MWNT synthesis. Thus amorphous carbon phases could be easily removed and better quality of MWNTs could be resulted by adding a small amount of NH_3. When the NH_3 concentrations are too high, however, the concentration of etching precursors is too high to grow MWNTs. It is to be noted that the variation of C_2H_2 intensities with NH_3 or CH_4 concentrations exhibits the opposite tendency to those of HCN and H_α intensities. In this study, hence, the intensities of C_2H_2 are utilized to estimate the quantity of C precursors for the MWNT synthesis. Fig. 5 shows ratios of C_2H_2 to H_α intensities at different NH_3 and CH_4 concentrations. Here C_2H_2 and H_α intensities were obtained from MS and OES spectra, respectively. Open circles indicate the deposition conditions where the highest density of MWNTs was grown at each series of experiments. MWNTs seem to be synthesized at the conditions indicated by the shaded box. In this low temperature growth region, the ratios of C_2H_2 to H_α intensities range from 1 to 3. No MWNT was deposited below the ratio of 1 and only amorphous carbon was formed over the ratio of 3. In this study, different methods were utilized to obtain the intensities of H_α and C_2H_2, and C_2H_2 intensity was resulted by neglecting the CN contribution to the mass 26 peak. Thus the ratios of C_2H_2 to H_α intensities for

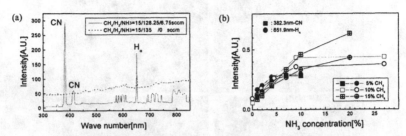

Fig. 2 (a) Optical emission spectra from the plasma of the CH_4/H_2 gas mixtures with and without NH_3 (intensity×10) and (b) intensities of CN and H_α radicals with the NH_3 concentrations for different CH_4 concentrations.

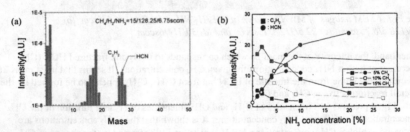

Fig. 3 (a) Mass spectrum from the plasma of a $CH_4/H_2/NH_3$ gas mixture during the deposition process and (b) intensities of HCN and C_2H_2 with the NH_3 concentrations for different CH_4 concentrations.

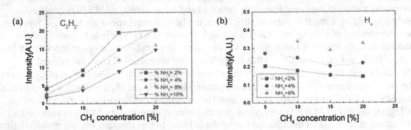

Fig. 4 MS intensities of C_2H_2 and OES intensities of H_α as a function of the CH_4 concentrations for different NH_3 concentrations.

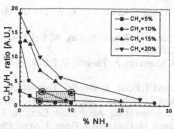

Fig. 5 Intensity ratios of C_2H_2 to H_α with the NH_3 concentrations for different CH_4 concentrations. Open circles represent the ratios where the highest density of MWNT was synthesized at each series of depositions, and the shaded area is considered to be an optimum region for the low temperature MWNT synthesis.

the MWNT synthesis may have a range of uncertainty. Outside the growth region, over 10% NH_3, there was no MWNT growth. It is thought that HCN and N_2 formed by plasma act as dilution gases, resulting in different kinetics of gas phase reactions. We analyzed the chemical species of the plasma for the low temperature MWNT synthesis using in-situ diagnostic tools such as OES and MS. It is worth quantitatively characterizing growth and etching parameters for low temperature MWNT synthesis.

CONCLUSION

We quantitatively diagnosed chemical species of the plasma of the $CH_4/H_2/NH_3$ gas mixtures for the MWNT synthesis in MPECVD using OES and MS. MWNTs were grown within a very restrictive range of gas compositions at low temperatures. From the OES and MS analyses, it was concluded that growth and etching processes were involved simultaneously in the synthesis of MWNTs. The nanotubes were deposited only at the intensity ratios of C_2H_2 to H_α being between 1 and 3.

REFERENCES

1. T. S. Iijima and T. Ichihashi, Nature (London), **363**, 603 (1993)
2. C. Journet, W. K. Maser, P. Bernier, A. Loiseau, M. Lamy de la Chapelle, S. Lefrant, P. Deniard, R. Lee, and J. E. Fisher, Nature (London), **388**, 756 (1997)
3. T. Guo, P. Nikolaev, A. Thess, D. T. Colbert, and R. E. Smalley, Chem. Phys. Lett., **236**, 419 (1995)
4. A. A. Puretzky, D. B. Geohegan, X. Fan, and S. J. Pennycook, Appl. Phys. Lett., **76**, 182 (2000)
5. M. Terrones, N. Grobert, J. Olivers, J. P. Zhang, H. Terrones, K. Kordatos, W. K. Hsu, J. P. Hare, P. D. Townsend, K. Prassides, A. K. Cheetham, H. W. Kroto, and D. R. M. Walton, Nature (London), **388**, 52 (1997)
6. Z. F. Ren, Z. P. Huang, J. W. Xu, J. H. Wang, P. Bush, M. P. Siegal, and P. N. Provenico, Science, **282**, 1105 (1998)
7. O. M. Kuttel. O. Groening, C. Emmernegger, and L. Schlapbach, Appl. Phys. Lett., **73**, 2113

(1998)
8. S. L. Sung, S. H. Tasi, C. H. Tseng, F. K. Chiang, X. W. Liu and H. C. Shih, Appl. Phys. Lett., **74**, 197 (1999)
9. Y. C. Choi, D. J. Bae, Y. H. Lee, B. S. Lee, I. T. Han, W. B. Choi, N. S. Lee and J. M. Kim, Synthetic Metals, **108**, 159 (2000).
10. M. Marinelli, E. Milani, M. Montuori, A. Paoletti, P. Paroli, and J. Thomas. Appl. Phys. Lett., **65**, 28 (1994)
11. R. Manukonda, R. Diollon, and T. Futak, J. Vac. Sci. Technol. A, **13**, 1150 (1995)
12. Jingbiao Cui, Yurong Ma, and Rongchuan Fang, Appl. Phys. Lett., **69**, 3170 (1996)
13. Jin-Jen Wu and Franklin Chua-Nan Hong, Appl. Phys. Lett., **70**, 185 (1997)
14. M. Ikeda, H. Ito, M. Hiramatsu, M. Hori, and T. Goto, J. Appl. Phys., **82**, 4055 (1997)
15. M. Okkerse, M. H. J. M. de Croon, C. R. Kleijn, H. E. van der Akker, and G. B. Marin, J. Appl. Phys., **84**, 6387 (1998)
16. X. Zhao, T. Okazaki, A. Kasuya, H. Shinoyama, and Y. Ando, Jpn. J. Appl. Phys., **38**, 6014 (1999)
17. S. H. Kim and I. T. Han, Diamond Films and Technology, **8**, 425 (1998)

SYMPOSIUM Q:
FLAT-PANEL DISPLAY
MATERIALS

Field Emission and
Display Applications

Mat. Res. Soc. Symp. Proc. Vol. 621 © 2000 Materials Research Society

Surface Modification of Si Field Emitter Arrays for Vacuum Sealing

M. Nagao, H. Tanabe[1], T. Kobayashi[2], T. Matsukawa, S. Kanemaru, and J. Itoh

Electrotechnical Laboratory, 1-1-4 Umezono, Tsukuba-shi, Ibaraki 305-8568, JAPAN
[1]Dai Nippon Printing Co.,Ltd, 250-1 Wakashiba, Kashiwa-shi, Chiba 277-0871, JAPAN
[2]Musashi Institute of Technology, 1-28-1 Tamazutsumi, Setagaya-ku, Tokyo 158-8557, JAPAN

ABSTRACT

Vacuum packaging is a very important issue for vacuum microelectronics devices, especially for field emission displays. Emission current from the field emitter array (FEA), however, is known to decrease significantly after the vacuum packaging process. The current decrease is caused by heating treatment in the vacuum sealing process. In the present paper, the effect of the heating treatment on Si FEA was investigated and CHF_3 plasma treatment was proposed for avoiding the problem. The Si FEA was exposed to plasma for 15sec and emission characteristics were measured before and after the vacuum sealing process using frit. It was confirmed that CHF_3 plasma treatment was very effective for avoiding the emission degradation of the Si FEA. Details of the heating damage and CHF_3 plasma treatment are described.

INTRODUCTION

Si field emitter array (FEA) is one of the most attractive devices in vacuum microelectronics because of its possibility in monolithic integration with various circuits. We have already developed a MOSFET-structured Si FEA[1], which offers stable emission and can be controlled by a relatively low voltage. But its superior performance was demonstrated only in a ultra-high vacuum (UHV) chamber. For practical application of such devices, especially for field emission displays (FEDs), vacuum packaging is important technology. It is well known that emission current of FEAs decrease after the vacuum packaging process.[2, 3] The vacuum packaging of the FED device is currently carried out using the frit sealing process. This process needs high temperature heating of about 450°C, and could damage the emitter surface seriously. Influence of the heating process on Spindt-type Mo-FEA has already been reported.[4] Influence on Si FEA has not, however, been investigated so much. In the present study, the effect of the heating process on Si FEAs was investigated and surface modification of Si FEA was proposed in order to avoid the degradation of emission characteristics.

EXPERIMENTS

Gated cone-shaped Si FEAs were fabricated by the conventional method using reactive ion etching (RIE) and thermal oxidation sharpening.[5] The FEAs having 1000 tips were mounted on TO-5 header. The FEAs were operated in DC bias mode at a pressure of 5×10^{-7} Pa for a few hours to stabilize the emission. After stabilization, current voltage curves were measured. Then the Si FEAs were taken out of the chamber to perform heating treatment at the temperature of 300°C for an hour in order to simulate the vacuum packaging process. The heating process was performed in 1 atm N_2 gas ambient. After that, the emission characteristics in UHV were measured again.

In Fig.1, the open circles and closed circles show the emission characteristics before and after heating treatment, respectively. As shown in this figure, emission current dropped significantly after heating. Although the temperature of this heating treatment is lower than that of the frit sealing process, Si FEA was damaged significantly. In order to investigate the origin of the emission drop, we dipped the Si FEA in 1% hydrofluoric acid (HF) solution for 15 sec and measured the emission characteristics in UHV again. As a result, emission current recovered to the initial level before heating, as shown in Fig.1 (triangles). This result indicates that the emission drop by the heating process is caused by oxidation of the Si tip surface.

The formation of the SiO_2 layer was confirmed by x-ray photoelectron spectroscopy (XPS). Figure 2 shows the Si 2p spectra obtained from the Si substrate before and after heating

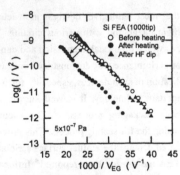

Figure 1. Emission characteristics of Si FEA before the heating treatment (open circles), after the heating (closed circles), and after HF cleaning (triangles).

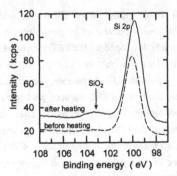

Figure 2. Si 2p spectra before and after heating treatment. Formation of SiO_2 after heating was observed.

treatment in the same condition as described above. The formation of SiO_2 was clearly shown in the spectra after heating treatment. Therefore, for less damaged vacuum packaging it is essential to prevent the emitter surface from oxidation.

For the above purpose, we tried surface modification of coating the emitter surface with carbon layer by exposing to CHF_3 plasma. Generally, CHF_3 plasma is used for etching SiO_2 in the RIE process, and it is well known that the Si surface is covered by carbon polymer after the plasma exposure. We applied this phenomenon to coat the Si tip with a thin carbon film. In the present study, the CHF_3 plasma treatment was conducted using the commercially available RIE system. The background pressure before plasma treatment was $1x10^{-3}$ Pa. The conditions of CHF_3 plasma treatment are summarized in table I.

Si surface exposed to CHF_3 plasma was investigated by XPS measurement. The photoelectron emission angle was changed to determine the distribution of composition corresponding to depth. XPS spectra measured at an angle of 80° indicated that the surface was fully covered by a thin film containing C and F. Figure 3 shows the C1s photoelectron spectra of CHF_3 plasma exposed Si surface measured in normal angle (0°). A typical C-C or C-H peak and three chemically shifted carbon peaks, which can be ascribed to CFx, were clearly observed.

Table I. Condition of CHF_3 plasma treatment for Si FEA.

Back ground gas pressure	$1x10^{-3}$ Pa
CHF_3 gas flux	80 sccm
Gas pressure	2.5 Pa
R.F. Power	80 W
Exposure time	15 sec

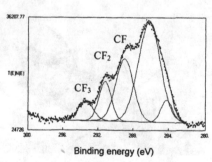

Figure 3. C1s photoelectron spectra of CHF_3 plasma exposed Si surface. Note that a minor peak around 284eV was required to fit the obtained spectrum.

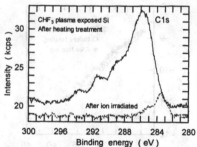

Figure 4. C1s spectra of plasma exposed Si surface after heating treatment. The dashed curve was obtained after Ar^+ ion irradiation.

Gaussian curves in Fig. 3 represent the various peak-fitted chemical components. A minor component around 284eV was required to fit the low energy side of C1s peak. This minor peak will be discussed later.

Figure 4 shows the C1s peak of plasma exposed Si surface after the heating treatment. The CFx component was decreased after the heating treatment. After measurement of this spectra, the Si substrate was irradiated by Ar^+ ion beam to etch the C,F-containing film. In Fig. 4, the dashed curve was obtained after Ar^+ ion irradiation. As shown in this figure, a minor peak around 283.5eV was clearly seen. Coyle et al. reported the existence of a similar minor peak in a CF_4/H_2 plasma etched Si surface and concluded that it is attributed to the SiC.[6] In our case, the situation is very similar to that of Coyle. This peak consequently indicates the existence of a SiC layer. This SiC layer and C,F-containing films are well known to be quite stable and inert against ambient, and therefore are possible for use as protection material in the vacuum sealing process.

The Si FEA was exposed to CHF_3 plasma in the same condition as described in Table I. Prior to the plasma treatment, Si FEA was cleaned by 1% HF solution for 15 sec. Emission characteristics of the plasma treated Si FEA in UHV was measured before and after heating treatment, as shown in Fig. 5. No decrease in emission current after the heating process was observed. It should be noted that there was no significant change in emission characteristics before and after plasma exposure.

Finally, we sealed the Si FEA with and without CHF_3 plasma exposure in a vacuum package by using the frit process. The maximum temperature during the present frit sealing

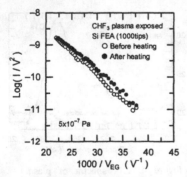

Figure 5. Emission characteristics of CHF_3 plasma treated Si FEA before and after heating treatment.

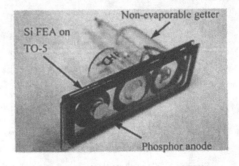

Figure 6. Photograph of frit sealed vacuum package with Si FEA mounted on a TO-5 header.

process was 430°C. Figure 6 shows the photograph of the frit sealed vacuum package with Si FEA. Non-evaporable getters were used to maintain the vacuum condition. The anode was phosphor coated ITO film biased to 200V. Figures 7(a) and (b) show the emission characteristics of Si FEA with and without CHF$_3$ plasma exposure, respectively. In this figure, the open circles show the characteristics in a UHV chamber and the closed circles show the characteristics in a vacuum package. Emission current of the normal Si FEA dropped significantly in the frit sealed vacuum package compared to UHV chamber. However, CHF$_3$ plasma exposed Si FEA was not damaged after the frit sealing process at all. Thus it was confirmed that the CHF$_3$ plasma treatment is very effective for restraining Si FEAs from the deterioration of emission characteristics during frit sealing process.

SUMMARY

It was confirmed that emission performance of Si FEA was significantly damaged by heating treatment in the frit sealing process and that the origin is oxidation of the Si tip. We tried surface modification by exposing Si FEA to CHF$_3$ plasma for only 15 sec in order to prevent it from oxidation. The heating test and frit sealing test revealed that the CHF$_3$ plasma exposed Si FEA was not damaged after the heating or the frit sealing process up to 430°C. XPS measurement indicated the formation of a silicon-carbide and C,F-containing layer on the plasma exposed Si surface. These carbon related thin films could cause endurance to the heating process.

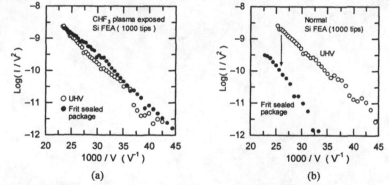

Figure 7. Emission characteristics of CHF$_3$ plasma exposed Si FEA (a) and normal Si FEA (b). Although emission current of normal Si FEA dropped significantly, that of plasma exposed FEA did not change.

Although details of the protection mechanism are not clear as yet, we can conclude that CHF_3 plasma exposure is very effective for avoiding the heating damage during the vacuum packaging process.

REFERENCE

1. T. Hirano, S. Kanemaru, H. Tanoue, and J. Itoh, Jpn. J. Appl. Phys. **35**, 6637 (1996).
2. J. W. Jeong, B. K. Ju, D. J. Lee, Y. H. Lee, N. Y. Lee, Y. W. Ko, Y. G. Moon, D. J. Choi, and M. H. Oh, in *Technical Digest of the 11th Int'l Vacuum Microelectronics Conf.* (Asheville, U.S.A., 1998), p.42.
3. H. Kim, B. K. Ju, K. B. Lee, M. S. Kang, J. Jang, and M. H. Oh, in *Technical Digest of the 11th Int'l Vacuum Microelectronics Conf.* (Asheville, U.S.A.,1998), p.67.
4. F. Ito, K. Konuma and A. Okamoto, J. Vac. Sci. Technol. **B16**, 783 (1999).
5. K. Betsui, in *Technical Digest of the 4th Int'l Vacuum Microelectronics Conf.* (Nagahama, Japan, 1991), p.26.
6. G. J. Coyle, Jr. and G. S. Oehrlein, Appl. Phys. Lett. **47**, 604 (1985).

Mat. Res. Soc. Symp. Proc. Vol. 621 © 2000 Materials Research Society

Nitrogen containing hydrogenated amorphous carbon prepared by integrated distributed electron cyclotron resonance for large area field emission displays

N. M. J. Conway, C. Godet, B. Equer
Laboratoire de Physique des Interfaces et Couches Minces, Ecole Polytechnique, 91128
Palaiseau-Cedex, FRANCE

ABSTRACT

The field emission properties of hydrogenated amorphous carbon containing up to 29at% nitrogen (a-C:N:H), grown in an integrated distributed electron cyclotron resonance (IDECR) reactor were studied using a sphere-plane geometry. All films were smooth in character and required a high field (20-70V/μm) activation process before emission, which created micron-sized craters in the emission region. Further analysis suggested that the emission originates from activation-created geometrically enhanced areas around the crater region. Upon low-level nitrogen incorporation (N/N+C≤0.2), the field required for activation decreased from 54V/μm to a minimum value of 20V/μm. The turn-on field required for 1μA of current also decreased, reaching a minimum of 11.3V/μm. The decrease in activation and turn-on field was related to the increase in conductivity observed with increasing nitrogen content. At higher nitrogen concentrations, the increase in activation energy and turn on field for emission may be due to changes in overall material structure, as indicated by the decreasing optical gap

INTRODUCTION

The field emission properties of carbon-based materials have attracted a great deal of attention with regards to their potential use as cold cathodes in field emission displays (FEDs) due to their ability to emit electrons from nominally flat, or intrinsically geometrically enhanced films at low electric fields. [1-5] It was initially thought that the good emission properties of diamond and diamond-like carbon could be related to the low, or even negative, electron affinity of the material. However, similar emission characteristics can be obtained from graphite-based, and hence high work function, nanotubes and nanoclusters [4,5]. Furthermore, it has been shown that electrons are actually emitted from the grain boundaries in diamond rather than the grains themselves[6] with a work function similar to that of graphite[7]. It is clear therefore that the emission mechanism is still not fully understood [8].

For carbon-based field emission displays to be feasible from an industrial perspective, it should be possible to grow them over large areas, at high deposition rates and at low temperatures (so that low cost substrates can be used). The majority of current deposition methods do not meet these criteria, often requiring high deposition temperatures or using ion beam or magnetic confinement techniques not easily scaleable to large areas. In contrast, the integrated distributed electron cyclotron resonance (IDECR) reactor used in this study satisfies all these criteria; its high density plasma produces a high deposition rate and the deposition area can be easily scaled up due to its unique design, as will be discussed in the experimental details. Furthermore, the IDECR is also able to produce high quality low temperature materials such as SiO_2[9], which could be used as a ballast layer or a gate insulator layer in a triode structure. The addition of nitrogen, which has been shown to reduce the field required for emission in both ta-C and a-C:H[2,3], can be achieved efficiently as the electron resonance condition results in high ionisation of the nitrogen gas. This paper therefore examines the emission properties of nitrogen containing hydrogenated amorphous carbon films (a-C:N:H) grown by IDECR and discusses the results in terms of the material properties and current understanding of the emission mechanism.

EXPERIMENTAL DETAILS

The IDECR is described in detail elsewhere[9]. Briefly, it consists of 4 microwave antennae containing $SmCo_5$ permanent magnets along the length of the antennae in a plane parallel to the substrate. Microwave power is supplied by a 2.45GHz continuous wave magnetron generator and is equally distributed to the antennae by a four-way splitter. The microwave frequency and magnetic field are chosen so that the electrons go into a cyclotron resonance condition which results in high ionisation and hence a high plasma density at low pressures(0.8mTorr). The multipolar magnetic geometry confines the primary electrons to magnetic lobes near the antennae whilst the ions are free to diffuse towards the substrate, resulting in a uniform diffusion plasma in the substrate region. The plane parallel geometry means that the system is easily scalable to any size by either increasing the length or number of antennae, without compromising plasma uniformity.

a-C:N:H films were grown using acetylene and nitrogen gas, inserted into the system(base pressure $<10^{-6}$Torr) just behind the antennae at flow rates of 8-30sccm and 0-70sccm respectively. At higher N_2 flow rates, the C_2H_2 flow rate was reduced so that the deposition pressure always remained below 2mTorr and the minimum nitrogen content corresponds to the film grown with no N_2. All samples were grown using 100W microwave power and an rf bias of 50W (corresponding to ~160Vdc bias) was applied to the substrate. Up to nitrogen concentrations of N/N+C=0.3, the deposition rate was 6-10Ås^{-1}, after which it decreased sharply, limiting the maximum possible concentration to N/N+C=0.37. All samples were 600-700Å thick.

The field emission measurements were performed on samples grown on p^{++} Si(0.01Ωcm) at pressures of $<5\times10^{-7}$Torr using a sphere-plane geometry, where the sample (cathode) was clipped to a stainless steel electrode and a 4mm diameter, polished stainless steel ball (anode) was held at a distance of 150μm above the film. A negative voltage in the range 0-10500V was applied to the cathode with a ramp rate of 5Vs^{-1} and the anode was connected to ground via a 1MΩ ballast resistor. In-situ UV-visible(1.5-5eV) ellipsometry was used to determine the film thickness and optical properties. Conductivity measurements were made using coplanar aluminium gap cells on samples deposited on Corning glass. The structural properties were examined using a Dilor Raman spectrometer at 633nm wavelength and a Philips XL40 scanning electron microscope(SEM). The concentration of nitrogen relative to carbon was determined using nuclear reaction analysis and a combination of Rutherford back-scattering and elastic recoil detection analysis was used to determine the absolute concentrations.

RESULTS AND DISCUSSION

The field emission properties of a-C:N:H films with N/N+C=0.007 to N/N+C=0.37 (29at%N) were examined. Fig 1 shows the field emission characteristics for N/N+C = 0.007,0.05,0.1 and 0.28. For all films, an activation process was required before a significant emission current was observed (see inset, Fig 1). This process consisted of increasing the electric field until a sharp increase in current was observed, after which the electric field was decreased

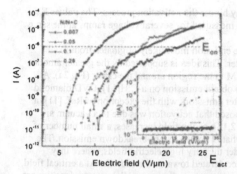

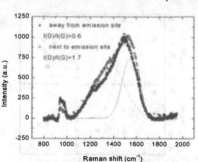

Figure 1 *I-E behaviour of 4 a-C:N:H films after activation (inset: activation process)*

relatively quickly to minimise damage caused by the high currents. Samples were then ramped up to a maximum of 25V/μm on subsequent measurements.

Before emission, all films were relatively smooth (<5nm rms roughness) as a good fit to the ellipsometric data was obtained without having to include a roughness layer in the model. However, SEM analysis of the films after activation found craters of ~5-10μm in diameter in the emission region in all cases. Fig 2(a) shows a crater formed on a sample activated at 29V/μm.

The effect of activation on local film structure was further examined using Raman spectroscopy. Fig 2(b) shows the spectra of the sample N/N+C = 0.007 next to and away from the emission site. The spectra consist of the usually observed 'graphite'(G) peak and the 'disorder'(D) peak of the sp^2 bonds in amorphous carbon[10]. Near the emission site, the position of the G and D peaks increase from 1506cm^{-1} to 1543cm^{-1} and 1295cm^{-1} to 1365cm^{-1} respectively. The ratio of the area of the D peak to that of the G peak (I(D)/I(G)) also increases. This suggests an increase in the number and/or size of the sp^2 clusters within in the film in the region of the emission site after activation.

Fig 3 shows the variation of the activation field, E_{act} and the turn-on field, $E_{on}(I=1\mu A)$ (see Fig 1) as a function of the N content. As the N content increases, E_{act} decreases from 55V/μm to 20-30V/μm at N/N+C~0.1. Similarly, E_{on} decreases from a maximum of 29.1±0.9V/μm to 11.3±0.5V/μm at N/N+C=0.1. Both tend to increase again at higher N contents. The values of E_{act} were on the whole reproducible, however it should be noted on several occasions, E_{act} was either found to be much higher than usual or the sample showed no signs of emission up to 70V/μm. This may be related to a difference in the surface structure of the film at these points or

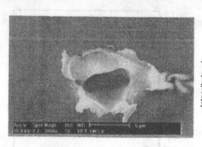

Figure 2(a) *SEM image of an emission site after activation at 29V/μm*

Figure 2(b) *Raman spectra for N/N+C =0.007. Also shown are the fits of the G(solid line) and D peaks(dashed line)*

some form of surface contamination, which may hinder the activation process. The values E_{on} were obtained by taking the mean of the values measured for several voltage ramps at the same emission site after activation.

The observation of craters suggests that the emission most likely originates from geometrically enhanced regions around the crater. This idea is supported by the greater amount of secondary electron emission observed by SEM in the region around the crater (Fig 2a). A similar activation process has also been used to obtain emission on a-C:H[11], CVD diamond [12] and tetrahedral amorphous carbon[13]. After emission, with the exception of Ref. [11], a crater formation was also observed. It was proposed that activation was due to a vacuum arc discharge between the cathode and the anode[12,13]; as the field increases, a small number of electrons are emitted from the carbon film in what is known as pre-breakdown emission. The electrons are accelerated towards the anode under the very high electric fields, where they promote the release of ions, which are in turn accelerated towards the cathode. At a critical field, the electrons will produce enough ions to give rise to an arc discharge. High energy densities are present during arcing[12], which create a local heating of the sample and hence cratering, cracking or even evaporation of carbon material. This model is consistent with the sharp increase in current and crater formation observed and the increase in sp^2 implied by the Raman measurements (Fig 2b) is consistent with a high local heating.

Fig 4 shows the variation in the conductivity and optical (E_{04}) gap with N concentration. E_{04} remains relatively constant at ~1.4eV up to N/N+C=0.1, after which it decreases. The conductivity shows a sharp increase from $10^{-9}Scm^{-1}$ to $10^{-5}S\ cm^{-1}$ between N/N+C=0.007 and 0.1. Within the region where E_{04} decreases, it initially increases slightly towards $10^{-4}Scm^{-1}$, then decreases. The increase in σ without a decrease in E_{04} below N/N+C=0.1 suggests that weak N doping may be occurring[15]. The decrease in E_{act} in Fig 3 corresponds to the region where there is an increase in conductivity, without a significant change in the material structure (as indicated by the constant optical gap), suggesting that improved conductivity results in discharge activation at lower electric fields. It is difficult to provide a precise explanation as to why this is the case as the exact nature of the vacuum arc discharge is not known. However, it could be due an increase in the number electrons available for emission upon ion impact due to the improved

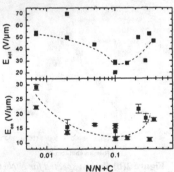

Figure 3 *Variation of the electric field required for activation (E_{act}) and 1μA of current (E_{on}) with N content*

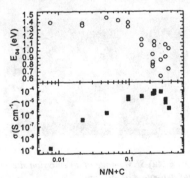

Fig 4 *The variation of optical (E_{04}) gap and the conductivity, σ, as a function of nitrogen content*

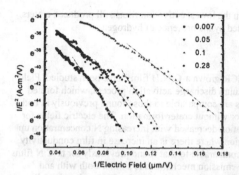

Figure legend values: 0.007, 0.05, 0.1, 0.28

Axis labels: I/E^2 (Acm^2/V) ; $1/Electric\ Field\ (\mu m/V)$

Figure 5 *FN plots of the data in Figure 1*

conductivity, or improved pre-threshold emission of electrons, both of which would result in a discharge occurring at lower electric fields. A decrease in the activation field with nitrogen inclusion was also observed in activation of ta-C films[13].

The similar behaviour of E_{on} with increasing N may also be explained by the increase in conductivity. The emission properties of the films at high currents will more likely be limited by the availability of emission electrons at the emission sites. The increase in conductivity will increase the ease of transport of electrons injected from the silicon back contact through the film to the emission sites and hence increase the emission current at higher fields.

The ratio $\Phi^{3/2}/\beta$, where Φ and β are the work function and geometrical enhancement factor respectively can be obtained by fitting the experimental data to the theoretically expected tunnelling behaviour governed by the Fowler Nordheim (FN) equation [16]. Fig 5 shows the FN plots of the samples in Fig 1. At low electric fields, a linear variation is obtained, indicating that in this range, emission is consistent with the FN law. If no geometrical enhancement is assumed($\beta=1$) an unrealistically low work function in the range 0.06-0.13eV is obtained from the gradient. These values are similar to those obtained in literature for carbon-based films, both with[7,13] and without[2,3] crater formation. Alternatively, if emission is assumed to be from sp^2 regions, a more realistic value of work function would be that of graphite(4.6eV). Using this value, the β values range from 200-500, suggesting a large geometrical enhancement, consistent with the emission originating from sharp points formed around the crater edges upon activation. These are similar to the values of β found for activated diamond and ta-C[7,13]. The area, A, obtained from the intersection of FN fit with the y-axis, was found to be around 10^{-2}-$10^{-1}\mu m^2$ in most cases, also consistent with the idea that emission is originating from enhanced regions around the micron-sized craters.

The results found here differ to those found previously for a-C:N:H films[2], where a continual improvement with increasing N content was found. This may be partially due to the fact that no crater formation was observed in their case. However, the maximum N content was smaller than here (14%) and further studies suggested that increased roughness at higher N contents could explain their improvement and therefore the behaviour of their films with increasing N content is different to that observed here. The improvement of E_{on} is however similar to the behaviour of the threshold field for $1\mu Acm^{-2}$ of current in ta-C:N [3], despite the fact that no crater formation was observed in their case. The similarity of the results in the two cases suggest a similar mechanism. Here, the emission is from the geometrically enhanced sp^2 sites. In ta-C it may be the case that sp^2 filaments embedded in the film experience field enhancement due to the their much greater conductivity compared to surrounding predominantly sp^3 regions. The crater formation in ta-C can be prevented by using very low ramp rates ($\leq 5Vs^{-1}$), which forms sp^2 filaments without creating a discharge[14]. A ramp rate of $5Vs^{-1}$ was used here, but even measurements at $1Vs^{-1}$ ramp rates in the lowest activation films still produced craters. This may be due to increased instability caused by the presence of hydrogen.

CONCLUSIONS

The electron emission properties of IDECR-grown a-C:N:H films have been studied. Field emission was observed after a high-field vacuum discharge activation process, which formed craters in the emission region. The properties are comparable to those found previously in both diamond and amorphous carbon films, with or without crater formation. The electric fields for activation and for $1\mu A$ of current after activation decreased with increasing N concentration up to N/N+C~0.2, which corresponds to the region where there is an increase in film conductivity before a decrease in the optical gap. This is comparable to that found previously for ta-C:N films with no crater formation, suggesting that the emission mechanism is similar both with and without cratering. The potential advantage of these films over other forms of carbon-based emission materials currently being studied is the compatibility of the IDECR deposition technique with large area applications. As the emission properties are stable after activation, with careful control of the activation process, perhaps using dummy, disposable anodes to pattern the crater formation, these films could find use in flat panel display applications.

ACKNOWLEDGEMENTS

Thanks go to C. Clerc and A. Grossman for the stoichiometry measurements, D. Caldemaison for SEM and S. Vernay for help with the commissioning of the field emission system.

REFERENCES

1. M.W. Geis, N. N Efremov, J. D. Woodhouse, M. D. McAleese, M. Marywka, D. G. Socker and J. F. Hochedez, IEEE Trans. Electron Device Lett. **12**, 456 (1991)
2. G.A.J. Amaratunga, S.R.P. Silva, Appl. Phys. Lett. **68**, 2529 (1996)
3. B.S. Satyanarayana, A. Hart, W.I. Milne, J. Robertson, Diam. Relat. Mater. **7**, 656 (1998)
4. W. A. DeHeer, A. Chatelain and D. Ugarte, Science, **270**, 1179 (1995)
5. B.F. Coll, J. E. Jaskie, J. L. Markham, E. P. Menu, A. A. Talin, P. vonAllmen, MRS Symp. Proc., **498**, 185 (1998)
6. C. Wang, W. Garcia, D. Ingram, M. Lake and M.E. Kordesch, Electron Lett. **27**, 1459 (1991)
7. O. Gröning, O.M. Küttel, P. Gröning, L Schlapbach, J. Vac. Sci. Technol. B **17**, 1064 (1999)
8. J. Robertson, J. Vac. Sci. Technol. B **17**, 664 (1999)
9. P. Bulkin, N. Bertrand, B. Drévillon, Thin Solid films, **296**, 66 (1997)
10. M.A. Tamor and W.C. Vassell, J. Appl. Phys. **76**, 3823 (1994)
11. R.D. Forrest, A.P. Burden, S.R.P. Silva, L. Cheah, X. Shi, Appl. Phys. Lett. **73**, 3784 (1998)
12. O. Gröning, O.M. Küttel, E. Schaller, P. Gröning and L. Schlapbach, Appl. Phys. Lett. **69**, 476 (1996)
13. A.A. Talin, T. E. Felter, T.A. Friedmann, J.P. Sullivan and M.P. Siegal, J. Vac. Sci. Technol. A **14**, 1719 (1996)
14. N. Missert, T.A. Friedmann, J.P. Sullivan, R.G. Copeland, Appl. Phys. Lett. **70**, 1995 (1997)
15. N. Conway, P. Bulkin, C. Godet, *to be published*
16. R.H. Fowler and LW Nordheim, Proc. R. Soc. London A **119**, 173 (1928)

Field Emission Display/
Cathodoluminescence

Mat. Res. Soc. Symp. Proc. Vol. 621 © 2000 Materials Research Society

LOW-TEMPERATURE CVD CARBON COATINGS ON GLASS PLATES FOR FLAT PANEL DISPLAY APPLICATIONS

YONHUA TZENG*, CHAO LIU*, CALVIN CUTSHAW *, AND ZHENG CHEN**
* Alabama Microelectronics Science and Technology Center, Department of Electrical and Computer Engineering, Auburn University, AL 36849
** Space Power Institute, Auburn University, AL 36849

ABSTRACT

Low-temperature chemical vapor deposition processes were studied for coating carbon films on metal-coated glass plates. Thermal CVD in hydrocarbon mixtures was used for carbon deposition at temperatures between 300°C and 550°C. Carbon deposited on metal coated glass plates were examined by SEM and analyzed using a pin to disk setup in an ultra high vacuum chamber for measuring the electron emission characteristics. Using a one-millimeter diameter tungsten rod with a hemispherical tip as the anode while the carbon coatings as the cathode, current-voltage characteristics of the carbon coatings were measured and used for calculating the electric field at which electron emission started as well as calculating the field enhancement factor of the carbon coatings. Field emission of electrons from carbon coatings starting from an electric field as low as 1.4 volts per micrometer has been achieved.

INTRODUCTION

In order to use carbon coatings as the electron field emitters for large-area plasma displays it is desirable that an inexpensive substrate such as a glass plate is used for carbon deposition. A conductive layer needs to be deposited on a glass plate before carbon is coated onto the substrate and used as a cold cathode. Because of the low melting point of glass and the large mismatch in the coefficient of thermal expansion between metal and glass, carbon deposition cannot be performed at too high a temperature that is usually considered optimal for carbon nanotube coating on metal substrates. The mismatch in coefficients of thermal expansion for metal coatings and glass makes the adhesion of carbon coatings on glass more difficult when the glass plate has to be heated to a high temperature and then cooled down to the room temperature. Various low temperature CVD techniques are being studied to achieve low threshold field emission of electrons at a high emission current density. In this paper, the results based on thermal CVD techniques are presented.

EXPERIMENTS

Shown in Figure 1 is the schematic diagram for the thermal CVD reactor that was used for this study. Mixtures of acetylene and argon or nitrogen [1] were fed into a quartz tubing chamber that was evacuated by a mechanical pump. Gas pressure was controlled by a throttle valve and monitored by a manometer pressure gauge. A resistive heater was used to heat the quartz tubing and the substrate inside the tubing to a preset temperature. Copper, nickel, and Cobalt were used as the catalysts for the carbon coatings.

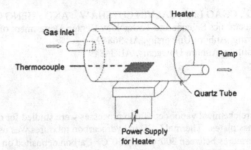

Figure 1. Schematic diagram for the thermal CVD reactor.

Carbon coated substrates were then loaded into a high vacuum chamber shown in figure 2 that was pumped down by a turbomolecular pump and an ion pump to a vacuum of about 1×10^{-7} Torr. The carbon coated substrate served as the cathode while a tungsten rod of one-millimeter diameter with the tip grounded into a hemispherical shape was used as the anode. The distance between the anode and the cathode was adjusted using a linear vacuum feedthrough to move the tungsten rod closer or farther from the carbon coated substrate. A desktop computer controlled the output of a high-voltage power supply for applying a voltage between the anode and the cathode. The electric field is calculated by dividing the applied voltage by the gap spacing between the anode and the cathode. The electron emission current was measured by a digital ammeter and recorded by the computer for further plotting and calculation.

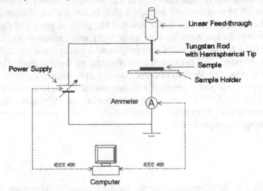

Figure 2. Experimental setup for electron emission measurement.

RESULTS

Shown in Figure 3 and 4 are the electron field emission characteristics for carbon deposited on a copper coated glass plate at a temperature of 500°C for six hours. The gas mixture for the carbon coating was 50% acetylene and 50% nitrogen with pressure controlled to be at 100 Torr. The gap spacing between the tungsten rod and the carbon coating for electron emission measurement was 614 micrometers. Figure 3 shows that field emission of electrons started at an electric field equal to about 1.4 volts per micrometer. Electron emission current rose exponentially at electric field higher than the turn-on field of 1.4 volts per micrometer. The slope decreased at the electric field about 2 volts per micrometer and then started to become saturated.

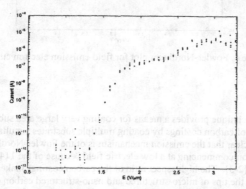

Figure 3. Field emission electron current vs. electric field.

The Fowler-Nordheim plot of the data is shown in Figure 4. A well-defined region of negative slope appears prior to the region of emission current saturation. The field enhancement factor was calculated from the negative slope using 5 eV emission barrier for graphite to be 1140.

Carbon films were deposited on copper coated glass plates at temperatures as low as 360°C. Unlike fiber-like appearance for carbon films deposited at the higher end of the substrate temperatures, at low substrate temperatures, the carbon coatings did not appear like nanotubes or carbon fiber. Instead, it showed under SEM a micro-structured rough surface. The electron field emission threshold electric field was higher for carbon films deposited at lower substrate temperatures. For example, a carbon film deposited at 360°C for 11 hours resulted in an electron field emission threshold electric field being 2.3 volts per micrometer with the field enhancement factor being 913.

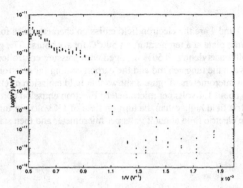

Figure 4. Fowler-Nordheim plot for field emission electron current.

DISCUSSION

Thermal CVD technique provides a means for coating very large area substrates as well as for mass production of carbon coatings by coating multiple substrates simultaneously. From Figure 3 and 4, it is clear that the emission mechanism is of the Fowler-Nordheim type with electron field emission commencing at a low electric field because of the 1140 times of field enhancement factor [2-3]. The high aspect ratio of the carbon coatings makes it possible for the local electric field at the tips of micro-structured and nano-structured carbon coatings to be more than one thousand times higher than that calculated by dividing the applied voltage by the distance between the anode and the carbon coating. Optimization of the coating conditions and the proper choice and applications of catalysts in order to achieve a high electron emission current exceeding 10m A per square centimeter at an electric field less than one volt per micrometer is desirable.

CONCLUSIONS

Micro-structured and nano-structured carbon films have been deposited on metal-coated glass plates at temperatures as low as 360°C. The threshold electric field for electron field emission from carbon films deposited on glass plates was measured to be as low as 1.4 volts per micrometer. The low threshold electric field and the capability of the thermal CVD technique for mass production of large-area carbon coated glass plates and other inexpensive substrates make this process a promising candidate for further study as an economic applications to flat panel displays. Field emission results from carbon coatings showed Fowler-Nordheim-like behavior. Further research is being conducted to optimize the carbon coating process for achieving high current density electron field emission.

ACKNOWLEDGEMENTS

The authors wish to acknowledge the support from the Office of Naval Research, the NASA, the Alabama Microelectronics Science and Technology Center, and Auburn University Space Power Institute.

REFERENCES

1. X. Xu and G.R. Brandes, Appl. Phys. Lett. **74**, 2549 (1999).
2. W. Zhu, C. Bower, O. Zhou, G. Kochanski and S. Jin, Appl. Phys. Lett. **75**, 873 (1999)
3. G. A. J. Amarutunga, M. Baxendale, N. Rupesinghe, M. Chhowalla, and I. Alexandrou, Proceeding of Applied Diamond Conference/ Frontier Carbon Technology Joint Conference 1999, eds. M. Yoshikawa, Y. Koga, Y. Tzeng, C.-P. Klages, and K. Miyoshi, MY, K.K., Tokyo, Japan (1999), pp. 335-339.

Mat. Res. Soc. Symp. Proc. Vol. 621 © 2000 Materials Research Society

Epitaxial Oxide Thin-Film Phosphors for Low Voltage FED Applications

Yong Eui Lee, David P. Norton, John D. Budai, Philip D. Rack[1], Jeffrey Peterson[1], and Michael D. Potter[1]
Solid State Division, Oak Ridge National Laboratory, Oak Ridge, TN 37831-6056
[1]Advanced Vision Technologies, Inc., W. Henrietta, New York 14586

ABSTRACT

Epitaxial $ZnGa_2O_4$ and Sr_2CeO_4 thin-film phosphors were successfully grown on (100) MgO, YSZ, and $SrTiO_3$ single crystal substrates using pulsed laser ablation. Cathodoluminescence efficiency was remarkably enhanced by adding lithium in the $ZnGa_2O_4$ and $ZnGa_2O_4$:Mn for both blue and green light emitting thin-film phosphors. The highest efficiencies, in this experiment, were 0.35 and 0.29 lm/W at 1kV for as-deposited blue and green zinc gallate phosphor films, respectively. In case of Sr_2CeO_4 films, the highest luminescence was 0.14 lm/W at 1kV and 0.26 A/m^2 for films annealed at 1000°C in air.

INTRODUCTION

Oxide thin-film phosphors are very attractive for field emission displays (FEDs) which operated with low voltage and high current density. Advantages include high resolution screens, enhanced lifetime at high current density, and excellent thermal and mechanical stabilities. Unfortunately, one of the fundamental problems for oxide thin-film phosphors in applications is inferior luminous efficiency and brightness as compared to powder systems.

Many researchers have studied luminescence for thin-film phosphors in an attempt to improve the properties through increasing light scattering centers using surface modification, reducing defect density by post-annealing at high temperatures, and identifying new phosphor materials.[1-3] Recently, Lee et al.[4] reported that epitaxial phosphor films exhibit superior photoluminescent (PL) intensity as compared to randomly-oriented polycrystalline films on glass substrates. This result indicates enhanced luminescence due to improved intragranular crystallinity via epitaxy. The study of epitaxial phosphor film is useful not only for achieving high luminescence performance, but also in understanding the fundamental properties of the materials. Most known phosphor materials are multi-component materials possessing complex crystallographic structures, making it difficult to find suitable deposition technique and/or substrate materials for epitaxial film growth.

Recently several interesting studies on low-voltage blue and/or green light emitting cathololuminescent(CL) oxide phosphors have been reported. 1. $ZnGa_2O_4$ and mixtures of $Li_2O/LiGa_5O_8$ and $ZnGa_2O_4$ having a spinel crystal structure have showed improved CL properties in powders and have been suggested as promising candidates in applications of FEDs.[5-6] 2. Sr_2CeO_4 have showed high efficiency in CL about 5.4 lm/W at 4 kV.[7] We have investigated the growth and luminescence characteristics of several oxide thin-film phosphors including $ZnGa_2O_4$, $Li_xZn_{1-x}Ga_2O_4$, $ZnGa_2O_4$:Mn, $Li_xZn_{1-x}Ga_2O_4$:Mn, and Sr_2CeO_4 for blue and green low-voltage CL.

EXPERIMENTAL DETAILS

Blue light emitting $Li_xZn_{1-x}Ga_2O_4(x=0-0.2)$, mixtures of $LiGa_5O_8$ and $ZnGa_2O_4$, and Sr_2CeO_4 thin-film phosphors were grown using pulsed-laser ablation. Polycrystalline 1 inch dia. ablation targets were prepared by mixing and pressing Li_2O [Alfa, 99.99%], ZnO [Alfa, 99.999%], Ga_2O_3 [Alfa, 99.9995%] powders, followed by sintering in air at 1250 °C for 24 hours. To compensate Zn loss in the films, all films were deposited using mosaic targets being composed of phosphor target/ZnO with an area ratio of 50/50.[4] A Sr_2CeO_4 target was prepared using CeO_2[Alfa, 99.994%] and $SrCO_3$[Alfa, 99.99%] powders. Approximately 2μm-thick films were grown using an excimer KrF laser with a wavelength of 248 nm. Detail deposition conditions are summarized in Table 1. Crystal structure was investigated using four-circle x-ray diffraction (XRD) with CuKα radiation (0.15406 nm wavelength). Surface microstructure of the films was observed using atomic force microscope.(AFM) Low voltage cathodoluminescence (CL) was measured with a Plasma Scan PSS-2 spectrometer, and CL intensity measurements were taken with an International Light 1700 photometer equipped with a photopic filter.

Table 1. Summary of film growth conditions.

Target	Substrate Temperature (°C)	Oxygen Pressure(mTorr)	Laser Fluence (J/cm^2)
$Li_xZn_{1-x}Ga_2O_4$/ZnO	730	100	3.5
Sr_2CeO_4	750	300	4.0

RESULTS AND DISCUSSION

1. $ZnGa_2O_4$ Thin-Film Phosphors

Figure 1 shows x-ray diffraction θ-2θ and φ scans for the films using a $Li_{0.25}Zn_{0.75}Ga_{2.25}O$:Mn/ZnO mosaic target. In this study, all films exhibited a $ZnGa_2O_4$ (400) diffraction peak. Some films revealed a Ga_2O_3 peak with very low intensity. In the case of Li_2O added $ZnGa_2O_4$, formation of secondary phase could be expected. No direct evidence for the formation of lithium gallate related secondary phase was found in this experiment. φ scans confirm that the films are well aligned in-plane indicating epitaxial films.

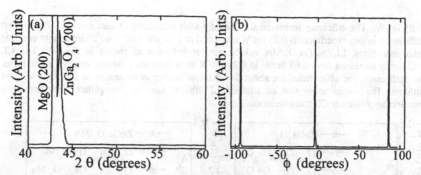

Figure 1. X-ray (a) θ-2θ and (b) φ scans for the films using $Li_{0.25}Zn_{0.75}Ga_{2.25}O:Mn/ZnO$ mosaic target.

Surface microstructures in the films were observed using AFM as shown in Fig. 2. $ZnGa_2O_4$ films show a smooth surface with root mean square(r.m.s.) roughness of about 2.3 nm while rather irregular-shaped surface features were observed in the lithium added films. The r.m.s. roughness of the $Li_{0.05}Zn_{0.95}Ga_2O_4$ and $Li_{0.25}Zn_{0.75}Ga_{2.25}O_4$ films were approximately 2.9 and 4.9 nm respectively. Mn-doped films show higher values in r.m.s. roughness relatively to the films grown without Mn. The values are 5.6 nm, 3.8 nm, and 8.0 nm for $ZnGa_2O_4:Mn$, $Li_{0.05}Zn_{0.95}Ga_2O_4:Mn$, and $Li_{0.25}Zn_{0.75}Ga_{2.25}O_4:Mn$ films, respectively. Consistent relationship between the surface microstructural features and lithium content could not be observed.

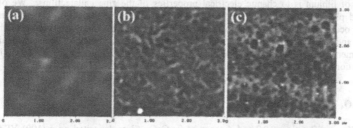

Figure 2. Surface morphology of the films deposited using mosaic targets of (a) $ZnGa_2O_4/ZnO$, (b) $Li_{0.05}Zn_{0.95}Ga_2O_4/ZnO$, and (c) $Li_{0.25}Zn_{0.75}Ga_{2.25}O_4/ZnO$. Scan area was 3μmx3μm.

The peaking wavelengths in CL spectra for blue light emitting lithium-doped $ZnGa_2O_4$ films ranged between 406 and 430 nm. CL emission spectra for green lithium-doped $ZnGa_2O_4:Mn$ showed peaking wavelength located from 507 to 509 nm which corresponds to the transition $^4T_1 \rightarrow {}^6A_1$ of the 3d electrons in the Mn^{2+} ion occupying the fourfold coordinated Zn position in the host lattice.[8]

Low voltage CL efficiency for the lithium zinc gallate films with current density at fixed voltage of 1kV are shown in Fig. 3. Without lithium, the film shows efficiencies lower than

0.05 lm/W. The efficiency increased significantly with increasing lithium content. The highest efficiency in this experiment is 0.35 lm/W measured at 1kV and 0.05 A/m^2. Green light emitting phosphor films, $Li_xZn_{1-x}Ga_2O_4$:Mn, exhibit similar behavior as shown in Fig. 4. The CL efficiency increased from 0.03 lm/W to 0.29 lm/W with increasing lithium content in the films. In both cases, the efficiencies are about 5 – 8 times higher as compared to the films without lithium. The results show that an addition of lithium ions in zinc gallate phosphor films remarkably enhances CL characteristics.

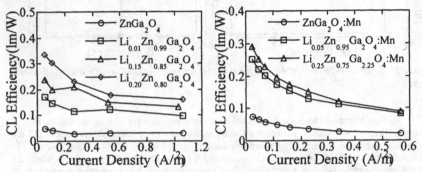

Figure 3. *CL Efficiency for blue films with lithium content measured at 1kV.*

Figure 4. *CL efficiency for green films with lithium content measured at 1kV.*

As described above, compatible effects of additive lithium on structural and surface feature of zinc gallate films were not observed in this experiment. One highly possible influence of the addition of lithium is to change electrical properties of the zinc gallate films. We observed enhanced photoconductivity with increasing lithium content in the films. The enhancement possibly occur during electron bombardment that may reduce charge build-up near phosphor surface and consequently, enhance CL brightness and efficiency. The mechanism on improvement of CL efficiency and brightness in zinc gallate films by addition of lithium is under investigation.

2. Sr₂CeO₄ Thin-Film Phosphors

X-ray diffraction θ-2θ and ϕ scans of Sr_2CeO_4 films deposited on a (100) YSZ single crystal substrates are shown in Fig. 5. The ϕ scans through the (110) Sr_2CeO_4 confirms in-plane aligned structure for the films. The film on (100) $SrTiO_3$ shows a near singular orientation with a very weak (111) diffraction peak. The films have orthorhombic crystal structure as determined by four-circle x-ray diffraction. High-resolution scans with a Ge analyzer crystal yields lattice parameters of the orthorhombic Sr_2CeO_4 film to be a=0.6115 nm, b=0.1036 nm, c=0.3596 nm on (100) $SrTiO_3$ and a=0.6119 nm, b=0.1035 nm, c=0.360nm on (100) YSZ. These values are in good agreement with the values reported by Danielson.[9]
The epitaxial relationships between film and substrate were (100)[010]Sr_2CeO_4//(100)[010]$SrTiO_3$ and (100)[010]Sr_2CeO_4//(100)[011]YSZ.

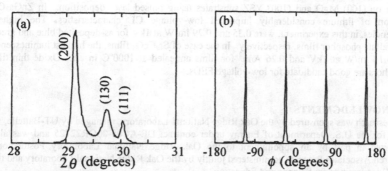

Figure 5. X-ray (a) θ-2θ and (b) φ scans through (110) for Sr₂CeO₄ films deposited on (100) YSZ single crystal substrate.

As-deposited films exhibit a broad-band CL emission, peaking at about 465nm as shown in Fig. 6. The small peaking at 505 nm may be due to contamination of Mn from the chamber. Figure 7 shows the CL efficiency versus voltage for as-deposited and subsequently post-annealed Sr_2CeO_4 films on (100) YSZ single crystal substrates. With increasing annealing temperatures up to 1000°C, significant enhancement in the CL efficiency is observed. A 2-μ m thick film annealed at 1000°C in air showed the highest luminous efficiency of about 0.14 lm/W at 1kV and 22μA/cm².

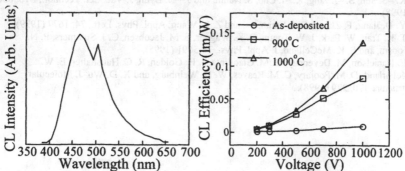

Figure 6. CL spectra of as-deposited Sr_2CeO_4 Thin-Film Phosphors.

Figure 7. Variation of CL efficiency for Sr_2CeO_4 films measured at 22μA/cm².

SUMMARY

Blue and green light emitting $Li_xZn_{1-x}Ga_2O_4$ and Sr_2CeO_4 thin-film phosphors were grown on (100) MgO and (100) YSZ substrates using pulsed laser deposition. In $ZnGa_2O_4$, addition of lithium considerably improved low-voltage CL characteristics. The highest efficiencies, in this experiment, were 0.35 and 0.29 lm/W at 1kV for as-deposited blue and green zinc gallate phosphor films, respectively. In the case of Sr_2CeO_4 films, the highest luminescence was 0.14 lm/W at 1kV and 0.26 A/m^2 for films annealed at 1000°C in air. Oxide thin-film phosphors are good candidate for low-voltage FEDs.

ACKNOWLEDGMENTS

This research was sponsored by the Oak Ridge National Laboratory, managed by UT-Battelle, LLC, for the U.S. Department of Energy under contract DE-AC05-00OR22725 and was also supported in part by an appointment to the Oak Ridge National Laboratory Postdoctoral Research Associates Program administered jointly by the Oak Ridge National Laboratory and the Oak Ridge Institute for Science and Education.

REFERENCES

1. S. L. Jones, D. Kumar, K. –G. Cho, R. Singh, and P. H. Holloway, Display, 19 151 (1999).
2. K. –G. Cho, D. Kumar, P. H. Holloway, and R. Singh, Appl. Phys. Lett., 73, 3058 (1998).
3. T. Minami, T. Maeno, Y. Kuroi, and S. Tanaka, Jpn. J. Appl. Phys., 34, L684 (1995).
4. Y. E. Lee, D. P. Norton, and J. D. Budai, Appl. Phys. Lett., 74, 3155 (1999).
5. S. Itoh, H. Toki, F. Kataoka, Y. Sato, K. Tamura, and Y. Kagawa, IEICE Trans. Electron, E82-C, 1808 (1999).
6. K. –S. Suh, S. –J. Jang, K. –I. Cho, S. Nahm, and J. –D. Byun, J. Vac. Sci. Technol., B16, 858 (1998).
7. Y. D. Jiang, F. Zhang, C. J. Summers, and Z. L. Wang, Appl. Phys. Lett., 74, 1677 (1999).
8. T. K. Tran, W. Park, J. W. Tomm. B. K. Wagner, S. M. Jacobsen, C. J. Summers, P. N. Yocom, and S. K. McCelland, J. Appl. Phys., 78 5691 (1995).
9. E. Danielson, M. Devenney, D. M. Giaquinta, J. H. Golden, R. C. Haushalter, E. W. McFarland, D. M. Poojary, C. M. Reaves, W. H. Weinburg, and X. D. Wu, J. Molecular Structure 470, 229 (1998).

Mat. Res. Soc. Symp. Proc. Vol. 621 © 2000 Materials Research Society

Mechanisms Affecting Emission in Rare-Earth-Activated Phosphors

David R. Tallant, Carleton H. Seager and Regina L. Simpson
Sandia National Laboratories, Albuquerque, NM, 87185-1411, U.S.A.

ABSTRACT

The relatively poor efficiency of phosphor materials in cathodoluminescence with low accelerating voltages is a major concern in the design of field emission flat panel displays operated below 5 kV. Our research on rare-earth-activated phosphors indicates that mechanisms involving interactions of excited activators have a significant impact on phosphor efficiency. Persistence measurements in photoluminescence (PL) and cathodoluminescence (CL) show significant deviations from the sequential relaxation model. This model assumes that higher excited manifolds in an activator de-excite primarily by phonon-mediated sequential relaxation to lower energy manifolds in the same activator ion. In addition to sequential relaxation, there appears to be strong coupling between activators, which results in energy transfer interactions. Some of these interactions negatively impact phosphor efficiency by nonradiatively de-exciting activators. Increasing activator concentration enhances these interactions. The net effect is a significant degradation in phosphor efficiency at useful activator concentrations, which is exaggerated when low–energy electron beams are used to excite the emission.

INTRODUCTION

Phosphors are the light-emitting components used in flat panel displays based on electroluminescent, plasma or field emission excitation technologies. Unlike cathode ray tube (CRT) displays, which use highly energetic electrons and substantial beam currents to excite phosphor emission, design constraints in flat panel displays severely limit the amount of excitation power available to excite phosphors. Consequently, the emission efficiency of phosphor materials becomes an issue of major concern in the design of flat panel displays. Unfortunately, the emission efficiency of rare-earth-activated phosphors appears to be limited by concentration quenching of activators (the dopants which produce characteristic light emission in phosphors) [1] and other effects [2]. The quenching effects are aggravated when low-voltage electrons are used for excitation [3]. It is, therefore, important to understand the mechanisms of quenching in phosphors and so determine whether quenching limitations of phosphor efficiency can be overcome and, if so, how. Our work has identified interactions between activators in excited states which have a major impact on phosphor efficiency. This paper presents evidence for these interactions (primarily) from photoluminescence (PL) experiments but also with data from cathodoluminescence (CL) which show concomitant effects.

EXPERIMENTAL DETAILS

Powder samples of phosphors from the systems Y_2O_3:Eu, Y_2SiO_5:Tb and Zn_2SiO_4:Mn were prepared by spray pyrolysis (Superior MicroPowders) in dopant concentrations from 0.2 to 10 atomic %. Photoluminescence (PL) emission spectra were obtained using a D_2 lamp/monochromator combination for excitation and a 0.6-m spectrograph with a charge-coupled device detector to disperse and record the emitted light. PL persistence (light intensity versus time) curves were obtained using a Nd:YLF pump laser and optical parametric oscillator to provide pulsed, ultraviolet and visible excitation and a 0.6-m spectrograph, photomultiplier tube and digital storage oscilloscope to disperse, detect and capture light intensity transients. Cathodoluminescence (CL) persistence measurements were obtained at accelerating voltages from 0.8 to 4 keV. Emission spectra were recorded using a small spectrograph and a charge-coupled device detector. Light transients were captured using wavelength-isolating optical filters, a photomultiplier tube and a digital storage oscilloscope.

RESULTS AND DISCUSSION

A primary limitation to increasing phosphor efficiencies is that the individual activators in essentially all phosphor systems become less efficient at converting input energy to light output as the activator concentration is increased. As shown in Figure 1, the output intensity of (red) light from the Eu $^5D_0$ energy level divided by the Eu activator concentration (a value proportional to the efficiency per Eu activator) decreases tenfold over the 0.2 to 9 atomic % range of Eu concentration for PL excitation at 260 nm. CL efficiency per activator decreases even more over the same activator concentration range for an accelerating voltage of 3 keV and degrades further if lower accelerating voltages are used. Both PL and CL efficiency per activator values were normalized to 1.0 for the Y_2O_3: 0.2 atomic % Eu powder

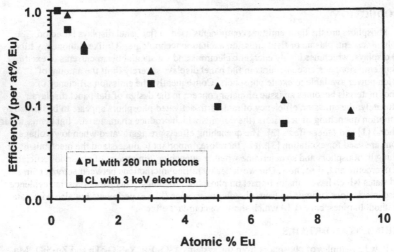

Figure 1. Normalized Y_2O_3:Eu $^5D_0$ (Red) emission per activator for PL excited at 260 nm and CL excited with 3keV electrons.

The decrease in efficiency per activator shown in Figure 1 as a function of activator concentration implies that the rate of a process which de-excites the activators nonradiatively, i.e., without photon emission, is increasing with increasing activator concentration. This increase in the nonradiative de-excitation rate increases the overall rate of depopulation of the excited states. Stated in other terms, the persistence lifetimes of the excited states of the Eu activators should decrease as the activator concentration increases, due to an increased rate of depopulation from nonradiative de-excitation events.

The energy level diagram for Eu^{3+} is shown in Figure 2. Eu^{3+} has a ground state of ^7F manifolds. The lowest excited states are of ^5D character, with the $^5D_0$ state having the longest lifetime and emitting the red light characteristic of Eu^{3+}-activated phosphors. Above the ^5D states are closely spaced states of f-orbital character and then charge transfer bands [3]. When a deep-ultraviolet (250 nm) photon is absorbed by Eu^{3+}, the Eu^{3+} is excited into a charge transfer band. For isolated Eu^{3+} ions (those incapable of transferring energy to another species in the phosphor lattice), the excitation relaxes sequentially through the various energy states of the ion, quickly to the $^5D_3$ level, then more slowly, but still sequentially through the ^5D levels to the $^5D_0$ level, which has a high probability of emitting a photon.

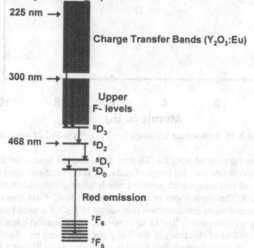

Figure 2. Energy levels for Eu^{3+} and the sequential relaxation model.

By exciting Eu^{3+} activators with ultraviolet photons at 225 nm from a pulsed source, we can excite detectable emission from the ^5D levels and obtain persistence curves from each of these levels. Relaxation within a single, isolated activator obeys first-order kinetics, and the persistence (emission intensity versus time) curve has an exponential, exp[-kt], dependence on time, t, and the total relaxation rate, k (S^{-1}). A semi-log plot of persistence is linear with slope, k, whose inverse is the lifetime of the state whose emission is plotted. PL persistence curves were thus obtained for the Y_2O_3:Eu powders whose emission efficiencies are shown in Figure 1. PL persistence lifetimes from the $^5D_0$, $^5D_1$ and $^5D_2$ states for these powders are shown in Figure 3.

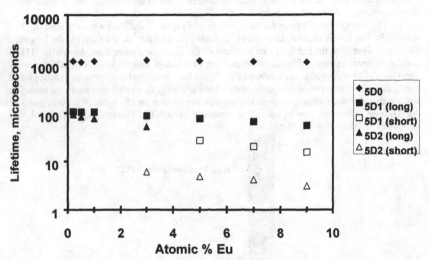

Figure 3. PL Persistence lifetimes for Y_2O_3:Eu with 225 nm excitation.

The semi-log persistence plots for $^5D_0$ state emission are linear, and the lifetimes derived from these plots are constant over the range of activator concentrations shown in Figure 3. The emission efficiency of this state, which provides nearly all of the characteristic Eu^{3+} light output, must also be constant. The implications of this result are twofold. First, loss of activator efficiency as a consequence of concentration quenching in Y_2O_3:Eu must be due to nonradiative processes operating on states above $^5D_0$. In other words, concentration quenching of Y_2O_3:Eu emission is not due to loss of efficiency of the $^5D_0$ state, but it must be related to how excitation is transferred to the $^5D_0$ state. Second, this result is not consistent with quenching related to resonant or near-resonant energy transfer [4,5,6] between activators. In this scenario excitation migration occurs until the activator emits light or the excitation reaches an activator near a "trap" site, which nonradiatively de-excites the activator. By this mechanism, quenching by trapping should most strongly affect $^5D_0$ state, since excitation will migrate the furthest in this, the longest-lived, excited Eu^{3+} state. Quenching by trapping will increase as increasing Eu^{3+} concentration enhances migration rates by reducing the average separation between Eu^{3+} ions. We conclude that such trapping effects (involving single, excited activators) are not strongly affecting the emission efficiency of Y_2O_3:Eu.

The semi-log plots of persistence from the $^5D_1$ and $^5D_2$ states are linear at low Eu^{3+} concentrations, but, as the Eu^{3+} concentration reaches 3 atomic %, they become nonlinear, indicating the advent of additional, non-first-order processes,. These nonlinear semi-log plots can be fit by equations involving more than one persistence lifetime. In Figure 3 these multiple lifetimes are plotted as long and short lifetime components of the $^5D_1$ and $^5D_2$ states. Both the long and short lifetime components decrease regularly with increasing Eu^{3+} concentration. The nonlinear appearance of the semi-log persistence plots from $^5D_1$ and $^5D_2$ emission and the decrease in $^5D_1$ and $^5D_2$ lifetimes with Eu^{3+} concentration suggest that non-first-order processes, which enhance the relaxation rates of these states, become increasingly important as the Eu^{3+} concentration increases.

We have carried out the same type of experiment for the Y_2SiO_5:Tb system and have obtained similar results. The Tb^{3+} $^5D_4$ state, which is the longest-lived and provides nearly all the characteristic Y_2SiO_5:Tb (green) emission, has linear semi-log persistence plots and shows little variation in its lifetime over the 0.2 to 10 atomic % range of Tb^{3+} concentrations. The (higher in energy) $^5D_3$ state has nonlinear semi-log persistence plots over nearly this entire Tb^{3+} concentration range, and its lifetimes decrease by two-to-three orders of magnitude by 10 atomic % Tb^{3+}.

The persistence plots in Figure 4 suggest the nature of the additional relaxation processes which are affecting the relaxation rates of the higher excited states in these phosphor systems. The figure includes the PL persistence curve obtained by exciting Y_2O_3: 1 atomic % Eu with 225 nm excitation and monitoring $^5D_0$ emission (611 nm). Figure 4 also includes calculations from model [7] based on sequential relaxation (see Figure 2) and the $^5D_0$, $^5D_1$ and $^5D_2$ lifetimes from Figure 3. The lifetimes of the states above $^5D_2$ are too short to significantly affect the shape of the persistence curve for the time scale depicted. The long tail of the data and the calculated curve, which represents the lifetime of the $^5D_0$ state, match well, but the rise of the data is much faster than that of the calculated curve. The corresponding persistence curve from CL excitation likewise matches the $^5D_0$ lifetime, but it rises even faster (deviates even more from the calculated curve) than the PL persistence curve. The fast rises of the PL and CL curves must be due to a process which provides excitation to the $^5D_0$ state without passing through one or both of the intermediate states ($^5D_1$ and $^5D_2$). Figure 5 depicts a process by which this might occur.

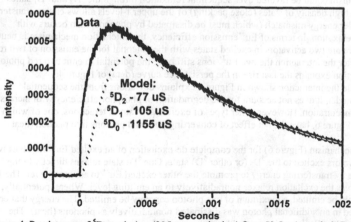

Figure 4. Persistence of $^5D_0$ emission (225 nm excitation) from Y_2O_3: 1atomic %Eu and curve calculated assuming sequential relaxation through excited states.

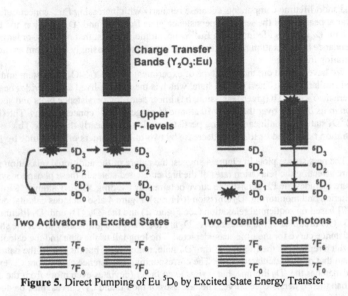

Figure 5. Direct Pumping of Eu $^5D_0$ by Excited State Energy Transfer

In Figure 5 two Eu^{3+} ions excited to the $^5D_3$ state transfer energy so that one Eu^{3+} ion is raised to an upper f-level while the other drops directly to the $^5D_0$ state, by-passing the $^5D_2$ and $^5D_1$ states and populating the $^5D_0$ state faster than it would be populated by sequential relaxation. This mechanism could be favored over sequential relaxation through the $^5D_2$ and $^5D_1$ states because the high density of states (close packing) of the upper f-levels allows energy transfer with a smaller energy mismatch (which must be dissipated by phonons) than occurs with sequential relaxation. In terms of Eu^{3+}emission efficiency, this interaction mechanism is neutral. Initially there are two activators in excited states with the potential for the emission of two red photons. After the interaction the two Eu^{3+} ions still have the potential to emit two red photons. This interaction explains the fast rise in the persistence curve (data) of Figure 4.

While the interaction shown in Figure 4 explains deviations from the sequential relaxation model, it does not account for the degradation of phosphor efficiency with increasing activator concentration. However, other types of excited activator interactions, one of which is depicted in Figure 6, have the net effect of converting potential photons into phonon (heat) energy.

A mechanism (Figure 6) for the complete de-excitation of an excited Eu^{3+} involves two nearby activators excited to the $^5D_3$ (or other ^5D) state. One $^5D_3$ state relaxes directly to the ground state by transferring energy to promote the other excited Eu^{3+} to an upper level. The Eu^{3+} which receives the excitation relaxes nonradiatively to an emitting level. Where, potentially, two photons could be emitted, a maximum of one photon can now be emitted. The energy that could have resulted in an additional photon was dissipated nonradiatively as phonons (heat). The remaining excited Eu^{3+} may interact with another excited activator in the same way, essentially acting as a catalyst for nonradiative de-excitation. Such repeated interactions will seriously degrade phosphor efficiency.

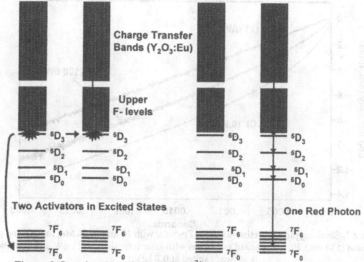

Figure 6. Complete de-excitation of a Eu^{3+} ion by excited state energy transfer

The two Eu^{3+} sites which interact as shown in Figure 6 form a type of "trap" - but not the usual concept of a trap, which involves de-excitation of a single excited activator, with the energy going to some other type of site. This type of trap requires two excited Eu^{3+} ions in close proximity. Individually, each Eu^{3+} in the "trap site" can act like a normal emitter.

One consequence of the excited state interaction mechanism is the introduction of a term involving second order kinetics (because the interaction depends on the concentrations of two activators in excited states) into the rate equations governing excited state relaxation. Experimentally, the addition of second order rate terms should result in nonlinear semi-log persistence curves and a dependence of the shape of the persistence curves on excitation intensity. We found, at most, a very weak dependence of the persistence curve shape on excitation intensity for intermediate activator concentrations in the Y_2O_3:Eu and Y_2SiO_5:Tb systems. Future experiments will measure the effect of excitation intensity on the persistence curves of powders with higher activator concentrations.

Transition metal activators might be expected to show enhanced excited state interactions compared to rare earth activators, since the valence-band d-orbitals of transition metals extend further from the activator ion than the f-orbitals of rare earths. Activators with d-orbitals should be able to interact with other (especially excited state) activators over longer distances. The shape of the persistence curve of Zn_2SiO_4 with 10 atomic % Mn was found to show a strong dependence on excitation intensity, as shown in Figure 7.

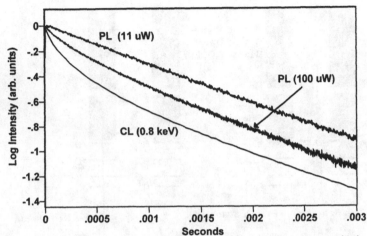

Figure 7. Semi-log plot of persistence from Zn_2SiO_4 with 10 atomic % Mn, monitoring green emission (530 nm). PL was excited at 232 nm with laser intensities of 11 uW and 100 uW. CL was excited at 0.8 keV.

The semi-log plot of persistence from Zn_2SiO_4 with 10 atomic % Mn excited at low (PL) excitation intensity (11 uW) is nearly linear and is consistent with first order (isolated activator) kinetics. The long tail of the persistence resulting from higher (PL) excitation intensity (100 uW) is also linear and has the same slope (lifetime) as the persistence resulting from lower excitation intensity. At shorter times the persistence resulting from higher excitation intensity is distinctly nonlinear. The nonlinear portion of this curve is modeled well if terms second-order in excited activator concentration (corresponding to excited activator interactions) are added to the rate equation for excited state relaxation. It is expected that a higher density of excited activators, resulting from higher excitation intensities, results in enhanced excited state interactions. The changes in the persistence curves due to nearly an order of magnitude difference in (PL) excitation intensity can be replicated by only a 20% change in manganese concentration in the host Zn_2SiO_4. At least for the excitation intensities used in PL excitation, the concentration at which activator is doped into the host phosphor matrix appears to have a stronger effect on the rate of excited state interactions than the excitation intensity. Increased activator concentration will both reduce the average distance between activators in the host matrix and increase the absorption of photons which cause excitation.

The semi-log plot of the persistence curve from Zn_2SiO_4 with 10 atomic % Mn obtained by electron beam excitation (Figure7, CL with 0.8 keV electrons) shows significantly more curvature, implying even higher rates of excited state interaction than was achieved for PL at the highest excitation intensity. The persistence curves of Zn_2SiO_4 with 10 atomic % Mn obtained with electron energies greater than 0.8 keV have less curvature and lie between the (0.8 keV) CL persistence curve and the (100 uW) PL persistence curve. CL apparently achieves higher local densities of excited activators than PL in these experiments.

CONCLUSIONS

We have presented evidence for the existence of interactions between excited state activators. Certain forms of these interactions explain why persistence curves deviate from the relaxation model, which assumes that excitation relaxes sequentially between excited states of an activator. Other forms of these interactions completely de-excite activators and can seriously degrade phosphor efficiency by converting energy which potentially can be emitted as photons into lattice phonons. The efficiency-degrading interactions explain the decrease in phosphor efficiency as activator concentration increases

Three factors potentially influence the rate at which these excited state interactions occur with excitation by photons (PL):
1. Activator concentration - determines the average distance between activators and, hence, the rate of excited state interactions and also affects the amount of excitation energy absorbed;
2. Resonant energy transfer - effects migration of excitation to activators in close proximity, i.e., where they can interact; and
3. Excitation intensity - determines the initial concentration of excited activators.

Throughout this paper we have presented data obtained by electron beam excitation (CL) comparison to that obtained by photon absorption (PL). The trends related to excited state interactions are uniformly exaggerated in the CL data compared to the PL data. We believe that a primary difference between CL and PL excitation is a much greater local density of excited activators along the track of an impinging electron than is achieved in the phosphor matrix by the (photon) excitation intensities used in our PL experiments. It is this high local density of excited activators which enhances excited state interactions in CL experiments.

Excited state interactions appear to intrinsically limit phosphor efficiency. However, there may be ways to ameliorate their effects. First, preparation techniques which minimize aggregation of activators into locally high concentrations in the phosphor matrix will minimize excited state interactions. Phosphors prepared by spray pyrolysis achieve higher efficiencies with low energy CL excitation than those prepared by conventional techniques. This improvement in efficiency may be due to less local aggregation of activators when spray pyrolysis is used to synthesize the phosphors. Second, there may be dopants, which, when inserted into the phosphor matrix, block interactions between excited state activators. Miller [8] has found that rare earth dopants with energy states near resonant with those of Eu^{3+} in Y_2O_3:Eu tend to quench its emission, but dopants with no resonant energy states may enhance its emission. We plan to investigate these means of enhancing phosphor efficiency.

ACKNOWLEDGMENTS

Funding was provided by DARPA and the U.S. DOE. The authors thank Superior MicroPowders for providing the phosphor materials used in this study. Sandia National Laboratories is a multiprogram laboratory operated by Sandia Corporation, a Lockheed Martin Company, for the United States Department of Energy under Contract DE-AC04-94AL85000.

REFERENCES

1. L. Ozawa, *J. Electrochem. Soc.* **126(1)**, 106-109 (1979).
2. U. Vater, G. Kunzler and W. Tews, *Journal of Fluorescence* **4**, 79-82 (1994).
3. C. H. Seager and D. R. Tallant, "The Role of Activator-Activator Interactions in Reducing Low-Voltage-Cathodoluminescence Efficiency in Eu and Tb doped Phosphors," to be published in *J. Appl. Physics*.
4. J. Heber and U. Kobler, *Izv. Akad. Nauk SSSR, Ser. F.*, **37(4)**, 772-777 (1973)
5. J. Heber, K. H. Hellwege, U. Kobler and H. Murmann, *Z. Physik* **237**, 189-204 (1970).
6. R. B. Hunt Jr. and R. G. Pappalardo, *Journal of Luminescence*, **34**, 133-146 (1985).
7. D. R.Tallant, M. P. Miller and J. C. Wright *J. Chem. Phys.* **65(2)**, 510-521 (1976).
8. M. J. Fuller, *J. Electrochem. Soc.*, **128(6)**, 1381-3 (1981).

Mat. Res. Soc. Symp. Proc. Vol. 621 © 2000 Materials Research Society

Modeling of Interface Scattering Effects during Light Emission from Thin Film Phosphors for Field Emission Displays

K. G. Cho, R. K. Singh, Z. Chen, D. Kumar, and P. H. Holloway
Department of Materials Science and Engineering, University of Florida, Gainesville, FL 32611.

ABSTRACT

It has been experimentally shown that the light trapping due to internal reflection from a smooth surface is reduced as the surface becomes progressively rougher. Although this phenomenon is qualitatively understood well, there has been a lack of detailed analysis of the scattering phenomenon which affects the light emission from thin film phosphors (TFPs). Factors which affect the light emission from the TFPs include electron beam-solid interaction (EBSI), film thickness, microstructure, surface recombination rate, surface roughness, and substrate (thus the interface formed). In many cases, they cannot be varied independently and thus making it difficult to interpret the results quantitatively. Furthermore, as the surface roughness is smaller or same as the wavelength of the emitted light, classical theories based on rectilinear propagation of the light cannot be used without gross simplification. A new theoretical model has been developed by incorporating diffraction scattering at the various interfaces and the factors mentioned above. The model provides an integrated solution to explain the cathodoluminescence (CL) properties of TFPs for field emission displays (FEDs).

INTRODUCTION

There have been intensive efforts to enhance the relatively low brightness of TFPs compared to that of powder phosphors by tailoring the surface roughness of the TFPs. The various methodologies include variation in processing conditions [1], modifying TFP-s surface [2], and adding a buffer layer between TFP and substrate [3]. Although their experimental results showed that light trapping due to internal reflection from a smooth surface is drastically reduced as the surface roughness increases to a certain level, it is hard to explain the empirical results quantitatively. Some simplified models have been suggested to predict the brightness of the TFPs as a function of the effect of bulk and surface structure [2,3]. However, their solutions include only a couple of affecting parameters mentioned above and do not encounter complex multiple scattering at the interfaces. If the dimensions of the facets of the crystallites are large compared with the wavelength of light, the reflectance of a surface in a given direction is determined entirely by geometrical optics and is only a function of the inclinations of the facets [4]. As the dimensions of the facets becomes smaller than the wavelength of light, diffraction effects become more important, and the reflectance is a function of both the inclination and the size of the facets. As the dimensions of the facets become very small compared with the wavelength of light, the reflectance of the surface is determined almost entirely by diffraction effects [4,5]. A new model based on diffraction scattering has been developed and incorporated most of the CL parameters such as EBSI, film thickness, optical interfaces, substrate, SRR, and so on. The model helps us to understand the effects of a single or multiple parameters controlling the luminance properties of TFPs with a programable theoretical model. This understanding will eventually lead to fabrication of TFPs for FEDs with optimum CL properties.

MODELING DETAILS

The model of light emission from Y_2O_3:Eu TFPs is based on the diffraction scattering of light from the various interfaces (air/film: interface 1, film/substrate: interface 2, and substrate/air: interface 3). When surface roughness is smaller or comparable to the wavelength of the incident light, the diffraction scattering theory explains two different scattering components of the reflected light, coherent (specular) which reflects at an angle equal to the incident angle and incoherent (diffuse) component which scatters in a space with some fraction of light propagating in the specular direction along with the specularly reflected light [5]. The generated light from each activator in the TFP, encounters absorption, diffuse and specular reflection, and diffuse and specular transmittance effects, as it traverses across the TFP. The observed brightness is the total sum intensity of the light escaping out of film surface (R-mode) or substrate surface (T-mode) through the multiple scattering occurring at the interfaces of 1, 2, and 3. Since the roughness of the interfaces 2 and 3 are very low, the diffuse effects from these interfaces are not considered. Also, the scattering effects at the grain boundaries are neglected, because significant changes in the refractive index are not expected to occur at the grain boundaries. Details of the modeling procedure are described in other publication [6].

Sinusoidal roughness was assumed for the surface roughness of the TFPs. From the definition of RMS roughness (σ), the relationship between the roughness height (h) and σ was derived as $h = 2\sqrt{2}\sigma$. By definition of the solid angle (Ω) with a surface area of s in a cone of emission, the derivative of the Ω (the solid angle change: $d\Omega$) which is the change of the cone of emission is defined as

$$d\Omega = 2\pi \sin\theta d\theta \tag{1}$$

where θ is the angle of propagating light with respect to normal. When light is generated from the activator center, the initial intensity of the light (I_0) in all directions has 4π solid angle. This defines the intensity of the light of an unit solid angle as $I_0/4\pi$. Therefore, the intensity change (dI_0) by an angular element ($d\theta$) is

$$dI_0 = \frac{I_0}{4\pi} d\Omega = \frac{I_0}{2} \cdot \sin\theta d\theta. \tag{2}$$

When propagating light scatters at the interface 1 having a RMS roughness of σ, the light is divided into four different components, such as specular reflection (R_s) and transmittance (T_s), and diffuse reflection (R_d) and transmittance (T_d) [7], and the intensity of the specular reflection (I_{Rs}) can be defined as

$$I_{R_s} = I_{R_o} \exp\left[-\left(\frac{2\sqrt{2}\pi\sigma}{\lambda}\cos\theta\right)^2\right] \tag{3}$$

where I_{R0} is the intensity of the internally reflecting light when $\sigma=0$, and λ is the wavelength of the light generated in the TFP.

Figure 1 illustrates the diffuse component of the propagating light after being generated from an activator. When the propagating light scatters at the interface 1, the light reflected back to the film with an angle less than $_c$ (R_d< $_c$) will transmit into the substrate, and the light reflected back to the film with an angle greater than $_c$ (R_d> $_c$) will internally reflect with another angle , in addition to R_s. For this case, the intensity of the total internal reflection when $\sigma\neq0$ ($I_{R \neq 0}$) is reduced by the presence of the intensity of the diffuse transmittance (I_{Td}) which is the light escaping through the interface 1 due to the presence of rough film surface. Then, $I_{R \neq 0}$ can be defined as the sum intensity of the specularly reflected light (I_{Rs}) and the diffusely reflected light (I_{Rd}):

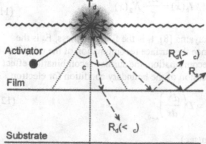

Figure 1. *When the light generated from an activator diffusely transmits (T_d) and reflects (R_d) at the rough film surface, the light reflected back to film with an angle less than $_c$ will transmit into substrate, and the light reflected back to film with an angle greater than $_c$ will reflect internally with an angle , in addition to specularly reflected light (R_s).*

$$I_{R_{\text{0}}} = I_{R_0} - I_{T_d} = I_{R_s} + I_{R_d}. \qquad (4)$$

I_{Td} is a part of the I_{Rd} since the internal reflection is reduced by the existing rough interface, thus it is defined as

$$I_{T_d} = \eta I_{R_d} \qquad (5)$$

where η is a constant between 0 and 1. Therefore, I_{Td} is an additional contributor to the total sum intensity of the light from the film side, in addition to the intensity of specularly transmitted light (I_{Ts}) which directly escapes from the film when θ is smaller than the critical angle θ_c. According to Equation (3) and (4), the I_{Rd} is also defined as

$$I_{R_d} = I_{R_0}\left[1 - \exp\left[-\left(\frac{2\sqrt{2}\pi\sigma}{\lambda}\cos\theta\right)^2\right]\right]. \qquad (6)$$

I_{Rd} is assumed to be uniformly distributed in all directions and is expressed again using Equations (4)-(6) as

$$I_{R_d} = \frac{I_{R_0}}{1+\eta}\left[1 - \exp\left[-\left(\frac{2\sqrt{2}\pi\sigma}{\lambda}\cos\theta\right)^2\right]\right]. \qquad (7)$$

Therefore, the intensity of the diffusely reflected ($I_{Rd\_}$) light related to $d\Omega$ in an unit solid angle with an angular unit $d\beta$ in an hemisphere is defined as

$$I_{R_d\_\beta} = \frac{I_{R_0}}{(\eta+1)}\left[1 - \exp\left[-\left(\frac{2\sqrt{2}\pi\sigma}{\lambda}\cos\theta\right)^2\right]\right]\sin\beta d\beta. \qquad (8)$$

CL is a part of the diffusion process of the electrons in the phosphor since electrons diffuse further into the matrix than the EBSI range creating new electron-hole (e-h) pairs. Under the steady state condition the number of photons emitted is equal to the number of carriers released by the incident electrons that recombine. Then, a simple diffusion equation can be applied to determine the carrier concentrations in a phosphor, as defined as [8]

$$D\frac{d^2}{dx^2}N - \frac{N}{\tau} + G = 0 \qquad (9)$$

where D is diffusivity of electron, N is the number of e-h pairs, x is the depth from surface, is the life time of the radiative recombination, and G is the generation rate of the e-h pairs by an incident electron-beam (e-beam). The number of e-h pairs created in the TFP can be obtained by solving Equation (9) with respect to N. Therefore, the initial intensity of the generated light (or brightness) can be obtained using the solutions below:

- when $0<x<t_1<R$ or $0<x<R<t_1$:

$$I_0(x) = \frac{E_e}{\tau} \cdot N_1(x) \qquad (10)$$

- when $0<R<x<t_1$:

$$I_0(x) = \frac{E_e}{\tau} \cdot N_2(x) \qquad (11)$$

where R is the EBSI range defined by Kanaya-Okayama [8], t_1 is the film thickness, E_e is the photon energy emitted in a TFP by an e-beam. Since the surface recombination of the e-h pairs reduces the number of e-h pairs of the radiative recombination, the surface recombination effect should be considered in the model using the expression of the boundary condition for electron diffusion to the surface [9],

$$[S \cdot N(x)]_{x=0} = D\left(\frac{dN}{dx}\right)_{x=0} \qquad (12)$$

where S is the surface recombination velocity (cm/sec).

Finally, the optical parameters of the materials are also included in the model and some of the representatives are listed in Table 1. The n, , t and $_c$ represent refractive index, absorption coefficient, thickness and critical angle with a wavelength of 611 nm, respectively.

Table 1. *Optical parameters of Y_2O_3:Eu TFP and substrates.*

Constant	Y_2O_3:Eu	Quartz	Sapphire	LAO	Silicon
N	1.93	1.46	1.77	2	3.91
(mm⁻¹)	5	0.07	0.3	0.55	>200
t (m)	0.07-0.5	1000	500	500	500
$_c$ (w/air)	0.545 (31.2)(	0.755 (43.2)(	0.600 (34.4)(	0.523 (30.0)(	0.258 (14.8)(
$_c$ (w/film)	1.570 (90.0)(	0.858 (49.2)(	1.160 (66.5)(	-	-
		= 611 nm,	n_1: 1.0 (air)		

MODEL AND EXPERIMENTAL RESULTS

Figure 2 shows the predicted e-h pair concentration profiles of Y_2O_3:Eu TFPs with different S values. The graphs indicate that the e-h pair concentration drastically decreases with increasing the penetration depth except for the very high S values; higher S (>1e5 cm/s) significantly reduces the number of e-h pairs which will participate the luminescent process; and smaller S (<1e3 cm/s) negligibly affects the e-h pair concentrations. Figure 3 predicts the total initial CL intensities (ΣI_0) generated inside the Y_2O_3:Eu TFP before propagation with three different diffusion coefficients of D=1, 5, and 10

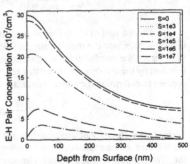

Figure 2. *E-h pair concentration profiles of Y_2O_3:Eu TFPs with different S values when D=5.0 cm²/s, =1e-9 sec, i_0=100 nA/cm², and V_0=2 kV.*

cm^2/sec as a function of film thickness. This figure shows that the intensities increase very rapidly and then saturate at around 1 μm film thickness for all the three cases. It also shows very small differences among the intensities before saturation and slightly higher intensities with increase of D value after 1 μm. Since the initial intensities are directly proportional to the number of charge carriers in a TFP, this calculation indicates that the charge carrier diffusion is an important part of CL generation process over the EBSI range. It also explains the reason why the Y$_2$O$_3$:Eu TFPs show rapid increase in CL brightness up to around 1 μm film thickness.

In order to verify the working model, Y$_2$O$_3$:Eu TFPs of various thicknesses (0.07 to 0.5 m) were deposited on various substrates such as

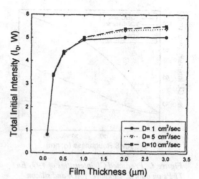

Figure 3. *Initial CL intensity profiles of Y$_2$O$_3$:Eu TFPs with different D vales when =0.3, S=1e4 cm/s, =1e-9 sec, i$_0$=100 nA/cm^2, and V$_0$=5 kV.*

quartz, sapphire, silicon, and lanthanum aluminate (LAO) under identical processing conditions using a pulsed laser deposition (PLD) technique and CL measurements were conducted. Figure 4 shows the calculated increase in relative brightness as a function of the surface roughness of the Y$_2$O$_3$:Eu TFPs deposited on quartz, sapphire, LAO, and silicon substrates. These calculations were conducted for a constant spatial generation rate of the emitted light. The figure shows that as the surface roughness increases the reflected brightness also increases. The rate of the increase is a function of the substrate properties and the roughness of the TFP-s surface. The figure also shows that the nature of the substrate plays a critical role in controlling the film brightness. The brightness values of the TFPs obtained from the quartz substrates are the highest, while those of the TFPs from the silicon substrates are the least. The difference in CL brightness is mainly attributed to the difference in their optical properties in terms of n and . The smaller n value of the quartz substrate with the Y$_2$O$_3$:Eu TFP provides a larger amount of total internal reflection inside the film and the trapped light can escape more easily through multiple internal reflection as the film surface is progressively rougher. Also, the smaller value gives less loss of the intensity of the propagating light during multiple internal reflection. The combined effects result in higher CL brightness from the Y$_2$O$_3$:Eu TFPs on quartz substrates. The experimental CL data obtained from the Y$_2$O$_3$:Eu TFPs on different substrates also showed the similar trends. Figure 5 shows one representative of the predicted and experimental CL intensities of the Y$_2$O$_3$:Eu TFPs deposited on quartz substrates in R-mode as a function of RMS roughness. The model predicts the CL properties of the TFPs precisely and experimental results can be explained using the model with any of the CL-determining parameters.

CONCLUSIONS

A new model of CL from Y$_2$O$_3$:Eu TFPs has been developed using the diffraction scattering theory. The number of e-h pairs created and involved in the CL process was obtained using the diffusion equation under the steady state condition. The Kanaya-Okayama equation for the EBSI range was used to solve the generation rate and the diffusion equation. Two different

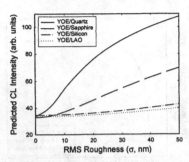

Figure 4. Predicted CL intensities of Y_2O_3:Eu TFPs on quartz, sapphire, LAO, and silicon substrates in R-mode as a function of RMS when $=0.3$, $S=1e4$ cm/s, $D=5.0$ cm^2/s, $=1e-9$ sec, $i_0=100$ nA/cm^2, and $V_0=2$ kV.

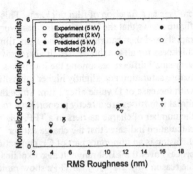

Figure 5. Experimental and predicted CL intensities of Y_2O_3:Eu TFPs deposited on quartz substrates in R-mode as a function of RMS when $=0.3$, $S=4e3$ cm/s, $D=5.0$ cm^2/s, $=1e-9$ sec, and $i_0=100$ nA/cm^2.

regimes of CL generation ($0<x$ R and R$<x$ t_1) were considered in a TFP due to the charge carriers reaching the activator centers outside the EBSI range by diffusion process. Since the light generated from each activator in a TFP, experiences absorption, reflection and transmittances at the various interfaces, the brightness was obtained integrating the total intensity of light escaping film surfaces through the multiple interfacial scattering. The new model satisfactorily predicts the brightness of the TFPs well being in good agreement with the controlled experiments. The model explains CL generation due to the charge carrier diffusion process in addition to the EBSI range, rapid increase of CL brightness of the Y_2O_3:Eu TFPs up to around 1 μm film thickness and saturation after that, higher CL intensities from the Y_2O_3:Eu TFPs on quartz substrates than the TFPs on the other substrates mainly due to smaller n and values. The model provides a general solution to predict the luminance properties of the TFPs for the CL-based technologies like FEDs and CRTs.

REFERENCES

1. S. L. Jones, D. Kumar, K. G. Cho, R. K. Singh, and P. H. Holloway, *Displays* **19**, 151 (1999).
2. Song Jae Lee, *Jpn. J. Appl. Phys.* **37**, 509 (1998).
3. K. G. Cho, D. Kumar, S. L. Jones, D. G. Lee, P. H. Holloway, and R. K. Singh, *J. Electrochem. Soc.* **145**, 3456 (1998).
4. H. E. Bennett and J. O. Porteus, *J. Opt. Soc. Amer.* **51** (2), 123 (1961).
5. A. S. Toporets, *Sov. J. Opt. Technol.* **46** (1), 35 (1979).
6. K. G. Cho, *Ph.D. Dissertation*, University of Florida, 2000.
7. A. Roos, and D. Ronnow, *Applied Optics* **33** (34), 7908 (1994).
8. J. S. Yoo and J. D. Lee, *J. Appl. Phys.* **81** (6), 2810 (1997).
9. B. G. Yacobi and D. B. Holt, *Cathodoluminescence microscopy of inorganic solids.* Plenum, New York, 1 (1990).

Mat. Res. Soc. Symp. Proc. Vol 621© 2000 Materials Research Society

The Effects of Microstructure on the Brightness of Pulsed Laser Deposited Y_2O_3:Eu Thin Film Phosphors for Field Emission Displays

K. G. Cho, R. K. Singh, D. Kumar, P. H. Holloway, H-J. Gao[1], S. J. Pennycook[1], G. Russell[2], and B. K. Wagner[2]

Department of Materials Science and Engineering, University of Florida, Gainesville, FL 32611.
[1]Solid State Division, Oak Ridge National Laboratory, Oak Ridge, TN 37831-6030.
[2]PTCOE, Georgia Institute of Technology, Manufacturing Research Center, Atlanta, GA 30332.

ABSTRACT

In order to investigate the effect of microstructure on the brightness of thin film phosphors for field emission displays, Y_2O_3:Eu thin film phosphors were prepared using pulsed laser deposition. To deconvolute the effects experimentally, the Y_2O_3:Eu films of controlled thickness and microstructure were prepared on various substrate materials such as amorphous quartz, (0001) sapphire, (100) lanthanum aluminate ($LaAlO_3$), and (100) silicon wafers. Cathodoluminescent brightness and efficiency of the films were obtained in both transmission and reflection modes. The Y_2O_3:Eu films deposited on the quartz substrates showed the maximum brightness followed by the films on (0001) sapphire, (100) lanthanum aluminate ($LaAlO_3$), and (100) silicon substrates. The role of interface scattering of the emitted light on the film brightness will be discussed together with changing surface roughness and film thickness.

INTRODUCTION

There has been significant interest in the development of thin film phosphors (TFPs) for field emission displays (FEDs). Since the field emitter tips are in close proximity to the phosphor screen in FEDs, there exist two major concerns using powder phosphors in terms of debonding and outgassing. Any failure of adhesion may cause debonding of powder particles from the bulk phosphor screen. The free particles debonded from bulk powder layer will block the gates of the field emitter tips. This blocking of emitter tips will eventually cause no light output from certain area of screen. Outgassing is another major concern of powder phosphors. The residual gases outcoming from the binding materials for powder particles and the relatively porous structure of powder layer itself will increase the gas pressure of the vacuum sealed device under energetic electron bombardments. Ionization, formation of plasmas, and arcing effects due to these gas phases with the electron beam (e-beam) will be detrimental to the emitter tips and phosphor dots. Therefore, TFPs are believed to be the only solution to solve these problems. However, the relatively low efficiency of the TFPs have been kept them behind the powder phosphors in FEDs so far. The relatively low efficiency of the TFPs compared to powder phosphors, can be significantly enhanced by tailoring the surface roughness of the TFPs using various methodologies including variation in processing conditions [1-3], modifying TFP-s surface [4,5], and addition of a buffer layer between TFP and substrate [6,7]. It has been experimentally shown that the light trapping due to internal reflection from a smooth surface is reduced as the surface becomes progressively rougher [4,5,8,9]. In this report, therefore, the correlations between microstructure and cathodoluminance (CL) characteristics of Y_2O_3:Eu TFPs will be discussed in detail, including the role of surface roughness, film thickness, and substrate properties.

EXPERIMENTAL DETAILS

Y_2O_3:Eu TFPs were deposited on amorphous quartz, (0001) sapphire, (100) $LaAlO_3$ (LAO), and (100) silicon wafers using pulsed laser deposition (PLD) with 248 nm KrF excimer laser at a temperature of 750°C. Y_2O_3:Eu target was ablated at 1.2 J/cm^2 laser energy density maintaining an oxygen pressure of 70 mTorr and a target-substrate distance of ~9 cm during the ablation experiments. Two sets of films from 70 nm to 500 nm in thickness were fabricated applying different number of pulses from 3000 to 24000 pulses with a repetition rate of 10 Hz. The absorption coefficient ( ) of the materials and refractive index (n) of LAO were measured using variable angle spectroscopic ellipsometry (VASE) and a Perkin-Elmer Lambda 9 UV-VIS-NIR spectrometer (UVNS). The temperatures of the substrates were controlled and monitored using a programable thermocoupled temperature controller and a pyrometer.

The Y_2O_3:Eu TFPs were characterized using X-ray diffraction (XRD), scanning electron microscopy (SEM), atomic force microscopy (AFM), and transmission electron microscopy (TEM) techniques. Film thicknesses were estimated using VASE, TENCOR Alpha-step 500 surface profilometer having a mechanical stylus, and SEM cross-sectional views of the fractured Y_2O_3:Eu TFPs. CL efficiency in both reflection mode (R-mode) and transmission mode (T-mode) was measured in the Phosphor Technology Center of Excellence (PTCOE) laboratory. TEM analysis was conducted in the Oak Ridge National Laboratory.

DISCUSSION

Figure 1 shows that the film thicknesses of Y_2O_3:Eu TFPs deposited on amorphous quartz, (0001) sapphire, (100) LAO, and (100) silicon substrates, increase with deposition time. The very small differences among substrates are believed to be induced by the experimental errors during the measurements due to different film morphologies and optical properties of the substrates. The average growth rate was ~140 Å/min (2.3 Å/sec or 0.23 Å/pulse). Figure 2 shows variations in the RMS roughness of the Y_2O_3:Eu TFPs with increase in film thickness. The

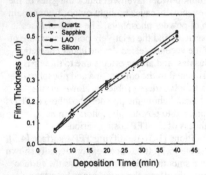
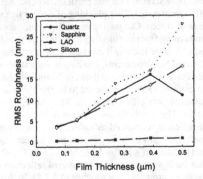

Figure 1. Film thickness of Y_2O_3:Eu TFPs deposited on quartz, sapphire, LAO, and silicon substrates with deposition time.

Figure 2. RMS roughness variation of Y_2O_3:Eu TFPs deposited on quartz, sapphire, LAO, and silicon substrates with film thickness.

Y$_2$O$_3$:Eu TFPs on sapphire substrates showed the highest roughness among the films while the Y$_2$O$_3$:Eu TFPs on LAO substrates showed lowest roughness. The Y$_2$O$_3$:Eu TFPs on quartz and silicon substrates showed less roughness except the film on quartz substrate which shows a rapid drop from 16 to 12 nm in RMS roughness at about 0.5 µm thickness.

The AFM micrograph studies on the surface morphologies of the Y$_2$O$_3$:Eu TFPs show visually the difference in roughness. Figure 3 shows a representative set of the AFM micrographs of the Y$_2$O$_3$:Eu TFPs deposited on (a) quartz, (b) sapphire, (c) silicon, and (d) LAO substrates for 18000 pulses (30 min, ~0.4 µm thick). The measured RMS roughness values were 16.00, 17.00, 13.60, and 1.00 nm from (a) to (d) in Figure 3, respectively. The surface morphologies of the Y$_2$O$_3$:Eu TFPs on quartz, sapphire, and silicon substrates from (a) to (c) in the figure show little difference while that of the films on LAO substrate show very smooth surface structures. The large difference in morphology between the Y$_2$O$_3$:Eu TFPs on the former three substrates and the Y$_2$O$_3$:Eu TFPs on LAO substrates is attributed to the different crystal structures of the substrate materials. Only ~1 % lattice mismatch between the Y$_2$O$_3$:Eu TFP and LAO substrate allows the films to grow epitaxially [10,11], which resulted in the very flat surface morphologies of the films. Figure 4 shows the XRD patterns of the Y$_2$O$_3$:Eu TFPs deposited on (a) quartz, (b) sapphire, (c) silicon, and (d) LAO substrates which are corresponding to the AFM micrographs in Figure 3. The Y$_2$O$_3$:Eu TFPs on quartz, sapphire, and silicon substrates exhibit only (111) orientation while the Y$_2$O$_3$:Eu TFPs on LAO substrates exhibit only (200) orientation.

Figures 5(a)-(b) are a low magnification TEM micrograph of a plan view and the corresponding selected area electron diffraction pattern (SAEDP) of the Y$_2$O$_3$:Eu TFP. The SAEDP indicates that the Y$_2$O$_3$:Eu TFP on the LAO substrate is a single crystalline film. The

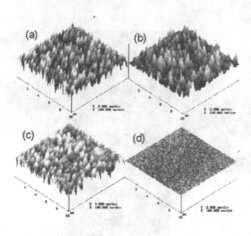

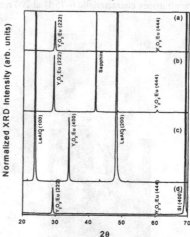

Figure 3 *Surface morphologies of the films fabricated with 18000 pulses on (a) quartz, (b) sapphire, (c) silicon, and (d) LAO substrates (x: 2µm/div, z: 100 nm/div).*

Figure 4. *XRD patterns of Y$_2$O$_3$:Eu TFPs deposited on (a) quartz, (b) sapphire, (c) silicon, and (d) LAO substrates for 18000 pulses (30 min, ~0.4 m thick).*

numerous small voids in the image suggest an island growth mechanism with incomplete coalescence of the islands. The electron projection is along the [001] zone axis of the Y_2O_3:Eu crystalline TFP. A cross section image of the sample is presented in Figure 5(c), showing the smooth surface, sharp interface and a uniform thickness of 300 nm maintained over the entire region. Figures 5(d)-(e) also show the SAEDPs of the TFP and the LAO substrate, showing the orientation relationship to be [110]Y_2O_3//[100]LAO and [-110]Y_2O_3//[010]LAO. A comparison of atomic positions in the layers of the film and the substrate materials suggests that a rotation of the film-plane (Y-layer or O-layer) by 45° results in epitaxial growth of the Y_2O_3:Eu TFPs on the LAO substrates. Under this atomic stacking order, the lattice mismatch which is given by $[\{(2a_s)^2 + (2a_s)^2\}^{1/2} - a_f]/[(2a_s)^2 + (2a_s)^2]^{1/2}$ between the TFP and the substrate is only ~1 %. This negligible lattice mismatch between the TFP and substrate promotes epitaxial film growth with better crystallinity. The columnar structure of the TFP is also apparent from the cross section image, with small rotations between neighboring grains giving the strong diffraction contrast. Each individual column however appears to be a good single crystal, which implies that the presence of the voids may avoid the need for a high density of dislocations between the grains to accommodate the rotations, and/or a high level of stress within the grains.

Table 1 shows the measured and calculated optical data for the Y_2O_3:Eu TFP and the substrate materials with an incident wavelength of 611 nm. The constants of n and t represent the refractive index and the thickness of films or substrates, respectively. The value of the critical angle (θ_c) is 0.545 arc degree which corresponds to 31.2°. According to the data in Table 1, the quartz substrate has the smallest optical absorption coefficient (α) for a wavelength of 611 nm. LAO is crystallographically a better substrate material for epitaxial Y_2O_3:Eu TFPs [10,11], but it is optically less advantageous than the other two transparent substrates because of its lowest transmittance characteristic (~75 % at 700 nm).

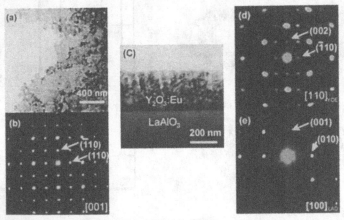

Figure 5. (a) Plan view TEM image and (b) corresponding electron diffraction pattern, and (c) cross sectional view TEM image and corresponding electron diffraction patterns of (d) Y_2O_3:Eu TFP and (e) LAO substrate showing [110]Y_2O_3// [100]LaAlO₃ and [-110]Y_2O_3// [010]LaAlO₃.

Table 1. Optical properties of Y_2O_3:Eu TFP and various substrate materials.

Constant	Y_2O_3:Eu	Quartz	Sapphire	LAO	Silicon
n	1.93	1.46	1.77	2.0	3.91
α (mm^{-1})	5	0.07	0.3	0.55	>200
t (:m)	0.07-0.5	1000	500	500	500
θ_c (w/air)	0.545 (31.2°)	0.755 (43.2°)	0.600 (34.4°)	0.523 (30.0°)	0.258 (14.8°)
θ_c (w/film)	1.570 (90.0°)	0.858 (49.2°)	1.160 (66.5°)	-	-
λ = 611 nm,		n_1: 1.0 (air)			

Figure 6 shows the CL brightness of the Y_2O_3:Eu TFPs deposited on quartz, sapphire, LAO, and silicon substrates with film thickness. The Y_2O_3:Eu TFPs on quartz substrates also showed higher CL brightness followed by the TFPs on sapphire, LAO, and silicon substrates. The differences in the CL brightness of the Y_2O_3:Eu TFPs were also caused by the different optical properties of the substrates. The quartz substrate with the smaller refractive index (n:~1.46) and less absorption of light, as shown in Table 1, gave higher CL brightness. Higher CL brightness of the TFPs on LAO in the smaller film thickness region where the roughness effects is small, is mainly due to better crystallinity of the TFPs on LAO substrates. Figure 7 shows the variations in the CL efficiency of the Y_2O_3:Eu TFPs on the quartz, sapphire, and LAO substrates for 30 minute deposition (~0.4 μm thick) in both R- and T-modes as a function of the accelerating voltage. These TFPs are the ones showing the best CL brightness and efficiency values in this set of experiments. The saturation and decrease of efficiency after ~4 kV was apparent through most of the Y_2O_3:Eu TFPs in the figure. The abrupt changes in T-mode CL efficiency were observed in the graphs due to experimental errors. The Y_2O_3:Eu TFPs measured in R-mode showed higher CL efficiencies than the Y_2O_3:Eu TFPs measured in T-mode.

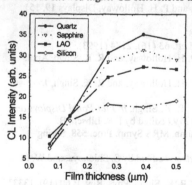

Figure 6. CL brightness of Y_2O_3:Eu TFPs on various substrate materials with different film thickness measured at 5 kV.

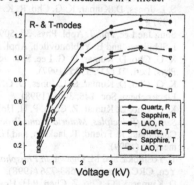

Figure 7. CL efficiencies of Y_2O_3:Eu TFPs on quartz, sapphire, and LAO substrates for 30 minutes (~0.4 m thick) in both R- and T-modes as a function of accelerating voltage.

CONCLUSIONS

Higher CL brightness and efficiency values were observed from the Y_2O_3:Eu TFPs on quartz substrates in R-mode. The higher CL brightness values of the Y_2O_3:Eu TFPs on quartz substrates than those on sapphire, LAO, and silicon substrates, were mainly associated with the advantageous optical properties of the quartz substrate in terms of smaller refractive index and low absorption coefficient. Higher CL brightness of the Y_2O_3:Eu TFPs on LAO substrates in the smaller film thickness region was mainly associated with the capability of better epitaxial growth of the Y_2O_3:Eu TFPs on the LAO substrates than the Y_2O_3:Eu TFPs on quartz, sapphire, and silicon substrates. TEM studies strongly support this explanation. The best CL brightness and efficiency was observed from the Y_2O_3:Eu TFPs deposited for about 30 minutes of PLD (18000 laser pulses) in this study. The Y_2O_3:Eu TFPs showed higher CL brightness and efficiency in R-mode than in T-mode.

ACKNOWLEDGEMENTS

This research was supported by an AFOSR Grant F49620-96-1-0026 and through the Phosphor Technology Center of Excellence by DARPA Grant No.MDA972-93-1-0030. The authors would also like to thank the Department of Energy (Grant No. DE-FG 05-95ER45533) for partial support to this research.

REFERENCES

1. S. L. Jones, D. Kumar, R.K. Singh, and P.H. Holloway, Appl. Phys. Lett. **71** (3), 404 (1997).
2. S. L. Jones, D. Kumar, R. K. Singh, and P. H. Holloway, *in Flat Panel Display Materials III* edited by R. T. Fulks, G.N. Parsons, D.E. Slobod, and T. H. Yuzuriha, MRS Symp. Proc. **471**, 299 (1997).
3. S. L. Jones, D. Kumar, K. G. Cho, R. K. Singh, and P. H. Holloway, Displays **19**, 151 (1997).
4. Song Jae Lee, Jpn. J. Appl. Phys. **37**, 509 (1998).
5. I. Schnitzer, and E. Yablonovitch, Appl. Phys. Lett. **63** (16), 2174 (1993).
6. K. G. Cho, D. Kumar, D. G. Lee, S. L. Jones, P. H. Holloway, and R. K. Singh, Appl. Phys. Lett. **71** (23), 3335 (1997).
7. K. G. Cho, D. Kumar, S. L. Jones, D. G. Lee, P. H. Holloway, and R. K. Singh, J. Electrochem. Soc. **145**, 3456 (1998).
8. K. G. Cho, D. Kumar, Z. Chen, P. H. Holloway, R. K. Singh, *in Flat-Panel Displays and Sensors (Principles, Materials, and Processes) 1999* edited by F.R. Libsch, B. Chalamala, R. Friend, T. Jackson, and H. Ohshima, MRS Symp. Proc. **558**, Spring meeting (1999).
9. T. Yoshioka, and M. Ogawa, *in Phosphor Handbook* edited by S. Shionoya, and W. M. Yen, CRC, New York, 683-726 (1998).
10. D. Kumar, K. G. Cho, Z. Chen, P.H. Holloway, R. K. Singh, Phys. Rev. B **60** (19), 13331 (1999).
11. H-J. Gao, D.Kumar, K. G. Cho, P. H. Holloway, R. K. Singh, X. D. Fan, Y. Yan, and S. J. Pennycook, Appl. Phys. Lett. **75** (15), 2223 (1999).

Organic Light Emitter Devices

Mat. Res. Soc. Symp. Proc. Vol. 621 © 2000 Materials Research Society

ORGANIC LIGHT-EMITTING DIODES USING TRIPHENYLAMINE BASED HOLE TRANSPORTING MATERIALS

Hisayoshi Fujikawa, Masahiko Ishii, Shizuo Tokito, and Yasunori Taga
TOYOTA Central Research and Development Labs., Inc.
Nagakute, Aichi, 480-1192, JAPAN.

ABSTRACT

The durability of the tris(8-quinolinolato) aluminum based light-emitting diode (LED) is related to the thermal stability of the hole transport layer. Several linear linkage triphenylamine oligomers were used for the hole transport layer. The thermal stability was clearly seen to depend on a glass transition temperature (Tg) of the hole transporting material, and a linear relationship between the Tg and the thermal stability was found. A lowering of "turn-on voltage" for light emission and an increase of luminous efficiency with increasing temperature was also observed. Excellent durability of the organic LED with a tetramer of triphenylamine was achieved at a high temperature of 120°C. Our results indicate that the linear linkage of triphenylamine leads to a high Tg and high device performance at high temperatures.

INTRODUCTION

Light-emitting diodes (LEDs) based on organic materials are attracting much attention as candidates for flat-panel displays and backlights for liquid crystal displays[1-6]. Over the past few years a considerable number of studies has been made on the improvements of the durability of the devices[7-9]. A half-decay lifetime of more than 10000 h from an initial brightness has been already realized at the present time. More recently, a practical multi-color display has been commercialized in small molecular organic materials. However, further research and development are required to improve the lifetime and the color tunability, especially materials developments are significantly important.

For the applications for automobile, good reliability at a high temperature is necessary. The most typical structure of the organic LED comprises two organic layers of a hole-transporting layer and an emitting layer. The thermal stability of the device characteristics is primarily limited by the morphological stability of the organic layers, especially, the hole-transporting layer[10,11]. For the hole-transporting layer aromatic amines are commonly used, because they have the excellent hole-transporting ability and good film-forming ability. By vacuum deposition, the aromatic amine forms a uniform amorphous film on any substrate. Therefore, the hole-transporting material based on aromatic amine structure which has a high glass transition temperature (Tg) is one of promising candidates to realize the high durability at a high temperature and at high driving current.

In our previous study, it was shown that the organic LEDs with a tetramer of triphenylamine can be operated in a continuous operation up to 130 °C without breakdown[12]. Furthermore, a systematic study of the temperature dependences of EL characteristics has been carried out in organic LEDs fabricated using trimer, tetramer and pentamer of triphenylamine in addition to the well-known dimer for the hole-transporting layer, and the Alq₃ for the emitting layer.   In this paper, we present the developments of hole transporting materials and the EL properties of the organic LEDs operated at a high temperature[13, 14].

EXPERIMENTAL DETAILS

The hole transporting materials, used in this study, which are based on a linear linkage of triphenylamine moiety, are shown in Fig. 1.   We use the terms "TPTR", "TPTE" and "TPPE" referring to trimer, tetramer and pentamer, respectively.   All the synthesized materials were purified by column chromatography through silica gel.   The Tg of each derivative was measured by differential scanning calorimetry (DSC).

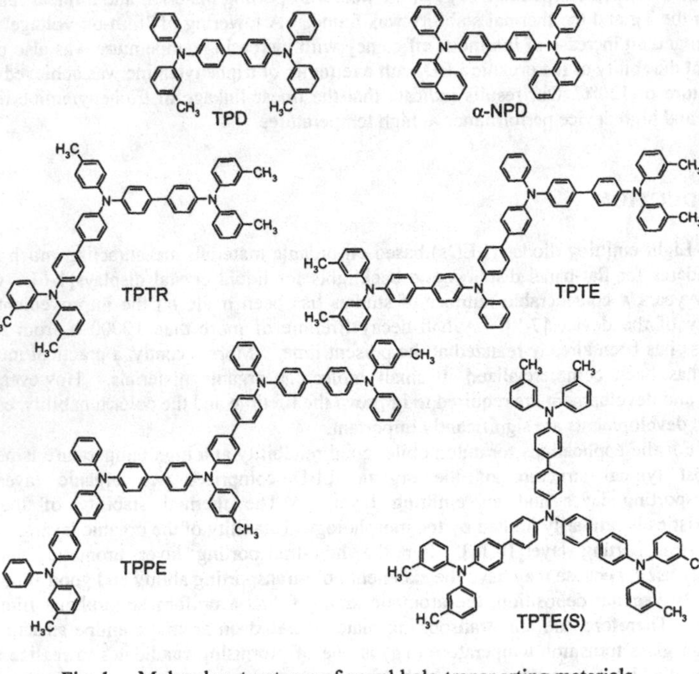

Fig. 1.   Molecular structures of novel hole-transporting materials.

Schematic structure of the organic LED and chemical structure of the emitting material are shown in Fig. 2.   The Alq$_3$ is a well-known emitting material which is one of most stable molecules in the class of metal chelates. The Alq3 was purified by a train sublimation method. The organic LEDs have been fabricated by a conventional vacuum deposition technique. The glass substrates coated with ITO (sheet resistance:15Ω/sq.) were cleaned ultrasonically in acetone and isopropyl alcohol, and dried in an oven kept at 60°C. The substrates were treated with a UV ozone chamber prior to use. At first, the hole-transporting layer (typically 60 nm thick) was deposited on the ITO electrode by vacuum evaporation from a graphite crucible at a pressure of 5 x 10^{-7} Torr. Then, the emitting layer (typically 60 nm thick) was deposited on the hole-transporting layer. The deposition rates were typically 3-5 nm/min for the transporting and emitting layers. Finally, the cathode electrode of Mg:Ag (9:1) mixed film (180nm) or LiF (0.5nm) / Al(160nm) was deposited from two metal sources under a vacuum of 5 x 10^{-7} Torr. An emitting area of the LEDs was typically 0.09 cm^2.

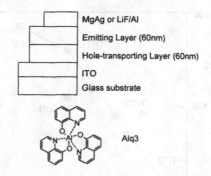

Fig. 2. Schematic structure of LED and molecular
structure of emitting material Alq3

The luminance-current and current-voltage characteristics were measured at a dry nitrogen atmosphere to prevent oxidation of the cathode with a programmable voltage/current source (TAKEDARIKEN TR-6150 or KEITHLEY 2400) and a Minolta photometer (LS-110). In the high temperature experiments, the LED was set in a small stainless steel chamber. The chamber was heated from room temperature to 160°C by heating tapes connected to a programmable temperature controller. Light output was measured through a glass viewport (ANELVA φ70) of the chamber.

The mobility measurements were carried out using a conventional time-of-flight (TOF) technique. Charge carriers were generated in a thin organic layer close to an ITO electrode by irradiation of a N$_2$ laser with a pulse width of 10 ns at a wavelength of 337 nm. A resistor with the value between 50 and 25KΩ was connected in series to the sample. The current response time was longer than 10 RC, where R was the series resistance and C the sample capacitance. The photocurrent was measured with a digitizing stragescope. All measurements were carried out at room temperature.

RESULTS AND DISCUSSIONS

The glass transition temperatures of the hole-transporting materials were found to be 60°C for TPD, 95°C for TPTR, 130°C for TPTE and 145°C for TPPE from DSC measurements[15]. The Tg of TPD is almost in agreement with that (63°C) reported in a literature[16], and for TPTR it's Tg is close to that (110°C) of a similar material reported by Adachi et al.[17].

Figure 3 shows the Tg versus the oligomerization number for the hole-transporting materials. The linear linkage of triphenylamine causes a rise in the Tg and shows a tendency of saturation of the Tg at high molecular weight above the tetramer[12]. This result indicates, as

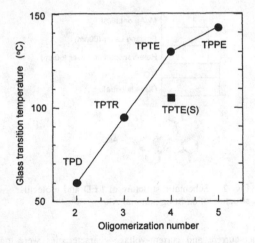

Fig. 3. Glass transition temperature vs oligomerization number for the hole-transporting materials.

we expected, that the linear linkage of triphenylamine is a promising way to create the high Tg hole-transporting materials. By comparing the TPTE with the TPTE(S), we have also found that the linear linkage is much more effective than the branch linkage in increasing Tg.

These materials were confirmed to have the ability to form good films on glass substrates by vacuum deposition. The resultant films were transparent and amorphous. The glassy state was very stable at room temperature, and no crystallization was observed, with the exception of TPD. The TPD film exhibited a partial morphological change accompanying crystallization after three months even at room temperature.

All the organic LEDs emitted green light with a peak wavelength of 530 nm, which was fully identical to the fluorescence of the Alq$_3$. The devices using TPTR, TPTE, and TPTE(S) exhibited the power efficiency, 0.94 - 1.0 lm/W, which is comparable to that of the TPD/Alq$_3$ device.

The thermal stability of the LEDs can be examined from the temperature dependence of the EL intensity measured under a constant current drive condition. The experiments were carried out at a constant current of $11mA/cm^2$ in ascending temperature of 1°C/min from room temperature. Figure 4 (a) shows the temperature dependences of the EL outputs for TPD/Alq₃ and TPTE/Alq₃ devices. The conventional LED with TPD exhibits a drastic drop of the EL output at 70°C, which is good correspondence with the Tg (60°C). The temperature of the intensity drop for the TPTE/Alq₃ device is over 100 °C, and it is also correspondence with the Tg (135 °C) of TPTE.

For all the LEDs, the EL outputs gradually decreased with increasing temperature. These decreases are primarily ascribed to the decrease of fluorescence quantum efficiency of Alq₃, although the temperature dependences are different. At a higher temperature, a significant difference in the thermal stability of the LEDs is clearly seen. The temperatures of the intensity drop for other LEDs are 110°C for TPTR, 155°C for TPPE, and 120°C for TPTE(S).

Figure 4 (b) shows the Tg's of the hole-transporting materials plotted against such critical temperatures of intensity drop. It is apparent that a linear relationship exists between the Tg and the critical temperature. This undoubtedly indicates that the thermal stability of the LEDs with Alq₃ emitting layer are dominated by the hole-transporting layers. The intensity drops above the Tg's are attributable to the morphological changes of the hole-transporting layers of each devices. Microscopic observations showed the marked morphological changes, from smooth to rough films, accompanying crystallization for the whole areas of the devices. We assume that a local molecular motion at around Tg brought about the morphological changes, and the electronic contacts between organic layers and electrodes became poor.

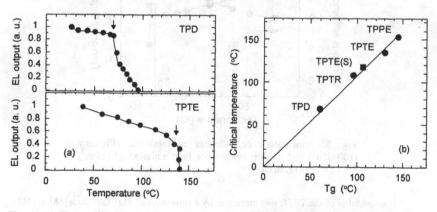

Fig. 4. (a) Temperature dependence of EL output for devices with TPD and TPTE under a continuous DC operation of $11mA/cm^2$. (b) Tg of the hole-transporting materials vs critical temperature of the LEDs.

Here, we describe the temperature dependence of EL characteristics in the device with TPTE (Tg=130°C). Figure 5 shows the temperature dependences of the turn-on voltage for light emission and the luminous efficiency. The turn-on voltage lowered from 3.1 V at room temperature to 2.1 V at 131°C with increasing temperature. The lowering of the turn-on voltage may be related to the increase of hole and electron mobilities in the hole transporting and emitting layers. It was reported that the hole mobility of triphenylamine derivative thin film significantly increased with increasing temperature[15].

Although the quantum efficiency gradually decreased, the luminous efficiency increased with increasing temperature, the luminous efficiency of 1.25 lm/W at 131°C was obtained. Improvement of the luminous efficiency is attributed to the lowering of the driving voltage. At the temperature above Tg, however, the turn-on voltage increased and the luminous efficiency dropped. These drastic changes of characteristics at the glass transition temperature of TPTE indicate that the Tg is the critical temperature for the LED based on such amorphous organic materials. Nevertheless, there is indication of the thermal stability. Even after the LED was heated up to 150°C without an applied voltage, the device showed almost the same emission intensity at room temperature as before.

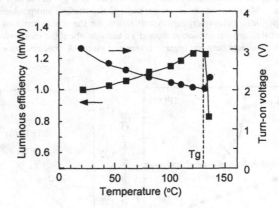

Fig. 5. Temperature dependences of luminous efficiency (100cd/m^2) and turn-on voltage for light emission (0.1cd/m^2) for a TPTE/Alq$_3$ device.

Hole mobility of the TPTE was measured by a conventional TOF technique[18]. ITO / 5μm-thick hole transporting material / Ag structure is used. Figure 6 (a) shows the transient photocurrent profile for hole in the TPTE film at 1.6×10^5V/cm. The transient photocurrent of TPTE shows dispersive behavior, the transit time could be determined from clear change of double linear plot of photocurrent versus time characteristics.

The mobility μ is calculated from transit time, film thickness, and applied voltage. Figure 6 (b) shows the logarithmic mobilities versus $E^{1/2}$ plot for TPTE compared with TPD and

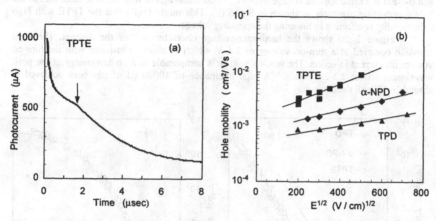

Fig. 6. (a) Transient photocurrent profile of a TPTE film at 1.6×10^5V/cm. (b) Hole drift mobilities for three hole-transporting materials.

α-NPD. The linear relation is obtained for all hole transporting materials, which means that the log(μ) of amorphous organic films is linear with a square root of the applied field [18]. The hole mobility for TPTE approaches on the range from 3×10^{-3} cm^2/Vs (4×10^4V/cm) to 8×10^{-3} cm^2/V s (2.5×10^5V/cm) with a weakly dependent on the electric field. In comparison with other hole transporting materials, the TPTE has the highest hole mobility, indicating that the TPTE has superior hole-transport capability.

We have also investigated the influence of hole transporting material on the EL property of the Alq$_3$ based device. The EL characteristics of the device with TPTE were compared with those with TPD and α-NPD devices. Here, the fabrication process of the devices was changed, the ITO surface was treated by r.f. plasma of argon/oxygen gas for 30 s at a pressure of 7×10^{-2} Torr for cleaning. The applied r.f. power was 10 W to the ITO-coated glass substrate and the power density was about 100 mW/cm^2. In addition, LiF/Al electrode was used instead of MgAg elecrode in order to enhance the electron injection from the cathode.

Figure 7 (a) shows the current-voltage (I-V) characteristics of the devices using different hole transporting materials. All the devices show similar current injection characteristics. These results imply that no difference in the energy barrier for hole-injection from the ITO electrode. The turn-on voltage for current injection was around 2.6 V, and the I-V characteristics follow the power-law over four orders of current density. This power-law dependence of the I-V data suggests that the current is dominated by space-charge injection effects [19]. And the low turn-on voltage suggests that the hole injection/ from the ITO at low voltage may be ohmic, regardless of the hole transporting material.

In addition, we can observe a slight difference of the I-V characteristics at a high voltage over 7V. The current density of the device with TPTE is larger than those with TPD and α-NPD at a same applied voltage over 7V. This result agrees with the hole mobilities of the hole transporting materials as shown in Fig. 6 (b). This might imply that the TPTE with high hole mobility is effective in lowering the operating voltage.

Figure 7 (b) shows the luminance-voltage characteristics of the devices. The light emission occurred at a turn-on voltage of 2.4 V which is almost consistent with the turn-on voltage for current injection. The turn-on voltage is comparable to a photon energy of the peak wavelength (about 2.3-2.4 eV). A high luminance of 10000 cd/m^2 has been achieved at around 10 V in all the devices.

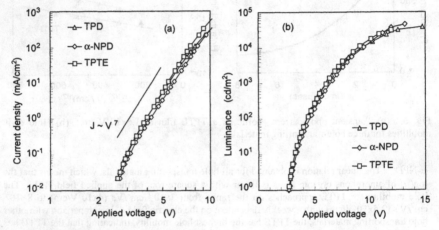

Fig. 7. (a) Current density-voltage characteristics, and (b) luminance-voltage characteristics for devices using three hole-transporting materials.

Table 1 summarizes the photometric, external quantum and luminous efficiencies of the devices. The luminous efficiency is also the another important efficiency for practical emitting

Table 1. EL efficiencies of devices using three hole transporting materials (at 300cd/m^2)

HTL	Photomeric Efficiency (cd/A)	Quantum Efficiency (%)	Luminous Efficiency (lm/W)
TPD	5.6	1.9	3.5
α-NPD	5.7	2.0	3.3
TPTE	4.9	1.7	3.1

devices. The devices gave a luminous efficiency in excess of 3 lm/W at a luminance of 300 cd/m^2. The high luminous efficiency is attributable to the high quantum efficiency and the low operating voltage which were achieved by the appropriate hole transporting material and the plasma treatment of ITO surface. The difference in the photometric and quantum efficiencies is attributable to the interaction such as exciplex between the hole transporting layer and the Alq$_3$ layer[14].

Even at room temperature, if the device is driven at a high current or a high luminance, the lifetime is influenced by the hole transporting material. The lifetime tests were carried out under a continuous operation of a constant current from an initial luminance of 3100 cd/m^2 as shown in Fig. 8. The half-decay lifetime was defined to be time for decreasing until 50% of the initial value. The current stress was around 100 mA/cm^2. Compared with α-NPD (Tg=95°C) and TPD (Tg=60°C), the device using TPTE (Tg=130°C) exhibited a longer lifetime: 22 h for TPTE, 8 h for α-NPD and 5h for TPD. This result is attributable to a rise of device temperature by the Joule-heat. The TPTE is a suitable hole transporting material for the devices operated at a high luminance.

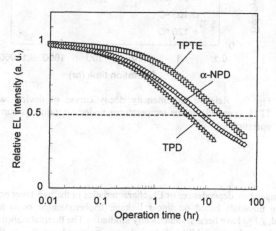

Fig. 8. Relative EL intensity decay curves of devices with three hole-transporting materials under a continuous current operation of 100 mA/cm^2.

The durability of continuous operation at a high temperature has been also examined in the LED with TPTE. Dimethylquinacridone was doped in the 20nm-Alq$_3$ layer close to TPTE layer for high efficiency green emission[20]. Figure 9 shows the luminance decay curves of the devices operated at various temperatures. The lifetime tests were carried out under a continuous operation of a constant current from an initial luminance of 300 cd/m^2. At room temperature, the lifetime is estimated to be over 15000 h. The lifetime of the LED operated at

85°C is 1700 h, 680 h at 105°C, and 270 h at 120°C. On the other hand, the LED with the low Tg material, TPD, exhibited a very short lifetime less than 1 min over 85°C. Such excellent durability at the high temperature is undoubtedly attributed to the excellent thermal stability of the TPTE layer. At a temperature above 130°C, however, the lifetime significantly became short, as the temperature approaches the Tg. Morphological change of the TPTE layer may gradually occur even at temperatures below the Tg.

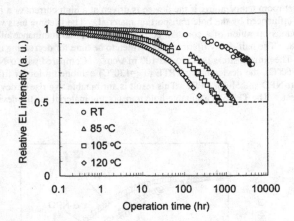

Fig. 9. Relative EL intensity decay curves of devices with TPTE operated various temperatures under a continuous current operation.

SUMMARY

The temperature dependences of EL characteristics in the two-layer organic LEDs using hole-transporting materials based on linear linkage triphenylamine oligomers and a typical emitting material, Alq_3, have been systematically studied. The thermal stability of the LED was found to be dominated by the stability of the hole-transporting layer, that is, the glass transition temperature. Our study clearly indicates that the linear linkage of triphenylamine provides high Tg materials which result in high device performances at high temperatures. Specially, triphenylamine tetramer, TPTE, was found to be a superior hole transporting material with thermal stability and highly electrical properties. Excellent durability of the organic LED with TPTE was achieved at a high temperature of 120°C.

REFERENCES

1. C. W. Tang and S. A. VanSlyke, Appl. Phys. Lett. **51**, 913 (1987)
2. C. Adachi, S. Tokito, T. Tsutsui, and S. Saito, Jpn. J. Appl. Phys. **27**, L269 (1988)
3. J. H. Burroughes, D. D. Bradley, A. R. Brown, R. N. Marks, K. Mackay, R. H. Friend, P. L. Burns, and A. B. Holmes, Nature **347**, 539 (1990)
4. J. Kido, M. Kimura, and K. Nagai, Science **267**, 1332 (1995)
5. H. Tokailin, M. Matsuura, H. Higashi, C. Hosokawa, and T. Kusumoto, Proc. SPIE 1910, 38 (1993)
6. Y. Fukuda, S. Miyaguchi, S. Ishizuka, T. Wakimoto, J. Funai, H. Kubota, T. Watanabe, H. Ochi, T. Sakamoto, M. Tsuchida, I. Ohshita, H. Nakada, and T. Tohma, SID 99 Digest, p.430 (1999)
7. Y. Shirota, Y. Kuwabara, H. Inada, T. Wakimoto, H. Nakada, Y. Yonemoto, S. Kawami, and K. Imai,   Appl. Phys. Lett. **65**, 807 (1994)
8. C. Adachi, T. Tsutsui, and S. Saito, Appl. Phys. Lett. **56**, 799 (1990)
9. T. Sano, Y. Hamada and K. Shibata, Inorganic and organic electroluminescence/EL96 Berlin, pp. 249-254, (Wissenschaft und Technik Verlag, Berlin, 1996)
10. E. Han, L. Do, Y. Niidome and M. Fujihira, Chem. Lett., pp. 969-972 (1994)
11. S. Tokito, and Y. Taga, Appl. Phys. Lett. **66**, 673 (1995)
12. S. Tokito, H. Tanaka, K. Noda, A. Okada, and Y. Taga, Appl. Phys. Lett. **69**, 878 (1996)
13. S. Tokito, H. Tanaka, K. Noda, A. Okada, and Y. Taga, IEEE Trans. Electron Devices, **44**, 1239 (1997).
14. S. Tokito, K. Noda, K. Shimada, S. Inoue, M. Kimura, Y. Sawaki, Y. Taga, Thin Solid Films **363**, 290 (2000)
15. H. Tanaka, S. Tokito, Y. Taga, and A. Okada, J. Chem. Soc., Chem. Commun., 2175 (1996)
16. M. Stolka, J. F. Yanus, and D. M. Pai, J. Phys. Chem. **88**, 4707 (1984)
17. C. Adachi, K. Nagai, and N. Tamoto, Appl. Phys. Lett. **66**, 2679 (1995)
18. S. Naka, H. Okada, H. Onnagawa, and T. Tsutsui, Appl. Phys. Lett. **76**, 197 (2000)
19. P. E. Burrows, S. R. Forrest, Appl. Phys. Lett. **64**, 2285 (1994)
20. J. Shi and C. W. Tang, Appl. Phys. Lett. **70**, 1665 (1997)

Mat. Res. Soc. Symp. Proc. Vol. 621 © 2000 Materials Research Society

Optical Properties of Microcavity Structures using the Organic Light Emitting Materials

Boo Young Jung, Nam Young Kim, Chang Hee Lee, Chang Kwon Hwangbo, and Chang Seoul[1]
Department of Physics, Inha University, Inchon 402-751, Republic of Korea [1]Department of Textile Engineering, Inha University, Inchon 402-751, Republic of Korea

ABSTRACT

We investigated the optical properties of Fabry-Perot microcavity with a tris(8-hydroxyquinoline)aluminum (Alq_3) organic film by measuring the photoluminescence (PL) and transmittance. An Alq_3 layer as an active layer was sandwiched between two mirrors, which were metal or ($TiO_2|SiO_2$) dielectric multilayer reflectors. An Alq_3 layer on glass, [air|Alq_3|glass], showed a PL peak around 513 nm and its full width half maximum (FWHM) was about 80 nm. Three types of microcavity, such as Type A [air|metal|Alq_3|metal|glass], Type B [air|dielectric|Alq_3|dielectric|glass], and Type C [air|metal|Alq_3|dielectric|glass], were fabricated. The result shows that the FWHMs of three Fabry-Perot microcavities were reduced to 15~27.5, 7~10.5 and 16~16.6 nm, respectively, and the microcavity structure is expected to improve the efficiency and tunability of emission spectrum in display.

INTRODUCTION

Tang and Van Slyke [1] introduced the organic light emitting devices (OLED) using Alq_3 as an emitting material in 1987. The organic light emitting material such as Alq_3 has been expected to play a significant role in future display technology, since it has the high fluorescence quantum efficiency in the visible spectrum and properties of semiconductor. Compared with the inorganic LED, the organic LED is to be fabricated at a low cost as well as in a flexible way. As the lifetime of OLED is improved, it is recently used in the car radio display.

A microcavity structure [2-5] has been expected to show the narrow spectral emission, the improvement of emission efficiency, the tunable emission spectrum, the fabrication of organic semiconductor laser [6], the application of optical switching and optical sensor, and the source of optical communication. The cavity is formed between top and bottom mirrors that are ($TiO_2|SiO_2$) distributed Bragg reflectors or metal.

In this study, we used Alq_3, a light emitting material in green, as an active layer and simulated the phase change on reflection at the interface between mirror and Alq_3 film. Also we showed two ways to control a resonant wavelength (RW) of microcavity. First, it is to adjust the cavity length as a common method. Second, it is to control the phase change on reflection [5].

EXPERIMENTAL

In order to obtain the maximum quantum efficiency, it is important that the mirrors have the minimum reflectance at an excitation wavelength and the wavelength of the maximum PL peak intensity is coincident with the RW. We designed our specimens to be in the optimum condition and they were three types: Type A is [air|Ag|Alq_3|Ag|glass], Type B [air|$(LH)^6$|Alq_3|$(LH)^6$|glass], and Type C [air|Ag|Alq_3|$(LH)^n$|glass]. In this paper L stands for the quarter wave optical thickness of low refractive index SiO_2 film and H for the quarter wave optical thickness of high refractive index TiO_2 film. The Ag and Alq_3 films were fabricated by a thermal source and the dielectric multilayer films by an electron-beam source. The base pressure of vacuum chamber was under 5×10^{-6} torr. TiO_2 and SiO_2 thin films were used as a high and a low refractive index material, respectively. Substrate temperature was $290\pm10°C$ for top mirror and $90\pm10°C$ for bottom mirror. The deposition rate for organic films and multilayers was about $1\sim2$ Å/s by a quartz crystal monitor. The PL spectrum was measured with a spectrofluorophotometer, RF-5301 (Shimadzu Corp.). Also the transmittance of specimens was measured with a spectrophotometer, Cary500 (Varian Corp.).

RESULTS AND DISCUSSION

Figure 1 is the PL spectrum as a function of excitation wavelength. The incident angle

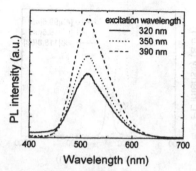

Figure 1. PL spectrum as a function of excitation wavelength for bare Alq$_3$, [air|Alq$_3$|glass] at the detecting angle of 45 °.

and detecting angle were 45°. The maximum PL peak was at 513 nm and the FWHM of the peak is about 80 nm. The ideal excitation wavelength is 390nm for the bare Alq$_3$. However, This excitation wavelength does not given the maximum PL peak intensity for all types due to the different reflection on surface. So, The excitation wavelengths for Type A, B, and C were 320, 390, and 350 nm, respectively, which were selected at the maximum quantum efficiency.

The mth order of the RW in the microcavity is given by

$$\lambda_m = \frac{4\pi Nd\cos\theta}{\Phi_a + \Phi_b - 2\pi m} \qquad (1)$$

where Φ_a and Φ_b are the phase changes on reflection at interfaces between mirrors and Alq$_3$ film. N is the refractive index of Alq$_3$ film and d is the thickness. From eq. (1), the RW shifts to the short wavelength region as the phase change on reflection increases, the incident angle increases or the thickness of Alq$_3$ film decreases.

Figure 2 shows the PL spectrum of p-polarized light of Type A at the detecting angle of 45 °. The RW of thick Ag film of [68|99.5|68] is shifted slightly to the short wavelength compared to thin one of [45|99.5|45] where the unit of each number is nm. The PL peak intensity of thick Ag film is remarkably reduced compared to thin one of [45|99.5|45]. The RW of thick Alq$_3$ film of [68|118.8|68] is shifted to the long wavelength compared to thin one of [68|99.5|68]. The FWHM of Type A is about 15~27.5 nm. The weak PL peak intensity of s-polarized light is not plotted.

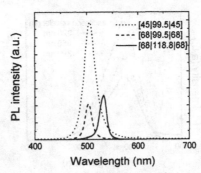

Figure 2. PL spectrum of Type A at the detecting angle of 45 °.

In order to reduce the FWHM at the RW, the dielectric multilayer mirrors were employed. Figure 3 is the RW of *s*-polarized light of Type B at the detecting angle of 45 °. When the thickness of Alq3 film increases from 112 nm of $[(LH)^6|112|0.92(LH)^6]$ to 128 nm of $[(LH)^6|128|0.92(LH)^6]$, the RW is shifted to the long wavelength. The simulation shows the phase change on reflection decreases as the optical thickness of dielectric multilayer increases. So, as shown in Fig. 3 the increase in the thickness of dielectric layer and Alq3 film leads the RW to the long wavelength. The FWHM of Type B is about 7 ~10.5 nm. The smaller FWHM of Type B than that of Type A is due to the lower absorption and higher reflection of the dielectric mirrors.

Fig. 4 is the RW of *p*-polarized light of Type C at the detecting angle of 45 °. The reflectance at the bottom mirror is enhanced as the number of pairs of dielectric layers increase from 5 of $[55|196|1.14(LH)^5]$ to 6 of $[55|196|1.10(LH)^6]$. Hence it gives rise to the increase of the PL peak intensity and the reduced FWHM of 16 ~ 16.6 nm.

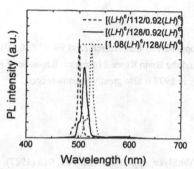

Figure 3. PL spectrum of Type B at the detecting angle of 45 °.

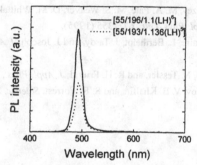

Figure 4. PL spectrum of Type C at the detecting angle of 45 °.

CONCLUSIONS

We have made the three types of organic microcavity structure employing Alq₃ film as an emitting layer. The resonant wavelength is shifted not only by the cavity length but also by the phase change on reflection. The FWHM of PL spectrum of Alq₃ film was drastically reduced by high reflection mirrors of microcavity and the narrowest FWHM was 7~10.5 nm for Type B. We expect that the organic microcavity is going to be used widely in future display due to the color tunability and the reduction in fluorescent spectrum.

ACKNOLEGMENT

This work was supported by the research fund 98-0702-03-01-3 of Korea Science and Engineering Foundation and the Brain Korea 21 Project, Republic of Korea. The financial support by Inha University in 1999 is also greatly acknowledged.

REFERANCES

[1] C. W. Tang and S. A. VanSlyke, *Appl. Phys. Lett.*, **51**, 913 (1987).

[2] A. Dodabalapur, L. J. Rothberg, R. H. Jordan, T. M. Miller, R. E. Slusher, and Julia M. Phillips, *J. Appl. Phys.*, **80**, 6954 (1996).

[3] T. A. Fisher, D. G. Lidzey, M. A. Pate, M. S. Weaver, D. M. Whittaker, M. S. Skolnick, and D. D. C. Bradley, *Appl. Phys. Lett.*, **67**, 1355 (1995).

[4] B. Masenelli, A. Gagnaire, L. Berthelot, J. Tardy, and J. Joseph, *J. Appl. Phys.*, **85**, 3032 (1999).

[5] H. Becker, S. E. Burns, N. Tessler, and R. H. Friend, *J. Appl. Phys.,* **81**, 2825 (1997).

[6] V. Bulovic, V. G. Kozlov, V. B. Khalfin, and S. R. Forrest, *Science.* **279**, 553 (1998).

Mat. Res. Soc. Symp. Proc. Vol. 621 © 2000 Materials Research Society

Experiment and Modeling of Conversion of Substrate-Waveguided Modes to Surface-Emitted Light by Substrate Patterning

Min-Hao M. Lu, Conor F. Madigan, and J. C. Sturm
Center for Photonic and Optoelectronic Materials, Department of Electrical Engineering,
Princeton University, Princeton, NJ 08544, U.S.A.

ABSTRACT

To predict the optical power that could be harvested from light emission that is waveguided in the substrate of organic light emitting devices (OLEDs), a quantitative quantum mechanical model of the light emitted into the waveguided modes has been developed. The model was used to compute the exact distribution of energy in external, substrate and ITO/organic modes as a function of the distance of the emission zone from the cathode. The results are compared to the classical ray optics model and to experiments in two-layer OLED devices. Classical ray optics is found to substantially over-predict the light in waveguided modes.

INTRODUCTION

OLEDs have received enormous interest because of their promise for cheaper and more efficient flat panel displays. A large amount of light is trapped in the substrate due to total internal reflection; therefore, substrate patterning can be used to increase external coupling efficiency [1-3]. The exact distribution of optical energy in all of the waveguide modes has not been calculated previously, except for by classical ray optics. The goal of this paper is to develop such a quantitatively accurate model of such waveguided light and verify it with experiments.

The radiative modes can be classified into external, substrate and ITO/organic modes (Figure 1a). External modes are those with angle to the surface normal in the organic layer less than the critical angle between air and Alq3, $\theta_{c1} = \sin^{-1} n_{air}/n_{alq}$; substrate modes are those with angle between θ_{c1} and the critical angle between glass and Alq3, $\theta_{c2} = \sin^{-1} n_{glass}/n_{alq}$; and

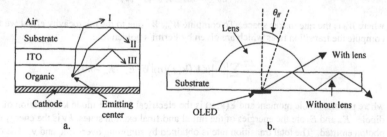

Figure 1. a. Three radiative modes in OLEDs: I. External modes, II. Substrate modes, and III. ITO/organic modes. B. Attaching a lens to the back side of OLED converts some light in substrate modes into external modes [3].

ITO/organic modes are those with angle greater than θ_{c2}. Substrate modes are emitted through the edges of the substrate, whereas the ITO/organic modes are heavily attenuated, and does not emit through the edge [5]. According to classical ray optics the amount of energy emitted into the external, substrate and ITO/organic modes are 18.9%, 34.2% and 46.9%, respectively, for refractive indices of $n_{alq} = 1.71$, $n_{glass} = 1.51$, and $n_{air} = 1$.

By laminating lens arrays on the backside of the OLEDs, some of the light waveguided in the substrate is allowed to emit externally (Figure 1b). For example, Figure 3a shows the far-field emission intensity with and without a lens attached to an OLED made with a single layer poly-(N-vinylcarbazole) (PVK) blend [3]. In one case, the OLED was 2.3 mm below the center of curvature of the lens, and there is a strong focusing effect. Intensity in the normal direction was increased by a factor of 9.5. In another case, the OLED was near the center of curvature of the lens and the total external emission was increased by a factor of 2.0. To understand how much light can be harvested by such techniques, ray optics is not adequate due to microcavity and other quantum mechanical effects in OLED structure. Therefore, in this paper we develop the full theory of light emission into waveguided modes and compare it with experiments.

THEORY

Field quantization in one-sided leaky microcavities was developed in an early paper [4]. Microcavity effects in OLEDs were first investigated by Bulovic et al. [5], whose treatment we follow closely.

The exciton recombination is modeled as a radiating dipole. The total energy transfer rate from the dipole is:

$$W_{Tot} = W_{ext} + W_{sub} + W_{ITO-org} + W_{DM-SP} + W_{NR}$$

where W_{ext}, W_{sub} and $W_{ITO-org}$ are the transfer rates into the external, substrate-wave guided and ITO/organic modes. W_{DM-SP} denotes the dipole-metal energy transfer, and W_{NR} the non-radiative energy transfer due to system-crossing. The external efficiency can be found as η_{ext}:

$$\eta_{ext} = \left(W_{ext}/W_0\right)\left(W_0/W_{Tot}\right)$$

where W_0 is the rate in free space. To compute W_{ext}/W_0 due to the microcavity effect we first compute the transition rates which are given by Fermi's golden rule:

$$f = \frac{2\pi}{h} \sum_n \left|\langle m|\mu \Box E(k,z)|n\rangle\right|^2 \delta\left(E_n - E_m - h\upsilon\right)$$

where μ is the dipole moment and $E(k, z)$ is the electrical field for mode k at location of the dipole. E_m and E_n are the energies of the initial and final exciton states. $h\upsilon$ is the energy of the photon emitted. The total transition rate is obtained by summing over all k, and υ. The electric filed is determined by the microcavity structure:

$$\mathbf{E}_k^{TE} = \mathbf{A}(\mathbf{k}) \text{Sin}^2 (k_{oz} l) \mathbf{x}$$

$$\mathbf{E}_k^{TM} = \mathbf{B}(\mathbf{k}) \text{Cos}^2 \theta_o \text{Sin}^2 (k_{oz} l) \mathbf{y} + \mathbf{C}(\mathbf{k}) \text{Sin}^2 \theta_o \text{Cos}^2 (k_{oz} l) \mathbf{z}$$

where $\mathbf{A}(\mathbf{k})$, $\mathbf{B}(\mathbf{k})$ and $\mathbf{C}(\mathbf{k})$ are functions of material constants and $\mathbf{k}$. k_{oz} is the z component of the wave vector in the emitting layer. l is the distance of the emitting center to the cathode. And θ_o is the angle of the mode in the emitting layer. For spontaneous emission, the electrical fields are normalized such that the energy in each mode is equal to that of a single photon. In our two-layer OLEDs, we assume the excitons are created at the PVK/Alq3 interface, and diffuse into Alq3 with a characteristic length of 20 nm [5].

The ratio W_0/W_{Tot} is calculated by considering the intrinsic PL efficiency and the quenching effect of a metal surface next to the dipole, which has been solved classically via the Green's function method [6,7]. It has been computed for a 250 nm Alq3/α-NPD/ITO waveguide, which we will use as an estimate for our OLEDs [5]. We note that the cathode quenching effect diminishes rapidly as the dipole moves away from the metal surface, and virtually disappears once the dipole is 60 – 80 nm away.

EXPERIMENTAL DETAILS

OLEDs were fabricated on O_2 plasma treated, indium tin oxide (ITO, 180 nm, n_{ITO} = 1.8) soda lime glass (n_{glass} = 1.51) purchased from Applied Films Co. [8]. The hole transporting layer was 80 nm PVK (n_{PVK} = 1.67) deposited by spin-coating. The electron transporting and emitting layer was aluminum *tris*(8-hydroxyquinoline) (Alq3, n_{alq} = 1.71) deposited by vacuum sublimation. Its thickness was varied from 20 to 80 nm. The cathode was 50 nm of Mg:Ag (10:1) followed by a thick Ag cap, deposited by evaporation through a shadow mask with 1.7 mm diameter holes. The energy band diagram of these devices is shown in Figure 2. The two-layer OLED structure was used to precisely control the location of the exciton creation and emission. This is because the model results depend critically on the cathode-exciton distance, and this distance is not well known in single-layer devices.

In some cases, a plano-convex lens of radius 3.4 mm was attached to the backside of the OLEDs by index matching gel. The resulting effective substrate thickness is 1.1 mm. The far-field intensity was measured as a function of polarization both with and without the lens. Emission out of the edge of the glass was measured by placing the substrate through the thin slit of a black diaphragm and directly against a photo detector. The substrates were diced, thus the

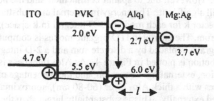

Figure 2. Energy band diagram of the OLEDs. The distance between the recombining exciton and the cathode, l, is controlled by the thickness of the Alq3 layer. Excitons are assumed to be created at the Alq3/PVK interface.

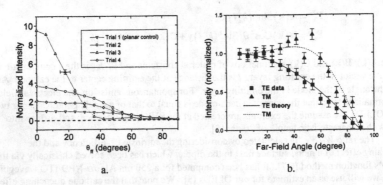

Figure 3. a. Far-field intensity patterns of single-layer OLEDs with planar substrate and with lenses of various geometry. b. Modeling and experimental data of far-field intensity pattern of a planar two-layer OLED with a 20 nm Alq₃ layer. Solid square and upper triangle: measured TE and TM intensity. Solid and dashed lines: theory.

edges are flat but not optically clear, acting as a diffusive scatterer. With this method the change in the optical energy in the substrate modes can be directly measured.

RESULTS AND DISCUSSION

The polarization resolved far-field intensity profile for a device with a 20 nm Alq₃ layer with a planar substrate (no lens) is plotted along with the theoretical prediction (Figure 3b). The fit between theory and data is excellent, reproducing the peak in the TM intensity profile which is a non-classical phenomenon due to the fact that TM radiation is enhanced for an exciton close to the cathode – a microcavity effect.

The QM microcavity model can be used to compute light emission into various modes. The external modes are a continuum. So are the substrate modes since the substrate in much thicker than the ITO/organic layer and the wavelengths in question. ITO/organic modes are discrete. For most of the wavelengths, there is only one TE and one TM mode in the ITO/organic waveguide. However, due to spatial confinement and the normalization condition, the energy in them is still significant. The energy in external and substrate modes is found by integrating over wave vector $\mathbf{k}$ for each mode (a 3-D integral in $\mathbf{k}$-space), the exciton profile, and the Alq₃ emission spectrum. The energy in ITO/organic modes is computed in the same fashion, except that the 3-D integral is replaced by a discrete sum and a 2-D integral in $\mathbf{k}$-space. The calculated energy distribution is plotted in Figure 4. As Alq₃ thickness is increased from 20 to 80 nm, total light emission, external light emission, and the percentage of external light emission all increases. For devices with a thick Alq₃ layer (60-80 nm), approximately 35% of the total light generated is emitted externally. This is substantially larger than the 18.9% external emission given by classical ray optics. The highest external quantum efficiency, η_{ext}, measured in an 80-nm-Alq₃ planar device was 0.9%, from which we infer that approximately 10% of injected electrons produce singlet excitons [5].

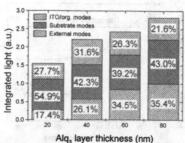

Figure 4. Distribution of energy into external, substrate and ITO/organic modes as a function of Alq3 thickness. The energy in the external modes is normalized according to measurement. The percentages are calculated from the model. The energy in the substrate and ITO/organic modes is computed from the ratio of the percentages.

We now look at simultaneous measurements of external and edge emission before and after a lens is attached. Since light in ITO/organic modes suffer from high propagation loss, all the light entering the edge detector are from the substrate modes [5]. We define r_1 and r_2 as the ratio of external and edge emissions before and after attaching the lens. The lens does not alter the spatial energy distribution of external modes which is predominantly in air. Furthermore, the substrate is much thicker than the wavelengths in question, any change at the backside of the substrate should not affect the ITO/organic microcavity; therefore, the total light emission is assumed to be the same before and after attaching the lens. Then the ratio of energy emitted into the external and substrate modes (which is equal to η_{ext}/η_{sub}) can be calculated directly from r_1 and r_2. For a sample with an 80 nm Alq3 layer, r_1 and r_2 were found to be 1.75 ± 0.12 and 0.30 ± 0.05 respectively, which implies a η_{ext}/η_{sub} of 93 ± 20%. In comparison, classical ray optics gives η_{ext}/η_{sub} = 55%, and our model gives η_{ext}/η_{sub} = 83%. The model clearly gives much better agreement with experimental data.

For display systems engineering, it is important to know how much light can be harvested by substrate patterning. Due to the finite size of the lenses, we examine light emission only in a 120° cone, because modes with larger angles run into edge of the lens. Total integrated light emission in this cone before and after attaching the lens is measured experimentally and computed by the model (Figure 5). For devices with thick Alq3 layer (60-80 nm), the agreement between data and modeling result is well within the experimental error. Both give an increase by a factor of 2.10. The model over-predicts the amount of increase at thinner Alq3 thickness. We believe the discrepancy originates from the estimate of cathode quenching effect, which is critical for excitons located near the cathode. On the other hand, classical ray optics predicts an increase by a factor of 3.2 in all devices. Again, our model fits the data better, especially at higher Alq3 layer thickness.

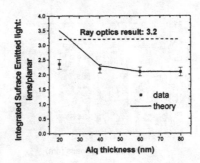

Figure 5. Modeling and experimental data for the improvement factor in total integrated light in the forward 120 ° cone as a result of attaching a lens as a function of Alq₃ layer thickness. Ray optics predicts an improvement of 3.2X.

CONCLUSIONS

We developed a quantum mechanical microcavity model for OLEDs to compute light emission into the external, substrate and ITO/organic modes. The validity of the model is verified by good fits to far-field intensity and edge emission data. The model also accurately predicts the increase of externally emitted light as a result of attaching a lens to an OLED with thick Alq₃ layers (60-80 nm). Ray optics was found to over-predict both the fraction of waveguided light and the increase due to attaching a lens.

ACKNOWLEDGEMENTS

M.-H. Lu would like to thank V. Bulovic for helpful discussions. This work is supported by DARPA and NSF.

REFERENCES

1. G. Gu, D. Z. Garbuzov, P. E. Burrows, S. Venkatesh, and S. R. Forrest, Opt. Lett., **22**, 396 (1997).
2. T. Yamasaki, K. Sumioka, and T. Tsutsui, *App. Phys. Lett.*, **76**, 1243, (2000).
3. C. Madigan, M.-H. Lu, J. C. Sturm, *App. Phys. Lett.*, **76**, 1650 (2000).
4. K. Ujihara, *Phys. Rev. A*, **12**, 148 (1975).
5. V. Bulovic, V. B. Khalfin, G. Gu, and P. E. Burrows, D. Z. Garbuzov, and S. R. Forrest, *Phys. Rev. B*, **58**, 3730 (1998).
6. R. R. Chance, A. Prock, and R. Sibley, *Adv. in Chem. Phys.*, **37**, 1 (1978).
7. D. Z. Garbuzov, V. Bulovic, P. E. Burrows, and S. R. Forrest, *Chem. Phys. Lett.*, **249**, 433 (1996).
8. C.-C. Wu, J. C. Sturm, R. A. Register, J. Tian, E. P. Dana, and M. E. Thomson, *IEEE Tran. Elec. Dev.*, **44**, 1269 (1997).

Mat. Res. Soc. Symp. Proc. Vol. 621 © 2000 Materials Research Society

New Dendritic Materials as Potential OLED Transport and Emitter Moeities

Asanga B. Padmaperuma[1], **Greg Schmett**[1], **Daniel Fogarty**[1], **Nancy Washton**[1], **Sanjini Nanayakkara**[1], **Linda Sapochak**[1], **Kimba Ashworth**[2], **Luis Madrigal**[2], **Benjamin Reeves**[2], and **Charles W. Spangler**[2]
[1]Department of Chemistry, University of Nevada, Las Vegas,
 Las Vegas, Nevada 89154-4003, USA.
[2]Department of Chemistry and Biochemistry, Montana State University,
 Bozeman, MT 59717, USA.

INTRODUCTION

Traditionally, organic light-emitting devices (OLEDs) are prepared with discrete layers for hole and electron transport.[1] Different materials must be used for these layers because most materials will preferentially transport one charge carrier more efficiently than the other. In most cases, the emitter material serves a dual purpose as both the emitter and the hole or electron transporter. One of the major failure modes of OLEDs results from thermal instabilities of the insulating organic layers caused by joule heating during device operation.[2,3] The problem is most pronounced for the hole transporting layer (HTL) material which are usually tertiary aromatic amines (i.e., TPD and NPD). This has been attributed to the relatively lower glass transition temperatures (Tg) and resulting inferior thermal stabilities compared to the other materials making up the device.[4,5] Many researchers have produced HTL materials with higher Tgs based on tertiary aromatic amine oligomers[6] and starburst compounds.[7] Starburst or model dendritic materials offer the advantages of high thermal stabilities and multi-functionality.

Model dendrimers based on bis-(diphenylamino)-E-stilbene subunits as the dendrimer repeat units have been shown to form exceptionally stable bipolaronic dications upon oxidative doping.[8,9] These charge states can also be photogenerated in the presence of electron acceptors on the picosecond time scale and are currently being evaluated as optical power limiters. Replacement of P for N reduces the capability for bipolaron formation, but substitution of only some of the N atoms in the dendrimer allows the preferential formation of polaronic radical cations.[10] Thus design of the dendrimer can predetermine which charge states dominate the photophysical processes, a key factor in OLED design. Thus we anticipate that dendrimers based on bis-(diphenylamino) and bis-(diphenylphosphino) substituents incorporated into the dendrimer repeat units can possibly function as not only efficient hole transport materials, but also as emitters in various OLED design configurations. We present a study of the luminescence and thermal properties of bis(diphenylamino- and diphenylphosphino)-E-stilbene oligomers and model dendrimers and will illustrate the efficacy of controlling the P/N substitution patterns in the model dendrimers in terms of device performance.

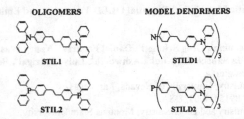

OLIGOMERS **MODEL DENDRIMERS**

STIL1   STILD1

STIL2   STILD2

Figure 1. *Structures of bis(diphenylamino)- and bis(diphenylphosphino)-E-stilbene oligomers and model dendrimers.*

EXPERIMENTAL

Materials

The bis(diphenylamino)- and bis(diphenylphosphino)-E-stilbene oligomers and model dendrimers are listed in Figure 1. Synthesis and structural characterization of all materials are reported elsewhere.[8,9,10]

Photophysical Characterization

Absorbance spectra were recorded with a VARIAN CARY 3BIO UV-Vis Spectrophotometer and photoluminescence spectra were obtained with a SLM 48000 Spectrofluorometer in $CHCl_3$ and as vapor-deposited films with an excitation wavelength of 365 nm.

Thermal Analysis

Thermal analysis was conducted utilizing a Netzsch Instrument Simultaneous Thermal Analyzer (STA-DSC/TGA) at a heating rate of 20°C/min.

Device Fabrication and Electroluminescence Characterization

Devices were grown on glass slides precoated with indium tin oxide (ITO). The model dendrimers STILD1 or STILD2 were deposited next as the HTL on the ITO substrate by thermal evaporation from a baffled Mo crucible at a nominal rate of 2-4 Å/s under a base pressure of <$2X10^{-6}$ Torr. A 550 Å emitter layer of aluminum tris-(8-hydroxyquinoline) (Alq_3) was then deposited on the HTL. A top Mg:Ag electrode consisting of circular 1 mm diameter contacts was subsequently deposited by thermal evaporation through a mask. The Mg:Ag alloy layer (1250 Å) was deposited by coevaporation of the two metals from separate Mo boats in a 10:1 Mg:Ag atomic ratio under a base pressure of 10^{-5} Torr, followed by a 300 Å Ag cap. Electrical pressure contact to the device was made by means of a 25µm diameter Au wire. Current-voltage characteristics were measured with a Hewlett-Packard HP4145 semiconductor parameter analyzer, and EL intensity was measured with a Newport 835 powermeter with a broad spectral bandwidth (400-1100 nm) photodetector placed directly below the glass substrate.

RESULTS AND DISCUSSION

Absorbance and Photoluminescence Data

The absorbance properties of all materials in $CHCl_3$ and as vapor-deposited films are reported in Table I. The oligomer containing only N-heteroatoms (STIL1) exhibits two well-separated absorbance peaks in both solution and in the solid-state. Upon substitution of P for N (STIL2) the longer wavelength peak is shifted to higher energy by 4874 cm^{-1} compared to STIL1 resulting in an overlap of the higher energy absorbance peak (see Figure 2). Furthermore, vibrational fine structure is observable for STIL2 only. Similar trends for absorbance are observed for solid-state films of STIL1 and STIL2. This suggests that delocalization of electron density through the diphenylphosphino-group is substantially lessened compared to the diphenylamino-group. This is not unusual due to the larger size of the P-heteroatom and the lack of overlap of the P- lone pair electrons with the extended π-system of the molecule.[11]

A comparison of the PL spectra of STIL1 and STIL2 in solution and as solid-state films is shown in Figure 3. The effect of P-substitution results in a shift of PL emission to higher energy in solution compared to the nitrogen-containing analogue, similar to what is observed for absorbance. However, in the solid-state film the PL emission of STIL2 is significantly shifted to lower energy compared to STIL1.

The absorbance and PL emission spectra of the model dendrimer, STILD1 containing only N-heteroatoms is similar to the oligomeric counterpart (STIL1) discussed above except that the longer wavelength band is shifted to lower energy. This may result from delocalization of electron density between the arms of the model dendrimer through the core N-heteroatom. Replacement of the core N-heteroatom with P gives a mixed heteroatom system (STILD2). The effect of P-substitution on the absorbance and PL emission energies is less dramatic for the model dendrimers (see Figure 4). Unlike the oligomer STIL2, which contains only P-heteroatoms, the mixed heteroatom system STILD2 exhibits two well-resolved absorbance peaks. Although the longest wavelength peak is shifted to higher energy compared to STILD1 in solution and in solid-state films, this shift is much smaller than what is observed for the oligomer STIL2. Furthermore, the PL emission spectrum of STILD2 in solution and the solid-state film are not significantly different, but are both shifted to longer wavelength compared to STILD1.

Table I. Absorbance and PL Emission Data

Compound	Solution Absorbance (nm)	Film Absorbance (nm)	Solution PL Emission (nm)	Film PL Emission (nm)
STIL1	307, **389**	309, **391**	439	464
STIL2	316, **327**, 348	321, **332**, 348	380	476
STILD1	309, **409**	309, **410**	464	471
STILD2	300, **392**	**302**, 389	471	480

* Highest intensity absorbance peak in bold.

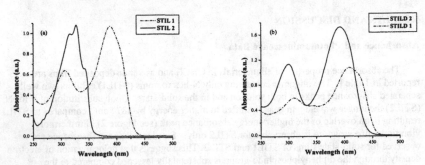

Figure 2. Absorbance spectra in $CHCl_3$ of a) oligomers and b) model dendrimers.

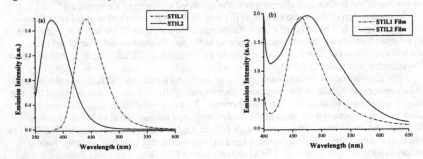

Figure 3. PL emission spectra of oligomers in a) $CHCl_3$ and b) vapor-deposited films.

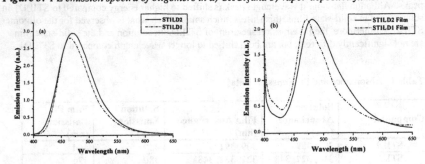

Figure 4. PL emission spectra of the model dendrimers in a) $CHCl_3$ and b) vapor-deposited films.

Table II. Thermal Properties

Compound	1st Heating	2nd Heating
STIL1	$T_m = 252°C$	No Change
STIL2	$T_m = 173°C$	$T_g = 41°C$
STILD1	$T_g = 96°C$	$T_g = 105°C$
STILD2	$T_g = 115°C$	$T_g = 121°C$

Thermal Properties

The DSC results for the oligomers and model dendrimers are shown in Table II. After one heating scan only the model dendrimers exhibited glass transitions (Tg's). However, after a cooling quench and a second heating scan of the P-containing oligomer the melting endotherm did not appear and a T_g at 41°C was observed. In this series of materials, substitution of P for N results in improved thermal properties.

Electroluminescence Data

The EL efficiencies and voltages at a constant drive current of 13 mA/cm² for devices prepared with STILD1 and STILD2 as HTL materials were 0.09% (8.5V) and 0.13% (11.8V) respectively. The current/voltage and optical power output/current plots are shown in Figure 5. Although both devices exhibit lower efficiencies and higher voltages than devices utilizing NPD as the HTL,[12] a significant difference in device performance is observed on P-heteroatom substitution of the HTL material. Efficient hole injection depends on the relative energies of the HOMO levels of the HTL and emitter material at the device interface. Therefore, the effect of P-substitution in the model dendrimers on device properties may result from a difference in the HOMO energy and/or charge-transport ability of the material. The relative energies of the HOMO levels can be accessed by cyclic voltammetry[6] and by x-ray spectroscopic methods.[13] Those experiments are in progress and the results will aid in understanding how P-substitution affects the properties of these materials as HTL's.

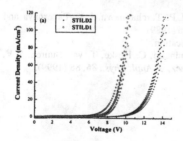 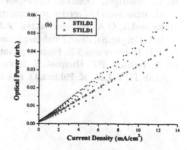

Figure 5. a) Current density/voltage plots, and b) Optical Power/current density plots for OLEDs utilizing STILD1 and STILD2 as HTL materials.

CONCLUSIONS

The absorbance, PL emission and thermal properties of oligomeric and model dendrimers based on bis(diphenylamino- and diphenylphosphino)-E-stilbene units were evaluated. Large energy shifts in absorbance and PL emission were observed for the oligomers when the N-heteroatoms were substituted with P. On the other hand, for the model dendrimers substitution of only the core N-heteroatom with P produced much smaller effects on the absorbance and emission properties. Finally, the model dendrimer STILD2 containing a P-heteroatom core exhibited superior thermal stability and higher EL efficiency than the corresponding material containing only N-heteroatoms when utilized as a HTL material.

ACKNOWLEDGMENTS

This research was supported in part by a NSF/CAREER-DMR-9874765 grant, Technical Management Concepts, Inc. under subcontract No. TMC-97-5405-0013-02 and BMDO/AOR. The authors would like to thank Dr. P.E. Burrows and Prof. S.R. Forrest, Princeton University for device fabrication and testing.

REFERENCES

1. C. W. Tang and S. A. VanSlyke, *Apl.Phys.Lett.*, **51**, 913 (1987).; C. W. Tang and S. A. VanSlyke, *J Appl. Phys.*, **65**, 3610 (1989).
2. Y. Kuwahara, H. Ogawa, H. Inada, N. Norma and Y. Shirota, *Adv. Mater.*, **6**, 677(1994).
3. C. Adachi, K. Nagui and N. Tamoto, *Appl. Phys. Lett.*, **66**, 2679 (1995).
4. C. Adachi, T. Tsutsui and S. Saito, *Appl. Phys. Lett.*, **56**, 799 (1990).
5. S. Tokito and Y. Taga, *Appl. Phys. Lett.*, **66**, 673 (1995).
6. B.E. Koene, D.E. Loy and M.E. Thompson, *Chem. Mater.*, **10**, 2235 (1998).
7. Y. Shirota, T. Kobata and N. Noma, *Chem. Lett.*, 1145 (1989).
8. E.H. Elandaloussi and C.W. Spangler, *Polymer Preprints*, ACS, **39(2)**, 1055 (1998).
9. K. Ashworth, C.W. Spangler and B. Reeves, *SPIE Proceedings on Third-Order Nonlinear Optical Materials II*, SPIE, **3796**, 191 (1999).
10. L. Madrigal, Doctoral Dissertation, Department of Chemistry and Biochemistry, Montana State University, in preparation.
11. Ratovskii, O.A. Shivernovskaya, A.M.Panov, L.P. Turchaninova, L.K. Papernaya and E.N. Deryagina, *Zhur. Obshchei Khimii*, **59**, 2026 (1989).
12. P.E. Burrows and S.R. Forrest, private communication.
13. R. Treusch, F.F. Himpsel, S. Kakar, L.J. Terminello, C. Heske, T. van Buuren, V.V. Dinh, H.W. Lee, K. Pakbaz, G. Fox and I. Jimenez, *J. Appl. Phys.*, **86**, 88 (1999).

Electroluminescence

Mat. Res. Soc. Symp. Proc. Vol. 621 © 2000 Materials Research Society

Electroluminescent Oxide Phosphor Thin Films Prepared by a Sol-gel Process

Tadatsugu Minami, Toshihiro Miyata, Tetsuya Shirai and Toshikuni Nakatani
Electron Device System Laboratory, Kanazawa Institute of Technology,
7-1 Ohgigaoka, Nonoichi, Ishikawa 921-8501, Japan

ABSTRACT

Ga_2O_3:Mn and (Ga_2O_3-SnO_2):Eu thin films have been prepared by a sol-gel process. High luminance emissions were obtained in TFEL devices: green using a Ga_2O_3:Mn and red using a Ga_2O_3:Eu or a SnO_2:Eu thin-film emitting layer. These sol-gel prepared devices were produced without using any vacuum processes and always exhibited higher luminances than equivalent devices prepared by r.f. magnetron sputtering: 40, 309 and above 1000 cd/m^2 using SnO_2:Eu, Ga_2O_3:Eu and Ga_2O_3:Mn phosphor thin films, respectively, in TFEL devices prepared by the sol-gel process and driven at 1kHz.

INTRODUCTION

High luminance multicolor emissions have been recently reported in thin-film electroluminescent (TFEL) devices using oxide phosphors[1,2]. In addition, a stable long term operation in the atmosphere has been demonstrated in an unsealed TFEL device with an oxide phosphor thin-film emitting layer[3,4]. Oxide phosphors feature a higher chemical stability than the sulfide phosphors which are widely used as the emitting layer of conventional TFEL devices. As a result, these sulfide phosphor thin-film emitting layers must be prepared in a vacuum using various processes such as electron beam evaporation, sputtering or atomic layer deposition methods[5,6]. In contrast, TFEL devices with an oxide phosphor thin-film emitting layer can be fabricated in the presence of water. We have recently reported that high luminance oxide phosphor TFEL devices could be fabricated by chemical deposition using a solution coating technique [7-9] which eliminates the need for vacuum processes. However, this deposition process using a solution of metallic complex salt source materials dissolved in methanol is unsuited for large area coatings since it requires a heat treatment featuring a rapid rise in temperatures.

In this paper, we describe newly developed high luminance TFEL devices using Ga_2O_3, SnO_2 and SnO_2-Ga_2O_3 phosphor thin films prepared by a sol-gel process which eliminates the need for a heat treatment with a rapid temperature rise.

EXPERIMENTAL

Electroluminescent (EL) characteristics were investigated using the ac single-insulating layer TFEL device structure [10] shown in Fig.1. The TFEL devices were fabricated by depositing Ga_2O_3 or SnO_2 phosphor thin films onto thick sintered $BaTiO_3$ insulating ceramic sheets (thickness, about 0.2 mm): gallium acethylacetonate ($Ga(C_5H_7O_2)_3$) or trimethoxy gallium ($Ga(OCH_3)_3$) as the Ga source and tetra-i-proxy tin ($Sn(O-i-C_3H_7)_4$) or tetra-n-buthoxy tin ($Sn(O-n-C_4H_9)_4$) as the Sn source with manganese chloride ($MnCl_2$) or europium chloride ($EuCl_3$) as the dopant source. For example, a Eu-activated SnO_2 (SnO_2:Eu) thin film was prepared by the following sol-gel process. $Sn(O-n-C_4H_9)_4$ and $EuCl_3$ were dissolved in methanol (CH_3OH) by stirring for 30 minutes at room temperature (RT) in a N_2 gas atmosphere. Subsequently, H_2O and HCl were added into the solution and stirred for 5 hours at 50°C in the atmosphere; as a result, the solution exhibited a pH of about 2.6, a Sn concentration of 0.05

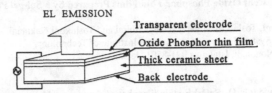

EL EMISSION

Transparent electrode
Oxide Phosphor thin film
Thick ceramic sheet
Back electrode

Fig.1 *Cross-sectional structure of a thick ceramic-insulating-layer-type TFEL device.*

molar% and a Eu content (Eu/(Sn+Eu) atomic ratio) of 0.5-60 atomic%. The BaTiO$_3$ ceramic sheets were first immersed in the solution and then dried about 5 min in air at RT before being heated in a furnace for 10 min in air at 600-1000℃. This heat-treatment temperature is referred to as the depositing temperature in this report. This procedure was repeated 20-30 times in order to obtain a film thickness of approximately 1.5-2 μ m. In addition to SnO$_2$:Eu, Ga$_2$O$_3$:Eu and Ga$_2$O$_3$:Mn phosphors were used as the thin film emitting layer. The Ga$_2$O$_3$ phosphor thin films were also deposited by the same sol-gel process described above: the solution exhibited a pH of about 3.4, a Ga concentration of 0.08 mol.% and a Mn content (Mn/(Ga+Mn) atomic ratio) of 0.3 at.%.

In order to improve crystallinity as well as luminescent properties, SnO$_2$:Eu and (SnO$_2$-Ga$_2$O$_3$):Eu thin films were postannealed in air for 1 h at temperatures of 950 to 1170℃, while Ga$_2$O$_3$:Eu and Ga$_2$O$_3$:Mn thin films were postannealed in Ar for 1 h at tempertaures of 850-1070℃. The heat-treatment temperature used in this annealing process is referred to as the annealing temperature. In the final procedure of TFEL device fabrication, an Al-doped ZnO transparent conducting film and an Al film were deposited as the transparent electrode and the back electrode, respectively. The electroluminescent (EL) characteristics of the TFEL devices driven by a sinusoidal wave voltage at 1 kHz were measured using a Sawyer-Tower circuit, an ac power meter and a conventional luminance meter.

RESULTS AND DISCUSSION

TFEL Devices using Ga$_2$O$_3$ Phosphors

Ga$_2$O$_3$ phosphors have recently attracted much attention as a new phosphor material for TFEL devices. For example, we reported that TFEL devices using Ga$_2$O$_3$:Mn exhibit a high luminance green emission which is suitable as the green primary color of full-color displays [1,2,11]. The EL characteristics of TFEL devices using Ga$_2$O$_3$:Mn thin films prepared by the sol-gel process using Ga(C$_5$H$_7$O$_2$)$_3$ or Ga(OCH$_3$)$_3$ source materials were strongly dependent on preparation conditions such as depositing and annealing temperatures. As an example, luminance (L)- and transferred charge density (Q)-applied voltage (V), L-V and Q-V characteristics, as functions of annealing temperature are shown in Fig.2 for TFEL devices with Ga$_2$O$_3$:Mn thin-film emitting layers prepared using Ga(OCH$_3$)$_3$ at a depositing temperature of 900℃. As can be seen in this figure, the effect of the annealing temperature on the L-V and Q-V characteristics were relatively correlated. It was found that the highest luminance was obtained in Ga$_2$O$_3$:Mn TFEL devices prepared at the depositing and annealing temperatures of 900 and 1020℃, respectively, irrespective of the Ga source materials. Figure 3 shows the L-V characteristics for TFEL devices using Ga$_2$O$_3$:Mn thin films prepared under optimal conditions:

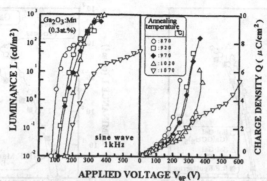

Fig.2 *L-V and Q-V characteristics as functions of annealing temperature for Ga$_2$O$_3$:Mn TFEL devices.*

sol-gel process using Ga(C$_5$H$_7$O$_2$)$_3$ or Ga(OCH$_3$)$_3$ as the Ga source material and r.f. magnetron sputtering. It should be noted that the devices prepared without a vacuum by the sol-gel process exhibited higher luminances than the device prepared by r.f. magnetron sputtering, which uses a vacuum process. Luminances above 1000 cd/m^2 were obtained in Ga$_2$O$_3$:Mn TFEL devices prepared by the sol-gel process under optimal conditions using either Ga(C$_5$H$_7$O$_2$)$_3$ or Ga(OCH$_3$)$_3$. However, Ga(OCH$_3$)$_3$ was unstable and not as easy to handle as Ga(C$_5$H$_7$O$_2$)$_3$.

In regard to red emission, we have already reported on TFEL devices using Ga$_2$O$_3$:Eu thin films prepared by r.f. magnetron sputtering[12]. Based on the sol-gel process techniques described above, we prepared Ga$_2$O$_3$:Eu TFEL devices. Figure 4 shows the L-V characteristics for TFEL devices using Ga$_2$O$_3$:Eu thin films prepared by both the sol-gel process using Ga(C$_5$H$_7$O$_2$)$_3$ under the optimal conditions described above and r.f. magnetron sputtering under optimal conditions; the sol-gel process resulted in higher luminances than r.f. magnetron sputtering with its vacuum process. Maximum luminances of 27 and 309 cd/m^2 were obtained in Ga$_2$O$_3$:Eu TFEL devices prepared by r.f. magnetron sputtering and the sol-gel process, respectively.

TFEL Devices using (Ga$_2$O$_3$-SnO$_2$):Eu and SnO$_2$:Eu Phosphors

SnO$_2$ has been widely used as transparent conducting films [13] and phosphor screens for vacuum fluorescent displays (VFD) [14] because its commonly available form has conduction electrons resulting from an oxygen vacancy. In contrast, when SnO$_2$ phosphors are used as the emitting layer of TFEL devices, a problem arises; it is very difficult to apply a high electric field to the phosphor thin-film emitting layer. In order to resolve this problem, we used high temperature annealing in air as well as impurity doping in the preparation of SnO$_2$ phosphor thin films. Figure 5 shows the L-V and Q-V characteristics as functions of Ga$_2$O$_3$ content for (Ga$_2$O$_3$-SnO$_2$):Eu TFEL devices prepared by the sol-gel process under the optimal conditions for SnO$_2$:Eu TFEL device preparation. The (Ga$_2$O$_3$-SnO$_2$):Eu thin-film emitting layers were prepared using Ga(OCH$_3$)$_3$ and Sn(O-i-C$_3$H$_7$)$_4$ as the Ga and Sn source materials, respectively. The conductivity of (Ga$_2$O$_3$-SnO$_2$):Eu thin-film emitting layers was almost unaffected by a

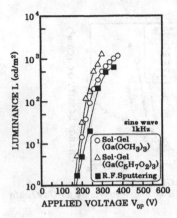

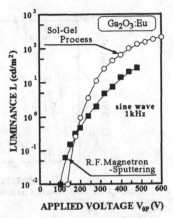

Fig.3 *L-V characteristics for TFEL devices using Ga₂O₃:Mn thin films prepared under optimal conditions by the sol-gel process (○ or △) and r.f. magnetron sputtering (●).*

Fig.4 *L-V characteristics for TFEL devices using Ga₂O₃:Eu thin films prepared under optimal conditions by the sol-gel process (○) and by r.f. magnetron sputtering (■).*

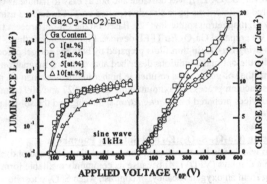

Fig.5 *L-V and Q-V characteristics as functions of Ga₂O₃ content for (Ga₂O₃-SnO₂):Eu TFEL devices prepared by the sol-gel process .*

Ga_2O_3 content up to 20 at.%, whereas with a Eu doping of 1 at.%, conductivity decreased.

Figure 6 shows the L-V and Q-V characteristics as functions of Eu content for SnO_2:Eu TFEL devices prepared by the sol-gel process under optimized conditions using $Sn(O-i-C_3H_7)_4$: depositing and annealing temperatures of 700 and 1070℃, respectively.   As can be seen in Fig.6, the conductivity of SnO_2:Eu thin films decreased as the Eu content was increased.

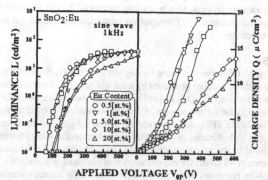

Fig.6 *L-V and Q-V characteristics as functions of Eu content for $SnO_2:Eu$ TFEL devices prepared by the sol-gel process.*

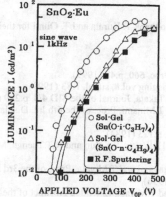

Fig. 7 *L-V characteristics for TFEL devices using $SnO_2:Eu$ thin films prepared under optimal conditions by the sol-gel process (O or △) and r.f. magnetron sputtering (■).*

However, the obtained maximum luminance remained almost unchanged as the Eu content was increased from 1 to 10 at.%: 30 cd/m² using $Sn(O-i-C_3H_7)_4$. A higher luminance was obtained in TFEL devices with a $SnO_2:Eu$ thin-film emitting layer prepared by the sol-gel process using $Sn(O-i-C_3H_7)_4$; however, $Sn(O-i-C_3H_7)_4$ proved to be unstable and not as easy to handle as $Sn(O-n-C_4H_9)_4$.

Figure 7 shows the L-V characteristics for TFEL devices using $SnO_2:Eu$ thin films prepared under the optimal conditions by the sol-gel process using $Sn(O-i-C_3H_7)_4$ or $Sn(O-n-C_4H_9)_4$ as the Sn source material and by r.f. magnetron sputtering; devices prepared by the sol-gel process exhibited higher luminances than the device prepared by r.f. magnetron sputtering with its vacuum process. Maximum luminances of 27 and 40 cd/m^2 were obtained in SnO$_2$:Eu TFEL devices prepared by r.f. magnetron sputtering and the sol-gel process using Sn(O-i-C$_3$H$_7$)$_4$, respectively.

CONCLUSIONS

Newly developed TFEL devices using Ga$_2$O$_3$:Mn and (Ga$_2$O$_3$-SnO$_2$):Eu phosphor thin films could be prepared without vacuum processes by a sol-gel process. Source materials composed of suitable combinations of Ga(C$_5$H$_7$O$_2$)$_3$ or Ga(OCH$_3$)$_3$, Sn(O-n-C$_4$H$_9$)$_4$ or Sn(O-i-C$_3$H$_7$)$_4$, EuCl$_3$ and MnCl$_2$ were used in the sol-gel process. A luminance above 1000 cd/m^2 for green emission was obtained in Ga$_2$O$_3$:Mn TFEL devices prepared by the sol-gel process, irrespective of the Ga source materials used. A luminance of 309 cd/m^2 for red emission was obtained in a Ga$_2$O$_3$:Eu TFEL device prepared by the sol-gel process using Ga(C$_5$H$_7$O$_2$)$_3$ as the Ga source material. In addition, a luminance of 40 cd/m^2 for red emission was obtained in a SnO$_2$:Eu TFEL device prepared by the sol-gel processes using Sn(O-i-C$_3$H$_7$)$_4$ as the Sn source material. Our tests showed that TFEL devices using Ga$_2$O$_3$ and SnO$_2$ phosphor thin films prepared by the sol-gel process always exhibited higher luminances than equivalent   devices prepared by r.f. magnetron sputtering with its vacuum process.

ACKNOWLEDGMENTS

The authors would like to thank T. Kurata and T. Ohmi for their technical assistance in the experiments.

REFERENCES

1. T.Minami, MRS Symp. Proc. **560**, p.47 (1999).
2. T.Minami, Display and Imaging **vol.8**, suppl. P.83 (1999).
3. T.Minami, Y.Kuroi and S.Takata, Journal of the SID **4/4**, p.299 (1995).
4. T.Miyata, T.Nakatani and T.Minami, Proc. of the 6th Int. Display Workshops, Sendai, p. 865 (1999).
5. P.M.Alt, Proc. of SID, **25**, p.123 (1984).
6. P.D.Rack, A.Naman, P.H.Holloway, S.S.Sun and R.T.Tuenge, MRS Bulletin **21**, p.49 (1996).
7. T.Minami, Y.Kuroi, Y.Sakagami and T.Miyata, Proc. of the 3rd Int. Display Workshops, Kobe, p.97 (1996).
8. T.Minami, Y.Sakagami and T.Miyata, Extended Abstract of the 3rd Int. Conf. on the Science and Technology of Display Phosphor, p.37 (1997).
9.   T.Minami, T.Miyata and Y.Sakagami, Surface and Coatings Tech., **108-109**, p.594 (1999).
10. T.Minami, S.Orito, H.Nanto and S.Takata, Proc. of the 6th Int. Display Research Conf. Tokyo, p.140 (1986).
11. T.Minami, H.Yamada, Y.Kubota and T.Miyata, Jpn. J. Appl. Phys., **36**, L1191 (1997).
12. T.Minami, Y.Kubota, H.Yamada, T.Miyata and S.Takata, Extended Abstract of the 3rd Int. Conf. on the Science and Technology of Display Phosphor, p. 207 (1997).
13. H.L.Hartnagel, A.L.Dawar, A.K.Jain, C.Jagadish, "Semiconducting Transparent Thin Films" Institute of Physics Publishing Bristol and Philadelphia, (1995)
14. "Phosphor Handbook", ed. S.Shionoya and Y.M.William CRC Press (1994).

Mat. Res. Soc. Symp. Proc. Vol 621 © 2000 Materials Research Society

Electroluminescent Devices with Nanostructured ZnS:Mn Emission Layer Operated at 20 V_{0-p}

Toshihiko Toyama, Daisuke Adachi and Hiroaki Okamoto
Department of Physical Science, Graduate School of Engineering Science, Osaka University
Toyonaka, Osaka 560-8531, Japan

ABSTRACT

We have developed a new type of a thin-film electroluminescence (TFEL) device with nano-structured (NS)-ZnS:Mn utilizing its enhanced luminescent efficiency due to the quantum confinement (QC) effects. As NS-ZnS:Mn, ZnS:Mn/Si$_3$N$_4$ multilayers with thicknesses of 1.9–3.5 nm for ZnS were prepared by a rf-magnetron sputtering method. From the results of grazing incidence X-ray reflectometry and X-ray diffractmetry, formation of ZnS:Mn nanocrystals in the ZnS layers are confirmed. With a decrease in the ZnS:Mn layer thickness, the photoluminescence (PL) efficiency associated with the Mn^{2+} transitions is increased, and the PL excitation spectrum is shifted toward higher energies, indicating appearance of the QC effects. As the results of the application of NS-ZnS:Mn to the emission layer of the TFEL device, we have successfully observed reddish-orange emission above the threshold voltage of 12 V_{0-p}, and the maximum luminance is 3.0 cd/m^2 operated with the 1-kHz sinusoidal voltage of 20 V_{0-p}.

INTRODUCTION

Luminescence properties of nanostructured (NS-) semiconductors such as nanometer-size multilayers or nanocrystals have been studied extensively in the last two decades, since they exhibit characteristics different from bulk semiconductors due to the quantum confinement (QC) effects [1]. The research interest on the QC effects on the luminescent properties has been extended from the interband transitions to the radiative transitions related to the transition metals or rare-earth elements doped in the NS-semiconductors. In particular, high photoluminescence (PL) efficiencies of Mn^{2+} $^4T_1(^4G)$-$^6A_1(^6S)$ transitions demonstrated in chemically-synthesized NS-ZnS:Mn has stimulated both basic research and application fields [2], because ZnS:Mn has been widely used as an emission layer of thin-film electroluminescence (TFEL) devices for about 30 years [3]. Many studies have been already reported about the QC effects on the PL properties of NS-ZnS doped with Mn [4], Cu [5,6], and Eu [6]. Moreover, some research groups have observed the EL from NS-ZnS:Cu [7,8], whereas, to our knowledge, there has been no report on EL from NS-ZnS:Mn.

In this article, we propose a novel NS-ZnS:Mn system, i.e., a multilayer with ZnS:Mn nanocrystals sandwiched by amorphous Si$_3$N$_4$ [9]. The Si$_3$N$_4$ layer is expected to act as a role in intercepting the crystal growth of ZnS:Mn, so that the resulting ZnS:Mn layer should contain ZnS:Mn nanocrystals. First, we report structural and PL properties of NS-ZnS:Mn with a stress on the influence of the crystal size as well as the lattice constant of ZnS on the PL properties, then describe the developed low-voltage-driven TFEL device with the NS-ZnS:Mn emission layer.

EXPERIMENTAL DETAILS

The ZnS:Mn/Si$_3$N$_4$ multilayers were prepared on quartz substrates or indium tin oxide (ITO) coated glass substrates with a multi-target rf (13.56 MHz)-magnetron sputtering system (Shimadzu HSR-552S). A mixture of ZnS and Mn powder was used as the ZnS:Mn target, and the Mn concentration was 1.0 mol%. The deposition was done in Ar atmosphere with a pressure of 10 mTorr, and the substrate temperature was 250°C. The ZnS:Mn/Si$_3$N$_4$ multilayer of 30 periods was formed by alternative deposition of ZnS:Mn and Si$_3$N$_4$ layers. For the EL device, Al back contacts with an area of 0.075 cm^2 were finally deposited by a vacuum evaporation technique. As a reference sample as bulk ZnS, we also deposited a 500-nm thick ZnS:Mn film with the identical deposition conditions with those for NS-ZnS:Mn mentioned above, and fabricated a conventional TFEL device consisting of the 500-nm thick ZnS:Mn emission layer sandwiched by 150-nm thick Si$_3$N$_4$ insulating layers.

Grazing incidence X-ray reflectivity (GIXR) and wide-angle X-ray diffraction (XRD) measurements with Rigaku RINT 2200 were utilized to characterize the multilayer structure and NS-ZnS:Mn crystallinity, respectively. Microscopic structural image was obtained by a transmission electron microscope (TEM, JEOL JEM 2000EX). The PL and PL excitation (PLE) spectra were measured with a conventional setup composed of a cooled GaAs(Cs) photomultiplier tube, a 20-cm monochromator, a current amplifier and a lock-in amplifier. A 325-nm line of He-Cd laser was used as the excitation light source for the emission spectrum and Xe lamp coupled with another monochromator was used for the PLE measurement. Luminance and EL spectra were measured for the EL devices driven by an ac sinusoidal voltage with a frequency of 1 kHz. All measurements were carried out at room temperature.

DISCUSSION

Figure 1(a) shows a cross-sectional view of typical NS-ZnS:Mn taken by the TEM. In Figure 1(a), light-gray parts corresponding to the Si$_3$N$_4$ layers and dark-gray parts corresponding to the ZnS:Mn layers are alternatively observed on the quartz substrate. The thicknesses of a ZnS:Mn layer, d_{ZnS}, and a Si$_3$N$_4$ layer, d_{SiN}, are directly estimated to be 1.7 nm and 0.6 nm, respectively. The thicknesses are also estimated from the GIXR spectra shown in Figure 1(b). In Figure 1(b), the GIXR spectra are plotted as a function of the deposition time of the ZnS:Mn layer. Except for the bottom spectrum with the shortest deposition time, the first and second diffraction peaks are clearly observed in the GIXR spectra. Alternatively, no diffraction peak is observed in the bottom spectrum, indicating mixing in the ZnS and Si$_3$N$_4$ layers. Employing the diffraction peak angles, the thickness of the ZnS:Mn/Si$_3$N$_4$ bilayer, $d_{ZnS} + d_{SiN}$, is evaluated from the Bragg law with a wavelength of X-ray radiation of 0.154 nm [10]. Applying $d_{SiN} = 0.6$ nm, d_{ZnS} is estimated as shown in Figure 1(c). The thickness of a ZnS:Mn layer is linear to the deposition time. From the XRD measurements, the diffraction peaks are observed at the diffraction peak angle, $2\theta = \sim29°$ corresponding to (111) β-ZnS [11], showing the presence of ZnS nanocrystals in the deposited ZnS:Mn layers. With a decrease in d_{ZnS} from 3.5 nm to 1.9 nm, the linewidth of the diffraction peak is monotonously increased, giving a decrease in the crystal size with the decreased d_{ZnS}. The estimated crystal sizes using Sherrer's formula are

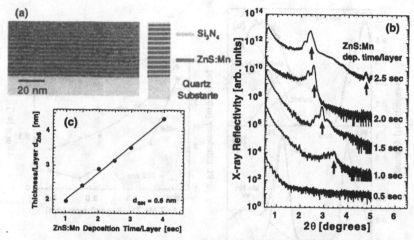

Figure 1. *Cross-sectional TEM image of NS-ZnS:Mn (d_{ZnS} = 1.7 nm, d_{SiN} = 0.6 nm) (a), GIXR spectra as a function of deposition time of ZnS:Mn layers (b), and ZnS:Mn layer thicknesses deduced from diffraction peaks in GIXR spectra plotted against deposition time of ZnS:Mn layer (c). In Figure 1(b), spectra are shown with logarithmic offset to clarify.*

slightly less than d_{ZnS}. Additionally, at d_{ZnS} < 2.8 nm, the diffraction peak angle tends to increase from 28.54° observed in the XRD patterns of bulk crystal ZnS, indicating a reduction of the lattice constant of ZnS, a_{ZnS}, for the ZnS nanocrystals in comparison with a_{ZnS} for bulk crystal ZnS.

In Figure 2, the PL and PLE spectra of NS-ZnS:Mn are displayed with those of bulk ZnS:Mn. The PLE spectra were detected at around the emission peak wavelengths of 610 nm for NS-ZnS:Mn and 585 nm for bulk ZnS:Mn, respectively. From the results of extra measurements (not shown here) on changes in the PL spectral lineshape associated with the measurement temperature, the excitation intensity, and the Mn-doping concentration as well as the temporal changes in the spectral lineshape, we have found the PL spectra consisting of two or three components, and assigned to their origins as follows.

The PL spectrum of bulk ZnS:Mn shown in Figure 2(b) is decomposed into two spectral components centered at 588 nm and 710 nm, respectively. The 588-nm band is assigned to the well-known Mn^{2+} $^4T_1(^4G)$-$^6A_1(^6S)$ transitions [12]. Alternatively, the 720-nm emission band is likely to be attributed to Mn-Mn pairs, of which luminescence have been found as PL, EL and cathodoluminescence spectra from highly Mn-doped bulk ZnS [13–15].

The PL spectrum of NS-ZnS:Mn shown in Figure 2(a) is decomposed into three components with the peak wavelengths of 520 nm, 610 nm and 700 nm, respectively. The emission centered at 520 nm with fast decay less than microseconds would be assigned to the recombination through defects located at the ZnS:Mn/Si$_3$N$_4$ interface, and must not be related to Mn ion. The emission centered at 700 nm would be assigned to the same origin with that of the emission band observed at 720 nm in bulk ZnS:Mn as mentioned above. The Mn^{2+} $^4T_1(^4G)$-$^6A_1(^6S)$ transitions

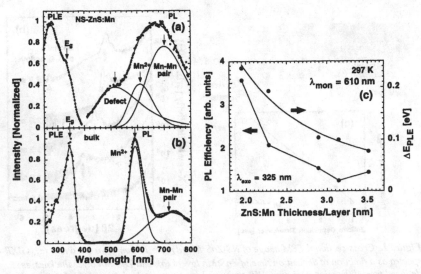

Figure 2. *Typical PL and PLE spectra of NS-ZnS:Mn (a), those of bulk ZnS:Mn (b), and PL efficiencies and PLE peak shifts, ΔE_{PLE}, of NS-ZnS:Mn plotted against thicknesses of ZnS:Mn layers (c). PL spectra shown in Figure 2(a) and (b) are decomposed into three and two spectral components shown by solid lines, respectively.*

would be responsible for the emission centered at 610 nm. However, in comparison with the Mn^{2+} transitions in bulk ZnS:Mn found at 588 nm, the Mn^{2+} emission peaks observed in NS-ZnS:Mn is shifted to lower photon energies, because the Mn luminescent center is strongly influenced by the surrounding ligand field, therefore, the emission peak energy tends to be decreased with a decrease in the lattice constant of ZnS [16]. As we have already described in the results of the XRD measurement, the lattice constant of NS-ZnS is smaller than that of bulk ZnS, so that the ligand field parameter Dq of NS-ZnS should be larger than that of bulk ZnS. Consequently, the red shift in the emission due to the Mn^{2+} $^4T_1(^4G)-^6A_1(^6S)$ transitions appears to be the result of the change in the ligand field induced by the decreased lattice constant of ZnS in NS-ZnS.

In the PLE spectrum of bulk ZnS:Mn shown in Figure 2(b), a sharp peak is observed at 346 nm (3.58 eV), which would correspond to the band gap of β-ZnS [17]. In contrast, two spectral features in the PLE spectra of NS-ZnS:Mn were observed at shorter wavelengths. The shoulder in Figure 2(b) would be assigned to the band gap of ZnS nanocrystals in NS-ZnS:Mn. Therefore, the band gap energy is increased in comparison with bulk ZnS:Mn, indicating appearance of the QC effects in the ZnS:Mn nanocrystals. In Figure 2(c), the shift in the PLE peak corresponding to the band gaps of ZnS in NS-ZnS:Mn, ΔE_{PLE}, is plotted against d_{ZnS}. With a decrease in d_{ZnS}, ΔE_{PLE} is monotonously increased.

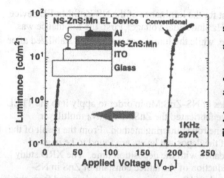
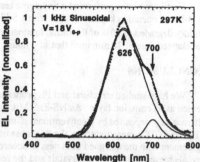

Figure 3. *Typical L-V characteristic of NS-ZnS:Mn TFEL device in comparison with that of conventional double insulating TFEL device. Inset shows schematic illustration of structure of TFEL device with NS-ZnS:Mn emission layer.*

Figure 4. *EL spectrum of NS-ZnS:Mn consisting of two spectral components shown by solid lines.*

Also shown in Figure 2(c) is the PL efficiencies plotted against d_{ZnS}. The PL efficiency, which is deduced from the PL intensity over the optical absorption measured at the excitation wavelength (325 nm) of the sample, is also increased with a decrease in d_{ZnS}. The similar tendency between the PL efficiency and ΔE_{PLE} against d_{ZnS} implies that the increase in the PL intensity would be also due to the QC effects as reported in the chemically-synthesized NS-ZnS:Mn [2]. Furthermore, the PL intensity of NS-ZnS:Mn with the 30 periods of the ZnS:Mn layer (d_{ZnS} = 1.7 nm) is 2–3 times larger than that of bulk-like ZnS:Mn of which thickness is 51 nm (= 1.7 nm x 30).

Utilizing the increased luminescence efficiency due to the QC effects, we have fabricated the TFEL device with NS-ZnS:Mn (d_{ZnS} = 2.5 nm) as the emission layer, and successfully observed light emission from the NS-ZnS:Mn TFEL device. The luminance-voltage ($L–V$) characteristics of the NS-ZnS:Mn TFEL device and its structure are shown in Figure 3. Above the threshold voltage of 12 V_{0-p}, the luminance curve rises steeply. We have achieved the maximum luminance of 3.0 cd/m^2 operated at 20 V_{0-p}. Also shown in Figure 3 is the $L–V$ characteristics of a typical conventional double-insulating TFEL device. A marked reduction in the threshold voltage of the NS-ZnS:Mn TFEL device is clearly observed.

Figure 4 shows the EL spectrum of NS-ZnS:Mn. The EL spectrum has a peak at 626 nm and a shoulder at 700 nm. In spite of little differences in the peak wavelengths between the EL spectrum and the PL spectrum shown in Figure 2(a), the two EL components observed at 626 nm and 700 nm would also correspond to the Mn^{2+} $^4T_1(^4G)$-$^6A_1(^6S)$ transitions and luminescence due to Mn-Mn pair, respectively, as assigned in the PL spectrum. In addition, the decay time of the integrated EL intensity is found to be on the order of milliseconds. These result imply that the Mn^{2+} $^4T_1(^4G)$-$^6A_1(^6S)$ related transitions dominate the EL emission as like the EL emission in the conventional TFEL devices. The emission color of the NS-ZnS:Mn TFEL device is reddish-orange corresponding to (0.59, 0.37) in the CIE color chart.

Finally, we have performed a tiny aging test for the luminance of the NS-ZnS:Mn EL device by 4-hours operation kept at 15 V_{0-p} with a 5-kHz sinusoidal voltage. First, the luminance was rapidly degraded to 70% of the initial luminance, while the luminance was almost stabilized after the elapsed time of 30 min until that of 4 hours.

CONCLUSIONS

We have studied structural and PL properties of NS-ZnS:Mn in order to apply it to the TFEL device as an emission layer. As NS-ZnS:Mn, we fabricated the ZnS:Mn/Si_3N_4 multilayer alternatively deposited by a conventional rf-magnetron sputtering method. From the result of the TEM and GIXR measurements, the thickness of the deposited layers are found to be in good agreement with the designed thickness in accordance with the deposition time. The XRD study reveals presence of ZnS nanocrystals and the reduction of the lattice constant of ZnS in NS-ZnS:Mn, which is responsible for the red shift in the 610-nm emission band originated from the Mn^{2+} $^4T_1(^4G)$-$^6A_1(^6S)$ transitions. The band gap widening and the increased emission intensity of the Mn^{2+} emission are found, implying appearance of the quantum confinement effects in NS-ZnS:Mn. Finally, we have developed the NS-ZnS:Mn TFEL device showing a reddish-orange broad band emission centered at 626 nm above the threshold voltage of 12 V_{0-p}. The maximum luminance of 3.0 cd/m^2 has been achieved with a 1-kHz sinusoidal voltage of at 20 V_{0-p}.

REFERENCES

1. A. D. Yoffe, Advances in Physics, **42**, 173 (1993) and references therein.
2. R. N. Bhargava, D. Gallagher, X. Hong, and A. Nurmikko, Phys. Rev. Lett. **72**, 416 (1994).
3. Y. A. Ono, in *Electroluminescent Displays*, (World Scientific, Singapore, 1995) p. 3.
4. A. A. Bol, and A. Meijerink, Phys. Rev. B **24**, 15997 (1998).
5. A. A. Khosravi, M. Kundu, L. Jatwa, S. K. Desphpande, U. A. Bhagwat, M. Sastry, and S. K. Kulkarni, Appl. Phys. Lett. **67**, 2702 (1995).
6. S. J. Xu, S. J. Chua, B. Liu, L. M. Gan, C. H. Chew, and G. Q. Xu, Appl. Phys. Lett. **73**, 478 (1998).
7. J. Huang, Y. Yang, S. Xue, B. Yang, S. Liu, and J. Shen, Appl . Phys. Lett. **70**, 2335 (1997).
8. W. Que, Y. Zhou, Y. L. Lam, Y. C. Chan, C. H. Kam, B. Liu, L. M. Gan, C. H. Chew, G. Q. Xu, S. J. Chua, S. J. Xu, and F. V. C. Mendis, Appl. Phys. Lett., **73**, 2727 (1998).
9. D. Adachi, T. Toyama, and H. Okamoto, submitted for the publication.
10. L. Nevot, B. Pardo, and J. Corno, Revue Phys. Appl. **23**, 1675 (1988).
11. JCPDS card, No. 5-0566.
12. M. Aven and J. S. Prenter, in *Physics and Chemistry of II-VI Compounds*, (North-Holland, Amsterdam, 1967), p. 471.
13. V. Marrello, and A. Onton, IEEE Trans. Electron Devices **27**, 1767 (1980).
14. O. Goede, and D. D. Thong, phys. stat. sol. (b) **124**, 343 (1984).
15. M. D. Bhise, M. Katiyar, and A. H. Kitai, J. Appl. Phys. **67**, 1492 (1990).
16. Y. Tanabe and S. Sugano, J. Phys. Soc. Jpn. **9**, 753 (1954).
17. Y. R. Wang, and C. B. Duke, Phys. Rev. B **5**, 2763 (1987).

Mat. Res. Soc.Symp. Proc. Vol. 621 © 2000 Materials Research Society

PHOTOLUMINESCENCE CHARATERISTICS OF SrTiO$_3$:Pr^{3+},Ga^{3+} SINGLE CRYSTAL

Jaedong Byun, Yongjei Lee, Boyun Jang[1], Youngmoon Yu[2], Sunyoun Ryou[3] and Kyungsoo Suh

[1]Dept. of Materials Science, Korea University,
Seoul 136-701, Korea.
[2]Korea Research Institute of Chemical Technology,
Taejon 305-343, Korea.
[3]Dept. of Materials Sci. & Engr., Sun Moon University,
Choongnam 336-840, Korea.
Electronics and Telecommunications Research Institute,
Taejon 305-350, Korea.

ABSTRACT

SrTiO$_3$ crystals containing varying concentrations of Pr and Ga were grown by a standard floating zone method and their photoluminescence characteristics were investigated. It was found that only a small fraction of Pr ions added are incorporated in the lattice. When the Ga ions are co-doped, the solubility of Pr ions is increased considerably. Comparing the ionic sizes of the host and dopants, it seems that the increase of solubility of Pr by Ga co-doping is due to charge compensation achieved by the additional Ga^{3+}. The addition of Ga in Pr - activated SrTiO$_3$ resulted also in a considerable enhancement of Pr^{3+} emission band at 615 nm. This result is attributed to the hole trapping effect of Ga^{3+} ions.

INTRODUCTION

A number of Pr-activated phosphor are known to emit blue-green or red color depending on the host lattice under ultraviolet (u.v.) or cathode ray (c.r.) irradiation. Pr-activated SrTiO$_3$ and CaTiO$_3$ show a single red emission band at 615nm which is assigned as $^1D_2 \rightarrow {}^3H_4$ intraconfiguration transition of Pr^{3+} ion. On addition of Ga[1] or Al[2] to SrTiO$_3$:Pr and monovalent ions[3] to CaTiO$_3$:Pr , the red emission intensity is considerably enhanced. Due to the strong emission intensity, SrTiO$_3$:Pr^{3+},Ga^{3+} is considered to be a very attractive phosphor for low-voltage applications such as field emission display[4]. However, the mechanism for the enhancement luminescence intensity with Ga^{3+} or Al^{3+} addition is not clearly understood. This is in part due to the lack of information related to the effect of Ga^{3+} or Al^{3+} addition on Pr^{3+} incorporation into $SrTiO_3$. Therefore, this work was initiated with an aim to understand the effects of Ga^{3+} addition on the Pr^{3+} incorporation into $SrTiO_3$ crystal and on the luminescence characteristics of $SrTiO_3:Pr^{3+}$. In addition the present authors are not aware of any previous reports on the luminescence characteristics of $SrTiO_3:Pr^{3+},Ga^{3+}$ single crystals

EXPERIMENTAL

SrTiO$_3$ crystals containing varying concentrations of Pr and Ga were grown by a standard floating zone method. The starting materials for preparation of the feed rods were, $SrCO_3$, TiO_2, Pr_6O_{11} and Ga_2O_3 of 99.99 % purity. The raw materials were mixed and calcined in air at 1200 °C for 2 hrs. The cylindrical rods were prepared by the cold isostatic pressing of the compounds and then sintered at 1200 °C in O_2 atmosphere for 2 hrs. Feed and seed shafts were rotated appositely at a rate of 30 rpm. The zone traveling rate was 15 mm/h. The crystal boules were cut into wafers of about 1 mm thickness and polished to an optical quality by successively finer diamond abrasives. The dopant concentrations in the crystal were determined by X-ray fluorescence spectroscopy.

Optical absorption measurements were made with a Cary5 double-beam spectrophotometer and luminescence measurements were carried out using a Aminco-Bowman luminescence spectrometer. A powder holder was loaded at a suitable angle to the excitation and emission beams to achieve maximum signal and minimum scattering. Samples were excited with appropriate radiations. The emission was monitored in the range of 400 to 800 nm to locate the wavelength of the maximum point in the emission spectra and the excitation wavelengths were scanned in the range of 220 to 500 nm.

RESULTS AND DISCUSSION

The amounts of Pr and Ga added to the starting materials and their concentrations in the crystals determined by XRF are summarized in Table 1. This result clearly shows that the concentration of Pr ions incorporated in the SrTiO$_3$ lattice is much lower than the added amount and it also indicated that when Ga ions are co-doped, the greater amount of Pr ions are incorporated into the crystal. Comparing the ionic sizes of the hosts and dopants it seems that Pr^{3+} and Ga^{3+} ions enter the Sr and Ti site, respectively, and the increase in solubility of Pr^{3+} is

Table I. Results of XRF analysis of the single crystals

Specimen number	Concentration in starting materials (mole fraction)		Concentration in crystal (mole fraction)	
	Pr	Ga	Pr	Ga
1	1.0×10^{-3}	—	9.0×10^{-5}	—
2	1.0×10^{-3}	1.0×10^{-3}	2.4×10^{-4}	8.7×10^{-4}
3	3.0×10^{-3}	—	4.5×10^{-4}	—
4	6.0×10^{-3}	—	9.5×10^{-4}	—
5	6.0×10^{-3}	6.0×10^{-3}	1.56×10^{-3}	5.26×10^{-3}

due to charge compensation achieved by the additional Ga^{3+}.

Fig. 1 is the absorption spectra of the single crystals of pure and doped $SrTiO_3$ investigated in this study. From the spectra, the band gap of ~ 3.2 eV was estimated by extrapolating the steep slope in the curve. The absorption spectra of pure $SrTiO_3$ and $SrTiO_3$ with low impurity concentration, # 1 and # 2 specimens, were essentially same without any detectable structure corresponding to absorption by Pr^{3+} ions. However, for the heavily doped specimens, # 3, # 4 and # 5, the absorption coefficient increases with increasing wavelength beyond the edge absorption wavelength with a shoulder at 510 nm. The specimens were characterized with dark bluish color due to the broad band absorption.   This result is consistent with that observed in highly reduced $SrTiO_3$ and the absorption is due to free carriers[5]. The bluish color suggests the existences of Ti^{3+} ions[6]. Practically no luminescence was observed in these crystals under u.v. or c.r. excitation, thus on further analyses were made on their specimens.

Fig. 2 presents the excitation and emission spectra of specimens # 1 and # 2 at room temperature. The energy spectral distributions of excitation and emission are consistent with those observed for $SrTiO_3$:Pr,Ga powder phosphor[1,2]. The excitation spectrum (Fig. 2. (a)) of the $^1D_2 \rightarrow {}^3H_4$ emission at 615 nm consists of a broad band with a relatively sharp edge on the long-wavelength side peaking at 357 nm and line groups between 440 and 500 nm which can be ascribed to the transitions between definite electronic levels within the 4f-shell as indicated in the figure. The shape of broad band of excitation spectrum is in good agreement with previously published results for excitation of intrinsic emission in $SrTiO_3$[7,8] at low temperatures. It is interesting to note that extrapolating the steep slope in the excitation curve to the long-wavelength side gives a value of ~ 3.2 eV which is close to the band gap energy of $SrTiO_3$. The similarity between absorption and excitation spectra at long-wavelength side seems to indicate that the excitation band is due to the intrinsic band to band or the charge transfer transition in the

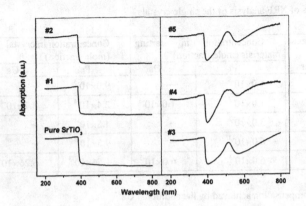

Figure 1. *Absorption spectra of SrTiO₃ containing Pr and Ga; pure SrTiO₃, specimens # 1, # 2, # 3, # 4 and #5.*

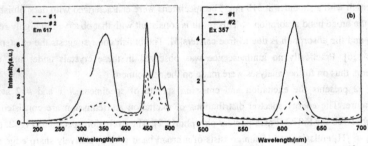

Figure 2. *(a) The excitation spectra of specimens # 2. (b) The emission spectra of specimens # 1 and # 2.*

host lattice[7,8].

The emission spectrum under broad band or line group excitation contains one strong red band peaking at 615 nm(Fig. 2 (b)). As noted from the figure, the intensity of this emission band was considerably enhanced by Ga^{3+} co-doping; the integrated area of the red emission bands is increased by a factor of ~ 20 as Pr concentration is increased by a factor 2.5. Similar enhancements have been previously reported for $SrTiO_3:Pr^{3+},Ga^{3+}$ [1] and $SrTiO_3:Pr^{3+},Al^{3+}$ [2] powder phosphors. The increased red emission by Al^{3+} doping has been attributed to the improvement of crystallinity[2]. However it is highly improbable that improvement of crystallinity alone should enhance the emission intensity since the same effect was observed in single crystals in this investigation. Therefore, the enhancement of emission intensity could be ascribed to the increase of Pr^{3+} concentration in $SrTiO_3$ crystal by the Ga^{3+} co-doping. As one can see from Table 1, in spite of the fact that the same amount of Pr was doped in the starting materials, the concentration of Pr in Ga co-doped crystal was higher by a factor of 2.5 than in crystal without Ga co-doping. However, since the Ga^{3+} addition increased the integrated emission intensity by a factor of 20, in spite of the fact that the Pr^{3+} concentration is increased only by a factor of 2.5, there must be another role played by the Ga^{3+} in enhancing the emission intensity besides increasing distribution constant for Pr^{3+} ions.

Since Ga at Ti-site has a negative effective charge, these ions prefer to be situated near the Pr^{3+} ion which has a positive effective charge (analogous to the di-vacancy formation). Because of the negative effective charge of Ga^{3+} in $SrTiO_3$, the hole created by excitation can be captured easily at Ga-site and subsequently recombine with an electron in conduction band. Thus, Ga^{3+} in $SrTiO_3$ is an effective hole trap center[9].

Summarizing the experimental results, the following model for the enhancement red emission by Ga addition is proposed. Excitation into the $SrTiO_3$ host lattice leads to the formation of electrons in the conduction band and holes in the valence band. The holes in valence band are captured by Ga^{3+} and the electrons in the conduction band recombine radiatively with the holes trapped at Ga^{3+} ion and the energy is then transferred to Pr^{3+} ion nearby and Pr^{3+} is ion excited to P_j levels. Since no direct emission from this level is observed, there seems to be a relaxation from P_j levels to $^1D_2$ level by a non-radiative transition producing finally the red $^1D_2 \rightarrow {}^3H_4$ emission[3,10].

CONCLUSION

Photoluminescence characteristics of $SrTiO_3:Pr^{3+},Ga^{3+}$ single crystals were investigated. It was found that only a small fraction of Pr ions added are incorporated in the lattice. Upon co-doping with Ga^{3+}, and the Pr emission intensity at 615 nm is also enhanced considerably. The enhancement of red emission is attributed to the increased Pr concentration and the hole trapping effect of Ga^{3+} ion.

ACKNOWLAGEMENTS

This work was supported by Korea Research Foundation Grant.(1998-017-E00130)

REFERENCES

1.   M. Lee, S. Nahm, M. Kim K. Suh and J. Byun, J. Kor. Cer. Soc. **35**. 757 (1998).
2.   S. Okamoto and H. Kobayashi, J. Appl. Phys. **86**. 5594 (1999).
3.   P. T. Diallo, P. Boutinaud, R. Mamou, and J. C. Cousseins, Phys. Stat. Sol. (a) **160**, 255 (1997).
4.   S. Jacobsen, J. SID, 4, **331** (1996).
5.   W. S. Baer, Phys. Rev. **144**, 734 (1966).
6.   O. Saburi, J. Phys. Soc. Jpn. **14**, 1159 (1959).
7.   L. De Haart, A. De Vries, and G. Blasse, J. Solid State Chem. **59**, 291 (1985).
8.   L. Grabner, Phys. Rev. **177**, 1315 (1969).
9.   Klasens. H.A, J. Electrochem. Soc., **100**, 72(1953)
10.   W. J. Schipper, M. F. Hoogendorp and G. Blasse, J. Alloys Comps., **202**, 283 (1993).

Mat. Res. Soc. Symp. Proc. Vol 621 © 2000 Materials Research Society

Stability in Electrical Properties of Ultra Thin Tin Oxide Films

Yuji Matsui[1], Yuichi Yamamoto and Satoshi Takeda
R&D Center , Asahi Glass Co., Ltd.
1150 Hazawa-cho, Kanagawa-ku, Yokohama-shi, Kanagawa 221, Japan
[1] New Products Development Center, Fabricated Glass General Div., Asahi Glass Co., Ltd.
426-1 Sumida, Aikawa-cho, Aikou-gun, Kanagawa, 243-0301 Japan

ABSTRACT

Tin Oxide films of less than 30nm in thickness were developed as transparent conductive electrodes. The films were deposited onto glass substrates by APCVD. Fluorine was used as a doping component. Stability to heat treatments in air at more than 500• was studied. Carrier concentration decreased and Hall mobility increased by the heat treatments. It was found relations between carrier concentration and mobility exhibited an exponential relation in extremely low fluorine concentration films. The exponent of the relation was close to −1.5. Resistivity decreased for the films, while it increased for films with high fluorine concentration. It was also found migration of sodium from the substrates into the films increased with increase of fluorine concentration in the films. Results suggest the sodium migration would affect grain growth and electrical properties of the films.

INTRODUCTION

Several kinds of materials, such as tin oxide(SnO_2), indium tin oxide (ITO) and zinc oxide(ZnO), are known as Transparent Conductive Oxide (TCO). Among these materials, tin oxide shows unique characteristics in chemical inertness, stability to heat treatment and mechanical hardness. In addition, tin oxide has industrial advantages about deposition processes. Atmospheric Chemical Vapor Deposition (APCVD) system is usually used as a production system [1]. Its instrument cost is lower and productivity is higher than vacuum evaporation or sputtering systems. Many studies have been reported about tin oxide films deposited by APCVD, especially as substrates for a-Si solar cells [2][3][4][5][6][7][8][9]. Almost all studies are about the films thicker than 100nm.

In recent years, industrial application of TCO substrates for what is called a pen touch screen is growing year by year. It is required the substrates show sheet resistance as high as 1k ohm/sq. and high optical transmission. In order to satisfy these requirements at the same time, film thickness of the tin oxide films is usually less than 30nm. We call the films ultra thin tin oxide films because electrical properties of the films in relation to deposition conditions are clearly different from those of bulk films.

Another requirement is stability to heat treatment, because manufacturing processes of pen touch screens sometimes include sintering of a silver paste at more than 500•. In this paper we report about the stability in electrical properties of ultra thin tin oxide films in relation to deposition conditions.

EXPERIMENTS

APCVD apparatus of an in-line type horizontal furnace is used for deposition. Heating up, film deposition and cooling down processes are done successively for moving substrates carried on a metal conveyer belt. Source gases supplied by injectors impinge onto a glass substrate in the form of curtain like down flow. Each Injector has 5 gas slits. Each of these supplies uniform gas flow across full width of the substrates. Since bottom edges of injector slits are apart from a substrate by 15mm, source gases mix together in the space between the injector and a substrate. Films were deposited through the hydrolysis reaction from stannic chloride and water by the following equation (1).

$$SnCl_4 + 2H_2O \rightarrow SnO_2 + 4HCl \qquad (1)$$

Methanol was added in order to regulate the reaction. Hydrofluoric acid was used to dope fluorine into SnO_2 films. Gas transportation of stannic chloride, water and methanol was performed by a bubbling method using nitrogen as a carrier gas. Storage tanks of the materials were controlled to a certain temperature. Molar flux of each source material can be calculated from temperature of the material and nitrogen flow rate. Substrate temperature on the deposition was 540•. Typical deposition conditions were summarized in a previous report [10].

Common Soda-Lime sheet glasses of 1.1mm thickness were used as substrates. It is generally known that electrical properties of TCO films are degraded by migration of alkali metal ions from glasses. To impede the migration, the glass was coated with a silica layer of 80nm as a migration barrier layer.

Film thickness was measured using a stylus instrument (Dektak3030). The electrical properties, the resistivity, Hall mobility and carrier concentration, were measured at room temperature by Van der Pauw method. Scanning Electron Microscope (SEM) was used for surface observation of the films. Secondary Ion Mass Spectrometry (SIMS) measurement was carried out to evaluate fluorine and sodium concentration in the films.

Typical conditions of heat treatments to evaluate the stability were at temperature of 530 or 600• for 2min.

RESULTS

Figure1 shows the changes in electrical properties of as deposited and 600• treated films as a function of fluorine concentration in the films. Film thickness is around 24nm.Carrier concentration of the films decreased at the treatment. The lowest fluorine concentration in the figure resulted from a contamination in the deposition system (virtually undoped). Resistivity was almost unchanged at the concentration, on the other hand it increased for films with more than 2wt% concentration.

It was found that changes in mobility depended on fluorine concentration. Namely, the mobility increased on the film with lowest concentration, while it deceased on the films with higher concentration.

Figure2 shows electrical properties of the virtually undoped films as a function of methanol to stannic chloride supply ratio.

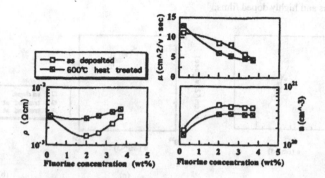

Figure 1 *Changes in resistivity(●), carrier concentration(n) and mobility(●) as a function of fluorine concentration. Films are deposited at source gas molar ratio of H20/SnCl4=10, CH3OH/H2O=0.8. Films Thickness is24nm.*

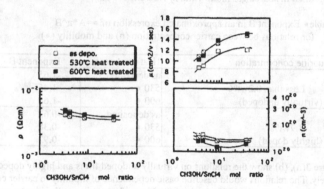

Figure 2 *Changes in resistivity(●), carrier concentration(n) and mobility(●) as a function of methanol to stannic chloride ratio on deposition. Films are deposited at source gas molar ratio of H20/SnCl4=10. Hydrofluoric acid was not supplied on deposition. Films Thickness is24nm.*

Electrical properties after heat treatments at 530 and 600● are shown in the figure together with as deposited values. Decrease in carrier concentration and increase in mobility took place by the heat treatment. Resistivity decreased as a result.

Changes in carrier concentration and mobility as a function of the ratio were also measured for films with 3.7wt% fluorine concentration (highly doped films). To understand the changes in mobility, we rearranged relations between carrier concentration and mobility for virtually undoped films and highly doped films.

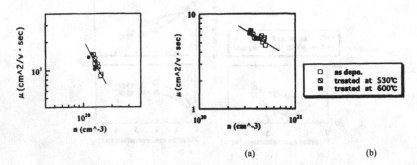

(a)   (b)

***Figure* 3** *Relations between carrier concentration and mobility of (a) virtually undoped films and (b) highly doped films (Fluorine concentration is 3.7wt) , respectively. Films are deposited at source gas molar ratio of H2O/SnCl4=10,*

Table• Exponent B in an approximated expression of •=A*n^B
for relations between carrier concentration (n) and mobility (•)

Fluorine concentration	Heat treatment	Exponent B
	as deposited	-1.65
Less than 0.2wt%	530 •	-1.45
(virtually undoped)	600 •	-1.68
	as deposited	-0.63
3.7wt%	530 •	-0.38
(highly doped)	600 •	-0.92

Figure 3(a), (b) show the relations on virtually undoped films and highly doped films, respectively. The relations would describe basic dependence of mobility to carrier concentration at a certain fluorine concentration. As shown in Figure 3(a), apparent linear relations were obtained for virtually undoped films on a logarithmic scale. The plots of as deposited and heat treated films are almost on a same line. Slopes of the lines were summarized in Table•. The values are close to -1.5 for virtually undoped films. This value is close to that which is represented by the ionized impurity scattering on electron transport [11]. On the other hand, the slope is larger than -1 on the highly doped films.

Figure 4 (a), (b) are SEM images of film surfaces of the virtually undoped and the highly doped film, respectively. Both films exhibit a grain structure. Grain sizes in each film are less than 50nm. The highly doped film exhibits smaller grains comparing to those in the actual non-doped film. Mobility decreased with fluorine concentration as shown in Figure 1. It can be considered that defects in the films increased with increase of fluorine concentration, since a crystalline film with low defect concentration generally shows high mobility [7].

(a)   (b)

Figure 4 SEM images of the film surface. (a)the virtually undoped film and (b)the highly doped film (Fluorine concentration is 3.7wt%), respectively. Source gas molar ratios are H20/SnCl4=10, CH3OH/H2O=0.8. Film thickness is 24nm.

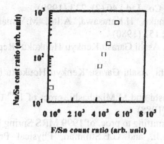

Figure5 Relations between fluorine concentration and sodium migration into the SnO2 films, estimated by SIMS. Source gas molar ratios are H20/SnCl4=10, CH3OH/H2O=0.8 Film thickness is 24nm.

Figure 5 shows determination of sodium migration into the films by SIMS measurements. Although a SiO_2 layer is deposited as a sodium migration barrier, the migration increases drastically as fluorine concentration increases. Because the migration occured during SnO2:F grain growth, there is a possibility that the incorporation of sodium impedes the grain growth.

Further research about relations between crystalline properties and the migration is awaited to understand the electrical property of the films.

CONCLUSIONS

The electrical stability of the tin oxide films to heat treatment was improved up to 600• by eliminating fluorine concentration. The relation between carrier concentration and mobility on the virtually undoped films were similar to that which is represented by the ionized impurity scattering. Sizes of crystalline grains decreased as fluorine concentration increased. Sodium incorporation during film growth seems to relate to decrease in grain sizes and degradation in the stability

ACKNOWLEDGEMENTS

We would like to thank to Mr. Y. GOTO, of Asahi Glass Co., Ltd. for valuable discussions.

REFERENCES

1. J.M.Blocher Jr.,Thin Solid Films. 77, 51 (1981)
2. R.N.Ghoshtagore, J. Electrochem. Soc. 125 (1), 110 (1978)
3. M.Mizuhashi, Y.Goto, and K.Adachi, Jpn J. Appl. Phys. 27 (11), 2053 (1988)
4. Y.Goto, K.Adachi and M.Mizuhashi, Asahi Garasu Kenkyu Houkoku (Rept. Res. Lab. Asahi Glass Co., Ltd.) 37 (1), 13 (1987)
5. K.Sato, Y.Goto, Y.Hayashi, K.Adachi, and H.Nishimua, Asahi Garasu Kenkyu Houkoku (Rept. Res. Lab. Asahi Glass Co., Ltd.) 40 (2), 233 (1990)
6. H.Iida, N. Shiba, T.Mishuku, H.Karasawa, A.Ito, M.Yamanaka, and Y.Hayashi, IEEE Electron Device Lett. ED-27, 157 (1980)
7. Y.Goto, and M.Mizuhashi, Asahi Garasu Kenkyu Houkoku (Rept. Res. Lab. Asahi Glass Co., Ltd.) 34 (2), 123 (1984)
8. K.Adachi and M.Mizuhashi, Asahi Garasu Kenkyu Houkoku (Rept. Res. Lab. Asahi Glass Co., Ltd.) 38 (1), 57 (1988)
9. K.Sato, K.Adachi, Y.Hayashi and M.Mizuhashi, proc. of 12th IEEE Photovoltaic Specialists Conference, Las Vegas, 267 (1988)
10. Y.Matsui and Y.Goto, submitting to proc. of 1999 MRS Spring Meeting 558, 393 (1999)
11. C.M.Wolfe, N.Holonyak,Jr. and G.E.Stillman, Physical Properties of Semiconductors (PRENTICE HALL, 1989) pp.175-201.

**Poster Session:
FPD**

Mat. Res. Soc. Symp. Proc. Vol. 621 © 2000 Materials Research Society

Photo- and Cathodoluminescence in Cerium-Activated Yttrium-Aluminium Borates

V. Z. Mordkovich
Intl Center for Materials Research, KSP East 601, Takatsu-ku, Kawasaki 213-0012, Japan

ABSTRACT

Ce^{3+} luminescence in Ce-activated yttrium aluminium borates was studied. Photoluminescnece was excited by continuous-wave UV or by pulse laser irradiation. Cathodoluminescence was excited at the acceleration voltage in the range 5 to 600 V. These phosphors show under continuous wave UV excitation three emission peaks: in near UV, deep blue and light blue ranges. The light blue peak at around 490 nm which may be connected to forbidden $^2D_{3/2} \rightarrow {}^2F_{7/2}$ transition appears in continuous wave-excited photoluminescence only and does not appear in cathodoluminescence nor under pulse excitation. The blue-emitting Ce-activated yttrium aluminium borates exhibit unusually strong activator ion-host lattice interaction. As a result a substantial red shift in the emission is induced by Ce concentration increase and complicated concentration dependence of the emission intensity is observed.

INTRODUCTION

Rare earth borate phosphors were first introduced by R.I.Smirnova et al. in 1969 [1] and have been of little or no interest for two decades. More recently due to rising demand for new efficient phosphors for flat panel display applications the rare earth borates' properties such as high efficiency, easy processibility, durability and often long afterglow have attracted attention. As a result, more publications appeared [2-4].

As for blue-emitting phosphors, they had been represented only by cerium-activated yttrium orthoborate with photo- and cathodoluminescence at ~410 nm i.e. of violet color [1]. Meanwhile, luminescence of Ce^{3+} is known to be strongly influenced by substituting some yttrium with aluminium [5]. Such substitution may lead to drastic changes in both emission wavelength and in efficiency either due to lower symmetry of Ce^{3+} ion surrounding or due to different contribution into crystal field by Y^{3+} and Al^{3+} ions. We recently introduced a Ce^{3+}-activated yttrium aluminium borate [6] as a blue-emitting phosphor which is more efficient than that of work [1].

The purpose of the present work was to study photo- and cathodoluminescence of the cerium-yttrium-aluminium borates with various Y/Al ratio and Ce^{3+} concentration.

EXPERIMENTAL DETAILS

The samples were prepared in the composition range $(Ce_xY_{1-x})_{1-y}Al_yBO_3$ where x=0.008 to 0.5 and y=0 to 0.75. The samples were synthesized at 900°C from stoichiometric amounts of yttrium chloride, cerium chloride and aluminium chloride mixed with excess boron oxide. After washing out the excess boron oxide the final product represents a powder with 2-3 μm particle size. X-ray diffraction showed a distorted hexagonal lattice typical for yttrium borate [7] with parameters slightly varying with Ce or Al concentration change.

Activation was performed by annealing the samples at the temperature in range 300-1200°C in a flow of a hydrogen/argon gas mixture (hydrogen concentration 10-14 %).

Photoluminescence was excited by continuous wave UV irradiation 200-400 nm. Also experiments with pulse excitation by 300 nm monochromatic irradiation (pulse frequency 4 MHz) irradiation were done. Cathodoluminescence was excited at the acceleration voltages 5 to 600 V and the current densities of up to 20 mA/cm^2. A fiber-optic probe was used to collect emission for the spectroscopic measurements. Reflection spectra were measured by Jasco V-570 spectrophotometer with integration sphere.

RESULTS AND DISCUSSION

Bright blue photoluminescence was observed with efficiency and chromaticity strongly depending on chemical composition and also on the temperature of activation.

A. Activation

Activation temperature strongly influences the efficiency of luminescence - see Figure 1. The optimum activation temperature was determined as 1000 °C. The resulting product of activation at 1000°C is a white powder while as-produced samples are yellow. It is reasonable to assume that the yellow color is determined by the presence of Ce^{4+} ions which usually show yellow color in oxide environment. The Ce^{4+} ions are reduced to Ce^{3+} by H$_2$ and this provides more Ce^{3+} emission centers. The drop in the luminescence efficiency at the temperatures higher than 1000°C may be explained by high-temperature phase transitions which is supported by X-ray diffraction results (annealing at 1200°C results in the change of peak intensity distribution).

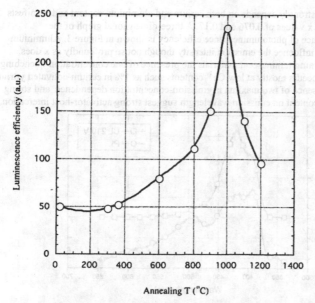

Figure 1. *Demonstration of the effect of activation by annealing in a hydrogen/argon gas mixture: luminescence efficiency vs activation temperature for $(Ce_{0.16}Y_{0.84})_{0.5}Al_{0.5}BO_3$.*

B. Photoluminescence

Continuous wave-excited photoluminescence (PL) spectra manifested three profound emission peaks as shown in Figure 2: near-UV peak with maximum at $\lambda_1 \approx 390$ nm; deep blue peak at $\lambda_2 = 410$ to 440 nm at different chemical compositions and light blue one at $\lambda_3 \approx 490$ nm. The low-cerium x<0.06 samples represented an exception - they showed only the first and the second peaks. Activated and as-synthesized samples showed similar spectra though all the features were more pronounced in case of activated samples.

The peak positions and intensity ratios in PL spectra are concentration-dependent. The near-UV λ_1 peak does not undergo position shifts but its contribution increases significantly at the Al content beyond y=0.33. The deep-blue peak position λ_2 appeared to be the most sensitive to the chemical composition variations. λ_2 increases with Ce content increase linearly in the range 0<x<0.25 and then stabilizes at the $\lambda_2 \approx 440$ nm. At the fixed rare-earth/Al ratio 1/1 this dependence fits the following equation:

$$\lambda_2 = 408 + (127 \pm 14)\, x$$

Al influence on λ_2 is more complicated but at any rate the increase in Al content results in a blue shift of λ_2. The biggest shift achieved by Al increase is 11 nm and was observed between the samples $(Ce_{0.17}Y_{0.87})_{0.67}Al_{0.33}BO_3$ ($\lambda_2 = 430$ nm) and $(Ce_{0.17}Y_{0.87})_{0.33}Al_{0.67}BO_3$ ($\lambda_2 = 419$ nm).

Emission intensity strongly depends on both cerium and aluminium content and manifests two distinct maxima at x values of 0.076 and 0.17. A three-dimensional graph of the concentration dependence of photoluminescence efficiency is shown in Figure 3. Aluminium content variations also influence the emission intensity though not so profoundly as x does.

For other Ce^{3+} containing luminescent materials the presence of one concentration quenching limit is typical and it occurs mostly at low Ce^{3+} contents such as 1% in cerium-activated yttrium silicate [8]. The presence of two maxima in emission-concentration dependence and strong influence of cerium content on emission wavelength suggest strong activator-host intercation.

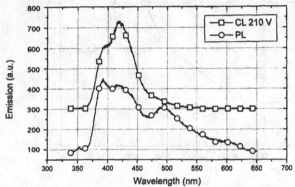

Figure 2. *Cathodoluminescence (CL) spectrum of activated $(Ce_{0.09}Y_{0.91})_{0.5}Al_{0.5}BO_3$ sample in comparison with photoluminescence (PL) spectrum of the same.*

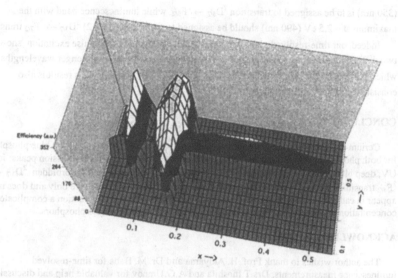

Figure 3. *Three-dimensional graph of concentration dependence of photoluminescence efficiency for $(Ce_xY_{1-x})_{1-y}Al_yBO_3$.*

C. Cathodoluminescence

Cathodoluminescence experiments showed relatively low threshold voltage value of 40 V. A sequence of cathodoluminescence spectra taken at different acceleration voltages is shown in Figure 4. Cathodoluminescence spectrum of $(Ce_{0.09}Y_{0.91})_{0.5}Al_{0.5}BO_3$ sample is shown in Figure 2 in comparison with the photoluminescence spectrum of the same. It can be seen from Figure 2 that though UV and deep-blue emission peaks stay the same in both cathodo- and photoluminescence, the third light-blue peak at λ_3 does not show in cathodoluminescence. This phenomenon was observed in the whole range of applied acceleration voltages. One shoud note that all our experiments were carried out far from saturation point as suggested by the observed linear dependence of emission intensity on anode current density.

To explain the luminescence spectra we observed, let us plot one of them along with a reflection spectrum of the sample in energy units (Figure5) and compare them with the scheme of energy levels of isolated Ce^{3+} ion, shown in the inset. States ^2D and ^2F split each into two levels thus making possible four transitions between excited and ground states. It is reasonable to assign two strongest luminescence bands to fully allowed transitions $^2D_{5/2} \rightarrow {}^2F_{7/2}$ and $^2D_{3/2} \rightarrow {}^2F_{5/2}$. The distance between two peaks in the reflection spectrum is ~ 0.9 eV. From the other hand, no distinctive peak has been observed in low energy (0.6 +1.5 eV) range of this spectrum. Therefore, the distance between levels $^2D_{5/2}$ and $^2D_{3/2}$ in the energy diagram of Ce^{3+} ion in the crystal lattice is higher than that between levels $^2F_{7/2}$ and $^2F_{5/2}$. So, we assign luminescence band with maximum at 390 nm to the transition from upper level $(^2D_{5/2} \rightarrow {}^2F_{7/2})$ and that with maximum at 420 nm to the transition $^2D_{3/2} \rightarrow {}^2F_{5/2}$. Shoulder with the maximum at ~ 3.5 eV

(350 nm) is to be assigned to transition $^2D_{5/2} \rightarrow {}^2F_{5/2}$, while luminescence band with the maximum at ~ 2.5 eV (490 nm) should be assigned to forbidden ($\Delta J = 2$) $^2D_{3/2} \rightarrow {}^2F_{7/2}$ transition.

Indeed, our time-resolved photoluminescence measurements with pulse excitation showed two distinct lifetimes: $\tau = 20$ ns at shorter wavelengths and $\tau = 60$ ns at longer wavelengths which correspond to two excited states: $^2D_{5/2}$ and $^2D_{3/2}$, respectively. This result is also consistent with Ce^{3+} lifetime data from [5].

CONCLUSIONS

Cerium-activated yttrium aluminium borates may be characterized as strong blue phosphors for both photo- and cathodoluminescence applications. They show three emission peaks: in near UV, deep blue and light blue ranges. The latter peak may be connected to forbidden $^2D_{3/2} \rightarrow$ $^2F_{7/2}$ transition and appears in continuous wave-excited photoluminescence only and does not appear in cathodoluminescence. As a result of strong activator-host interaction a complicated concentration dependence of the emission intensity is observed in these phosphors.

ACKNOWLEDGEMENTS

The author wishes to thank Prof. H. Akiyama and Dr. M. Baba for time-resolved luminescence measurements, Drs T.Inoshita and A.G.Umnov for valuable help and discussions.

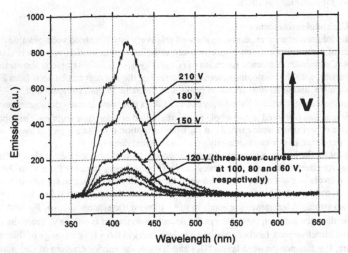

Figure 4. *Cathodoluminescence spectra of $(Ce_{0.09}Y_{0.91})_{0.5}Al_{0.5}BO_3$. The curves from bottom to top refer to the following acceleration voltages: 60 V; 80 V; 100 V; 120 V; 150 V; 180 V; 210 V.*

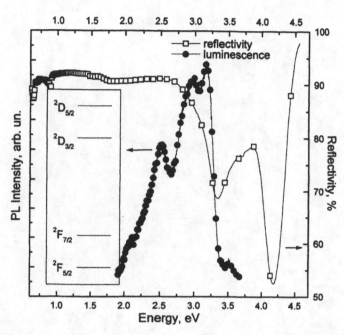

Figure 5. *Reflection and photoluminescence spectra of* $(Ce_{0.09}Y_{0.91})_{0.5}Al_{0.5}BO_3$. *See explanations in the text.*

REFERENCES

1. R.I.Smirnova, L.A.Konopko and L.A.Ivanenko, *Otkrytiya Izobreteniya Bulletin*, **46**, 74 (1969).
2. M.V.Ryzhkov, V.P.Dotsenko and V.A.Gubanov, *Opt.Spektrosk.*, **67**, 988 (1989).
3. Tsutomu Chikada and Takashi Hase, *Jap. Pat. 06096736* of 08.04.1994.
4. L.Zhang, C.Madej, C.Pedrini, C.Dujardin, I.Kamenskih, A.Belsky, D.A.Shaw, J.Mesnard and C.Fouassier, *Proc.-Electrochem.Soc,* **97-29** (Luminescent Materials), p.342 (1998).
5. *Phosphor Handbook*, edited by Shigeo Shionoya and William M. Yen (CRC Press, Boca Raton, 1998) 608 p.
6. Vladimir Mordkovich, *Jap. Pat.Appl. 11-27421* of 04.02.99.
7. A.Newnham, Redman and Santero, *J.Amer.Ceram.Soc.*, **46**, 253 (1963).
8. E.J.Bosze, G.A.Hirata, J.McKittrick and L.E.Shea, in *Flat Panel Display Materials*, edited by Gregory N.Parsons et al. (Mater. Res. Soc. Proc. **508**, Warrendale, PA 1998) p.269.

Mat. Res. Soc. Symp. Proc. Vol. 621 © 2000 Materials Research Society

Very low field electron emission from Hot Filament CVD grown microcrystalline diamond.

B.S.Satyanarayana[a], **X.L.Peng**[*b], **G.Adamopoulos, J.Robertson , W.I.Milne, & T.W.Clyne***
Electronic Materials & Devices, Dept of Engineering, Cambridge University, Cambridge. CB2 1PZ. U.K.
*Material Science dept, Cambridge University, Cambridge. U.K.

ABSTRACT.

Very low threshold field emission from undoped microcrystalline diamond films grown by the hot filament chemical vapour deposition process (HFCVD) is reported. The effect of crystal size, methane concentration and the temperature has been studied. The microcrystalline diamond films grown using 3% methane (CH_4) / hydrogen (H_2) gas mixture ratio under varying deposition temperatures exhibit very low emission threshold fields. The threshold fields varied from 0.4 V/ μm to 1 V/μm for an emission current density of 1 μA/cm^2. A correlation between the emission characteristics and the material properties is presented. These films exhibit an emission site density of ~ 10^4-10^5/cm^2 at an applied field of 3 V/μm.

INTRODUCTION

Low electron affinity was one of the prime factors for interest in diamond as a candidate for field emitters, besides its other physical properties like high thermal stability, mechanical hardness and chemical inertness. Recently low field electron emission has been reported from many carbon based materials including nanocrystalline diamond [1-8], nanotubes [9-11] and nanocluster or nanostructured carbons [12-15]. This clearly shows that electron affinity is not the only criteria but there are other factors which also influence the emission. The emission mechanism from these materials is still not clear. For efficient display applications the need is for high emission site densities ~ 10^6 and current preferably in the range of 10 mA / cm^2 [8] at reasonably low fields.

Undoped nano-crystalline / polycrystalline diamond films with threshold fields as low as 3-5 V/μm [4-6] and doped polycrystalline diamond with threshold fields as low as 0.5 V/μm [3] have been reported. In the case of doped diamond films, the doping efficiency is still poor and the films tend to be still resistive even after heavy doping.[3.16] It has been observed that the phosphorus doped films tend to show a decrease in threshold field with temperature while boron doped films seem to be relatively stable[16]. It has also been reported that nitrogen has a tendency towards phase transformation in carbon systems[17]. Therefore it is still not clear if the doped diamond cathode would be stable over a long period of time. Further the low field emitting graphitic materials [9-15] and the evidence that even in diamond the emission is possibly from the grain boundaries [1,2,18] suggest that there could be an optimum crystal dimension and sp^2/sp^3 bonding ratio for low field emission. This could possibly be done even without doping. Hence an effort has been made to systematically study the effect of crystal size, methane concentration and growth temperature on field emission from undoped diamond films grown using the hot filament chemical vapour deposition [HFCVD] process.

EXPERIMENTAL CONDITIONS

Prior to deposition, the heavily n-doped silicon substrate surface was pretreated by polishing with a 1μm diamond paste for ten minutes to increase diamond nucleation. The diamond thin films were grown using a hot filament chemical vapour deposition [HFCVD] system. The tantalum filaments [0.25 mm diameter] used in the system were precarburized in a 1% CH_4/H_2 atmosphere for 2 hours. The conditions used for deposition were :

Filament temperature : 2200 ~ 2300 °C,   Substrate temperature : 725 °C to 925 °C
Filament to substrate gap : 7 mm.   Gas flow rate   : 150 ~ 200 sccm
Chamber pressure   : 15 - 20 torr CH_4/H_2 ratio   : 0.5 % to 3 %

First the effect of crystal size on field emission was studied. Next the effect of methane concentration on emission from the nanocrystalline diamond films was studied by varying the methane to hydrogen ratio from 0.5 % to 3 %. Finally the effect of the substrate temperature was investigated while using a 3% CH_4/H_2 gas ratio by varying the temperature between 725 °C to 925 °C. Typical thickness of the films grown were 500nm.

The field emission measurements were carried out in a parallel plate configuration with an anode–cathode spacing of 100 μm in a test chamber maintained at 10^{-7} Torr.[19] The polycrystalline diamond film needed conditioning, before initiation of emission while the nanocrystalline/microcrystalline diamond films did not require any forming or conditioning.[19] The small dimension diamond crystals with dimension varying from 100 – 700 nm have been generally referred to as microcrystalline diamond in this paper. The experimental conditions for the emission site density measurement have been explained earlier [13]. The Raman measurements were carried out in a Spex 1401 double spectrometer, using a tuneable, 488 nm argon ion laser.

Results and discussion :

In Figure- 1 we show the field emission characteristics of three types of films namely polycrystalline diamond, defective diamond and microcrystalline diamond. These films were grown at 825°C with 1% CH_4/H_2 gas ratio. The different materials including polycrystalline diamond and microcrystalline diamond were grown by varying the flow rate from 150 Sccm to 200 Sccm and the corresponding deposition pressure from 15 torr to 20 torr. The thickness of the polycrystalline diamond film was ~1.5 μm, the defective and microcrystalline diamond films were nearly 500nm in thickness. The grain size vary from 2 μm to 200nm. It can be seen from the figure 1, that the threshold field reduces from around 18 V/μm for a ~ 2 μm sized polycrystalline diamond to nearly 1 V/μm for the nano-crystalline/microcrystalline diamond. The threshold field is defined as that field at which an emission current density of 1 μA/cm² is obtained. The reduction in the threshold field could be partly attributed to the reduction in thickness as observed by Lacher et.al [5]. The other factors being the reduction in the grain size and the effective increase in the conducting grain boundaries and other carbon phases.[4-7] The above results show that with the reduction in crystal size, it is possible to have low threshold fields and high emission currents. More details about the diamond crystal size affects have been presented earlier.[20] To further improve the surface coverage of diamond nuclei and the current density, the effect of methane concentration and substrate temperature was studied.

Figure –2 shows the SEM pictures of microcrystalline diamond films grown at different methane concentration. It can be seen, that with increase in methane concentration, there is an increase in the nucleation density and a corresponding increase in the coverage and hence increase in the emission current. The crystal size also seems to increase. Figure – 3 shows the effect on field emission due to variation in methane concentration during the growth of nano-crystalline diamond films. It can be seen that with increasing methane concentration there is a corresponding fall in the threshold field and increase in the current density. The films grown at 3% and 1.5 % both have a threshold fields less than 1 V/μm.

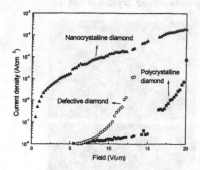

Figure - 1 *J-E emission characteristics of three different types of diamond thin films*

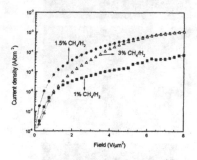

Figure – 3 *J-E emission characteristics of nano-crystalline/microcrystalline diamond films grown at different [CH₄/ H₂] methane concentration*

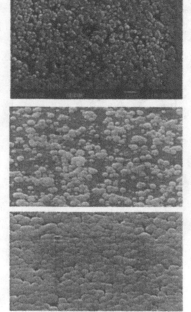

Figure – 2 *SEM pictures of nano-crystalline/microcrystalline diamond films grown at different [CH₄/ H₂] methane concentration*

Next the effect of temperature was studied while maintaining the methane concentration constant at 3%. Figure - 4 shows the effect of the substrate temperature on the emission current and the threshold field. It can be seen that the threshold field decreases from 5.25 V/μm to 0.4 V/μm as the temperature of growth is varied from 925 °C to 725 °C. The current density shows a two order change in magnitude at 5 V/μm applied field. The SEM micrographs of these materials are shown in figure 5. It can be seen that with increasing temperature there is further modification in the morphology of the microcrystalline material. From a crystal size of nearly 600- 700nm at 725 °C the average size of the crystals reduces to about 400 -500nm at 925 °C. The morphology also seems to change from cauliflower like structures, with fibers or whiskers on the top for 725 °C to more relatively more refined crystals at 925 °C.

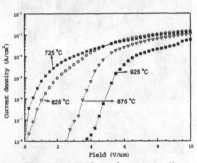

Figure 4 *J-E Plots of nanocrystalline diamond thin films grown at different temperatures*

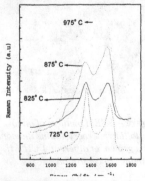

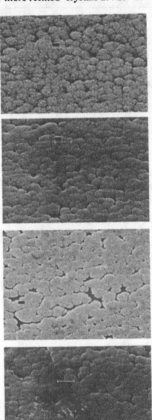

Figure 6. *Raman spectra of microcrystalline diamond films grown at different temperatures.*

Figure 5 *SEM micrographs of microcrystalline diamond filmsgrown at different temperatures from top a) 725 C, b) 825 C, c) 875 C & d) 925 C.*

The mobility of atomic hydrogen decides the growth of diamond crystals. Depending on the temperature during growth, graphite or amorphous carbon deposits result on the films[21-23]. Hence with increase in temperature from 725 °C to 925 °C, there seems to be a reduction in the embryonic graphitic nuclei due to the enhanced presence of atomic hydrogen. This leads to a corresponding enhancement in the quality of the diamond crystal growth. The improvement in diamond quality affects the emission process, leading to higher threshold fields and relatively lower currents. The removal of the embryonic graphitic nuclei from the carbon matrix, could be attributed to be the cause for the observed gaps in surface coverage of the films in the SEM micrographs with rise in temperature.

Shown in figure 6, are the Raman curves of the microcrystalline diamond films grown at various temperatures. It can be seen that the $1140 \sim 1150$ cm^{-1} peak associated with the disordered or nanocrystalline diamond decreases in intensity with decrease in temperature from 925 °C to 725 °C. The D and G peaks associated with the graphitic clusters or carbon inclusions become prominent with decrease in temperature. Thus the lowest threshold field and the highest emitted current is associated with the films with least crystalline features and the highest I(D)/I(G) ratio. The I(D)/I(G) ratio seems to increase linearly with decrease in threshold field in the case of microcrystalline diamond films. the same is represented in figure 7. This seems to be applicable for other carbon based nanostrucutred carbon films too.[20,24] The results show that good emission currents at low threshold fields require an optimum blend of diamond particles in a graphitic mesh spread uniformly over the surface. The additional surface morphological features due to the graphitic phases seem to further enhance the emission characteristics. Thus leading to emission threshold field as low as 0.4 V/μm, the lowest value reported in literature for 1μA/cm2 current density in the case of undoped microcrystalline diamond films. These diamond films exhibit an emission site density of $\sim 10^4$-10^5 at an applied field of 3 V/μm.

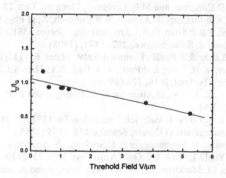

Figure 7. *Relation between Threshold field and I(d)/I(g) ratio derived from Raman spectra.*

CONCLUSION

We have shown that the undoped microcrystalline diamond films grown under optimum conditions emit at very low applied fields. A systematic study on the effect of crystal size, methane concentration and the temperature has been reported. The results show that with decrease in grain size of the diamond crystal there is an increase in the emission current and decrease in the threshold field from 18 V/μm to 0.75 V/μm. With increase in methane concentration there seems to be an increase in the nucleation density and a corresponding increase in the surface coverage leading to more uniform emission The microcrystalline diamond films grown using 3% methane (CH_4) / hydrogen (H_2) gas mixture ratio under varying deposition temperatures exhibit very low emission threshold fields. The threshold fields varied from 0.4 V/μm to 1 V/μm for an emission current density of 1 $\mu A/cm^2$. The better emitting nano/micro-crystalline diamond film seems to have greater graphitic inclusions, leading to higher emission current at lower applied fields. In the present case an emission threshold field of 0.4 V/μm (lowest value reported in literature for 1μA/cm2 current density) is observed. Graphitic inclusion and clustering of the nuclei is also confirmed by the Raman curves and the morphological features observed in the SEM curves. The I(D)/I(G) peaks ratio derived from Raman measurements seems to increase linearly with decrease in threshold field in the case of microcrystalline diamond films. The low threshold emission microcrystalline diamond films exhibit an emission site density of $\sim 10^4$-$10^5/cm^2$ at an applied field of 3 V/μm.

Acknowledgment :

The authors would like to thank Mr. B.Breton for the SEM pictures.

References:

1. C.Wang, A.Garcia, D.C.Ingram, and M.E.Kordesch, Electron . Lett. **27**, 1459 (1991).
2. A A Talin, L S Pan, K F McCarty, H J Doerr, R F Bunshah, Appl Phys Lett **69**, 3842 (1996).
3. K.Okano, S.Koizumi, S.R.P Silva, G.A.J.Amaratunga, Nature, **381**,140 (1996).
4. W. Zhu, G P Kochanski & S Jin, Science, **282**, 1471, (1998).
5. F.Lacher, C.Wild, D.Behr & P.Koidl , Diamond Relat. Mater. **6**, 1111(1997).
6. K.H.Park, Soonil Lee, K.H. Song, J. Il Park, K.J. Park, S.Y. Han, S.J. Na, N.Y. Lee, and K.H. Koh, J. Vac. Sci. Technol. **B 16**, 724 (1998).
7. M.Q. D. M. Gruen, A. R. Krauss, O.Auciello, T. D. Corrigan and R. P. H. Chang, J Vac Sci Technol **B 17**, 705 (1999).
8. Z.Li Tolt, R.L.Fink & Z.Yaniv, J. Vac. Sci. Technol. **B 16**, 1197, (1998).
9. W.A.de Heer, A.Chatelain, and D.Ugrate, Science **270**, 1179 (1995).
10. O.M.Kuttel, O.Groening, C. Emmenegger & L.Schlapbach, App Phys Lett, **73**, 2113 (1998).
11. Y.Chen, S.Patel, Y.Ye, D.T.Shaw & L.Guo, App. Phys. Lett, **73**, 2119 (1998).
12. B.F.Coll, J.E. Jaskie, J.L.Markham, E.P.Menu, A.A.Talin, P.von Allmen, MRS. Sym Proc. **Vol 498**, 185 (1998).
13. B S Satyanarayana, J Robertson, W I Milne, J. App. Phys. **87**, 3126, (2000)
14. O.N.Obraztsov, I.Yu.Pavlovsky and A.P.Volkov, J. Vac. Sci. Technol. **B 17** , 674, (1999).
15. A.C.Ferrari, B.S.Satyanarayana, P.Milani, E.Barborini, P.Piseri, J.Robertson and W.I.Milne. Europhys. Lett, **46**, 245 (1999).

16. T.Sugino, K.Kuriyama, C.Kimura & S.Kawasaki, App. Phys. Lett **73**, 268 (1998).
17. H.Sjostrom, S.Stafstrom, M.Boman & J-E Sundgren, Phys. Rev. Letts, **75** 1336 (1995).
18. A.V.Karabutov, V.D.Forlov, S.M.Pimenov & V.I.Konov, Diamond Relat. Mater. **8**, 763 (1999).
19. B S Satyanarayana, A Hart, W I Milne & J Robertson, App Phys Lett **71**, 1430 (1997). (b) Diamond & Related Materials **7**, 656 (1998).
20. B.S.Satyanarayana, Ph.D thesis, University of Cambridge, Cambridge, UK, (1999).
21. B.Lux & R.Haubner, in Diamond and Diamond like films and Coatings, Ed. R.E.Clausing, J.C.Angus, L.L.Horton, & P.Koidl , Plenum Press USA, p-579, (1990).
22. Y.Hirose, Japan. J. Appl. Phys, **25** ,L519, (1986).
23. B.B.Pate, Surface Science, **165**, 83, (1986).
24. A.Ilie, T.Yagi, A.C.Ferrari and J Robertson, to be published in App Phys Letts.

Note :

a) Electronic mail : satya@ele.kochi-tech.ac.jp
Present address : Sistec & Dept of Electronic &Photonic Systems Engineering, Kochi University of Technology, 185, Miyanokuchi, Tosayamada, Kochi 782-8502, Japan.

b) Present address : Dr. XiLin Peng,
Research Associate, Physics Department, 1150 University Ave.
University of Wisconsin-Madison, Wisconsin, WI53706, USA

Mat. Res. Soc. Symp. Proc. Vol. 621 © 2000 Materials Research Society

Green Emission from Er-Doped AlN Thin Films Prepared by RF Magnetron Sputtering

V. I. Dimitrova, F. Perjeru, Hong Chen and M. E. Kordesch
Physics & Astronomy Department, and Condensed Matter & Surface Science Program,
Ohio University, Clippinger Labs, Rm. 251
Athens, OH 45701, U.S.A.

ABSTRACT

Thin films of Er doped AlN, ~ 200 nm thick, were grown on indium tin oxide/aluminum titanium oxide/glass substrates using RF magnetron sputtering in a pure nitrogen atmosphere. To optically activate Er all films were subject to post-deposition annealing in flowing nitrogen atmosphere at atmospheric pressure at temperatures between 1023-1223 K for 10-60 minutes. The visible cathodoluminescence (CL) in the green was detected at both 11 K and 300 K. The strongest CL peaks were observed at 558 nm and 537 nm (11 K), which correspond to the transitions from $^4S_{3/2}$ and $^2H_{11/2}$ to the $^4I_{15/2}$ ground level. Electroluminescence (EL) studies of AlN:Er alternating-current thin-film electroluminescent (ACTFEL) devices were performed at 300 K. The turn-on voltage was found to be around 80-100 V for our ACTFEL devices. The intensity of the EL emission rapidly increases with the voltage increasing in the investigated range of 110-130 V.

INTRODUCTION

Recently, intense research on rare-earth (RE)-doped III-nitride semiconductor thin films has been performed because of their potentially significant new application in display technologies [1]. Considerable work has been done mainly on GaN doped with a variety of RE elements for a light-emitting diodes (LEDs) fabrication. Visible, room temperature photoluminescence (PL) and cathodoluminescence (CL) emission from various RE elements (Er, Tm, Eu, Dy, Pr and Tb) have been observed in GaN [2-6]. In addition to PL and CL, Steckl and coworkers have observed strong visible room temperature green and blue electroluminescence (EL) emission from Er and Tm-doped GaN-based Shottky LEDs and electroluminescent devices (ELD) [7,8]. These GaN films were prepared by molecular beam epitaxy (MBE) on p- Si (111) substrates, and doped by ion implantation.

Visible green CL emission at low temperature and at 300 K from 3.4 at. % Er-doped amorphous AlN thin films deposited by dc reactive magnetron sputtering have also been observed [9].

The purpose of this work is to report a novel approach for obtaining strong green EL emission from amorphous AlN phosphor material doped with Er impurities. Sputtered AlN:Er films may be a promising new material as a green light emitting phosphor for use in alternating-current thin-film electroluminescent (ACTFEL) device applications.

EXPERIMENTAL DETAILS

Thin films of AlN:Er were prepared by rf magnetron sputtering of an Al target of 99.999 % purity and an Er target of 99.9 % purity in pure nitrogen atmosphere. The background vacuum in the chamber was $\leq 3\times10^{-5}$ Torr. All films were deposited onto 1"x1" glass substrates coated with layers of indium tin oxide (ITO) and aluminum-titanium oxide (ATO), which serve as the bottom transparent electrode and the bottom insulator, respectively. The deposition pressure and the sputtering power during the film growth were 6×10^{-4} Torr and 199 W, respectively. The substrates were kept at room temperature ($T_{sub}\leq 403$ K) to obtain amorphous AlN: Er films. The growth rate was between 0.04 and 0.05 nm/s and the total thickness of the films as determined from the transmission IR spectrum of AlN:Er films was in the range between 180 and 200 nm.

The as-deposited films were annealed at 1023-1223 K up to 60 min in N_2 atmosphere using a tube furnace. To complete the ACTFEL device structure, some of the samples were covered by an Al_2O_3 top insulator layer. On the top aluminum dots, ~100 nm thick, were evaporated which serve as the second contact.

The film stoichiometry was measured by x-ray fluorescence in a scanning electron microscope (SEM) and wavelength-dispersive electron microprobe analysis (EPMA). The microstructure was examined by x-ray diffraction (XRD). The optical transmittance was recorded as a function of a wavelength (190-820 nm) by using an ultraviolet-visible spectrophotometer. The band gap (Eg) was determined as described in Ref. 9.

For all samples, CL spectra were recorded at 11 and 300 K at the following excitation conditions: 5 keV, ~ 75 μA at 11 K, and 3.8 keV at 300 K. EL data were collected with a SPEX monochromator and THORN EMI photomultiplier detector. For the EL measurements ac power supplies operating at frequencies of 1 kHz and 60 Hz were used.

RESULTS

XRD of as-deposited and activated AlN:Er films showed that there are no peaks observed due to crystalline AlN indicating that films are amorphous.

SEM and EMPA analyses of Er doped AlN films, ~200 nm thick, confirmed that films are stoichiometric (Al:N=1:1). The Er concentration in AlN:Er films was found to be about 1 wt. % (0.25 at. %), as determined by EMPA. Experimentally it was found that this amount of Er in the AlN films is the optimal one in order to get bright luminance.

The investigated films are transparent with a high optical transmittance of 98 % at 490 nm. The band gap (Eg) was estimated to be ~5.92 and ~5.55 eV for undoped and doped with 1 wt. % Er films, respectively.

In general, the thermal annealing is an important process used in the device phosphor preparation. The as-deposited AlN:Er films do no show any CL emission. It was found that to activate the Er^{3+} ions the films have to be annealed in the temperature range between 1023-1223 K. If the samples were annealed for a long time (30, 45 or 60 min), in a N_2 flow at temperatures 1073 and 1223 K, no luminescence or very weak luminescence was observed.

The optimum annealing temperature and time in a N_2 flow were found to be 1073 K and 15 min for a reproducible bright green CL emission at 11 K (Fig. 1). An identical CL spectrum to that shown in Fig. 1 is observed for the samples annealed at a temperature of 1223 K for 15 min.

Also, films deposited onto ATO/ ITO/ glass substrates and annealed at 1023 K in N_2 for 10 min gives strong green CL at 300 K (Fig. 2). In order not to destroy the ITO conductive layer used in our device structure in our experiments, all samples were annealed at 1023 K. From Fig. 1 and 2 it can be clearly seen that both spectra recorded at 11 K and 300 K contain four major lines.

emission spectra (Fig. 1 and 2) also show two intense lines in the violet and blue part of the visible spectral region at 408 and 378 (11 K) nm, respectively. The luminescence from these lines corresponds to $^2H_{9/2}$ and $^4F_{7/2}$ transitions to the ground level, which are nearly independent of temperature between 11 K and 300 K respectively.

Fig. 2 shows dependently of the EL emission of the applied voltage-voltage curve of an operating EL display of 9 Hz for the fabricated AlN:Er/ACTFEL device with a top insulator layer. The optical turn-on voltage for these devices is found to be around 140 V. This value is a compatible to the typical ones (100-200 V) for ACTFEL devices (e.g. intensity of the EL emission rapidly increases with the voltage increasing in the universal range of 170-190 V).

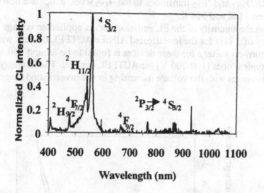

Figure 1. CL emission spectrum of amorphous $AlN:Er^{3+}$ films excited by electrons with 5 keV energy at 11 K. Films are annealed at 1073 K in N_2 for 15 min.

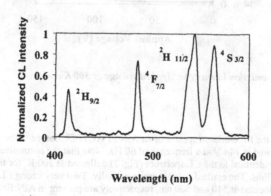

Figure2. CL emission spectrum of amorphous $AlN:Er^{3+}$ excited by electrons with 3.8 keV energy at 300 K. Films are annealed at 1023 K in N_2 for 10 min.

Two very strong lines are observed in the green wavelength region of the spectrum, at 558 nm (11 K) or 564 nm (300 K) and 537 (11 K) nm or 542 nm (300 K). These lines are due to $^4S_{3/2}$ and $^2H_{11/2}$ transitions to the $^4I_{15/2}$ ground state. Their intensity has temperature dependency. The emission spectra (Fig. 1 and 2) also show two intense lines in the violet and blue part of the visible spectral region at 408 and 478 (11 K) nm, respectively. The luminescence from these lines corresponds to $^2H_{9/2}$ and $^4F_{7/2}$ transitions to the $^4I_{15/2}$ level, which are nearly independent of temperature between 11 K and 300 K.

Fig. 3 shows the intensity of the EL emission versus applied ac voltage curve at an operating frequency of 1 kHz for the investigated AlN:Er ACTFEL devices with a top insulator layer. The optical turn-on voltage for these devices is found to be around 100 V. This value is a compatible to the typical ones (100-200 V) for ACTFEL devices. The intensity of the EL emission rapidly increases with the voltage increasing in the investigated range of 110-130 V.

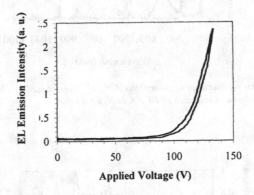

Figure 3. Visible EL emission intensity vs. Applied voltage at 300 K and frequency of 1 kHz.

Fig. 4 shows the EL emission spectra of an AlN:Er ACTFEL device without a top insulator layer measured at 144 V at a frequency of 60 Hz. Note that the normalized AlN:Er EL spectrum is virtually identical to the CL spectrum (Fig. 2) collected at 300 K for the same amorphous AlN:Er films. The emitted color is green, visually. Two very strong EL peaks $^2H_{11/2}$-$^4I_{15/2}$ and $^4S_{3/2}$-$^4I_{15/2}$ located at 540 and 560 nm, respectively are present in AlN:Er spectrum. That means the hot electrons are energetic enough to excite $^2H_{11/2}$ and $^4S_{3/2}$ excited states, which cause the green emission. The turn-on voltage for the devices without a top insulator layer is lower, around 80 V.

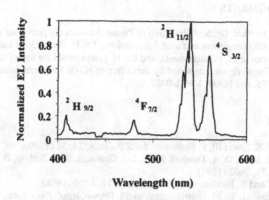

Wavelength (nm)

Figure 4. Room temperature EL emission spectrum of AlN:Er ACTFEL device without a top insulator layer at applied voltage 144 V and frequency 60 Hz.

DISCUSSION

Based on our experimental results, it was found that the long time annealing leads to decreasing of Er luminescence. XRD of AlN:Er films annealed to different temperatures in a N_2 atmosphere showed the annealing does not crystallize any of the investigated films. It is not known why the Er activity decreases with further heating.

Both CL and EL peaks show narrow peaks widths. These sharp characteristic emission lines observed in both luminescence spectra correspond to Er^{3+} intra-4f shell transitions. The temperature dependence of these lines is not so large, which is very important for prospective electronic devices operating at ambient temperatures. The emission lines at 540 nm and 560 nm in the EL spectra could be assigned to more than one transition. Therefore, more detailed investigations at this wavelength region of the spectrum will be necessary. We believe that excitation of the Er^{3+} ions in CL and EL involves direct impact excitation by injected hot (energetic) electrons.

CONCLUSIONS

In conclusions, we have demonstrated Er^{3+}-related EL emission from a sputter deposited amorphous AlN:Er ACTFEL device at 300 K. RF magnetron sputtering is a promising method to make good quality phosphor films, in particular, because the process can easily be scaled up for production. The potential for useful devices needs to be further evaluated.

ACKNOWLEDGMENTS

We want to thank Dr. Sey-Shing Sun of Planar America for providing substrates. Also, Dr. J. Wager for encouragement and useful discussions, Dr. H. Richardson and Dr. P. G. Van Patten for assistance in EL measurements, and Dr. H. Lozykowski for use of equipment for CL measurements. The work was supported by the Office of Naval Research through grants N00014-99-1-0975, and N00014-96-1-0782.

REFERENCES

1. J. Ballato, J. S. Lewis III, P. Holloway, *MRS Bulletin*, **24**, 51 (1999).
2. S. Kim, S. J. Rhee, D. A. Tumbull, X. Li, J. J. Coleman, S. G. Bishop, B. Klein, *Appl. Phys. Lett.*, **71**, 2662 (1997).
3. A. J. Steckl and R. Birkhahn, *Appl. Phys. Lett.*, **73**, 1700 (1998).
4. H. J. Lozykowski, W. M. Jadwisienczak and I. Brown, *Appl. Phys. Lett.*, **74**, 1129 (1999).
5. R. Birkhahn, M. Garter and A. J. Steckl, *Appl. Phys. Lett.*, **74**, 2161 (1999).
6. H. J. Lozykowski, W. M. Jadwisienczak and I. Brown, *Appl. Phys. Lett.*, **76**, 861 (2000).
7. A. J. Steckl, M. Garter, R. Birkhahn and J. Scofield, *Appl. Phys. Lett.*, **73**, 2450 (1998).
8. A. J. Steckl, M. Garter, D. S. Lee, J. Heikenfeld and R. Birkhahn, *Appl. Phys. Lett.*, **75**, 2184 (1999).
9. K. Gurumurugan, Hong Chen, G. R. Harp, W. M. Jadwisienczak and H. J. Lozykowski, *Appl. Phys. Lett.*, **74**, (3008) 1999.

Mat. Res. Soc. Symp. Proc. Vol. 621 © 2000 Materials Research Society

Effect of TiO₂ Addition on the Secondary Electron Emission and Discharge Properties of MgO Protective Layer

Younghyun Kim, Rakhwan Kim, Hee Jae Kim[1], Hyeongtag Jeon and Jong-Wan Park
Dept. of Metallurgical Engineering, Hanyang University, Seoul 133-791, Korea
[1]Dept. of Ordnance Engineering, Korea Military Academy, Seoul, KOREA

ABSTRACT

Secondary electron emission from a cathode material in AC PDP (Plasma Display Panel) is dominated by potential emission mechanism, which is sensitive to band structure of a protective layer. Therefore, the secondary electron emission property can be modified by a change in the energy band structure of the protective layer. $Mg_{2-2x}Ti_xO_2$ films were prepared by e-beam evaporation method to be used as possible substitutes for the conventional MgO protective layer. The oxygen content in the films and in turn, the ratio of metal to oxygen gradually increased with the increasing TiO₂ content in the starting materials. The pure MgO films exhibited the crystallinity with strong (111) orientation. The $Mg_{2-2x}Ti_xO_2$ films, however, had the crystallinity with (311) preferred orientation. The stress relaxation, when the [TiO₂/(MgO+TiO₂)] ratio in the evaporation starting materials was 0.15, seems to be related to inhomogeneous film surface due to an excessive addition of TiO₂ to MgO. When the [TiO₂/(MgO+TiO₂)] ratios of 0.1 and 0.15 were used, the deposited films exhibited the secondary electron emission yields improved by 50% compared to that of the conventional MgO protective layer, which resulted in reduction in discharge voltage by 12%.

INTRODUCTION

The AC plasma display panel (AC PDP) has recently received much attention in research and development as an alternative to the cathode ray tube (CRT). The AC PDP is also considered as a promising wall hanging TV set because there have been growing demands for the large-sized TV screens which are thin and light. The current MgO protective layer for the promising wall hanging TV, AC plasma display panel is thought to be still not one of the best materials to meet the demand of advanced high vision PDP with good light emission efficiency, low power consumption, low production cost and high image fidelity [1]. Therefore, to improve the overall performance of PDP, it is indispensable to develop new protective materials that can reduce the required voltage for discharging, and which will then be compatible with integrated circuits [2]. If other variables such as gas composition, electrode gap, electrode width and barrier height are constant, the operation voltage of ACPDP is closely related to secondary electron emission from protective layer for dielectric layer [3]. The secondary electron emission property of the protective layer is affected by its energy band structure such as electron affinity, work function, bandgap energy and intraband level [4]. In this study, $Mg_{2-2x}Ti_xO_2$ protective layers evaporated from the starting materials with different [TiO₂/(MgO+TiO₂)] ratios were introduced, and then the effects of TiO₂ addition on physical and electrical characteristics were investigated.

EXPERIMENTAL DETAILS

MgO and TiO_2 powders of 99.95% purity purchased from Cerac were used to prepare powder mixtures of desired compositions, which were milled and then dry-pressed into disks in one direction at the pressure of 2,000 Pa. Heat treatment at 1000 °C were performed for 3 h in a vacuum furnace for densification of the cold-pressed mixture pellets using the solid-state reactions [5]. Evaporation is performed at base pressure of 5×10^{-4} Pa without elevating the substrate temperature and introducing the oxygen gas. Input power and voltage of electron gun were 1.2 kW and 4 kV, respectively. The surface morphologies of the films were obtained by field emission scanning electron microscopy (FESEM). Film stress was calculated from measured wafer curvature based on Stoney's equation using a Tencor FLX-2908 thin film stress measurement system. Surface morphology of films was analyzed by an atomic force microscopy. The secondary electron emission properties of $Mg_{2-2x}Ti_xO_2$ films were characterized using the measurement system that was built at Seoul National University. The apparatus consists of an ion gun system, a measuring part for secondary electrons and a vacuum system, being similar to that of Lin *et al* [6]. The detail description of the measurement system was already reported elsewhere [7]. In this study, He ion was chosen to be the ion species and its kinetic energy was fixed at 200 eV, because secondary electron emission from films depended upon the type and kinetic energy of ion species [8]. The voltage characteristics of the panels such as firing voltage and minimum sustaining voltage were then evaluated using He gas of 300 Torr at a driving frequency of 20 kHz.

RESULT

Film composition obtained by energy dispersive spectrometry (EDS) is shown in Table I. As the [$TiO_2/(MgO+TiO_2)$] ratio in the starting materials increased from 0 to 0.3, cation ratio i.e., [$Ti/(Mg+Ti)$] in the films also increased from 0 to 0.117. The deposited films were thus depleted to some degree in Ti content in comparison with the initial Ti content in the starting materials. It was well anticipated from the fact that TiO_2 has much lower vapor pressure than that of MgO[9]. XRD was used to determine the structural changes of the Mg-Ti-O films by addition of the TiO_2 to the MgO.

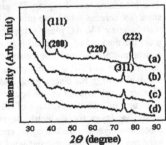

Table I. Composition of $Mg_{2-x}Ti_xO_{2x}$ films using energy dispersive spectrometry (EDS)

[$TiO_2/(MgO+TiO_2)$] in Starting Materials	Cation Ratio [$Ti/(Mg+Ti)$]	Metal to Oxygen Ratio [$(Mg+Ti):O$]	Deposition Rate (nm/min)
0	0	1 : 1.02	159
0.05	0.026	1 : 1.16	120
0.1	0.054	1 : 1.19	105
0.15	0.069	1 : 1.19	96
0.2	0.092	1 : 1.20	63
0.3	0.117	1 : 1.23	45

Figure. 1 XRD patterns of MgO-TiO_2 films as a function of 2θ for different [$TiO_2/(MgO+ TiO_2)$] ratio in sources for evaporation; (a) 0, (b) 0.1, (c) 0.2 and (d) 0.3

Figure 1 shows the XRD patterns of 5000 ±500 Å thick MgO-TiO$_2$ films evaporated on soda lime glass substrates. The XRD pattern of the pure MgO film shows strong (111) peak at the diffraction peak position of 2θ =36.95° and weak (200) and (220) peaks (Fig. 1a). With addition of TiO$_2$ to MgO, the structural change is such that a new peak appears at 2θ value of 71.4° as shown in Figs. 1b ~ 1d.

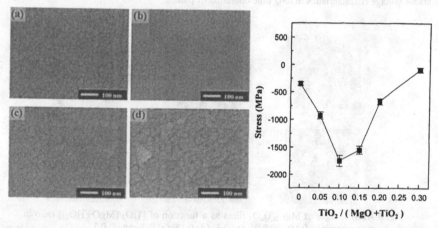

Figure 2 FESEM images of Mg$_{2-2x}$Ti$_x$O$_2$ thin films as a function of [TiO$_2$/(MgO+TiO$_2$)] ratio in the starting materials; (a) 0, (b) 0.1, (c) 0.2 and (d) 0.3

Fig. 3 Residual stress of Mg$_{2-2x}$Ti$_x$O$_2$ thin films as a function of [TiO$_2$/(MgO+TiO$_2$)] ratio in the starting materials

The surface morphologies of 5000Å thick Mg$_{2-2x}$Ti$_x$O$_2$ films analyzed by FESEM are shown in Fig. 2. The surface morphology of the film with x of 0.1 (Fig. 2b) is very fine and uniform compared to that of the pure MgO film (Fig. 2a). It is shown that a film with a proper composition can be very uniformly deposited by e-beam evaporation and in turn, this might help improve the voltage stability of a panel in long time operations. However, an excessive addition of TiO$_2$ to MgO, such as the case when the [TiO$_2$/(MgO+TiO$_2$)] ratio in the starting materials is as high as 0.2 or 0.3 (Figs. 2c and 2d), tends to make film surface rough and increase overall particle size and crack size probably due to agglomeration of particles. Cracks in the films are speculated to be related to stress release induced by introducing large amount of TiO$_2$ into MgO matrix. That is, to release the stress in accordance with excessive addition of TiO$_2$, the film surface ends up with formation of cracks, which leads to the decrease in the compressive stress as shown in Fig. 3. These cracks may also lead to instability in electron emission from the film surface and furthermore, reduce the lifetime of PDP because the dielectric layers cannot be effectively protected during long time operation.

The surface morphology of Mg$_{2-2x}$Ti$_x$O$_2$ films was observed by atomic force microscopy

(AFM). Figure 4 (a) shows the topography of the pure MgO film, x=0, which has rounded particles of about 0.1 μm in mean diameter. The particles look very small when x is 0.1 or 0.15, compared to those of the pure MgO. However, the overall morphologies of films are not greatly affected by addition of TiO_2.

Figure 5 shows the surface roughness of $Mg_{2-2x}Ti_xO_2$ films as a function of $TiO_2/(MgO+TiO_2)$ ratio in the starting materials obtained by AFM. $Mg_{1.8}Ti_{0.1}O_2$ film has the lowest surface roughness probably due to smaller particle size, which is expected to result in stable voltage characteristics in long time operation of panels.

Fig. 4 AFM Images of $Mg_{2-2x}Ti_xO_2$ films as a function of $[TiO_2/(MgO+TiO_2)]$ ratio in the starting materials; (a) 0, (b) 0.05, (c) 0.1, (d) 0.15, (e) 0.2 and (f) 0.3

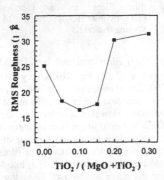

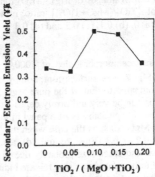

Fig. 5 Surface roughness of $Mg_{2-2x}Ti_xO_2$ films as a function of $[TiO_2/(MgO+TiO_2)]$ ratio in the starting materials

Fig. 6 Secondary electron emission yields of $Mg_{2-2x}Ti_xO_2$ films as a function of $[TiO_2/(MgO+TiO_2)]$ ratio in the starting materials

Low operation voltage is needed to reduce the cost of driving ICs, which can be obtained by increasing the secondary electron emission yield of a cathode material since the firing voltage is dependent on γ [10]. Thus, it is desirable to measure the secondary electron emission yield of $Mg_{2-2x}Ti_xO_2$ films as a function of TiO_2 content in the starting materials. Figure 6 shows the measured secondary electron emission yield of each film. Despite of a small variation in Ti cation ratio from 0 to 0.092 in film, the corresponding change in the secondary electron emission coefficient turned out to be very large. The secondary electron emission yield of the pure MgO film is 0.33. The films with the $[TiO_2/(MgO+TiO_2)]$ ratio in the starting materials of 0.1 and 0.15 (cation ratio of 0.054 and 0.069) exhibit the secondary electron emission yield of about 0.495 and 0.482, respectively. When the $[TiO_2/(MgO+TiO_2)]$ ratio is 0.2, the yield is, however, decreased to 0.349. Electrical instability was observed when the ratio was 0.3. This instability is probably due to the inhomogeneous morphology, as shown in Fig. 2d, which made the measurement of the secondary electron emission yield unsuccessful.

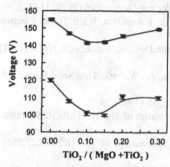

Fig. 7 Firing voltage and minimum sustaining voltage of panels with $Mg_{2-2x}Ti_xO_2$ protective layers as a function of $[TiO_2/(MgO+TiO_2)]$ ratio in the starting materials

The firing voltage (V_f) is the voltage amplitude for plasma ignition without wall charge and the minimum sustaining voltage (V_{sm}) is the minimum voltage amplitude which will sustain discharge sequence. Both firing and minimum sustaining voltages limit the bistable range of the sustain waveform. [2] Bistability, that is to sustain a discharge at a voltage lower than that required to initiate it, is a characteristic feature of AC PDP. Therefore, the larger the difference between the firing voltage and minimum sustaining voltage, defined as memory margin, the more stable the operation can be. The results of electrical measurement as a function of the $[TiO_2/(MgO+TiO_2)]$ ratio in the starting materials is shown in Fig. 7. Because other variables that can affect voltage characteristics of panels were kept constant, changes in voltage of panels are expected to depend entirely on the composition of protective layers in this work. The firing voltage and minimum sustaining voltage of panels with the conventional MgO protective layer are 155 V and 120 V, respectively. With addition of TiO_2 to the pure MgO, both voltages are drastically decreased. The lowest firing voltage, which is smaller than that of MgO by 13 V, is obtained when the $[TiO_2/(MgO+TiO_2)]$ ratio is 0.1. The minimum sustaining voltage is 101 V when the $[TiO_2 / (MgO+TiO_2)]$ ratio in the starting materials is 0.15.

CONCLUSION

$Mg_{2-2x}Ti_xO_2$ films were prepared by e-beam evaporation method to be used as a possible substitute for the conventional MgO surface protective layer for dielectric materials in AC PDPs. When the $[TiO_2/(MgO+TiO_2)]$ ratio in the starting materials was increased up to 0.3, the cation ratio in the film increased to 0.117 and the oxygen content in the films also gradually increased. By increasing the $[TiO_2/(MgO+TiO_2)]$ ratio in the starting materials the variation trends of the surface roughness of $Mg_{2-x}Ti_xO_{2x}$ films were found to be consistent with that of the voltage characteristics. The $Mg_{2-x}Ti_xO_2$ films yielded a secondary electron emission coefficient as high as 0.497 when the $[TiO_2/(MgO+TiO_2)]$ ratio in the starting materials was 0.1, while the secondary electron emission yield from a pure MgO film was 0.33. The firing and minimum sustaining voltages could be drastically decreased to 142V and 101V, respectively, when the $[TiO_2/(MgO+TiO_2)]$ ratio in the starting materials was 0.1 or 0.15.

ACKNOWLEGEMENT

The authors wish to thank the PDP Research Division (G7 project) in KOREA for financial support.

References

1. S. Mikoshiba, Society for Information Display 15 (2/1999) 28
2. T. Shinoda, H. Uchiike and S. Andoh, IEEE Trans. Electron Devices, 26 (1979) 1163
3. H. Uchiike, K. Miura, N. Nakayama, T. Shinoda, Y. Fukushima, IEEE Trans. Electron Devices, 23(1976) p.1211
4. M. O. Aboelfotoh, Omesh Sahni, IEEE Trans. Electron Devices, Vol. ED-28, No. 6 (1981) p.645
5. J. Cho, R. Kim, K-. W. Lee, G. –Y. Yeom, J. –Y. Kim, J. –W. Park, Thin Solid Films, Vol. 350, (1999) 173
6. H. Lin, Y. Harano and H. Uchiike, ASID Technical Digest (1995) 70
7. K. S. Moon, J. W. Cho, J. Lee, K. –W. Whang, Proceedings of The 18th IDRC/ISSN-0098-966X (1998) 401
8. A. Shiokawa, Y. Takada, R. Murai and H. Tanaka, Proceedings of The 18th IDRC/ISSN-0098-966X (1998) 519
9. G. V. Samsonov, in "The Oxide Handbook 2nd edition" (IFI/Plenum Data Corporation, New York,1982) p.206
10. A.Shiokawa, Y. Takada, R. Murai and H. Tanaka, in the Proceedings of International Display Workshops 98, (1998) 519

Mat. Res. Soc. Symp. Proc. Vol. 621 © 2000 Materials Research Society

Evaluation of Magnesium Oxide Films by Spectroscopic Ellipsometry

Hiroshi Kimura
Device Analysis Technology Laboratories, NEC Corporation,
1753 Shimonumabe, Nakahara-ku, Kawasaki, Kanagawa, 211-8666, JAPAN

ABSTRACT

Spectroscopic ellipsometry was used to evaluate the density and surface roughness of MgO films. The density of a film affects the dispersion of the film's refractive index, and the refractive index can be evaluated by spectroscopic ellipsometry. Among the various fitting models applied to all of the deposited films, the two-layered model gave the best results. For the two-layered model, we found that the thickness of the upper layer was strongly correlated with surface roughness. Thus, spectroscopic ellipsometry can be used to evaluate density and roughness.

INTRODUCTION

The plasma display panel (PDP) is a promising devices for use in large-area displays. Its required characteristics are long lifetime and low operating voltage [1]. The lifetime of a PDP is determined from the sputtering yield of MgO thin film [2], and this yield is directly affected by the film's density [3]. The firing and sustaining voltage depend on the secondary electron emission coefficient of MgO [4], and both voltages are lower when the surface is smooth [5]. The properties of a MgO film, such as its density and surface roughness, depend strongly on the deposition condition. Thus, to manufacture good PDPs, the MgO deposition process should be monitored continuously, but there is no suitable method to evaluate the process monitoring technique. The monitoring method should be simple, fast, and reliable while being non-destructive.

The dispersion of a film's refractive index reflects the density and surface roughness of the film. For MgO, they should be determined by measuring the film's refractive index. This refractive index has been evaluated by spectroscopic ellipsometry [6], which is a non-destructive, simple, and fast tool for analyzing thin films, so it can be applied to process monitoring. In this study, we used spectroscopic ellipsometry to evaluate quantitatively the density and surface roughness of MgO films deposited under various conditions.

EXPERIMENTAL

About 1 μm of MgO was deposited on a glass substrate by electron beam evaporation. Sample films were prepared by using various oxygen fluxes and substrate temperatures, which are both critical parameters affecting film quality [7, 8, 9]. They ranged from 0 to 100 sccm and 50 to 290°C, respectively, as shown in Table I.

Spectroscopic ellipsometry measures the polarization of light reflected from film, where polarization is described by the amplitude ratio of p- to s-component reflectance Ψ and the phase shift Δ. The ratio is described by

Table I. Deposition conditions.

Sample number	Oxygen flux [sccm]	Substrate temperature [°C]
#1	0	50
#2	65	50
#3	100	50
#4	0	200
#5	65	200
#6	100	200
#7	0	290
#8	65	290
#9	100	290

$$\frac{R_P}{R_S} = \tan\Psi \cdot \exp i\Delta,$$

(2)

(1)

where R_P and R_S indicate the reflection coefficients for p- and s-polarized light, respectively. Measurements were done on a spectroscopic ellipsometer (MASS-102, FiveLab) in the range from 350 to 830 nm, at an incident angle of 70 degrees. MgO film was also checked by atomic force microscope (TMX-1010, Topometrix) and x-ray diffractometer (RINT2000, Rigaku).

RESULTS AND DISCUSSION

Measured values of Ψ and Δ for a MgO film deposited without oxygen flux are shown in Figure 1. Experimental data was fitted with the dispersion data of bulk MgO [10], and the reliability of the model was comfirmed by a chi squared (χ^2) test:

$$\chi^2 = \frac{1}{2N-M}\sum_{i=1}^{N}\left[\left(\Psi_i^{mod} - \Psi_i^{exp}\right)^2 + \left(\Delta_i^{mod} - \Delta_i^{exp}\right)^2\right],$$

where N is the number of measuring points and M is the number of parameters in the applied model. A small χ^2 means that the assumed model was appropriate. Assuming that the film was a single crystal of MgO, the calculated Ψ and Δ did not agree with the observed ones, as shown in Figure 1(a), and χ^2 was very large for all samples. A large χ^2 suggests that the deposited film was not dense or that the stoichiometry of the MgO was different in the film. Two models has been proposed for a complex layer, the Cauchy function model and the effective medium approximation (EMA) model [11].

The lattice constants estimated from the (111) and (222) peaks in an x-ray diffraction pattern were coincident with the value of bulk MgO, as shown in Figure 2. The width of each

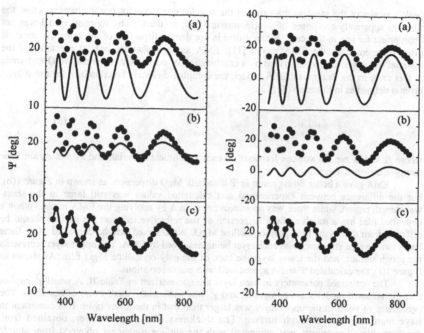

Figure 1. Typical Ψ and Δ from spectroscopic ellipsometry spectra and curve fitted by (a) bulk MgO, (b) one-, and (c) two-layered EMA models (sample #5). The circles indicate observed polarizations, and the solid curves are calculated ones.

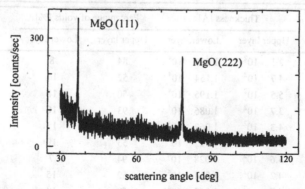

Figure 2. X-ray diffraction pattern of MgO film (sample #5).

peak was almost the same as the value of the angular resolution for our instrument. Thus, the film was apparently composed of stoichiometric MgO, so the Cauchy function model was not appropriate for use in this case. When a film is not dense, dispersion of the refractive index is often represented by the EMA model [11]. EMA assumes that the optical constants of the material of interest can be described by a combination of constituent materials, MgO and voids in this case. In the Bruggman EMA model, the complex dielectric function of ε for the subject layer is defined as in Equation (3) [12].

$$ f_A \frac{\varepsilon_A - \varepsilon}{\varepsilon_A + 2\varepsilon} + f_B \frac{\varepsilon_B - \varepsilon}{\varepsilon_B + 2\varepsilon} = 0, \tag{3} $$

where f_A and f_B are the volume fractions of each constituent material, and ε_A and ε_B are their complex dielectric functions.

EMA gave a better fitting result in Ψ than bulk MgO dispersion, as shown in Figure 1(b), but the difference between experimental and calculated values was still large in the short-wavelength region, while there was no improvement in Δ by applying the EMA model. When a deposited film has a rough surface, dispersion of the refractive index at the surface should be different from that in the bulk poly-crystalline MgO. We assumed that the deposited MgO films had a two-layered structure, with each layer being described by EMA. The upper layer represents the rough surface, and the lower layer the bulk of the poly-crystalline MgO film. As shown in Figure 1(c), the calculated Ψ and Δ agreed well with our observations.

The optimized parameters in each layer are summarized in Table II. A small χ^2 implies that the assumed model was appropriate, and χ^2 was sufficiently small in all of our films. The percentage of voids in the upper layer was larger than that in the lower layer. It is consistent to have more voids on a rough surface. The thickness of the upper layer, obtained from spectroscopic ellipsometry, was compared with the surface roughness obtained from atomic

Table II. Fitted values of thickness and percentage of voids (density) using two-layered EMA model.

Sample number	Thickness [Å]		Percentage of voids [%]		χ^2
	Upper layer	Lower layer	Upper layer	Lower layer	
#1	2.1 10^2	1.090 10^4	84	8	0.98
#2	4.7 10^2	1.154 10^4	52	16	0.80
#3	5.5 10^2	1.195 10^4	50	18	1.60
#4	3.7 10^2	1.086 10^4	93	10	0.73
#5	4.3 10^2	1.062 10^4	54	15	0.92
#6	5.2 10^2	1.122 10^4	53	19	1.55
#7	3.6 10^2	1.028 10^4	94	7	2.20
#8	4.2 10^2	1.014 10^4	52	15	0.96
#9	4.7 10^2	1.053 10^4	52	14	0.97

force microscopy (AFM). As shown in Figure 3, the maximum height range obtained by AFM increased for a film deposited by using oxygen flux, which was the same tendency demonstrated for the thickness of the upper layer obtained by spectroscopic ellipsometry. The dependence of the thickness in the upper layer of films on the substrate temperature was also the same as that for the maximum height range obtained by AFM. Thus, the thickness of the upper layer closely correlates with the surface roughness. The use of spectroscopic ellipsometry thus makes it possible to estimate surface roughness quickly in addition to evaluating the film density.

CONCLUSION

The density and surface roughness of MgO thin films vary depending on its deposition conditions. Their densities can be evaluated quantitatively by using spectroscopic ellipsometry. Compared with AFM results, the roughness obtained by spectroscopic ellipsometry is reliable when the two-layered EMA model is used. Since spectroscopic ellipsometry is an optical technique, rapid and non-destructive measurements can be done for a large area. All of these features are essential for process monitoring. Spectroscopic ellipsometry is thus likely to become a powerful tool for process monitoring.

ACKNOWLEDGEMENTS

The author thanks Mr. Ken Ito for providing the samples, and Dr. Ichiro Hirosawa, Mr.

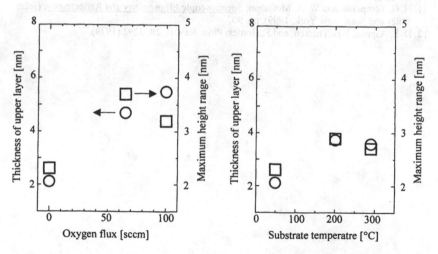

Figure 3. Copmarison of the results of spectroscopic ellipsometry and AFM. The circles and squares indicate thickness of the upper layer and maximum height range, respectively.

Toshiya Hayashi, Mr. Keiji Shiotani, and Dr. Tohru Tsujide for their helpful discussions and encouragement throughout this work.

REFERENCES

1. T. Urade, T. Iemori, M. Osawa, N. Nakayama, and I. Morita, IEEE Trans. Electron Devices, **23**, 313 (1976).
2. S. Hidaka, M. Ishimoto, N. Iwase, K. Betsui, and H. Inoue, IEICE Trans. Electron E, **82**, 1804 (1999).
3. P. Sigmund, Phys. Rev., **184**, 343 (1969).
4. E. H. Choi, J. Y. Lim, Y. G. Kim, J. J. Ko, D. I. Kim, C. W. Lee, and G. S. Cho, J. Appl. Phys., **86**, 6525 (1999).
5. C. H. Park, Y. K. Kim, B. E. Park, W. G. Lee, and J. S. Cho, Mat. Sci. Engineering B, **60**, 149 (1999).
6. S. Charvet, R. Madelon, F. Gourbilleau, and R. Rizk, J. Appl. Phys., **85**, 4032 (1999).
7. C. H. Park, W. G. Lee, D. H. Kim, H. J. Ha, and J. Y. Ryu, Surf. Coating Tech., **110**, 128 (1998).
8. E. Fujii, A. Tomozawa, H. Torii, R. Takayama, M. Nagaki, and T. Narusawa, Thin Solid Films, **352**, 85 (1999).
9. J. S. Lee, B. G. Ryu, H. J. Kwon, Y. W. Jeong, and H. H. Kim, Thin Solid Films, **354**, 82 (1999).
10. E. D. Palik ed., Handbook of Optical Constants in Solids II, (Academic Press, New York, 1991), pp. 919.
11. H. G. Tompkins, and W. A. McGahan, Spectroscopic Ellipsometry and Reflectmetry, (John Wiley and Sons, New York, 1999), pp. 90.
12. D. E. Aspnes, J. B. Theeten, and F. Hottier, Phys. Rev. B, **20**, 3292 (1979).

Mat. Res. Soc. Symp. Proc. Vol. 621 © 2000 Materials Research Society

Intense Photoluminescence and Photoluminescence Enhancement upon Ultraviolet Irradiation in HYydrogenated Nanocrystalline Silicon Carbide

M.B.Yu, Rusli and S.F.Yoon
School of Electrical and Electronic Engineering, Nanyang Technological University,
Nanyang Avenue, Singapore 639798, Republic of Singapore
S.J. Xu
Department of Physics, The University of Hong Kong, Pokfulam Road, Hong Kong, China
K.Chew, J. Cui, J. Ahn and Q. Zhang
School of Electrical and Electronic Engineering, Nanyang Technological University,
Nanyang Avenue, Singapore 639798, Republic of Singapore

ABSTRACT

Hydrogenated nanocrystalline silicon carbide (nc-SiC:H) films were deposited in an electron cyclotron resonance chemical vapor deposition (ECR-CVD) system using silane (SiH_4) and methane (CH_4) as source gases. It was discovered that under the deposition conditions of strong hydrogen dilution and high microwave power, nanocrystalline grains embedded in an amorphous matrix can be obtained, as confirmed by TEM study. Steady state and time-resolved photoluminescence (PL) from these films were investigated. The films exhibit intense visible PL at room temperature under laser excitation. The PL emission peaks at 2.64 eV, which is higher in energy compared to the bandgap of cubic SiC. Temporal evolution of the emission peak exhibits a double-exponential decay. Two distinct decay times of 179 ps and 2.88 ns were identified, which are at least 2 orders of magnitude faster than that of bound excition transitions in bulk 3C-SiC at low temperature. It was found that upon ultraviolet irradiation using an Ar^+ laser (351nm) the PL intensity of the films was enhanced. After 20 minutes irradiation, the PL intensity increased by about three times. This result suggests that the UV light may lead to modification of nonradiative recombination centers in the films. These nc-SiC:H films are promising for application in large area flat panel displays and in optoelectronic storage technology.

INTRODUCTION

Nanocrystalline semiconductor materials have attracted much attention due to their potential application in novel optical and electro-optical devices[1-2]. In recent years, many techniques, such as sputtering, laser-chemical vapor deposition, ion implantation, and plasma enhanced CVD, have been adopted for the preparation of films consisting of nanocrystalline Si embedded in a matrix of silicon nitride or oxide. These films have been shown to exhibit visible photoluminescence (PL), attributed to quantum size effects in the Si nanocrystallites[3-6]. Apart from nanocrystalline Si, fabrication of other similar films consisting of nanocrystallites of C[7], Ge[8], GaAs[9], GaN[10] and diamond[11] embedded in silicon oxide or other matrixes have also been reported.

Crystalline SiC, being a wide band gap semiconductor, exhibits weak blue PL at low temperature, and blue light emitting diodes (LED) based on this material have been demonstrated[12]. However, due to its indirect band gap characteristic, the intensity of the LED is very low, with a quantum efficiency of only about 10^{-4} [13]. To improve its luminescence intensity, blue light emitting porous SiC has been fabricated on crystalline 6H-SiC substrates and C^+ implanted silicon substrates by the same technique used for forming porous Si[14,15]. In this work we report the fabrication of nanocrystalline SiC (nc-SiC) films using the electron cyclotron resonance chemical vapour deposition (ECR-CVD) technique. With the attractive plasma characteristics such as high ion density and high electron temperature, coupled with the flexibility of being able to independently control the degree of plasma ionization and the ion energy, the ECR-CVD technique is conducive to thin film growth with high nucleation density. It was found that under suitable deposition conditions, nc-SiC films could indeed be obtained, as confirmed by the results obtained from different characterization techniques. Interestingly, strong visible PL with peak emission energy higher than the band gap energy of 3C-SiC has also been observed from these films at room temperature. Moreover, it was found that the PL intensity was enhanced upon ultraviolet irradiation.

EXPERIMENT

The nc-SiC films were deposited by the plasma decomposition of pure methane (CH_4) and silane (SiH_4) (10% diluted in H_2) in our ECR-CVD system. Further information on the system used can be obtained elsewhere[16]. The films were deposited on Si (100) and 7059 corning glass substrates for different structural and optical characterizations. During the deposition, the microwave power at a frequency of 2.45GHz was set at 850W, and the total gas pressure in the chamber was 10 mTorr. The reactant gases flow ratio (H_2:CH_4:SiH_4) was fixed at 100:2:10 sccm. The large microwave power and strong hydrogen dilution conditions were adopted to promote higher nucleation density. For all the depositions, the substrates were not heated and there was no appreciable increase in the substrate temperature due to the heating by the plasma.

The structures of the films were studied using a JEM-2010 FEG-TEM field emission gun electron microscope operating at 200 kV with an attached selected-area electron diffraction system. For the PL measurements, an Ar^+ ion laser with a wavelength of 351 nm and an output power of 10 mW was used as the excitation source, and the PL signals were dispersed by a Digikrom DK242 double grating 0.25 meter monochromator and detected by a Hamamatsu photomultiplier. For time-resolved PL measurement, the excitation source has a wavelength of 400 nm and consists of approximately 2 picoseconds pulse with a repetition rate of 82 MHz. This light source was derived from the frequency doubling of a 800 nm line obtained from a Spectra-Physics mode-locked Ti:Sapphire laser. The transient luminescence signal was dispersed in a Chromax 0.25 meter monochromator and captured by a Hamamatsu C4334 streak-camera. The overall temporal resolution of the whole system is about 20 picoseconds.

RESULTS AND DISCUSSION

In this paper, we focus on the microstructural and PL properties of those samples deposited under the conditions where nc-SiC films were obtained. For the HRTEM measurement, the films were stripped by etching the Si substrate in a mixture of HF and HNO_3 (HF:HNO_3 =1:3), and thinned by Ar^+ ions sputtering. A bright-field plan view TEM image and the corresponding selected area diffraction (SAD) are shown in Figure 1. There are a large number of crystalline grains within which the atoms are periodically arranged. Comparing the grains with a ruler of length 10 nm, the grain size was determined to be of several nanometers. It is noted that the direction of the crystalline grains in space is irregular and the interface area between the grains consists of irregularly arranged atoms. The SAD pattern indicated that the electron diffraction pattern has several rings. It should be emphasized that all the diffraction rings, except the first inside diffraction spot, are composed of many diffraction spots, which further reinforce the fact

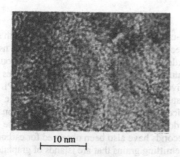

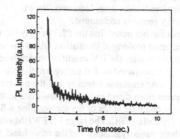

|← 10 nm →|

Figure 1(a). *HRTEM photograph of the nc-SiC sample.*

(b) *TEM diffraction pattern of the nc-SiC sample.*

that the crystalline grains are disorderedly oriented in the films. The first spread diffraction spot also shows that there exist amorphous components in the films.

Further evidence on the presence of SiC nanocrystallites in the film can be obtained through x-ray photoelectron spectroscopy (XPS), infrared absorption and Raman scattering[17].

Figure 2 shows the room-temperature PL spectrum of the nc-SiC films. The PL has a peak at 2.64 eV and a FWHM of about 0.56 eV. It is noted that the PL peak energy is higher than the room-temperature optical bandgap energy of 3C-SiC, which has a value of 2.2 eV[18]. Assuming that the PL derives from the SiC nanocrystallites, the peak emission energy being higher than that of bulk 3C-SiC can be a result of quantum size effect in these nanocrystallites.

As in the case of light-emitting Si nanocrystallites[19-20], valuable information on the nature of the radiative recombination, such as whether it involves direct optical transitions can be obtained through the study of PL decay lifetime. Figure 3 shows the time evolution of the room temperature PL intensity at the peak emission energy of 2.64eV. A best fit to the time evolution

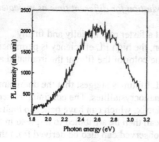

Figure 2 *Room temperature PL spectrum of the sample excited using an Ar+ laser at an emission wavelenth of 351 nm.*

Figure 3 *The time evolution of the PL intensity at the peak emission energy of 2.64 eV.*

of the luminescence was obtained by assuming a bi-exponential decay process. The smooth solid line shows the bi-exponential fitting results. Two decay time constants of 179 ps and 2.88 ns were obtained, which are noted to be at least two orders of magnitude faster than that of bound excitonic transitions in bulk 3C-SiC at low temperature[21]. The bi-exponential decay implies that multilevel transitions may be involved in the radiative recombination. It is noted that the PL decay lifetimes of the nc-SiC films at room temperature are comparable to that of excitonic transitions in high-quality bulk direct bandgap semiconductors[22-23] and their low-dimensional structures such as quantum wells[24-25] and quantum dots[26-27] at low temperature. It should be mentioned that short PL lifetimes of several nanoseconds have also been observed for carbon rich a-SiC films, attributed to the presence of light emitting grains that are islands of graphite-like sp^2 phase[28]. Further experiments are in progress to study the detailed relations between the radiative lifetime and the nanostructures of these nc-SiC films.

In the PL measuring experiment, visually it was found that the PL intensity of the samples increased under UV light irradiation. To investigate the effect of the UV light irradiation, the sample was put on the sample holder of the PL measurement system where both UV light irradiation and PL measurement can be carried out at the same area of the sample using the same laser beam of 3.53 eV. Higher power laser beam (50 mW) was used for the irradiation and lower power laser beam (10 mW) was used for the PL excitation. Figure 4 shows the PL spectra under different irradiation time for the nc-SiC sample. It is clear that the PL intensity of the as-grown nc-SiC sample was greatly enhanced with increasing irradiation time. However, the PL peak energy remains unchanged.

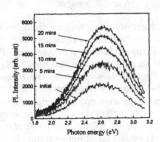

Figure 4 *PL spectra of the sample at room Temperature under continuous UV light Irradiation for different time as indicated*

It was further noted that the PL intensity increased at a faster rate initially and finally saturated upon prolonged irradiation. After the saturation, the high PL efficiency is permanently preserved even after the UV irradition is removed, and the color of the film at the irradiated area has changed compared to that before irradiation.

The strong emission intensity and the nanosecond PL lifetimes suggest that the dominant PL peak derives from direct radiative transitions in the SiC nanocrystallites. The confining materials in this case could be voids or the surrounding a-SiC:H matrix, which can have bandgap higher than the crystalline SiC. The large PL FWHM observed can then be attributed to a spread in the size of these nanocrystallites. On the other hand, the PL observed can also be derived from the hydrogenated amorphous SiC components in the films, the presence of which was confirmed from the electron diffraction patterns. Further work such as depositing a range of nc-SiC films with varying grain sizes, and correlating them to the PL characteristics observed will be useful. Besides, studies on the electric field quenching of the PL will also be helpful in verifying the origin of the strong PL observed.

On the assumption that the PL derives from the SiC nanocrystallites, there is an issue of how direct optical transitions can be possible in this indirect gap material. Given the nanosize of these SiC crystallites and the fact that they are surrounded by an amorphous SiC matrix, their energy band structures are expected to be very different from that of 3C-SiC. It is possible that the change in the band structure may be in a way that favours direct radiative transitions. This, couples with the strong confinement of carriers in the nanocrystallites, can lead to exciton-like recombination and account for the strong PL observed.

The exact origin of the UV irradiation enhanced PL of the nc-Si:H films is not well understood at present. Since the films have thickness in the range of about 200-300nm, which are much smaller than the absorption depth of the UV irradiation ($1/\alpha$) of about 700nm, where α is the absorption coefficient at the UV wavelength, it can be concluded that the PL enhancement seen is not a film surface effect. In view of the fact that all the samples showed PL enhanced effect unpon UV irradiation with the PL peak emission energies remain unchanged, it is suggested that there are some nonradiative recombination centers in the bulk of the nc-SiC:H films sensitive to UV light. It is likely that the UV light irradiation supplies energy and modifies the disordered atoms on the surface of the nanocrystalline gains or the defects which are located in the regions of grain boundaries and amorphous matrix. The light-induced passivation of the nonradiative recombination centers could be responsible for the enhancement of the PL intensity without any new emission energy involved.

CONCLUSION

In summary, nanocrystalline SiC (nc-SiC) films have been deposited using the ECR-CVD technique. Microstructural characterizations showed that the crystalline grains have irregular orientations and grain dimension of several nanometers. These nc-SiC films emit intense visible light at room temperature under UV laser excitation. Steady-state and ps time-resolved PL spectra of these samples were measured at room temperature. It was found that the photon energy of the dominant emission is higher than the bandgap of bulk 3C-SiC crystal. More interestingly, the PL decay follows a bi-exponential process with very short lifetimes of about 179 ps and 2.88 ns. These, being at least two orders of magnitude faster than that of bound exciton transitions in bulk 3C-SiC at low temperature, are attributed to direct optical transitions. It is postulated that the strong PL and the short lifetimes observed are likely to be a result of quantum confinement effect in the SiC nanocrystallites. It has been found that UV irradiation leads to PL intensity enhancement at room temperature, which can be preserved even after UV light irradiation was removed. These nc-SiC films emit strong visible light and are promising materials for the fabrication of large area LEDs.

REFERENCES

1. M.C.Schlamp, X.Peng and A.P.Alivisators, J.Appl.Phys. **82**, 5837 (1997)
2. K.D.Hirschman, L.Tsybeskov, S.P.Duttagupta and P.M.Fauchet, Nature (London) **384**, 338 (1996)
3. K.S.Min, K.V.Shcheglov, C.M.Yang, H.A.Atwater, M.L.Brongersma and A.Polman, Appl. Phys. Lett. **69**, 2033 (1996)
4. Yoshihiko Kanemitsu, Phys. Rev. B **48**, 4883 (1993)
5. S.Furukawa and T.Miyasato, Jpn. J. Appl. Phys., Part 2 **27**, L2207 (1988)

6. B.H. Augustine, E.A. Irene, Y.J.He, K.J.Price, L.E.Mcneil, K.n.Christensen and D.M.Maher, J.Appl. Phys. **78**. 4020 (1995)
7. S.Hayashi, M.Kataoka and K.Yamamoto, Jpn. J. Appl. Phys., Part2 **32**, L274 (1993)
8. A.K.Dutta, Appl. Phys. Lett. 68, 1189 (1996)
9. C.W.White, J.D.Budai, J.G.Zhu, S.P.Withrow, R.A.Zuhr, D.M.Hembree, Jr., D.O. Henderson, A.Ueda, Y.S.Tung, R.Mu and R.H. Magruder, J.Appl. Phys. **79**, 1876 (1996)
10. J.A.Wolk, K.M.Yu, E.D.Bourret-Courchesne and E.Johnson, Appl. Phys. Lett. **70**, 2268 (1997)
11. M.Zarrabian, N.Fourches-Coulon, G.Turban, C.Marhic and M.Lancin, Appl. Phys. Lett. **70**, 2535 (1997)
12. M.Ikeda, T.Hayakawa, S.Yamagiwa, H.Matsunami and T.Tanaka, J.Appl.Phys. **50**, 8215 (1979)
13. L.Hoffman, G.Ziegler, D.Theis and C.Weyrich, J. Appl. Phys. **53**, 6962 (1982)
14. T.Matsumoto, J.Takahashi, T.Tamki, T.Futagi and H.Mimura, Appl. Phys. Lett. **64**, 226 (1994)
15. Liang-Sheng Liao, Xi-Mao Bao, Zhi-Feng Yang and Nai-Ben Min, Appl. Phys. Lett. **66**, 2382 (1995)
16. S.F.Yoon. H.Yang, Rusli, J.Ahn, Q.Zhang and T.L.Poo, Diamond Relat. Mater. **6**, 1638 (1997)
17. M.B. Yu, Rusli, S.F.Yoon, S.J. Xu, K. Chew, J. Ahn and Q.Zhang, International conference on metallurgical coatings and thin films, April 10-14, 2000, San Diego, USA
18. Y.S. Park, *SiC Materials and Devices* (Academic Press, Lond, UK, 1998), Semiconductors and Semimetals Vol. 52, P14.
19. J. P. Wilcoxon and G. A. Samara, Appl. Phys. Lett. 74, 3164 (1999).
20. T. Toyama, Y. Kotani, A. Shimode, and H. Okamoto, Appl. Phys. Lett. 74, 3323 (1999)
21. J. P. Bergman, E. Janzén, S. G. Sridhara, and W. Choyke, *Silicon Crabide, III-Nitrides and Related Materials: Part I* (Trans Tech Publications Ltd, Switzerland, 1998), p485.
22. G. W. 't Hooft, W. A. J. A. van der Poel, and L. W. Molenkamp, Phys. Rev. B 35, 8281 (1987)
23. O.Brandt, J.Ringling, K.H.Ploog, H-J Wünsche, F.Henneberger, Phys. Rev. B 58, R15977 (1998)
24. J. Feldmann, G. Peter, E. O. Göbel, P. Dawson, K. Moore, C. Foxon, and R. J. Elliott, Phys. Rev. Lett. 59, 2337 (1987)
25. P. Lefebvre, J. Allègre, B. Gil, A. Kavokine, H. Mathieu, W. Kim, A. Salvador, A. Botchkarev, and H. Morkoc, Phys. Rev. B 57, R9447 (1998)
26. M. Colocci, A. Vinattieri, L. Lippi, F. Bogani, M. Rosa-Clot, S. Taddei, A. Bosacchi, S. Franchi, and P. Frigeri, Appl. Phys. Lett. 74, 564 (1999)
27. J. Bellesa, V. Voliotis, R. Grousson, X. L. Wang, M. Ogura, and H. Matsuhata, Phys. Rev. B 58, 9933 (1998)
28. S. V. Chernyshov, E. I. Terukov, V. A. Vassilyev and A. S. Volkov, J. Non-Cryst. Solids 34, 218 (1991)

Mat. Res. Soc. Symp. Proc. Vol. 621 © 2000 Materials Research Society

Nitride Phosphors for Low Voltage Cathodoluminescence Devices

Hisashi Kanie, Takahiro Kawano, Kose Sugimoto, Ryoji Kawai
Dept of Applied Electronics, Science Univ of Tokyo, Noda, Chiba, 278-8510, JAPAN.

ABSTRACT

Undoped and Zn doped InGaN microcrystals were synthesized by a two-step method. The InGaN microcrystals have a wurtzite structure and brownish body color. The InGaN samples prepared at 900 °C did not contain a metal In phase. The InGaN:Zn microcrystals showed blue photoluminescence (PL) at 77K different from that of GaN:Zn. Reflectivity and photoluminescence excitation (PLE) measurement showed that the fundamental absorption edge of the InGaN:Zn phosphors is 3.47 eV, which implies that the In content in the InGaN:Zn phosphors is less than 0.2%. GaN:Zn and InGaN:Zn showed a Zn related PLE peak at 3.34 eV. InGaN and InGaN:Zn showed an In related PLE peak at 3.14 eV. When the InGaN:Zn samples were selectively excited at 3.15 eV, an In-related emission band centered at 2.2 eV emerged. The InGaN:Zn phosphors mounted on vacuum fluorescent displays (VFDs) showed room-temperature blue cathodoluminescence (CL) and the CL peak shifted slightly toward the low energy compared to that of the GaN:Zn phosphors because of the superimposed In related band. The InGaN:Zn phosphors had a luminance of 50 cd/m^2 and a luminance efficiency of 0.03 lm/W at an anode voltage of 50 V.

INTRODUCTION

GaN is a wide band gap semiconductor and has low resistivity, high luminescence efficiency, and chemical stability [1]. GaN:Zn or InGaN is used for an active layer in a highly efficient blue light emitting diode [1-4]. We synthesized microcrystaline GaN and InGaN by a reaction between the mixture of Ga_2S_3 and In_2S_3 and NH_3 at 1050 °C [5-7] and reported that GaN:Zn microcrystals are applicable to phosphors for low-voltage CL devices, such as a field emission display (FED) or a VFD [5]. The current vs. anode voltage characteristics of VFDs using the GaN:Zn phosphors showed a threshold voltage as low as ZnO:Zn phosphors because of low resistivity of GaN. The CL spectra showed that a band in the violet region was dominant and a luminance efficiency in the blue region was not enough for practical use.

To shift the emission band from the violet to the blue region and to increase the intensity of emission band we prepared InGaN:Zn with better crystalline quality. Structural and optical properties of InGaN:Zn phosphors were studied. Structural properties of synthesized GaN, GaN:Zn and InGaN:Zn were measured using a conventional x-ray diffractometer. Low temperature PL and PLE spectra of InGaN:Zn phophors were recorded using a Xenon lamp. CL spectra and electric properties of InGaN:Zn phosphors mounted in VFDs were measure at room-temperature.

EXPERIMENTAL

The improvement of the InGaN crystalline quality requires synthesis at higher temperatures while preventing the InGaN decomposition. To counter the effect of decomposition, which caused from high temperature synthesis, we introduced a two step growth method. In the first step, undoped or Zn doped GaN microcrystals were synthesized at 1150°C. In the second step, InGaN(:Zn) was synthesized from a mixture of the GaN(:Zn) microcrystals and In_2S_3 at 800 °C or higher temperatures. GaN was synthesized by the nitridation of Ga_2S_3 under an NH_3 flow of 0.6 l/min for 1 or 5 hours. Zn was doped by using ZnS as a Zn source during synthesis. InGaN:Zn was synthesized by the nitridation of a mixture of the synthesized GaN:Zn and In_2S_3 with NH_3 for 5 hours. The Ga to In molar ratio was varied from 1:0.35 to 1:0.5.

Synthesized GaN:Zn has a pale grayish body color and InGaN:Zn has a brownish body color. X-ray diffraction patterns showed that hexagonal InGaN was grown and that the grown InGaN crystals contain metal In phase at lower synthesis temperatures, as shown in figure 1. InGaN synthesized at 900 °C did not contain metal In phase. Lattice constants of InGaN were comparable to GaN, which indicated that In concentration is very low.

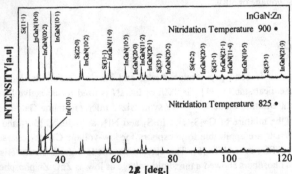

Figure 1. X-ray diffraction patterns of InGaN synthesized at different temperatures. They have wurtzite structure. In phase exists in the sample synthesized at 825°C and it does not exist in the sample at 900°C.

OPTICAL PROPERTIES

Reflectivity spectra showed that the band gaps of the GaN:Zn and InGaN:Zn samples are located at 3.4 eV. We estimated the In content to be less than 0.2% in the InGaN samples by taking account of the resolution of spectrometer. Corresponding to the brownish body color, strong absorption is observed for the InGaN:Zn samples.

PL spectra at 77K were measured under an excitation of a He-Cd laser 325 nm line. PL spectra showed that the synthesized GaN:Zn and InGaN:Zn have a blue luminescence band and the peak of the blue band shifted toward low energy with the synthesis temperature, as shown in figure 2. The results indicated that In concentration is higher in the InGaN samples synthesized at higher temperatures.

PLE spectra were recorded with the emission intensity monitored at the photon energy of emission band peaks while the excitation light wavelength was being scanned. Selectively excited PL spectra were recorded under an excitation with the photon energy corresponding to the PLE peaks. PLE and PL spectra of the InGaN:Zn samples showed that three kinds of optical transition exist in the samples. The PLE peak at 3.46 eV in figure 3 is assigned to the fundamental absorption of InGaN. The shoulder peak at 3.37 eV is assigned to the Zn absorption band, because the peak appears at comparable energy in PLE spectra of Zn doped GaN. When the InGaN:Zn samples were excited at 3.18 eV, a PL band with a peak at 2.20 eV emerged from the low energy side of the blue emission band excited by a He-Cd laser 325 nm line, as shown in figure 3. The PLE peak at 3.14 eV was observed for both undoped and Zn doped InGaN phosphors prepared by the two step method. We assigned the PLE band peak at 3.14 eV for undoped InGaN with a band gap energy of 3.47 eV to In isoelectronic centers [6,7]. The PL peak at 2.2eV for the InGaN:Zn phosphors is also assigned to the In centers.

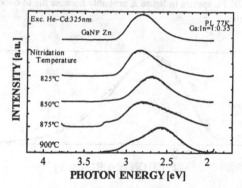

Figure 2. *Low temperature PL spectra of InGaN :Zn prepared at various temperatures. PL spectrum of GaN:Zn is shown in the figure.*

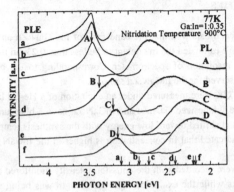

Figure 3. PLE and selectively excited PL spectra of InGaN:Zn. PLE spectra labeled a to f were recorded by monitoring the PL intensity with photon energies of 3.10 eV (indicated by the arrow labeled a), 2.89eV (b), 2.78 eV(c), 2.42eV (d), 2.20 eV (e), and 2.18 eV (f). PL spectra labeled A to D were recorded under an excitation with photon energies of 3.47 eV (indicated by the arrow labeled A), 3.37 eV (B), 3.20 eV (C), and 3.18 eV (D). PLE peaks were observed at 3.47 eV, 3.37 eV, and 3.15 eV. Shift of the PL bands toward low energy was observed with a decrease in excitation photon energies of the peaks in the PLE spectra.

Room temperature CL spectra were measured for VFDs using GaN:Zn or InGaN:Zn phosphors. The CL peak of InGaN:Zn phosphors shifts slightly toward low energy compared to that of GaN:Zn phosphors, as shown in figure 4. The amount of the shift was smaller than we expected from the 77K PL spectra in figure 2, probably because the In concentration in InGaN is very low in the samples.

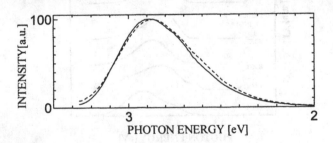

Figure 4. CL spectra of GaN:Zn and InGaN:Zn. The peak shifts toward low energy for InGaN:Zn.

The threshold voltage of luminance vs. an anode voltage characteristics is as low as 10 V, suggesting that the InGaN:Zn phosphor has low resistivity comparable to the value of ZnO:Zn. The luminance of the InGaN:Zn phosphors for VFDs working at an anode voltage of 50 V was 50 cd/m^2 , which is one third times higher than that of the GaN:Zn phosphors [8], and the luminescence efficiency 0.03 lm/W. To increase the luminance, it is necessary to decolorize the body color, increase the In concentration, and improve the crystal quality of InGaN.

CONCLUSION

InGaN:Zn phosphors were prepared by the nitridation of a mixture of the previously synthesized GaN and In$_2$S$_3$ with NH$_3$ in the range of 800 to 900 °C. The InGaN:Zn phosphors without a metalic In phase was obtained at 900 °C. The InGaN:Zn phosphors has a wurtzite structure and brownish body color. Reflectivity spectrum and PLE spectrum showed that InGaN:Zn has a band gap energy comparable to that of GaN, which indicated the synthesized InGaN:Zn had low In contents of 0.2%. The InGaN:Zn phosphors showed an broad blue PL with a peak at 2.74 eV. InGaN:Zn has a PLE peak related to In centers at 3.14eV, a peak corresponding to the fundamental absorption at 3.47 eV, and a peak related to Zn centers at 3.34 eV. When the InGaN:Zn phosphors were excited at 3.18 eV, they emitted an In-center-related luminescence at 2.2 eV. A room temperature CL peak of the InGaN:Zn phosphors mounted in a VFD shifted slightly toward the lower energy compared to that of the GaN:Zn phosphors. Luminance of 50 cd/m^2 and luminance efficiency of 0.03 lm/W was obtained for a VFD using the InGaN:Zn phosphors at an anode voltage of 50 V. The threshold voltage of luminance was as low as 10 V, which is comparable to ZnO:Zn phosphors.

ACKNOWLEDGEMENT

This work is partially supported by the Futaba Natural Science Foundation.

REFERENCES
1. Shuji Nakamura and Gerhard Fosal, *The blue laser diode*, (Springer-Verlag, 1997).
2. M. Ilegems, R. Dingle and R. A. Logan, *J. Appl. Phys.*, **43**, 3797 (1972).
3. J. I. Pankove, J. E. Berkeyheiser and E. A. Miller, *J. Appl. Phys.*, **45**, 1280 (1974).
4. M. Boulou, M. Furtado, G. Jacob and d. Bois, *J. Luminescence* **18/19**, 767 (1979).
5. H. Kanie, T. Kawano, and H. Koami, *Proc. the 2nd Intern. Symp. on Blue Laser and Light Emitting Diodes*, Kazusa, Japan, 552 (1998).
6. H. Kanie, H. Koami. T. Kawano, and T. Totsuka, *Mat. Res. Soc. Symp. Proc.*, **482**, 731 (1998).
7. H. Kanie, N. Tsukamoto, H. Koami, T. Kawano, and T. Totsuka, J. Crystal Growth, **189/190**,

52(1998).

8. F. Kataoka, Y. Satoh, Y. Suda, K. Honda, and H. Toki, *Electrochem. Soc. Proc.*, **99-40**, 17 (2000).

TFT I: Deposition
and Crystallization

Mat. Res. Soc. Symp. Proc. Vol. 621 © 2000 Materials Research Society

SOLID PHASE CRYSTALLIZATION (SPC) BEHAVIOR OF AMORPHOUS Si BILAYER FILMS WITH DIFFERENT CONCENTRATION OF OXYGEN:
Surface vs. Interface-nucleation

Myung-Kwan Ryu, Jang-Yeon Kwon, and Ki-Bum Kim,
School of Materials Science and Engineering, Seoul National University, Seoul, KOREA

ABSTRACT

Solid-phase crystallization (SPC) behavior of a-Si film [a-Si(II)] in which oxygen concentration (C_O) is higher at the a-Si/SiO$_2$ interface ($C_O=5 \times 10^{21}$ /cm^3) than at the film surface ($C_O=3 \times 10^{20}$ /cm^3) has been investigated. The results were also compared with that of a-Si single layer [a-Si(I), 600 Å] with $C_O=3 \times 10^{20}$ /cm^3. It has been found that the interface-nucleation was suppressed in the a-Si(II) and the surface-nucleation occurred to make a poly-Si/a-Si (300 Å/300 Å) bilayer structure. Many equiaxial grains with sizes of 1~2 µm were formed in the surface-nucleated poly-Si layer. Compared with the results of conventional SPC poly-Si (600 Å-thick) in which elliptical grains with sizes of 0.5~1 µm were formed by the interface (a-Si/SiO$_2$)-nucleation, we concluded that the poly-Si/a-Si bilayer scheme is a method to improve the microstructure of SPC poly-Si film.

INTRODUCTION

Solid-phase crystallization (SPC) of amorphous Si (a-Si) has been attracted many interests for the application to polycrystalline Si (poly-Si) thin-film transistors (TFTs) in a liquid-crystal displays (LCDs)[1,2] Many efforts on improving the microstructure of poly-Si film by reducing the grain boundary area and crystalline defects due to their detrimental effects on the performances of TFT.

In recent work, we proposed a novel crystallization scheme in which the crystallization start from a film surface, i.e. the *surface-nucleation*.[3] In that study, the *interface-nucleation* phenomenon, dominantly occurring in a conventional annealed a-Si, was suppressed by making an oxygen-rich region near the interface between a-Si film and SiO$_2$ substrate with an oxygen-blowing process during initial a-Si deposition period. Thus, we could observe a surface-nucleated SPC. However, the oxygen-blowing process still remains to be optimized since it may cause a problem such as a particle formation which is not practical for TFT fabrication process.

In this work, we utilized residual oxygen in our deposition chamber for making an oxygen-rich region at near the a-Si/SiO$_2$ interface in order to suppress the interface-nucleation. Thus. there is no extra oxygen-blowing process. From the work of crystallization behavior of a-Si films with different C_O, we found out the C_O at which the crystallization is nearly suppressed. Based on this result, we have deposited an a-Si/oxygen-rich a-Si bilayer and comparatively investigated its SPC behavior with that of a conventional a-Si single layer.

EXPERIMENTS AND RESULTS

Deposition and annealing of amorphous Si films with different oxygen concentration

Amorphous Si films at 425 °C with 1 SCCM of Si$_2$H$_6$ and no carrier gas. Deposition pressure was 0.05, 0.1, 0.2, and 0.4 Torr, which was adjusted by varying the conductance of throttle valve connected to a mechanical pump. After the depositions, C_O in each as-deposited a-Si film was measured by secondary ion mass spectrometry (SIMS). Figure 1 shows the C_O as a function of deposition pressure. The C_O significantly increases as the deposition pressure

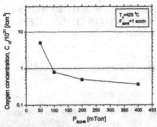

Fig. 1. Variation of C_O in a-Si films as a function of deposition pressure.

Fig. 2. PTEM image of a-Si films annealed at 650 °C for 3 hours. The C_O in each a-Si is (a)5×10^{21}, (b) 8×10^{20}, and (c) 5×10^{20} /cm^3, respectively.

decreases. C_O was measured as 5×10^{20}, 8×10^{20}, and 5×10^{21} /cm^3 for the deposition pressure of 200, 100, and 50 mTorr, respectively.

These a-Si films were annealed at 650 °C in nitrogen ambient for 3 hours. Figure 2(a)~(c) show the plan-view transmission electron microscope (PTEM) image of above a-Si films. As the C_O increases, the crystallization fraction is decreased. Particularly, a-Si with $C_O = 5 \times 10^{21}$ /cm^3 was not crystallized.

Deposition and crystallization of a-Si films with different C_O at a-Si/SiO$_2$ and a-Si surface.

By using the above results, we deposited an a-Si film [a-Si(II)], consisting of two a-Si layers. A lower a-Si layer (300 Å) with $C_O = 5 \times 10^{21}$ /cm^3 was firstly deposited at the deposition conditions of Fig. 1 and then, an upper a-Si layer (300 Å) with $C_O = 3 \times 10^{20}$ /cm^3 was deposited *in-situ*. For the comparison of crystallization behavior, we also deposited an a-Si (600 Å) [a-Si(I)] single layer with $C_O = 3 \times 10^{20}$ /cm^3 as a reference sample. Deposition scheme of these a-Si films were shown in Fig. 3(a) and (b).

Figure 4(a) and (b) are cross-sectional TEM (XTEM) images of a-Si(I) and a-Si(II) annealed at 600 °C for 11 hours. As shown in Fig. 4(a), a-Si(I) was fully crystallized. It has been well known that the crystallization process of conventional a-Si [a-Si(I)] starts from the nucleation of crystalline phases at the interface between a-Si and SiO$_2$ substrate.[3-8] For the case of a-Si(II), only 300 Å-thick of upper Si layer is polycrystalline while the lower part (300 Å) still remains amorphous. This result indicates that the crystallization was successfully suppressed and thereby we could observe the surface-nucleated SPC process in the a-Si(II).

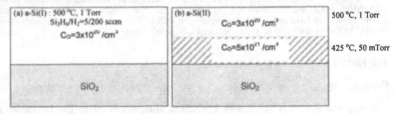

Fig. 3. Deposition scheme for (a) a-Si(I) (interface-nucleation sample) and (b) a-Si(II) (surface-nucleation smaple).

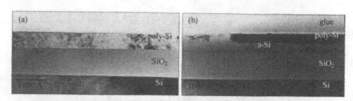

Fig. 4. XTEM image of crystallized (a) a-Si(I) and (b) a-Si(II) at 600 °C for 11 hours.

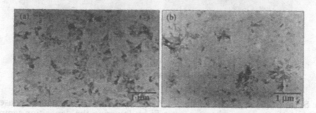

Fig. 5. PTEM image of crystallized (a) a-Si(I) and (b) a-Si(II) at 600 °C for 11 hours.

Unlike the small grains in Fig. 4(a), many large grains expanded laterally, such as the one in Fig. 4(b) were observed in the upper poly-Si layer of a-Si(II). The formation mechanism of the grains is deeply related with twins parallel to the film surface, which enhance the lateral growth process.[3,9] Those grains are corresponding to the equiaxial ones in the following plan-view TEM (PTEM) image of crystallized a-Si(II).

Plan-view TEM images of a-Si(I) and a-Si(II) annealed at 600 °C for 11 hours are shown in Fig. 5(a) and (b), respectively. Figure 5(a) is a typical microstructure of an conventional SPC poly-Si. High density of elliptical grains with a size of 0.5~1 μm are formed in the poly-Si.

The microstructure of annealed a-Si(II) is somewhat different from that of a-Si(I). Firstly, the film was partially crystallized indicating that the crystallization kinetics of a-Si(II) is slower that that of a-Si(I). Although many elliptical grains are also observed, what we focus on in Fig. 4(b) is the large equiaxial grains with sizes of 1~2 μm. In our previous report, it has been found that most of equiaxial grains has a strong {111}-orientation unlike the random orientations of elliptical grains.[3,9] Due to the twins parallel to the surface in the equiaxial grains, which enhance the lateral growth process,[3] they can grow upto 2 μm even in a half of a-Si(I) thickness.

DISCUSSION

Mechanism of interface-nucleation and suppression the interface-nucleation

As mentioned above, it has been well known that the conventional SPC process starts from the nucleation at the a-Si/SiO$_2$ interface. Thus, in order to observe the surface-nucleation phenomenon, it is necessary to suppress the interface-nucleation process. First of all, it is essential to investigate the mechanism which enhances the nucleation mechanism at the interface.

We attributed the origin of interface-nucleation in the conventional SPC to the microstructural difference in the as-deposited a-Si at between the a-Si/SiO$_2$ interface and film surface. Figure 6(a) and (b) show the XTEM images and schematic drawings of as-deposited

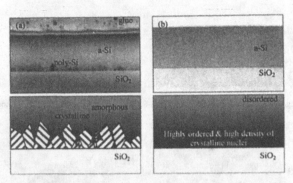

Fig. 6. XTEM image and its schematic drawing for (a) m-Si and (b) a-Si.

mixed phase Si (m-Si)[4] and a-Si. As shown in Fig. 6(b), most of crystalline phases are located at the Si/SiO_2 interface. Thus, when we get a-Si by lowering the deposition temperature from that of m-Si, it can be thought that the amorphous phase must have higher structural orderness and density of crystalline nuclei at the $a-Si/SiO_2$ interface than at the film surface. We believe that the high density of crystalline nuclei promotes the nucleation process at the $a-Si/SiO_2$ interface during the following SPC process.

As a method to suppress the interface-nucleation, we try to utilize the crystallization-preventing agent such as oxygen. It has been well known that even with 0.5 at.% of oxygen the crystallization kinetics can be retard by a factor of ten.[10] The results of Fig. 2(a)~(c) also demonstrate the effect of oxygen on the crystallization kinetics.

Crystallization kinetics

Figure 7 shows the crystallization behavior of the interface-nucleated [a-Si(I)] and surface-nucleated [a-Si(II)] SPC at 625 °C. It is found that the interface-nucleated SPC kinetics is faster that the surface-nucleated one. This difference is mainly caused by the microstructural difference at the nucleation site in between the a-Si(I) and a-Si(II). In the a-Si(I), the nucleation site is the $a-Si/SiO_2$ interface having high density of crystalline nuclei. For the case of a-Si(II), however, the density of crystalline nuclei at the surface is much smaller than that at the $a-Si/SiO_2$ interface of a-Si(I) from the explanation of Fig 6(a) and (b).

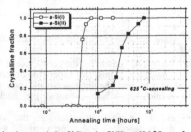

Fig. 7. Crystalline fraction in annealed a-Si(I) and a-Si(II) at 625 °C as a function of annealing time.

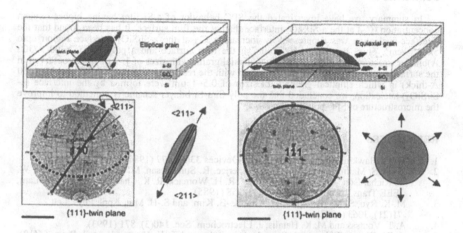

Fig. 8. Schematic diagram and its stereographic projection for (a) elliptical and (b) equiaxial grain. An {110}-oriented grain, a typical form of elliptical ones, is shown in (a). In this case, two <211>-directions normal to <110> are major growth directions for the lateral grain growth, which make the elongated grain. Other two <211>-directions facilitate the vertical growth of the grain. Thus, their role is not important to form the elongated shape. As for {111}-oriented grain having parallel twin plane, six <211>-directions normal to <111> are placed on the {111}-twin plane. Thus, the grain growth proceeds isotropically along the six <211>-directions to make the equiaxial shape. Also, in this case, other <211>-directions (open-circle) does not play an important role to make the equiaxial shape.

Formation mechanism of equiaxial and elliptical grain

On the microstructural viewpoint, it is desirable to make SPC poly-Si consisted of equiaxial grains since grain boundary area of equiaxial grain is smaller than that of elliptical one. Thus, the surface-nucleation scheme is a method to fabricate a high quality of poly-Si. However, even with the surface-nucleation scheme, there were still many elliptical grains although the density of elliptical ones was remarkably decreased by the surface-nucleation scheme. In order to further improve the microstructure of SPC poly-Si formed by the surface-nucleation, it is necessary to investigate the formation mechanism of these two types of grains.

Since the grain growth is enhanced along the six <211>-directions on {111}-twin plane,[3-9] the major factor determining the shape of grain is the location of twin plane in a grain. Except {111}-oriented grain having parallel twin to the film surface, all other oriented grains have oblique twin planes. In those grains, grain growth occurs anisotropically along certain <211>-directions to form elongated shape.[Fig. 8(a)] As for {111}-oriented grain with parallel twin plane to the film surface, the grain growth can occur isotropically toward six <211>-directions to make an equiaxial grain.[Fig. 8(b)]

SUMMARY

In summary, we have investigated the SPC behavior of a-Si film having higher oxygen concentration (C_O) at the a-Si/SiO$_2$ interface than at the film surface. It has been found that the interface-nucleation was suppressed by increasing C_O at the a-Si/SiO$_2$ interface region upto 5×10^{21} /cm^3. Thus, the nucleation occurs at the film surface and a poly-Si/a-Si (300 Å/300 Å)bilayer structure was obtained. Many equiaxial grains with sizes of 1~2 µm were observed in the surface-nucleated poly-Si layer. Compared with the results of conventional SPC poly-Si (600 Å-thick) in which elliptical grains with sizes of 0.5~1 µm were formed by the interface (a-Si/SiO$_2$)-nucleation, we concluded that the poly-Si/a-Si bilayer scheme is a method to improve the microstructure of SPC poly-Si film.

REFERENCES

1. W. G. Hawkins, IEEE Trans. Electron Devices **33**(4), 477 (1986).
2   S. D. S. Malhi, H. Shichijo, S. K. Banerjee, R. Sundaresan, M. Elahy, G. P. Pollack, W. F. Richardson, A. H. Shah, L. R. Hite, R. H. Womack, R. K. Chatterjee, and H. W. Lam, IEEE Trans. Electron Devices **32**(2), 258 (1985).
3. M.-K. Ryu, S.-M. Hwang, T.-H. Kim, K.-B. Kim, and S.-H. Min, Appl. Phys. Lett. **71**(21), 3063 (1997).
4. A. T. Voutsas and M. K. Hatalis, J. Electrochem. Soc. **140**(3), 871 (1993).
5. I.-W. Wu, A. Chiang, M. Fuse, L. Öveçoglu, and T.-Y. Huang, J. Appl. Phys. **65**(10), 4036 (1989).
6. K. Nakazawa, J. Appl. Phys. **69**(3), 1703 (1991).
7. L. Haji, P. Joubert, J. Stoemenos, and N. A. Economou, J. Appl. Phys. **75**(8), 3944 (1994).
8. J.-H. Kim, J.-Y. Lee, and K.-S. Nam, J. Appl. Phys. **79**(3), 1794 (1996).
9. T. Noma, T. Yonehara, and H. Kumomi, Appl. Phys. Lett. **59**(6), 653 (1991).
10. E. F. Kennedy, L. Csepregi, J. W. Mayer, and T. W. Sigmon, J. Appl. Phys. **48**(10), 4241 (1977).

TFT II: Laser Crystallization

Mat. Res. Soc. Symp. Proc. Vol. 621 © 2000 Materials Research Society

Excimer Laser Crystallisation of Poly-Si TFTs for AMLCDs

S D Brotherton, D J McCulloch, J P Gowers, J R Ayres, C A Fisher and F W Rohlfing
Philips Research Laboratories,
Cross Oak Lane,
Redhill,
Surrey, RH1 5HA, England

ABSTRACT

There is interest in reducing the shot number in the poly-Si laser crystallisation process in order to improve its throughput. Two distinct shot number dependent effects have been identified, which are both laser intensity dependent. The critical laser energy density is that which causes full film melt-through, and the major issue occurs at energies greater than this, where there is a considerable degradation in device uniformity with reducing shot number. The cause of this is non-uniform recovery of the full-melt-through fine grain poly-Si, and it is demonstrated that by extending the trailing edge of the beam, the material uniformity at reduced shot number can be improved. For energies less than this, the issue is not so much uniformity, as a general degradation in overall device properties with reducing shot number, which has been correlated with reducing grain size.

In more demanding, future applications (such as system-on-panel), it will be necessary to improve circuit performance and approach that of current MOSFET devices. This will require short channel, self-aligned (SA) TFTs, and some of the issues with this architecture, particularly lateral ion implantation damage beneath the gate edge and drain field relief are discussed.

1. INTRODUCTION

Excimer laser crystallisation is now the preferred technique for the formation of thin films of poly-Si which are used as the active layer in poly-Si TFTs. Whilst this technique, under optimum conditions, is able to form very high performance TFTs [1], there remain a number of issues in its application to commercial production. The broad objectives in applying the technique in a production environment are the formation of high quality films in a reproducible manner, with a large process window, and the maintenance of good uniformity under conditions of high throughput. Unfortunately, these conditions may be mutually exclusive: for instance, the highest quality and most uniform material is most easily produced with a large number of laser shots, but the high shot number would result in low throughput. Hence, in order to optimise these potentially conflicting requirements, it is necessary to develop a good understanding of the crystallisation process and to identify the key factors determining the resulting material properties.

In this paper we discuss the main causes of non-uniformity in material which has been crystallised by a line beam excimer laser. The fundamental issues with the laser and beam shaping optics are:

a) the pulse to pulse variation in output intensity from the laser, which, under critical conditions, can lead to large variations in material properties in the sweep direction of the beam,

b) the beam profile along its short axis, which will influence the impact of (a),

c) non-uniformities of the beam intensity along its long axis, which due to the high sensitvity of material properties to laser intensity, can be amplified in the resulting poly-Si film.

In addition to these fundamental issues with the laser, the final outcome will be strongly influenced by the number of shots employed per unit area in the crystallisation process.

In section 2, our background understanding of the crystallisation process is briefly reviewed in order to provide the framework for section 3, which describes the influence of short axis beam profile, and shot number, on device and material uniformity.

Even with high quality, uniform material, if the subsequent device processing degrades the material, the resulting TFTs will have sub-optimal performance. The fundamental influence of device architecture (self aligned or non-self-aligned) on these effects is discussed in section 4, and implantation damage, particularly lateral damage beneath the gate, in self-aligned structures, is demonstrated to be a significant issue. For the first time, the damage has been directly imaged by TEM and ways of controlling its influence are discussed.

2. REVIEW OF CRYSTALLISATION PROCESS

We have already published a full description of the crystallisation process, as inferred from TFT measurements [1], and demonstrated that it is consistent with all the key features in the SLG model [2], so only the main features will be summarised here as background to the following section.

Figure 1 consists of a sequence of mobility profiles [1], measured through the gaussian distribution of energies, in samples irradiated with a semi-gaussian KrF laser, and illustrates the important distinction between partial melt-through, near-melt-through and full-melt-through conditions [2]. The peak intensity in figure 1a results in near-melt-through, SLG conditions and gives the highest performance TFTs. In figure 1b, the peak intensity has been increased to full-melt-through conditions, resulting in fine grain material and an order of magnitude degradation in performance. However, as shown by figure 1c, it is possible to re-irradiate this fine grain material under near-melt-through conditions and convert it back to the high performance SLG material [1]. This recovery from full-melt-through will occur, to a greater or lesser extent, with the trailing edge of the beam in the conventional top-hat, swept beam TFT process. However,

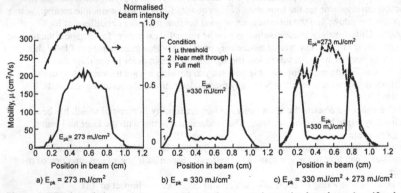

Figure 1. TFT degradation and recovery with a stationary semi-gaussian laser beam (t_{si} ~40nm).

achieving this recovery in a controlled fashion, in a low shot process, is one of the major issues in the laser crystallisation process and is discussed below.

3. LOW SHOT NUMBER AND SHORT AXIS BEAM SHAPE EFFECTS

The device architecture used in this part of the work has been the non-self-aligned (NSA) TFT [1], because, as discussed below, there are potentially serious artefacts in the operation of self-aligned TFTs due to the presence of implantation damage [3]. Hence, the NSA TFT yields the most transparent means of monitoring basic material properties as determined by the laser crystallisation process itself. The results reported have been mainly from ~40nm thick PECVD pre-cursor films, and thicker films gave broadly similar results, albeit at greater energies, which scaled with film thickness.

As is apparent from figure1, there is a narrow process window for achieving the high mobility TFTs under near-melt-through conditions. For 40nm thick films, the size of this window is ~30-40mJ/cm^2 at an optimum irradiation intensity of ~300mJ/cm^2. Ideally, it would be desirable to precisely set the energy intensity to give the peak mobility value, but in practice the energy setting is only accurate to about 5%. In addition, the film thickness across a plate may also vary by a comparable amount, and, finally, the maximum pulse to pulse fluctuation, in commercial crystallisation systems, has been quoted to be up to 15%. Hence, it will be very difficult to both maintain the optimum energy density, at near-melt-through, and to avoid unintentional excursions into full melt-through. When the latter happens, then in order to maintain good material uniformity, it will be necessary to recover the material with the trailing edge of the beam.

In order to investigate this regime experimentally, energy densities were used which deliberately caused full-melt-through conditions to occur, so that the consequences could be clearly seen. In addition, particular attention was paid to the uniformity of the results in the beam sweep direction, as this will be the direction in which maximum variation in material properties can be expected to occur.

A typical example of what happens to TFT properties when the shot number is reduced is shown in figure 2. In this figure, the mean electron mobility is plotted as a function of laser energy density for 100, 40, 20 and 10 shot top-hat beam irradiations. On the right hand axis, the normalised peak-to-peak scatter in the mobility values, $\Delta\mu_{max}$ ($=\mu_{max}$-μ_{min}), is also plotted, but, for clarity, only the values for the 100 and 10 shot processes are included. A number of features are apparent in these curves:

- There is a peak mobility value, which decreases with decreasing shot number, but occurs at approximately the same energy density, E_m (where E_m is the threshold energy density for full melt-through). The scatter in the mobility values for both the 100 and 10 shot process are less than ±10% at E_m.
- At energy densities greater than E_m, there is both a larger reduction in mobility values and increasing scatter in these values with reducing shot number, such that with the 10 shot process the scatter increases to ±50%, whereas it stays below ±10% for the 100 shot process. This is interpreted as a consequence of going into full-melt-through at energies greater than E_m, with progressively worse resetting of the material back into the SLG regime with reducing shot number. The details of this are discussed in section 3.1 below.

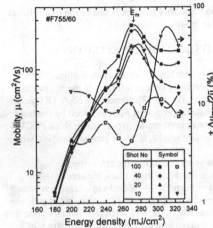

Figure 2. Dependence of electron field effect mobility, and maximum scatter in mobility, on laser intensity and shot number for top-hat beam irradiations (t. ~35nm).

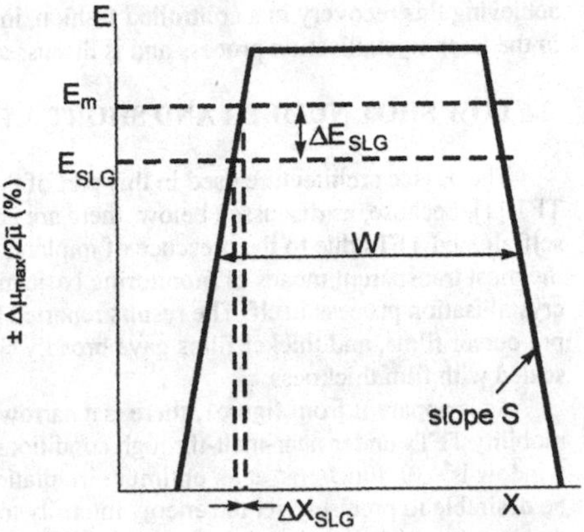

Figure 3. Schematic illustration of top-hat laser beam shape.

- At energy densities below E_m, there is also a reducing mobility with decreasing shot number, the origin of which is quite different from the phenomena at energies greater than E_m. Hence these two regimes of $E<E_m$ and $E>E_m$ need to be distinguished, and the former regime is discussed in section 3.2

3.1 Energy Densities, $E > E_m$

3.1.1 Trailing Edge Analysis

The detail at issue here is the reduction in average mobility, and the increase in scatter, with reducing shot number in samples irradiated with the conventional top-hat beam. As discussed qualitatively above, this is related to the recovery of the material following full-melt-through, and the details of this can be more easily appreciated with reference to figure 3. This is a schematic illustration of a top-hat beam, with the critical energy densities for the SLG regime and full-melt-through indicated. Once the material has been exposed to an energy density greater than E_m, it will need to be re-irradiated at an energy within the SLG window, ΔE_{SLG}, which has a spatial width of ΔX_{SLG}. Hence, to ensure that each point on the surface is re-irradiated within the SLG window, the size of the incremental beam step across the surface will have to be $< \Delta X_{SLG}$. This defines the minimum number of shots in the process to be $W/\Delta X_{SLG}$, where W is the FWHM of the beam. This can be related to the slope, S, of the trailing edge by:

$$\Delta X_{SLG} = \Delta E_{SLG}/S \qquad (1)$$

Hence, the minimum number of shots, N, is:

$$N > SW/\Delta E_{SLG} \qquad (2)$$

For a typical SLG window of 40mJ/cm^2 and a top-hat beam with a 50µm wide edge, the number of shots required is >75, which is consistent with the experimental data in figure 2.

3.1.2 Ramped Edge Data

As indicated in equation 2, to reduce the number of shots, it is necessary to reduce the slope of the trailing edge of the beam, and, in the course of this work, the influence of edge shape has been investigated. Schematic illustrations of a top-hat and a ramped edge beam are shown in figure 4. Figures 5a and b compare the experimental results obtained from a top-hat and an appropriately shaped ramped edge beam, for a 20 shot and 100 shot process, respectively.

The key result is in figure 5a, for the 20 shot process, in which the broadened beam shape produces a more constant value of average mobility into full-melt-through and the scatter remains below ±10% over a larger energy interval. In other words, as predicted, a beam with a ramped trailing edge does indeed produce more uniform material, within a process window enlarged from 15mJ/cm^2 to 50mJ/cm^2. The 100 shot results in figure 5b are included to demonstrate consistency with the basic model, by showing that when the shot number is high the beam shape has very little effect upon the results. Although, even here, it will be seen that the scatter is slightly larger with the top hat beam.

Whilst figure 5a illustrates the improvements in window size and uniformity achieved at low shot number, with the ramped edge beam, it should be appreciated that reducing the shot number does lead to an overall degradation in all device properties, irrespective of the particular beam shape. This was clearly seen for mobility with the top hat beam in figure 2, and these effects are discussed more fully in section 3.2. In spite of the overall degradation in device characteristics, all device parameters still have improved uniformity with the ramped edge beam, once full melt-through occurs.

3.2 Energy Density, $E < E_m$

In the preceding section, we discussed the shot number dependence of crystallisation at energy densities greater than for full melt-through, where the primary issue was recovery of the fine grain material. There is also a shot number dependence to the crystallisation phenomena at E

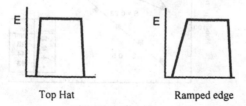

Top Hat   Ramped edge

Figure 4. Schematic illustration of top-hat and ramped edge beams

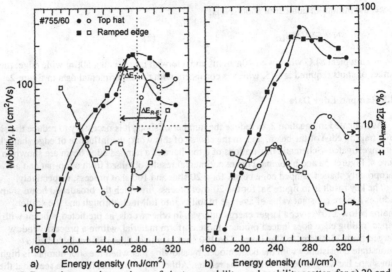

Figure 5. Dependence on beam shape of electron mobility, and mobility scatter, for a) 20 and b) 100 shot irradiations

$< E_m$, which is clearly unrelated to the full melt-through situation. It is primarily observed under near-melt-through conditions, when the large SLG grains are being produced [2]. The dependence of carrier mobility on shot number is plotted in figure 6 for films of different thickness. Samples showing similar properties to the 44nm thick films have been examined using TEM to determine grain structure, and this identified an increase in mean grain size of ~40%,

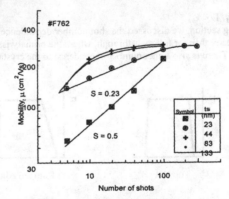

Figure 6. Variation of mobility (for $E<E^*$)

from 160nm to 220nm for 10 shot and 100 shot irradiations, respectively. The grain quality was generally high, with the most visible features being grain bridging stacking faults and micro-twins, for which the bounding dislocations are contained within the grain boundaries themselves. Hence, the major scattering features are concluded to be the grain boundaries, and the increasing mobility with shot number in figure 6 is assumed to result from reduced grain boundary scattering caused by the increase in grain size.

The other key features in this figure are the saturation in mobility after ~100 shots, in all but the thinnest films, and the more rapid saturation in the thicker films. Assuming that the mobility increase is a result of grain growth, these curves indicate a corresponding saturation in grain size. Increasing grain size with shot number has been reported by other workers [2], but the mechanism for this growth has not been fully clarified, beyond noting that it is likely to be driven by grain boundary melting and the preferential growth of grains having a particular orientation [2]. With this model of preferential melting of grain boundaries in mind, the saturation may be understood from the SEM micrograhs in figure 7. Figures 7a and b were obtained after 100 shot irradiations in the SLG regime at 320 and 340mJ/cm^2, and show clear grain delineation and increasing spatial ordering, respectively. The ordering is believed to be caused by a roughness induced interference effect, which has been widely reported in laser/solid interactions, and referred to as Laser Induced Periodic Surface Structures (LIPSS) [4,5]. This ordering may contribute to grain size stabilisation in multi-shot irradiations, and, moreover, the asperities (resulting from Si pile-up at grain boundaries [6]) will eventually be too thick to melt and thus prevent the preferential melting of the grain boundaries. Hence the driving force for grain growth will be terminated.

(It is interesting to note from figure 7c, that once full melt-through and recovery occurred, pre-existing grain structures were no longer dominant in mediating the re-growth and the ordering seen in 7b was lost. The grain size was also more irregular than at 340mJ/cm^2, and TEM examination identified the presence of dislocations within grains, contributing to a general degradation of grain quality. There was also a greater density of intra-grain asperities, which may be a surface topography residue of previous grain boundaries.)

The field effect mobility data in figure 6 were taken from 60μm channel length TFTs,

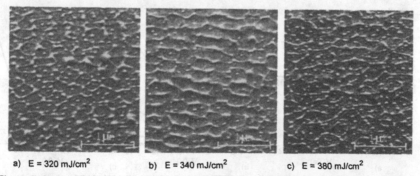

a)   E = 320 mJ/cm^2 b)   E = 340 mJ/cm^2 c)   E = 380 mJ/cm^2

Figure 7. SEM micrographs illustrating surface structure for different laser intensities (100 shot irradiations): a), b) E<E$_m$ and c) E>E$_m$. (Samples tilted at 45° about horizontal axis in the SEM).

because, with increasing film thickness, there were increasing errors in the estimation of mobility in short channel TFTs due to channel shortening. The channel shortening resulted from lateral diffusion of the source and drain dopants in the melt, and was directly indicative of increasing melt duration with increasing film thickness. Hence there was a shorter melt duration and more rapid cooling with reducing film thickness, which would influence the grain growth. In particular, the more rapid cooling of the thinner films would result in more rapid random nucleation in the melt, limiting the initial grain growth per shot.

4. DEVICE ARCHITECTURE

4.1 Self Aligned TFTs

In the preceding section, the TFTs all employed the NSA architecture, in which the source and drain regions were ion implanted before laser crystallisation. Hence, the ion implantation damage will be removed during film melting. In contrast, self-aligned (SA) TFTs have the source and drain regions implanted after the crystallisation of the film, and we have reported striking differences in the characteristics of phosphorus implanted n-channel NSA and SA TFTs, having had the same initial crystallisation stage [3]. When the phosphorus is thermally activated, the implantation conditions have to be tightly controlled to avoid full amorphisation of the poly-Si film [3]. Should this occur, the film will not re-crystallise in a low thermal budget process. For laser activation of the dopant, this failure to crystallise does not occur, but, nevertheless, there is still a limited range of implantation conditions within which successful TFT operation can be obtained. This is illustrated in figure 8a, in which we show the transfer characteristics of TFTs fabricated with two different implantation schedules. The implantation conditions are listed in table 1, together with the resulting sheet resistance values, which were almost independent of the change in conditions. However, as a result of increasing the implantation energy from 80 to 110keV, device (ii) in figure 8a failed to turn-on, and, as shown by the output characteristic in 8b, there was severe current crowding. However, Table 1 confirms that this effect cannot simply be due to high series resistance. The key result in figure 8c, of an anomalous increase in drain

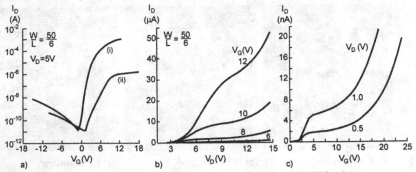

Figure 8. a) transfer characteristics for TFTs with (i) 80 and (ii) 110keV P implants, (b) output characteristics for 110keV TFT and (c) transfer characteristics for 110keV TFT.

current at large values of V_G, demonstrates that there is a current flow barrier at the channel/n^+ boundary, which can be modulated by the gate.

Table 1
Variation of sheet resistance with P implantation conditions

Implantation Energy (keV)	P Ion Dose (x 10^{15} cm^{-2})	Rs (Ω/)
80	7	500
110	3.5	650

We have previously reported higher leakage currents in laser activated n-channel TFTs, which we ascribed to lateral implantation damage beneath the gate edge [3]. We believe that the 110keV TFT results in figure 8 represent an extreme example of this lateral damage effect, and, for the first time, the damage region has been directly confirmed by cross-sectional TEM. The micrograph is shown in figure 9, in which the arrow indicates a residual amorphous region at the gate edge. This unambiguously confirms our previous interpretation of SA TFT data in terms of lateral implantation damage beneath the gate, but also identifies it extending beyond the gate edge, which may be due to scattering of the laser beam at the gate edge.

Under the less damaging implantation conditions of 80keV, TEM examination has not identified an extensive amorphous region, so much as a region of micro-defects beneath the gate edge, which are responsible for the higher leakage currents and lower mobilities in these structures compared with NSA TFTs. In order to improve the performance of the laser activated SA TFTs, it is necessary to either remove the defects or suppress their influence. We have reported that gate overlapped LDD regions (GOLDD [7]) very effectively suppress the influence of the defects on leakage current, by shifting the electron-hole pair generation site from the damaged n^+-channel junction to the damage-free GOLDD-channel junction [3].

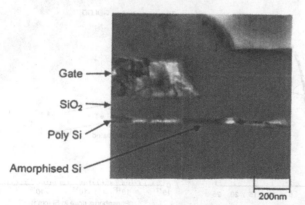

Gate
SiO$_2$
Poly Si
Amorphised Si

200nm

Figure 9. XS TEM micrograph of 110keV SA TFT, showing damage at gate edge.

Alternatively, the damage can be directly removed by off-setting the edge of the n⁺-region from the channel edge, so that the defects are no longer produced beneath the gate. In this situation, the end-of-range defects will be directly exposed to the laser during the dopant activation process, and will be removed together with all the other implantation damage. Moreover, the presence of this gap will also permit some lateral diffusion of the n^+ junction edge during the laser induced melting of the film; this will laterally broaden the junction and, thereby, reduce the drain field [7]. The off-set gap has been accomplished by a controlled etch-back of the gate, by ~0.5μm, following the phosphorus implant. The consequent removal of the end-of-range damage during the laser activation of the dopant is confirmed by the TFT characteristics in figure 10. In this figure, we compare the etch-back gate TFT with the standard SATFT, and the reduction in leakage current and its reduced sensitivity to gate bias are clearly seen. The leakage current reduction can be directly attributed to the removal of the implantation damage, but the reduction in the field-assisted component of the leakage current is unlikely to be due to junction broadening alone. Even 100 shot irradiated NSA TFTs (which will have √10 times greater junction broadening), do not display an off-state characteristic as flat as that shown in figure 10. Hence, part of the reduction in gate field sensitivity is due to the reduction in gate-drain coupling induced by the gap.

4.2 Drain Field Relief

Even with junction broadening, hot carrier stability remains an issue in both SA and NSA TFTs, and further field relief is required. From an architecture point of view, field relief can be achieved with an LDD region, which can either be overlapped by the gate (GOLDD [7]) or, more conventionally, interposed between the gate and the drain. In the case of GOLDD, the conductivity of the lightly doped region will be modulated by the gate in the on-state, and this

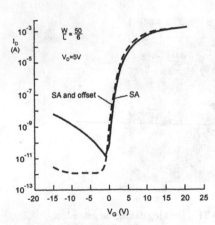

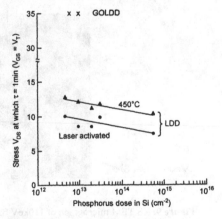

Figure 10. Effect of gate off-set in removing damage in SA TFT.

Figure 11. Maximum hot carrier stress bias for SA TFTs with GOLDD and LDD field relief.

will minimise series resistance. In contrast, the resistance of the LDD region, with the conventional architecture, can result in significant on-current loss. These differences are clearly illustrated by the apparent field effect mobilities of 70 and 200 cm^2/Vs measured in SA LDD and SA GOLDD TFTs, respectively, with LDD region widths of 3μm and doping levels of 6×10^{12} cm^{-2}. The low, apparent mobility measured in the SA LDD TFT, is entirely due to the resistance of the LDD region.

The differences in hot carrier stability are even more striking when the two architectures are compared, as shown by the data in figure 11. The LDD devices could barely sustain the application of 13V to the drain, whereas the GOLDD devices, at comparable doping level, were stable up to 35V. As these particular LDD devices did not have an off-set gate, there was some end of range damage present and the junctions were abrupt, both of which would increase the drain field. However, comparable stability values were obtained in the off-set gate structures [8]. Hence, there are fundamental differences in the stability conferred by the GOLDD and LDD field relief architectures. Some of the key advantages of the GOLDD structure are the positioning of the peak field beneath the gate [9], and the diversion of the electron avalanche current to the back of the film [10]. Further details of the instability phenomena in the LDD TFTs will be discussed elsewhere [8,11].

5. CONCLUSION

In order to improve the throughput of the laser crystallisation process, it is necessary to reduce the number of laser shots per unit surface area, and the influence of shot number on the resulting poly-Si TFT properties are discussed in this paper. Two distinct shot number dependent effects are identified, which are themselves laser intensity dependent. The critical laser energy density is that which causes full film melt-through, and for energies greater than this, the major issue is the uniformity of the device properties. It is demonstrated that the conventional top-hat beam is not an optimum shape, and that by deliberately extending the trailing edge of the beam, the material uniformity at reduced shot number can be improved. For energy densities less than full-melt-through, there was a general reduction in overall device properties with reducing shot number, which has been correlated with reducing grain size. However, the magnitude of this effect at low shot number is film thickness dependent and diminishes with thicker films, due to the increased melt duration and slower cooling of these films.

TFT architecture plays a major role in determining the final device characteristics, and differences in key aspects of the performance of self-aligned and non-self-aligned TFTs, such as leakage current and mobility, are confirmed to be artefacts of the fabrication process. The most important effect is lateral ion implantation damage, beneath the gate edge, in n-channel SATFTs. The presence of this damage region was unambiguously demonstrated with cross sectional TEM images, and various ways of removing it, or reducing its influence, were discussed. Finally, the GOLDD field relief architecture was demonstrated to offer superior hot carrier stability compared with conventional LDD regions.

6. REFERENCES

1. S D Brotherton, D J McCulloch, J P Gowers, J R Ayres and M J Trainor. J Appl Phys, 82, 4086 (1997).

2. H J Kim and J S Im. Mater Res Soc Symp, 321, 665 (1994).
3. S D Brotherton, J R Ayres, C A Fisher, C Glaister, J P Gowers, D J McCulloch and M J Trainor. Electrochem Soc Proc, 98-22, p25 (Boston, 1998).
4. J F Young, J S Preston, H M van Driel and J E Sipe. Phys Rev B 27, 1155 (1983).
5. D J McCulloch and S D Brotherton. Appl Phys Lett, 66, 2060 (1995).
6. D K Fork, G B Anderson, J B Boyce, R I Johnson and P Mei. Appl Phys Lett, 68, 2138 (1996).
7. J R Ayres, S D Brotherton, D J McCulloch, and M J Trainor. Jpn J Appl Phys, 37, 1801 (1998).
8. F W Rohlfing, J R Ayres, S D Brotherton, C A and D J McCulloch. To be published Proc Fall SID, 2000.
9. R Izawa, T Kure and E Takeda. IEEE Trans ED-35, 2088 (1988).
10. A Pecora, F Massussi, L Mariucci, R Carluccio, G Fortunato, J R Ayres and S D Brotherton. Proc AMLCD'99, p165 (1999).
11. A Valletta, A Bove, A Pecora, L Mariucci, G Fortunato, F W Rohlfing, J R Ayres and S D Brotherton. To be published in Proc AMLCD 2000.

Mat. Res. Soc. Symp. Proc. Vol. 621 © 2000 Materials Research Society

Single-Shot Excimer-Laser Crystallization
of an Ultra-Large Si Thin-Film Disk

Mitsuru NAKATA *, Chang-Ho OH and Masakiyo MATSUMURA
Department of Physical Electronics, Tokyo Institute of Technology,
2-12-1 O-okayama, Meguro-ku, Tokyo 152-8550, Japan
E-mail : nakata@silicon.pe.titech.ac.jp   URL : http://silicon.pe.titech.ac.jp/

Abstract

An excimer-laser-induced large-grain growth method has been proposed which utilizes non-uniform heat diffusion along the Si thin-film and also along the underlayer. A single-shot of KrF excimer-laser light pulse with uniform intensity could crystallize a circularly pre-patterned Si thin-film of 20µm in diameter, much larger than TFT feature size in present AM-LCD panels.

Introduction

Studies on polycrystalline silicon (poly-Si) crystallized by excimer-laser annealing (ELA) have been pursued so as to implement high driving-current and low leakage-current in thin-film transistors (TFTs). To produce single-crystal Si TFTs, following three conditions should be satisfied. The one is to grow grains larger than the TFT feature size [1,2], the second is to control position of these large grains [3], and the third is to keep the excess-process cost low, for example, by a single shot and uniform irradiation using a simple ELA system.

ELA-induced lateral growth can potentially satisfy these requirements, since long lateral growth can take place when there is a sufficient retardation of a start time of nucleation along the Si thin-film. There are two dominant ways in elongating the retardation, that is, introduction of a non-uniform light intensity distribution [4,5,6] or introduction of a non-uniform sample structure to initiate non-uniform heat out-diffusion from the uniformly molten Si thin-film. And we have studied here the latter way. Namely, a nucleation controlled ELA (NCELA) method has been proposed where the unmolten amorphous-Si (a-Si) region is localized to a very small volume by lateral heat diffusion in the Si thin-film. The NCELA method has been combined with pre-patterned ELA (P^{2}ELA) method where the molten Si film is heated up non-uniformly by the underlayer. By this IP^{2}ELA (improved P^{2}ELA) method, a Si disk of as large as 20µm in diameter was crystallized by a single shot and uniform irradiation, and a central part of a Si disk of 10µm in diameter was single-crystallized.

Proposal of IP^{2}ELA

In the conventional ELA method, Si grains start to grow from a lot of nuclei born simultaneously in the unmolten a-Si region. Since collision with neighboring grains interrupts their growth along the lateral direction, grown grains are column-like-shaped with a diameter much smaller than the TFTs feature size. The long lateral grain growth can take place, however, when the nucleation time is delayed along the molten Si layer. Retardation of the nucleation time along the Si layer, a necessary condition for the sufficiently-long lateral growth, may be introduced by several ways. One of them is to generate a temperature gradient after the uniform laser light irradiation by a non-uniform sample structure.

A proposed sample structure suitable to this approach is shown in Fig.1. There is a small light reflecting plate attached on the circularly patterned Si thin-film on the heat-tolerant light absorption layer, and the sample is irradiated by a uniform laser light pulse. The Si layer outside the reflector is completely molten and the Si layer under the reflector remains in an original amorphous state. There appears, however, a large temperature gradient in the Si layer near the edge of the reflector that generates the lateral heat diffusion in the Si thin-film. Thus the unmolten a-Si region is shrunken to a small value under the center of the reflector. Due to limited volume of the unmolten a-Si region, number of nuclei born there is reduced to only unity. The nucleus then starts to grow laterally by the help of steep temperature gradient around it.

Since the heat-tolerant underlayer outside of the Si disk is heated up to an extremely high temperature by absorbing the laser light, thermal energy given to the irradiated underlayer diffuses not only vertically to the substrate but also laterally to the non-irradiated underlayer of low temperature. Then, the vertical heat out-flow from the molten Si disk is reduced in magnitude especially around the disk edge compared with its central region. Therefore, the temperature gradient is produced in the Si disk from the edge to the center, and the lateral crystallization having started from the center can continue to whole the Si disk, resulting in a single crystal Si of large diameter.

Since there are two-steps in elemental mechanism of the lateral growth for the proposed IP^2ELA method, we will divide them and discuss each individual growth mechanisms, i.e. lateral growth (P^2ELA method) around the Si disk edge and limited nucleation and lateral growth (NCELA method) near the reflector edge. And then we will combine them.

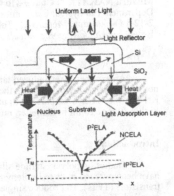

Fig.1. Schematics of IP^2ELA

Lateral Growth by P^2ELA Method

For the P^2ELA method no light reflector is placed at the center as shown in Fig.2. The SiON underlayer is used as the light absorbing layer since SiO_2 is transparent but Si_3N_4 is not transparent to the KrF excimer-laser light, and since both of these films are heat-tolerant. We have assumed the SiON layer to have the optimum light absorption coefficient distribution with the form of $(\lambda-x)^{-1}$ by adjusting the O/N ratio in the film, where x is the distance from the top SiON surface and λ the SiON layer thickness [7]. At the beginning of irradiation, the Si disk is uniformly heated up. Temperature in the Si disk stays once at the melting point T_M due to its large latent heat, and then restarted to be heated up again.
Temperature in the SiON layer outside the Si disk, on the other hand, goes up steadily, resulting in the extremely high temperature such as 3000°C. After irradiation, thermal energy given to the irradiated SiON region diffuses not only vertically to the substrate but also laterally along the SiON layer. Thus there appears steep temperature gradient in the SiON layer under the Si disk, and this gradient reflects to the non-uniform vertical heat flow from the molten Si disk. As a result, temperature gradient is generated in the Si disk for more than 4.5μm in length at the time when temperature at the center of the Si disk reduces to T_M, resulting in the lateral growth. It is worthy to note that whole Si region is completely molten, and temperature is the lowest and takes a uniform distribution near the center.

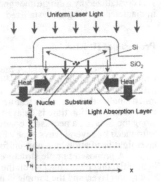

Fig.2. Schematics of P^2ELA

A SEM photograph after Secco ething is shown in Fig.3 for the P^2ELA sample. Thickness of the Si thin-film and λ were 200nm and 1.6μm, respectively. A substrate of the 850nm thick thermal SiO_2 on Si was heated at 500°C during ELA. For Si disks of as large as 12µm in diameter, there were no laterally grown grains, when the light intensity I was as low as 400mJ/cm² and only spots were observed, because thermal energy given to the SiON layer outside the Si disk was so low in density that the satisfactory temperature gradient could not be generated near the edge of the Si disk. The crystallization kinetics did not differ so much from those for the conventional and high intensity ELA. There appeared no lateral growth and the surface morphology was perfectly the same to that for the conventional ELA sample. For I=500mJ/cm², grains grew laterally but only near the edge of the disk. There were many small grains near the center as in case of I=400mJ/cm². Higher the laser light intensity was, larger the lateral growth in length became until the SiON layer was broken. At I=700mJ/cm² all Si disks were laterally crystallized. However, a single crystal was not grown and there appeared spoke-like-shaped lines of grain boundaries as shown in the figure. And there was a clear hole of a few tenth µm in diameter at the center. This is because temperature near the center of the Si disk took the minimum and uniform value. Nucleation site is not limited to a sufficiently narrow volume at the center, and many nuclei were born spontaneously as schematically shown in Fig.2. Since grown grains at the center were small and dense they can be removed completely by Secco ething, resulting in a hole. And grains started to grow laterally from nuclei having born in the vertically of the center. Then there appeared many grains.

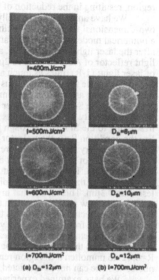

Fig.3. SEM photographs of the Si films crystallized by P^2ELA

Number Control of Nuclei by NCELA Method

Since average grain size for the conventional ELA method is at most a few tenth µm, number of nuclei may be reduced to unity in P^2ELA when there remains the unmolten a-Si region of a few tenth µm in diameter. To localize the a-Si region to such a small size, a light reflector of a few µm is placed just on the sample as shown in Fig.4. And uniform laser light irradiates the Si region outside the light reflector without the light diffraction effects, since the reflector is placed just on the sample surface. The narrow and unmolten a-Si region is then remaining under the light reflector. When temperature in this a-Si region of a few µm wide is higher than the critical value T_N for nucleation but lower than T_M, then a limited number of nuclei are born there. There appears also the steep temperature gradient near the light reflector edge, that initiates lateral grain growth from a nucleus having born at the center under the light reflector.

It should be taken into account that the steep temperature gradient near the edge of the light reflector introduces a large heat flow from the molten Si region outside the reflector to the unmolten a-Si region under the reflector. This flow will shrink further the unmolten a-Si

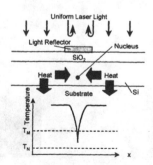

Fig.4. Schematics of NCELA

region, resulting in the reduction of number of nuclei.

We have analyzed numerically time-dependent two-dimensional temperature distribution. Figure 5 shows a numerical model and the temperature distribution just after the laser light irradiation, i.e., at a time t=21ns. The light reflector of Al of as large as 2μm may absorb a part of laser light; (1-R) times as much as the intensity to the Si region outside the reflector. Thus in the numerical model, we have assumed that excimer-laser light is absorbed by the Si thin-film "under the reflector", with the light intensity I(1-R), where I is the light intensity irradiating to the Si region through the capping SiO$_2$ layer. For the condition R=1, the reflector reflects perfectly the laser light, a very sharp temperature gradient is formed near the center of the reflector just at the finish of laser light irradiation. The high temperature region in the Si thin-film under the reflector extending from the edge is less than 0.3μm. This means that in order to prepare a sufficient narrow nucleation site in the unmolten a-Si region, the reflector should be less than 0.6μm wide, that is unacceptable from a production view point. As R decreases, the temperature slope near the edge becomes gentle, and thus the reflector can be widen. However, for R=0.42 no unmolten a-Si region remains at the center and nucleation site can not be restricted.

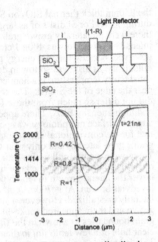

Fig.5. Temperature distributions during NCELA for several light absorption coefficients

We have examined experimentally the effects of the light reflector. Samples were of an Al(800nm)/SiO$_2$(100nm)/a-Si(200nm) structure Si wafer with 850nm-thick SiO$_2$. Only the top Al film was patterned circularly as a light reflector. Samples were irradiated by a KrF excimer laser pulse at the substrate temperature of 500 °C. Figure 6 shows the surface morphology after Secco ething for several values of I and of the Al disk diameter (D$_{Al}$). For D$_{Al}$=5μm, higher the

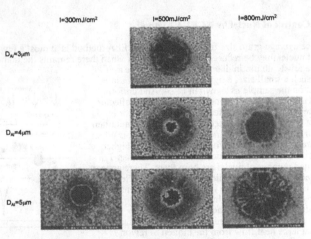

Fig.6. SEM photographs of the Si films crystallized by NCELA

light intensity was, larger the grains were grown at the edge of the light reflector. For I=800mJ/cm², the a-Si region at the center disappeared and grains started to grow from the center. When I was fixed at 500mJ/cm², the a-Si region became narrower as D_{Al} was reduced. And grains started to grow from the center when D_{Al} was reduced to 3μm. It will be important to make a comment that there was no clear hole at the center, different from the case of P²ELA. And sometimes the number of nuclei was as few as unity, and then single crystal grains with a few μm in diameter were obtained. A single crystal Si as large as 5μm in diameter was grown successfully by using a light reflector with D_{Al}=4μm at I=800mJ/cm² as shown in the figure.

Theoretical and Experimental Results of IP²ELA

By combining the NCELA method and the P²ELA method i.e., by using the IP²ELA method we have investigated a possibility to crystallize larger Si disks with a small number of grains. Prospected temperature distribution is shown in Fig.1. Since there is a sharp dip in temperature at the center due to NCELA component, number of nuclei born at the central region is very few and lateral growth starts from these nuclei. This lateral grain growth is then transferred to the P²ELA-induced lateral growth for the outside of the light reflector.

Figure 7 shows the time-dependent temperature distribution in the Si disk for the IP²ELA method where R was assumed 0.8, and the width of the light reflector 2μm, the thickness of the Si disk 200nm and λ 1.6μm respectively. The light intensity given to the SiON layer is assumed 1/T times higher than that to the Si layer by the reflection effect of light at the Si surface (where T=0.4). The steep temperature gradient appears at the center due to the effect of the light reflector and the unmolten a-Si region is as narrow as 1.4μm with the maximum temperature of about 1000°C at the center. Thus only a few nuclei can be born there. As the time advances, the region having sharp temperature gradient and growth temperature near T_M moves steadily to the disk edge by the effect of the SiON underlayer. Thus long lateral growth can take place.

Figure 8 shows a melt/solid interface from the center after the laser light irradiation for D_{Al}=2μm. For D_{Si}=16μm, since the Si region is very large, there appears a gentle temperature gradient region between 3μm and 6μm from the center. Therefore nuclei may be born there while grains grow laterally. As D_{Si} decreased, the growth speed becomes slow, for D_{Si}=8μm grains can grow with a constant low speed, resulting in a single crystal.

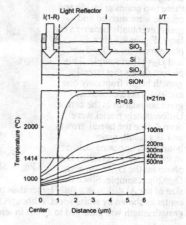

Fig.7. Temperature distribution during Improved P²ELA for several times after laser irradiation

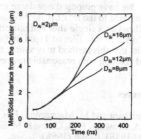

Fig.8. Melt/solid interface from the center after the laser light irradiation

A SEM photograph after Secco ething is shown in Fig.9 for the crystallized sample of an Al(800nm)/SiO₂(100nm)/a-Si(200nm)/SiO₂(100nm)/SiON(1.6m) structure on Si wafer with 850nm-thick SiO₂. Figure 9(a) was photographs at several I values for D_{Al}=10μm and D_{Si}=20μm, respectively. At I=400mJ/cm², since lateral growth due to the SiON under layer did not take place near the Si disk edge, there were large grains only around the light reflector (by NCELA). At I=600mJ/cm², the grains can grow around the Si disk edge and at the same time lateral growth length for the light reflector edge was enlarged. At I=800mJ/cm² these two grains at inside and outside were merged and grains eventually became very large. Figure 9(b) is a photograph for a D_{Al}=5μm and D_{Si}=11μm sample. This picture shows that lateral growth began from only one nucleus. There existed no grain boundaries at the center. Unfortunately nuclei were born during the lateral growth, and as a result, grain boundaries appeared outside in the region. Figure 9(c) is a photo for the D_{Al}=6μm and D_{Si}=20μm sample. Since the

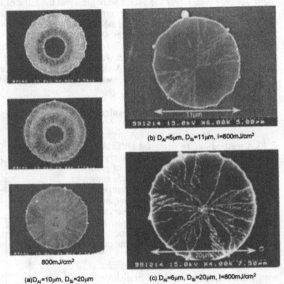

(a)D_{Al}=10μm, D_{Si}=20μm

(b) D_{Al}=5μm, D_{Si}=11μm, I=800mJ/cm²

(c) D_{Al}=6μm, D_{Si}=20μm, I=800mJ/cm²

Fig.9. SEM photographs of the films crystallized by IP²ELA

size of the Al disk was a little larger than the optimum value there are several nuclei near the center as shown by white spots. Although several grains started to grow from these nuclei, lateral growth length was enlarged to 10μm in length.

Conclusions

We have proposed new large grain growth methods named P²ELA and NCELA. And they are combined to the IP²ELA method. A disk-shaped Si thin-film of 20μm in diameter was crystallized by a single shot and uniform KrF excimer laser light. And a center region of the disk-shaped Si thin-film of 11μm in diameter was single crystallized. These results indicate the potential that this method may satisfy the requirements of large and single grain growth at desired position with reasonable process cost aiming at ultra high performance TFTs.

References

1. R. Ishihara and M. Matsumura, Jpn. J. Appl. Phys., **36**, p. 6167 (1997)
2. M. A. Crowder, P. G. Carey, P. M. Smith, R. S. Sposili, H. S. Cho, and J. S. Im, IEEE ED. Lett., 19 (8), p. 306 (1998)
3. M. Matsumura, Phys. Stat. Sol. (a) **166**, p. 715 (1998)
4. C. H. Oh, M. Ozawa and M. Matsumura, Jpn. J. Appl. Phys., **37**, p. L492 (1998)
5. C. H. Oh and M. Matsumura, Jpn. J. Appl. Phys., **37**, p. 5474 (1998)
6. M. Matsumura and C. H. Oh, Thin Solid Films, **337**, p. 123 (1999)
7. M. Ozawa, C. H. Oh and M. Matsumura, Jpn. J. Appl. Phys., **38**, p. 5700 (1999)

Mat. Res. Soc. Symp. Proc. Vol. 621 © 2000 Materials Research Society

Location-Controlled Large-Grains in Near-Agglomeration Excimer-Laser Crystallized Silicon Films

Paul Ch. van der Wilt, Ryoichi Ishihara and Jurgen Bertens
Delft Institute of Microelectronics and Submicron Technology (DIMES), P.O. box 5053, 2600 GB Delft, The Netherlands.

ABSTRACT

Large grains in thin silicon films were grown by controlling the location of unmolten islands, which are left after near-complete melting of the film during excimer laser crystallization. As the initially amorphous film was first transformed in small grain polycrystalline silicon, these islands contain seeds for crystal growth. To get a single large grain, either the number of seeds was reduced to one or a single one was selected from the seeds by a 'grain filter'. Former was achieved by making a small indentation in the isolating layer underlying the silicon film so that seeds remain embedded in the indentation. Latter was achieved by making a small diameter hole in the underlying isolating layer, which was filled with amorphous silicon. The lateral growth is preceded by a vertical growth phase during which a single grain is filtered from the initial set of seeds present at the bottom of the hole. In the experiment described, highest yield was achieved for samples in which the melt-depth to hole-diameter ratio was largest.

INTRODUCTION

Excimer-laser crystallization (ELC) of thin silicon films is an important technology for fabrication of thin-film transistors (TFTs) on low-temperature-budget substrates such as glass. The goal to make c-Si TFTs [1] requires that large crystalline islands can be grown on a predetermined position. Artificially controlling the super lateral growth (SLG) [2] phenomenon allows for this. The location of the solid seeds, which remain when the film is nearly completely melted, can be controlled either by patterning the beam [e.g. 3], by making a structure in or underlying the silicon film [e.g. 4], or by a combination of both [e.g. 5]. The issues in these methods are process window, process complexity and compatibility with existing processes. For TFT fabrication, issues such as crystal quality and crystal orientation are also very important.

In this work the focus is on structures, whereas the ELC process is kept as simple as possible i.e. a single square-shaped pulse, single beam process. By locally increasing the heat-extraction rate from the film, we have been able to control the location of the solid seeds and by that of the growth of large grains [4]. The number of seeds, however, was difficult to control and the yield of *single* large grains was at the most 8%. In an alternative method in which solid seeds remain in small indentations in the underlying SiO_2, this control could be achieved. A 30% yield of single large-grain growth was obtained with these embedded seeds [6]. As no heat sinks are needed, no highly thermal-conductive layer has to be deposited and process is kept simple.

The research presented in this report, further investigates the embedded-seed location-control method and shows how yield can be increased to 100%. Rather than making a small indentation, a small diameter (ca. 100 nm) hole is made with a depth several times the film thickness. This hole is filled during a-Si deposition, and subsequently the column-like filling is partially melted during ELC. Because of this, the lateral grain growth is preceded by a vertical growth phase, during which competitive grain-growth filters out a single grain; i.e. crystal

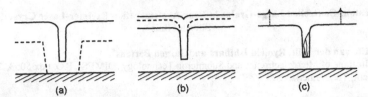

Figure 1. *Schematical overview of the process. (a) Small diameter holes are etched and subsequently filled by a second oxide deposition. (b) Before a-Si deposition, the surface is planarized to sharpen the edges of the hole. (c) During ELC, large grains grow laterally from the hole until they collide with grains grown from the undercooled melt.*

growth first goes through a 'grain filter' before the single surviving grain laterally grows into a large c-Si island.

EXPERIMENTAL

Two experiments will be discussed. In the first experiment, the film thickness dependence of the material-transformations during ELC was determined by visual inspection afterwards. The samples were made by 1 μm plasma-enhanced chemical vapor deposition (PECVD) using tetraethoxysilane (TEOS) at 350°C on crystalline silicon (c-Si) wafers. Subsequently, amorphous silicon (a-Si) layers were deposited with a thickness varying from 7 to 214 nm. Deposition was performed by low-pressure chemical vapor deposition (LPCVD) using silane at 570°C.

In the second experiment (Figure 1), a first 700 nm PECVD TEOS SiO_2 deposition was performed on c-Si wafers. Small holes defined by standard lithography with a diameter of 1.0 μm (the minimal *reproducible* feature size) were etched in a $CHF_3 - C_2F_6$ plasma. Due to horizontal etching and/or redeposition during etching, the sidewalls of these holes were slightly outwardly sloped [7]. Subsequently, a second PECVD TEOS layer was deposited to reduce the diameter. The thickness of this second layer was 763 nm, 814 nm, or 863 nm. Step coverage of the deposition was sufficient to retain the outwardly sloped side walls, necessary to prevent void formation during the subsequent a-Si filling. A plasma planarization was performed to sharpen the edges of the holes. The structure covered with SPR3017 photoresist were etched back in an $O_2 - CHF_3 - C_2F_6$ plasma with equal etch rates for the photoresist and the oxide. The 90 nm, 181 nm, or 272 nm LPCVD a-Si deposition had very uniform step coverage.

For both experiments, ELC was performed by a single pulse of a XeCl (308 nm wavelength) excimer laser with a FWHM pulse duration of 66 ns, with energy densities ranging from 0.15 J/cm^2 to 1.40 J/cm^2. The substrates were irradiated either at room temperature or at ca. 400°C in a vacuum chamber (10^{-5} Torr). Although after the filling a small indentation is observed at the surface of the a-Si, after melting and regrowth of the film, the surface is planarized (apart from the hillocks, characteristic for SLG). The films were etched in a diluted Schimmel-etch solution (4 parts of 40% HF solution : 5 parts of 0.3 M CrO$_3$ solution) to visualize grain boundaries and other defects. Analysis was performed with a SEM.

RESULTS AND DISCUSSION

The visually observed material transitions are plotted in Figure 2. At a certain energy density, the film is completely transformed into highly defective small-grain poly-Si by explosive crystallization (XC) [8]. For a film thickness less than 100 nm, this transition appeared very abrupt. Above this film thickness, however, more energy was required to reach complete XC, and the transition no longer appears abrupt. This is indicated in the figure.

The minimal energy density required to completely melt the poly-Si film (E_{CM}) is a strong function of film thickness. The energy density at which the film starts to agglomerate is also a strong function of film thickness. At this energy density, the surface gets roughened and at some locations holes with high ridges appear in the film, but no silicon is removed or evaporated. Agglomeration is related to the melt duration of the film combined with the wetting properties of the melted film. The energy density at which holes were visible for the first time (E_{Agg}) is indicated in Figure 2. For further increasing energy density, the density and the diameter of the holes increase up to a point where the film ablates, i.e. silicon is actually removed. For the 214 nm film this energy density is about two times E_{CM}, whereas the agglomeration starts at about 1.5 times E_{CM}.

As the goal in the second experiment is to melt a significant part of the silicon column embedded in the oxide, a much higher energy density than E_{CM} for the surrounding film will be used. The upper limit is of course the agglomeration energy density. An indication of the melt depth at this energy density can be obtained from Figure 2: looking at a certain energy density, the film thickness which will be melted just completely is about 2.5 times the film thickness at which agglomeration will occur. When horizontal heat flow is neglected (both going inwards from the film towards the hole as going outwards from the hole into the oxide) this means that the melt depth in the hole is about 1.5 times the film thickness.

Structured Films

As the deposition rate of oxide is dependent on the arrival angle of the gas molecules, the horizontal deposition rate is less than the vertical deposition rate [9]. Furthermore, it decreases for decreasing diameter of the hole to be filled. A result of this is that close to completely filling the hole, horizontal deposition rate reduces to less than half the vertical deposition rate. Cross sectional SEM images showed that after 763 nm second oxide deposition, hole-diameter was reduced to about 140 nm and after 814 nm second oxide deposition to about 90 nm. Thus, extrapolating, the holes after 863 nm deposition were also open and diameter was about 40 nm. Figure 3 shows the resultant structure after the subsequent filling with a-Si, before and after ELC. From these cross sections it can be seen that no void formation occurred during a-Si deposition and that the surface of the silicon film is planarized after ELC. Unfortunately, the edges of the holes were hardly sharpened,

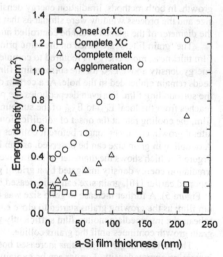

Figure 2. The various material transitions seen in ELC as a function of film thickness.

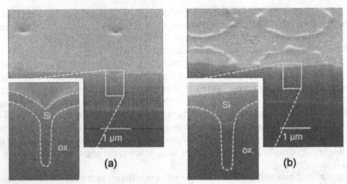

Figure 3. *Cross sectional SEM images of four positions for large grain growth: (a) before irradiation showing the indentations at the surface and the complete filling of the holes and (b) after irradiation, showing the edges of the large grain growth seen as hillocks at the surface and the planarization through melting of the film.*

due to micro loading-effects during the plasma planarization process.

In the SLG regime, solidification starts from the few unmolten islands left after nearly completely melting the film. In earlier work we presented two methods with which we managed to control the location of these unmolten islands [4, 6]. As the a-Si film was first transformed into columnar poly-Si during XC, these islands contain one or several seeds for large-grain growth. In both methods, irradiation energy density was slightly above E_{CM} and both the grain size and the process window were similar as that seen in the SLG regime. In the second method, the diameter of the islands could be controlled and average number of seeds could be reduced.

The 'grain filters' are based on the same principle as latter method: by locally increasing the film thickness, more energy is required to completely melt the silicon. When irradiated at an energy density above E_{CM} for the surrounding film, but below E_{CM} for the silicon column, solid seeds remain embedded in the hole. At a certain depth of the hole, the local E_{CM} equals E_{ABL} of the surrounding film. For even deeper holes the process window for large grain growth now reaches from the local E_{CM} to E_{ABL} of the surrounding film. As more heat is introduced in the film, the cooling rate at the onset of solidification of the film will be reduced. As a result, the lateral growth continues longer before it collides with grains growing from the undercooled melt. A doubling in grain size can be observed, as can be seen in Figure 4. This is further illustrated in Figure 5, which shows the diameter of lateral overgrowth as a function of film thickness and irradiation energy density instigated by a grid of grain filters with an 8 μm interspacing. As was reported earlier [10], grain size can be increased by heating the substrate during ELC (black dots in Figure 5). A further increase in grain size was observed for 1.40 J/cm² irradiation of the 272 nm film: as the growing grains started to close each other, the heat released during solidification reheats the intermediate molten film sufficiently to suppress copious nucleation, so that lateral grain growth continues until the grains collide.

The yield of *single* large grains increased both for decreasing hole diameter and for increasing energy density. Former can be explained by the fact that a smaller diameter reduces the number of seeds. Latter can be explained by looking at the ratio between the melt-depth and

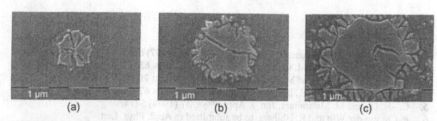

| (a) | (b) | (c) |

Figure 4. SEM images showing grain growth in a 181 nm silicon film instigated by isolated grain filters with second oxide deposition of 863 nm: ELC (a) at 0.64 J/cm², slightly above SLG; (b) at 0.87 J/cm², slightly below oxide melting; and (c) at 1.13 J/cm², slightly below ablation.

the hole-diameter. Assuming the average number of seeds for large grain growth is independent of melt-depth, the observed increase in yield of single grains must be due to occlusion of some of the grains, which started to grow. This indicates that grains were indeed filtered out during the vertical growth phase, and that filtering was more successful as melt-depth was increased. Melt-depth could not only be increased by increasing energy density (note that increasing film thickness allows for higher energy densities), but also by introducing heat from the backside by substrate heating. Figure 6 shows the grain growth in 272 nm films instigated by a grid of grain filters with smallest diameter and an interspacing of 3 μm. Without substrate heating yield was reasonable, however, cracks were formed, especially at highest energy density, due to stress induced by ELC. With substrate heating, on the other hand, no cracks were observed and yield was increased considerably.

CONCLUSIONS

It was shown that with grain filters the regime for large-grain growth was drastically increased compared to SLG and to methods developed earlier to control the SLG [4, 6]. Furthermore, when the melt-depth to hole-diameter ratio was sufficient, a high yield of single large grain growth could be achieved. In this work, the melt-depth to hole-diameter ratio was increased by introducing heat from the backside to decrease vertical temperature gradient. Another method would be to further sharpen the edges of the grain filter by planarizing the oxide surface after second oxide deposition. This way, the opening of the grain filter has the same diameter as the bottom and filtering will be more effective. Also, if the ablation is related to agglomeration, replacing it by a material with better wetting properties might further increase the maximum irradiation energy density.

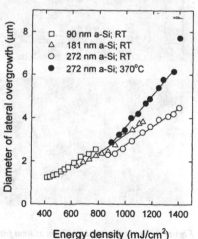

Figure 5. Diameter of the laterally overgrown area as a function of film thickness and substrate temperature.

REFERENCES

1. R. Ishihara and M. Matsumura, Jpn. J. Appl. Phys. **36** (1997) 6167.
2. J. S. Im, H. J. Kim, and M. O. Thompson, Appl. Phys. Lett. **63** (1993) 1969.
3. J. S. Im, R. S. Sposili, and M. A. Crowder, Appl. Phys. Lett **70** (1997) 3434.
4. P. Ch. van der Wilt and R. Ishihara, Solid State Phenomena **67-68** (1999) 169.
5. C. –H. Oh and M. Matsumura, Jpn. J. Appl. Phys. **37** (1998) 5474.
6. P. Ch. van der Wilt and R. Ishihara, to be submitted to Appl. Phys. Lett.
7. S. M. Sze, VLSI Technology (McGraw-Hill Book Co., New York, USA) 2nd ed. (1988) 223.
8. H. Watanabe, H. Miki, S. Sugai, K. Kawasaki, and T. Kioka, Jpn. J. Appl. Phys. **33** (1994) 4491.
9. S. M. Sze, VLSI Technology (McGraw-Hill Book Co., New York, USA) 2nd ed. (1988) 253.
10. H. Kuriyama, et al., Jpn J. Appl. Phys. **30** (1991) 3700.

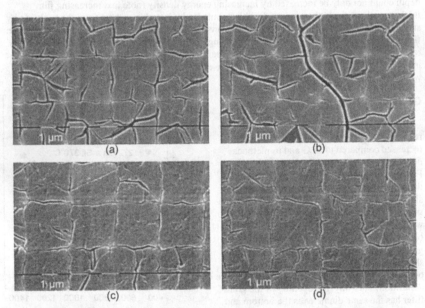

Figure 6. Grain growth in 272 nm silicon films instigated by a grid of smallest diameter grain filters with a interspacing of 3 μm. (a) and (b) without substrate heating at 1.05 and at 1.25 J/cm², respectively. (c) and (d) with 370°C substrate heating at 1.09 and at 1.26 J/cm², respectively. Note: The etch pits are not stained point defects. They were absent in other experiments and were caused by surface contamination during ELC as vacuum was not optimal.

Mat. Res. Soc. Symp. Proc. Vol. 621 © 2000 Materials Research Society

Defect population and electrical properties of Ar+-laser crystallized polycrystalline silicon thin films

S. Christiansen[1], M. Nerding[1], C. Eder[1], G. Andrae[2], F. Falk[2], J. Bergmann[2], M. Ose[2], H. P. Strunk[1]*

[1] Universität Erlangen-Nürnberg, Institut für Werkstoffwissenschaften-Mikrocharakterisierung, Cauerstr. 6, D-91058 Erlangen, Germany

[2] Institut für Physikalische Hochtechnologie, Winzerlaer Strasse 10, D-07745 Jena, Germany

*e-mail: sechrist@ww.uni-erlangen.de

Abstract

We crystallize amorphous silicon (a-Si) layers (thicknesses: ~300nm and ~1300nm for comparison) that are deposited on glass substrates (Corning 7059) by low pressure chemical vapor deposition using a continuous wave Ar+-laser. We scan the raw beam with a diameter of ~60μm in single traces and traces with varying overlap (30-60%). With optimized process parameters (fluence, scan velocity, overlap) we achieve polycrystalline Si with grains as wide as 100μm. The grain boundary population is dominated by first and second order twin boundaries as analyzed by electron backscattering analysis in the scanning electron microscope and convergent beam electron diffraction in the transmission electron microscope. These twins are known not (or only marginally) to degrade the electrical properties of the material. In addition to twins, dislocations and twin lamellae occur at varying densities (depending on grain orientation and process parameters). The recombination activity of the defects is analyzed by EBIC and according to these measurements crystallization receipts are defined that yield the reduction of electrically detrimental defects.

Introduction

The preparation of thin polycrystalline silicon (poly-Si) films on low cost amorphous substrates such as glass has been subject of intensive research with a variety of techniques [1-7] over the past few years, motivated by the perspective to use poly-Si thin film transistors (TFTs) in active matrix liquid displays [8] and for thin film solar cells on glass which could be an attractive low-cost alternative to conventional solar cells [9]. For both fields of applications it is the task to realize large grained material, a control of the defect population and especially in the case of TFTs a location control of the extended defects. These defects may affect the electrical characteristics of the material in different ways [10-15]. Thus, knowledge about the nature of the grain boundary and dislocation population in laser crystallized poly-Si is not only an important structural information but is also a prerequisite for a clear understanding of the electrical characteristics of the material. Presently laser crystallization is considered to be one of the most promising approaches to bring a thin crystalline Si layer on the amorphous, temperature sensitive glass substrate without damaging the substrate thermally. Different types of laser annealing procedures are tested, that all base on melting and re-solidification of an amorphous Si layer. The most common and frequently used laser is the short pulse excimer laser [1,2,3], but also other lasers like e.g. the frequency doubled Nd:YAG [4] or the Ar^+- ion laser are used [5]. Receipts to obtain micro-crystalline material rather than nanocrystalline one are available for all laser processes, they however lead to significantly different large grained material as will be pointed out in the following.

It is well known that the grain size is directly related to the supercooling of the generated melt [16,17]. Low (high) supercooling results in low (high) nucleation rates in the melt [16,17,18]. Low supercooling is for example realized in bulk Czochralski growth [29] as well as in zone melt recrystallization (ZMR) [6]. The required melt duration times are however too large to be compatible with the glass substrate. The excimer laser represents the other extreme, very low melt duration times (50-150 nsec at pulse durations between 15 and 50nsec) result in large supercooling thus in high nucleation rates and consequently comparably small grains (hundreds of nm up to few micro-meter at best). We report a laser annealing process that is intermediate between the two extremes: the raw beam of ~60μm in diameter of a continuous wave Ar^+-ion laser is scanned at high velocity of ~5cm/sec over the substrate. Melt duration times in the micro-second region result. These are still compatible with the glass substrate, the comparably low supercooling leads to small nuclei densities and thus comparably large grains of several tens to 100μm across, as has already been shown in earlier work [5]. With grains of this size, the grain boundary and dislocation population and thus the structure in laser crystallized poly-Si thin films can thoroughly be evaluated. For this purpose we use a combination of transmission electron microscopy (TEM) techniques and electron backscattering diffraction analysis (EBSD). We will consider the dependence of the defect population on layer thickness and overlap between adjacent laser crystallized traces that are formed in the scanning process.

Experimental

Laser crystallization (on areas of $3 \times 3 cm^2$) was performed by scanning the Gaussian shaped raw beam of an Ar^+- ion laser (operated at a wavelength of 514 nm, power 1-10Watt, intensity of about $20kW/cm^2$) at a velocity of up to 5cm/sec on an amorphous Si (a-Si) layer, that is deposited on a borosilicate glass. Crystallization was performed at 550°C substrate temperature. The a-Si is deposited by conventional rf-PECVD process using SiH_4 as a precursor. Two a-Si thicknesses of 300nm and 1300nm are compared as concerns the microstructure that develops under optimized crystallization parameters that yield the largest grains. Using the obtained optimized fluence we vary the overlap between adjacent scans between 30% and 60%. We focus in this paper on these two extreme overlaps.

Transmission electron microscopy (TEM) of the laser crystallized films is carried out using a Philips CM30 analytical microscope operated at 300kV and a conventional Philips CM20 microscope operated at 200kV. To preserve the as-grown surface the samples are thinned by conventional grinding and polishing techniques from the back side. They are finally thinned to electron transparency in an ion mill under an Ar^+-beam at 13°, 5kV in a liquid nitrogen cooling stage. Utilizing convergent beam electron diffraction (CBED) in the TEM, Kikuchi patterns [19] are obtained that contain a 3D information of the exact grain orientation. From Kikuchi patterns of adjacent grains the grain boundary types between the grains are determined utilizing a non-commercial analysis software developed in our group [20]. Unfortunately, it is rather difficult to obtain with the CBED technique a good statistics on the grain boundary population, since only a few grain boundaries can be evaluated on each TEM sample. Therefore, pseudo-Kikuchi patterns are obtained in addition using electron backscattering diffraction (EBSD) analysis in a scanning electron microscope (SEM). The pseudo-Kikuchi patterns are taken in steps of 0,5-2 μm as dependent on the magnification used. For each sample a matrix of several thousand measurements is obtained describing the grain orientations and the boundaries between grains. The grain boundary type is then evaluated by using the software of the commercial Oxford EBSD setup. The combined use of CBED, EBSD and TEM provides good statistics of grain boundary types and permits the evaluation of the fine structure of the grain boundaries, e.g. secondary defects such as dislocations or dislocation networks.

We use the following classification of grain boundaries. Firstly two main groups are distinguished; low and high angle grain boundaries. Low angle grain boundaries consist of dislocations that arrange in networks [21]. In case of high angle grain boundaries we distinguish between random grain boundaries, that can be described by the misorientation they cause, and so-called coincidence grain boundaries, which can be defined in a lattice common to both adjacent grains (coincidence site lattice [19]). In this purely geometric notion coincidence orientations are characterized by the Σ-parameter which is essentially the reciprocal value of the coincidence sites occupied by the atoms of the common lattice [22]. Special Σ-boundaries are the class of twin boundaries, such as first order ($\Sigma=3$), second order ($\Sigma=9$), third order ($\Sigma=27$) and fourth order ($\Sigma=81$) twins [23]. In general Σ-boundaries represent low energy configurations compared to random grain boundaries. Calculated energies of different Σ-boundaries can be found elsewhere [24,25]. Besides grain boundaries, films can also contain twin lamellae. These twin lamellae consist of few atomic layers in twin orientation relationship with respect to the surrounding matrix.

Electrical characterization of the observed defects, related to structural characterization is carried out by electron beam induced current measurements (EBIC) in a Jeol 6400 scanning electron microscope (SEM). The electron beam (acceleration voltage 10-20keV, ~1nA beam current) is scanned slowly over the sample and the resulting EBIC current is measured in parallel, thus revealing as dark features the recombination active defects such as dislocations and grain boundaries.

Results

Using optimized process conditions (for each layer thickness) as given by a specific fluence at the fixed scanning velocity of 5cm/sec we obtain in any single trace of the scanning process large grained material with grains that are as wide as half the trace width, i.e. around 30μm. Fig.1 shows a TEM micrograph of the rim of such a single trace. At the very rim we observe nanocrystals caused by explosive crystallization [25]. When these grains form the melt in the trace interiors still exists. Further solidification occurs in that these grains grow perpendicular to the solidification front towards the interior. These grains having the fast growth direction <110> perpendicular to the solidification front overgrow all the others grains [26]. As a result elongated grains form with width of 5-10μm and lengths of around 30μm. The long axis of the grains is oblique to the rims of the trace by an angle of 40° due to the rapid scanning process (without scanning solidification is perpendicular to the rims e.g. [4]).

Figure 1:
Plane view TEM micrograph of an Ar⁺- laser crystallized polycrystalline Si. Single scanned trace on a glass substrate. The grains, that grow fastest and laterally overgrow others are 2-10µm in width. The grain boundaries are oblique, 40° to the scan direction. Nanocrystals at the rim from explosive crystallization are visible.

The grain boundaries formed between the large grains in Fig.1 are essentially twins of first and second order (verified by CBED in the TEM). Fig.2 shows the grain boundary location, distribution and type with statistically relevant data volume in a EBSD analysis. Fig.2a shows an EBSD mapping. We use color coding to denote the sample normals: blue indicates {111} orientation, red {100}, and green {110}. Note, that the EBSD orientation analysis confirms the TEM result that the grains (seen as patches of the same color) are roughly 2-10 µm wide and 30µm in length. The middle of the picture coincides with the middle of the scanning trace. There the grains that start to grow from both rims meet. The surface normals of 2604 orientation mapping measurements are visualized in the standard triangle in Fig.2b. It is obvious, that this sample has no texture, the surface normals cover the whole standard triangle. The summary of the analysis of the grain boundary population in Fig.2a is shown in Fig.2d. Most of the grain boundaries are twin boundaries of first (Σ=3) or second (Σ=9) order. In Fig.2 roughly 65% of all measurements belong to Σ=3 boundaries, roughly 15% to Σ=9 boundaries and 20% belong to other coincidence boundaries and random grain boundaries. The evaluation of other areas of the sample and other samples yield similar data: in general 50%-70% of all grain boundaries are Σ=3 boundaries and 10%-15% are Σ=9 boundaries. Σ=27 boundaries (3rd order twins) occur, but less frequent. As already expected from CBED in the TEM, the Σ=3 and Σ=9 boundaries lie parallel to the fast grain growth direction. The grain boundary locations are shown in Fig.2c. The middle of the trace is free of any Σ- boundaries, i.e. random grain boundaries form as expected. It is well known that twin boundaries are electrically comparably harmless (no recombination activity, no level in the band gap) as long as the grain boundaries are clean (no precipitates or dislocations superimposed the grain boundary plane) [14,27] whereas random grain boundaries are recombination active and cause a barrier [10]. First EBIC measurements of the recombination activity of the grain boundaries in a single trace are shown in Fig.2e. The oblique grain boundaries are the twins. They are visible as faint darker lines. The middle of the trace, where the random grain boundaries lie shows strong EBIC contrast, indicating strong recombination activity of these grain boundaries and associated dislocations (according to TEM measurements that are not shown here) that reside in these regions.

In case of overlapping scans (overlap: 30% (low) and 60% (high)) we find essentially the same grain boundary population, dominated by Σ=3 and Σ=9 twins. At low overlaps the single scan traces can still be distinguished since the grains terminate at the scan trace rims. I.e. only the explosive crystallization region is re-melted compared to a single trace. Samples with high overlap (60%) however show grains that are extended across several adjacent scan traces due to lateral epitaxial growth. Grains can in principle be extended practically infinitely by this technique as long as the scanning procedure is carried out. Grains, compared to the single scan trace and low overlap traces, also vary in width. Using an evaluation software (Image Tool, share ware) we determine grain widths from large area EBSD mappings (300μmx500μm: 30% sample, 800μm-1800μm: 60% sample). For the 30% overlap sample most grains are 10-50μm in width as visible from the bar chart in Fig.3a. Grains in widths 100μm and larger do not exist. For the 60% overlap sample grains are generally larger in width and >100μm grains cover more than 1/3 of the whole scanned area as shown in Fig.3b. Fig.3c shows an optical micrograph of such a 60% overlap sample. Grain boundaries are made visible by Secco [28] etch as topographical features.

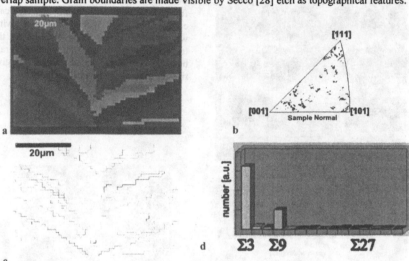

Figure 2:

a) Color coded orientation mapping with EBSP. Each color represents a different surface normal orientation; red: {100}, blue: {111} and green: {110}. Note that one can clearly distinguish the different grains (patches of single color).

b) Surface normal orientations of the 2604 independent EBSP measurements visualized in the standard triangle. The different orientations cover the whole standard triangle, i.e. this sample has virtually no texture.

c) The grain boundary location is shown. The color code of 2d is taken. It is obvious that essentially the green and pink $\Sigma=3$ and $\Sigma=9$ twins occur. They lie inclined by 40° to the scan direction comparable to the boundaries

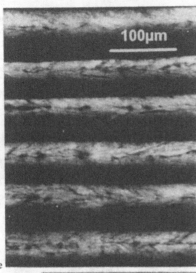

e

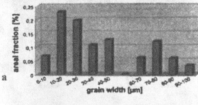

a

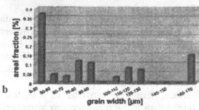

b

in Fig.1.

d) Grain boundary population in a bar chart as evaluated from Kikuchi patterns of the measurement in 2a. Color coded are the different grain boundary types. Green are the Σ =3 twins, pink are the Σ =9 twins.

e) EBIC measurement of a individual scan traces. The faint dark lines stem essentially from Σ=3 and Σ=9 twins and indicate comparably low recombination activity, whereas the strong dark contrast at the trace's center stems from random grain boundaries that are located there and give rise to strong recombination activity.

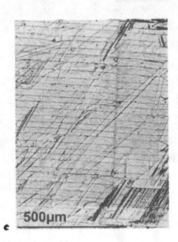

c

Figure 3:
a) Bar chart: percentages of areal coverage with grains of certain sizes (given by the grain width); 30% overlap scan sample.
b) Bar chart: percentages of areal coverage with grains of certain sizes (given by the grain width); 60% overlap scan sample.

c) Optical micrograph: Secco etched large grained Si layer (thickness: 1300nm, 60% overlap scan)

d) TEM micrograph showing a bundle of dislocations produced by a Frank Read source having operated at the rim of a scan trace.

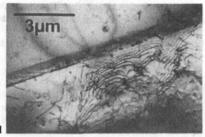

d

Discussion

We presented essentially two important findings:

i) The grain boundary population is dominated by twin boundaries of first and second order independent of process conditions. This is important since twin boundaries are known to be electrically harmless (no barrier, no recombination activity) [10,14,27].

ii) Process conditions could be found that result in grains that are not only large in scanning direction (this is achieved with almost all sequential laser melting processes, independent of the laser type [3,4]) but also in grain width (>100µm). The corresponding arrangement in the laser crystallized thin layers is reminiscent of the microstructure of cast silicon [29]. In addition, also compared to cast silicon, we observe even lower intra grain dislocation densities (<10^5cm^{-2}) in a vast number of grains (TEM analysis, to be published elsewhere). Grains with a high dislocation density reside at the rims of the overlapping traces together with twin lamellae that form there preferably. The twin lamellae form at high densities in certain grains. This density is high enough for the twins to act as pinning points for dislocations that then can multiply by a Frank Read mechanism [30] at these twin lamellae (Fig.3d).

Acknowledgements

We acknowledge access to the EBSP equipment in the Institute for General Materials Properties (Prof.H. Mughrabi) at the University of Erlangen. Part of the electron microscopy work was performed in the Central Facility for High Resolution Electron Microscopy, Friedrich Alexander Universität Erlangen Nürnberg.

References

[1] S. D. Brotherton, D. J. McCulloch, J. B. Clegg, and J. P. Gowers, IEEE Trans. Electr. Dev. **40** (2), 407 (1993)

[2] P. Mei, J. B. Boyce, M. Hack, R. Lujan, R. I. Johnson, G. B. Anderson, D. K. Fork, and S. E. Ready, Appl. Phys. Lett. **64** (9), 1132 (1994)

[3] J. S. Im, H. J. Kim, and M. O. Thomson, Appl. Phys. Lett. **63**, 1969 (1993); J. S. Im, M. A. Crowder, R. S. Sposili, J. P. Leonard, H. J. Kim, J. H. Yoon, V. V. Gupta, H. Jin Song, and H. S. Cho, phys. stat. sol. (a) **166**, 603 (1998)

[4] G. Aichmayr, D. Toet, M. Mulato, P.V. Santos, A. Spangenberg, S. Christiansen, M. Albrecht, H.P. Strunk, J.Appl.Phys. 85, 4010 (1999); J. R. Köhler, R. Dassow, R. B. Bergmann, J. Krinke, H. P. Strunk, and J. H. Werner, Thin Solid Films 337, 129 (1999);

[5] G. Andrä, J. Bergmann, F. Falk, E. Ose, H. Stafast, phys.stat.sol. (a) **166**, 629 (1998); G. Andrä, J. Bergmann, F. Falk, and E. Ose, Thin Solid Films **318** (1-2), 42-45 (1998)

[6] P.V. Evans, D.A. Smith, C.V. Thompson, Appl.Phys.Lett. 60, 439 (1992)

[7] M. Bonnel, N. Duhamel, L. Haji, B. Loisel, J. Stoemenos, Electron. Device Letters 14, 551 (1993)

[8] T. Sameshima, M. Hara, and S. Usui, Jpn. J. Appl. Phys., Part 1 28, 1789 (1989).

[9] R. B. Bergmann, J. Köhler, R. Dassow, C. Zaczek, and J. H. Werner, phys. stat. sol. (a) 166, 587 (1998)

[10] J. H. Werner, in *Structure and Properties of Dislocations in Semiconductors*, edited by S. G. Roberts (Institute of Physics, Bristol, 1989), Vol. 104, pp. 63; J. H. Werner and N. E. Christensen, "Classification of Grain Boundary Activity in Semiconductors," in *Polycrystalline Semiconductors II*, edited by J. H. Werner and H. P. Srunk (Springer Verlag, Berlin, 1991), Vol. 54

[11] S. Serikawa, IEEE Trans. Electron Devices 36, 1929 (1989)

[12] B. Cunningham, H. Strunk, and D. G. Ast, Appl. Phys. Lett. 40 (3), 237 (1982)

[13] A. Voigt, *Blockgegossenes Silizium für die Photovoltaik: Struktur und elektrische Eigenschaften von Defekten* (Vol. 2 Ser. Mikrostrukturelle Materialforschung, H.P. Strunk ed., Verlag Lehrstuhl für Mikrocharakterisierung, Erlangen, Germany, 1996; ISBN3-932392-01-9)

[14] A. Bary, B. Thercey, G. Poullain, J. L. Chermant, and G. Nouet, Revue Phys. Appl. 22, 597 (1987)

[15] F.. Liu, M. Mostoller, V. Milman, M.F. Chisholm, T. Kaplan, Phys.Rev.B 51, 17192 (1995); T.S. Fell, P.R. Wilshaw, M.D. DeCoteau, phys.stat.sol.(a), 138, 695 (1993)

[16] H. Watanabe, H. Miki, S. Sugai, K. Kawasaki, T. Kioka, Jpn.J.Appl.Phys. 33, 4491 (1994)

[17] G. Götz, Appl.Phys.A 40, 29 (1986)

[18] K.A. Jackson, B. Chalmers, Canadian Journal of Physics 34, 173 (1956)

[19] W. Bollmann, *Crystal Defects and Crystalline Interfaces* (Springer-Verlag, Berlin, 1970)

[20] A. Voigt, Orientation Analysis Software (University Erlangen,1996)

[21] D. Hull, D.J. Bacon, Introduction to dislocations, in: Series on Materials Science and Technology Vol. 37, 1984

[22] A full description of this type of grain boundary would additionally require the identification of the contact planes in the adjacent grains [19].

[23] Higher order twins form by subsequent twinning operations or by reaction of lower order twins, e.g. a Σ=3 and a Σ=9 boundary form a Σ=27 boundary or a Σ=27 and a Σ=3 boundary form a Σ=81 boundary.

[24] M. Kohyama and R. Yamamoto, Phys. Rev. B 49, 17102 (1994) ; M. Kohyama and R. Yamamoto, Phys. Rev. B 50, 8502 (1994)

[25] M.O. Thompson, G.J. Galvin, J.W. Mayer, P.S. Peercy, J.M. Poate, D.C. Jacobson, A.G. Cullis, N.G. Chew, Phys.Rev.Lett. 52, 2360 (1984); J.C.c. Fan, J.H. Zeiger, R.P. Gale, R.P. Chapman: Phys.Rev.Lett. 36, 158 (1980)

[26] J. L. Batstone, Phil. Mag. A 67, 51 (1993)

[27] A. Rocher, C, Fontaine, C. Dianteill, Inst. Phys. Conf. Ser. 60, 289 (1981)

[28] F. Secco d'Aragona, J.Electrochemical Soc.119, 948 (1972)

[29] H.J. Möller, Semiconductors for Solar Cells (Artech House Inc., Norwood MA, 1993

[30] F.C. Frank, W.T. Read, in: Symp. Plastic Deformation of Crystalline Solids, Carnagie Inst. Technol., Pittsburgh 1950 (p.44)

Mat. Res. Soc. Symp. Proc. Vol. 621 © 2000 Materials Research Society

Effect of Excimer Laser Fluence Gradient on Lateral Grain Growth in Crystallization of a-Si Thin Films

Minghong Lee, Seungjae Moon, Mutsuko Hatano[1], Kenkichi Suzuki[2], and Costas P. Grigoropoulos
Department of Mechanical Engineering, University of California,
Berkeley, CA 94720-1740, U.S.A.
[1]Hitachi Laboratory, Hitachi Ltd.,
Tokyo 185-8601, JAPAN
[2]Electron Tube & Devices Division, Hitachi Ltd.,
Mobara 297, JAPAN

ABSTRACT

In order to clarify the relationship between excimer laser fluence gradient and the length of lateral grain growth, the laser fluence is modulated by a beam mask. The fluence distribution is measured by using a negative UV photoresist. The lateral growth length and the grain directionality are improved with increasing fluence gradient. Lateral growth length of about 1.5 μm is achieved by using a single laser pulse without substrate heating on a 50 nm-thick a-Si film by enforcing high fluence gradient. Electrical conductance measurement is used to probe the solidification dynamics. The lateral solidification velocity is found to be about 7 m/s.

INTRODUCTION

Conventional excimer laser crystallization (ELC) can produce grains of hundreds of nanometer in size depending on the a-Si film thickness [1]. However, the processing window for the conventional technique is narrow because large grains can only be obtained in the so-called superlateral growth (SLG) regime [2]. In addition, the grain size produced by the conventional ELC is highly non-uniform with randomly oriented grain boundaries which results in non-uniform device characteristics. Therefore recent research efforts on ELC have been focusing on developing spatially and directionally controlled crystallization methods. Several methods have been shown to give laterally oriented grain growth. These methods include the use of beam mask [3], diffraction mask [4], anti-reflective coating [5], phase shift mask [6], and the interference effect induced by a frequency-doubled Nd:YAG laser [7]. The working principle in all these techniques relies on shaping the laser energy profile that is irradiated onto an a-Si sample. In order to induce lateral grain growth, a fluence gradient must be enforced such that the a-Si film completely melts at the area exposed to higher laser fluence and partially melts at the adjacent area exposed to lower laser fluence. Under this condition, grains grow laterally towards the completely molten region. However, relationship between fluence gradient and lateral growth length has not been demonstrated. The purpose of this study is therefore to experimentally clarify this relationship by shaping the excimer laser beam via a beam mask to produce the desired fluence gradient. Electrical conductance measurement was also performed to probe the solidification dynamics and estimate the lateral solidification velocity. A solidification model is proposed to qualitatively explain the grain microstructure produced by the crystallization process.

EXPERIMENT

The excimer laser used in the crystallization experiment is a KrF excimer laser operating at the wavelength of 248 nm. The laser light was directed into a 2:1 ratio projection system as shown in figure 1. An Al beam mask consists of 20 μm-wide line patterns separated by 20 μm spacings was used. The projection system included a fly's-eye type homogenizer that ensures energy uniformity to the level of ±5% or better. The patterns were imaged onto an a-Si sample placed on the X-Y translation sample stage. The a-Si thin films were deposited by Low Pressure Chemical Vapor Deposition (LPCVD) at 550°C onto quartz substrates. The laser crystallized samples were secco-etched for characterizing the grain microstructure by using Scanning Electron Microscope (SEM).

A negative photoresist was used to quantify the laser fluence distribution after passing through the 20 μm line patterns on the mask. The photoresist has a characteristic that the thickness of the developed photoresist after laser exposure will exhibit a linear relationship to the local laser fluence [8]. The photoresist that has been spin-coated onto a quartz wafer was placed onto the sample stage. The patterned excimer laser was pulsed through the quartz substrate onto the photoresist. After excimer laser exposure, the photoresist was developed such that the unexposed photoresist was stripped-off, but the laser-exposed photoresist remained on the quartz

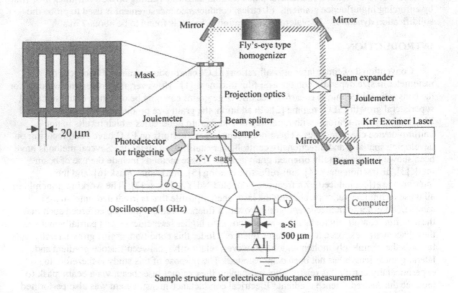

Sample structure for electrical conductance measurement

Figure 1. Schematic of experimental setup for laser crystallization and electrical conductance measurement

surface with thickness linear to the local laser fluence. The photoresist thickness across the pattern was then measured by an alpha step profiler. Because of the linear characteristic of the photoresist, the fluence distribution can be mapped by using the measured thickness distribution and the measured laser pulse energy by a joulemeter as shown in figure 2.

Time-resolved electrical conductance measurement was performed to probe the solidification dynamics and to yield the lateral solidification velocity (see figure 1). Electrical conductance measurement is based on the difference in electrical conductivity between solid silicon and liquid silicon. The electrical conductivity of molten silicon (75 $\Omega^{-1}cm^{-1}$) is higher than that of Si (0.3 $\Omega^{-1}cm^{-1}$) [9]. Therefore the conductance signal, which is captured by a fast digitizing oscilloscope, increases upon melting of the Si layer. This signal was then used to analyze the solidification dynamics.

RESULTS AND DISCUSSIONS

An example of crystallization results is also shown in figure 2. The circled regions (the dark lines) are the lateral growth regions. The fluence gradient at the origin of the lateral grain growth (identified as C in the figure) was found by measuring the distance L from the SEM picture and identifying location C on the mapped fluence distribution with the length L. The slope at the identified location C of the mapped fluence distribution corresponds to the fluence gradient at the lateral growth. By applying different excimer laser pulse energies and measuring the lateral growth lengths from the SEM pictures, the dependence of lateral grain growth length on fluence gradient is obtained. Figure 3 shows that the lateral growth length is almost constant at about 500 nm for fluence gradients below 80 mJ/cm^{2}μm, but increases rapidly as the laser fluence gradient increases further. The results also show grain directionality improvement as fluence gradient increases. It is noted that lateral grains of about 1.5 μm in size are obtained on a 50 nm-thick a-Si film by a single excimer laser pulse without any substrate heating under high fluence gradient.

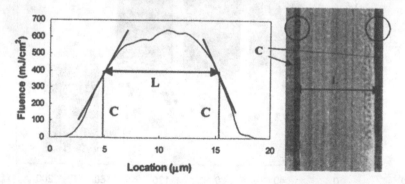

Figure 2. An example of a mapped fluence distribution and the illustration of fluence gradient calculation.

Figure 4 shows the results of the electrical conductance measurement. The solidification modes for excimer laser crystallization without a beam mask (the conventional technique) have been studied in detail by both electrical conductance measurement and a number of optical probing techniques [10]. On the other hand, laser crystallization with a beam mask takes a substantially longer solidification time, which starts at point "a", changes in slope on the conductance signal at point "b", and solidification ends at point "c".

To understand this conductance signal, a qualitative solidification model depicted by a series of one-dimensional temperature profiles is proposed as shown in figure 4. Figure 4 (a) represents the temperature profile at times between "a" and "b". The solid-liquid interface is at the highest temperature "T_i" because of the release of latent heat during solidification. The bulk liquid temperature is supercooled at "T_s", but above the spontaneous nucleation temperature "T_n". As time progresses to "b", the supercooled bulk liquid temperature drops below "T_n" and triggers spontaneous nucleation (see figure 4 (b)). As spontaneous nucleation and solidification continue in the bulk liquid, a second solid-liquid interface with interface temperature "T_{i2}" is produced due to the release of latent heat. The resulting temperature profile is shown in figure 4 (c) with lateral grains on the left due to lateral solidification, fine grains on the right due to spontaneous nucleation, and liquid silicon in between. Finally, the two interfaces meet together and solidification ends at time "c". The final grain arrangement is shown in figure 4 (d). Lateral grain 1, which grows from left to right in the schematic, takes place over time interval "c-a". Fine grains are produced from time "b" to "c". Lateral grain 2, which grows from right to left in the schematic, takes place over a time interval less than "c-b". Using the measured lateral solidification time, "c-a", and the lateral growth length from SEM pictures, a lateral solidification velocity of about 7 m/s is obtained. The lateral solidification velocity does not vary with fluence gradient.

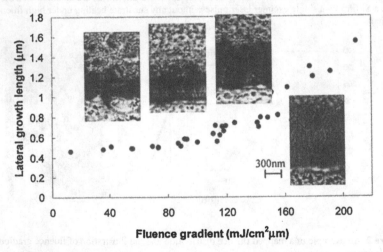

Figure 3. The dependence of lateral growth length on fluence gradient for 50 nm a-Si thin film.

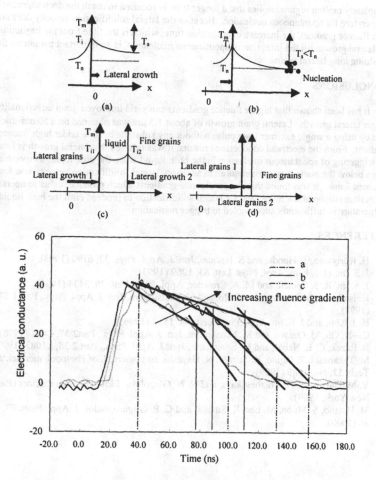

Figure 4. Electrical conductance signal for excimer laser crystallization with a beam mask and a proposed solidification (grain growth) model.

With the above results and analyses, lateral grain growth is limited by spontaneous nucleation in the bulk liquid. In order to produce long lateral grains, the spontaneous nucleation in the bulk liquid must be suppressed or delayed so that lateral growth can proceed to a longer distance. In the case of high fluence gradient, the temperature increases rapidly from the partially molten region to the completely molten region. The higher local temperature in the completely molten region implies that a longer time is required to reach the deep supercooling temperature for spontaneous nucleation. Because the lateral solidification velocity does not vary with fluence gradient, the increase in nucleation time, which is the time from the beginning of the lateral growth till the inception of spontaneous nucleation, is an important parameter for obtaining long lateral grains.

CONCLUSIONS

It has been shown that higher fluence gradients can yield improved grain directionality and longer lateral growth. Lateral grain growth of about 1.5 μm was obtained on a 50 nm-thick a-Si film by using a single excimer laser pulse without any substrate heating under high fluence gradient. From the electrical conductance results, it was inferred that lateral growth is limited by the triggering of spontaneous nucleation in the bulk liquid when the bulk liquid temperature drops below the nucleation temperature. In addition, lateral solidification velocity was found to be about 7 m/s. It was found that steeper fluence gradients which produce higher temperature in the molten silicon allow more time for lateral solidification to proceed until the bulk liquid temperature is sufficiently supercooled to trigger nucleation.

REFERENCES

1. H. Kuriyama, T. Honda, and S. Hakano, Jpn. J. Appl. Phys. **32**, 6190 (1993).
2. J. S. Im, H. J. Kim, Appl. Phys. Lett. **63**, 1969 (1993).
3. J. S. Im, R. S. Sposili, and M. A. Crowder, Appl. Phys. Lett. **70**, 3434 (1997).
4. K. Ishikawa, M. Ozawa, C. –H. Ho, and M. Matsumura, Jpn. J. Appl. Phys., Part 1 **37**, 731 (1998).
5. H. J. Kim, and J. S. Im, Appl. Phys. Lett. **68**, 1513 (1996).
6. C. –H. Oh, M. Ozawa, and M. Matsumura, Jpn. J. Appl. Phys., Part2 **37**, L492 (1998).
7. B. Rezek, C. E. Nebel, and M. Stutzmann, Jpn. J. Appl. Phys., Part 2 **38**, L1083 (1999).
8. M. Takahashi, T. Ogino, K. Suzuki, N. Hayashi, and others, J. of Photopolymer Sci. and Tech. **12**, No. 1, 89 (1999).
9. V. M. Galzov, S. N. Chizhevskays, and N. N. Glagoleva, Liquid Semiconductors (Plenum, New York, 1969).
10. M. Hatano, S. Moon, M. Lee, K. Suzuki, and C. P. Grigoropoulos, J. Appl. Phys. **87**, No. 1, 36 (2000).

TFT III:
Devices

Mat. Res. Soc. Symp. Proc. Vol. 621 © 2000 Materials Research Society

UNIFORM, HIGH PERFORMANCE POLY-SI TFTs FABRICATED BY LASER-CRYSTALLIZATION OF PECVD-GROWN a-SI:H

D. TOET,*,a T.W. SIGMON,a T. TAKEHARA,b C.C. TSAI,b,† W.R. HARSHBARGER.b
aLawrence Livermore National Laboratory, 7000 East Ave., L-271, Livermore CA 94550.
bAKT, 3101 Scott Blvd., M/S 9155, Santa Clara, CA 95054.
*e-mail: toet1@llnl.gov
†now at: Quanta Display, 188 Wen Hwa 2nd Rd., Kuei Shan Hsiang, Tao Yuan Shien, Taiwan.

ABSTRACT

Polycrystalline silicon thin film transistors (TFTs) were fabricated using laser crystallization of thin amorphous Si films grown by plasma-enhanced chemical vapor deposition. The films were exposed to a scanned XeCl excimer laser beam at 350 mJ/cm². At this fluence the Si film completely melted and crystallized in the form of uniformly distributed grains with an average size of 39 nm. One of the films was then subjected to a low fluence laser scan (250 mJ/cm²), which resulted in the melting of the top part of the film and lead to an increase in grain size. The TFTs fabricated without the partial melt method had good electrical properties and uniformities. The partial melt method lead to substantial improvements in most device characteristics, while the uniformity remained good.

INTRODUCTION

The on-panel integration of addressing and switching circuitry in active matrix liquid crystal displays (AMLCDs) will lead to a drastic reduction in production costs, as well as to improved ruggedness and performance [1]. Driver circuits, however, require a much higher performance than can be provided by amorphous silicon (a-Si), the material traditionally used for the fabrication of pixel switches. As a result, the past decade has seen extensive research into the development of low-temperature polycrystalline silicon (poly-Si) processes, which allow the manufacturing of high-mobility thin film transistors (TFTs). Poly-Si TFTs also provide sufficient currents to drive organic light-emitting diodes, enabling further improvements in display performance and ruggedness, as well as a reduction in the power consumption.

Short pulsed excimer laser-induced crystallization (ELC) of a-Si [2] is emerging as the leading technique for the fabrication of poly-Si TFTs [3]. This is mainly due to the fact that ELC is compatible with inexpensive low-temperature substrates such as plastic and glass. ELC involves ultrafast melting and resolidification of the near-surface region of the sample and consequently, minimal heating of the substrate [2,4]. Moreover, ELC yields poly-Si films consisting of defect-free grains that can reach sizes of several microns. In contrast, solid phase crystallization (SPC) [5], the most common alternative to ELC, requires temperatures above 600 °C, precluding the use of low-temperature substrates. Furthermore, SPC results in poly-Si films with a high density of intra-grain defects.

In this report, we present the characteristics of TFTs made by ELC of hydrogenated amorphous silicon (a-Si:H) thin films grown on glass substrates by plasma-enhanced chemical vapor deposition (PECVD). This technique, which is used by most AMLCD manufacturers, is not generally considered to be an obvious choice for the deposition of precursor films for ELC due the high hydrogen content (10%) of PECVD a-Si:H. Upon laser irradiation, hydrogen effuses explosively from the film, resulting in damage to the surface. However, we have recently developed a PECVD process yielding a-Si:H films that are compatible with ELC [6]. N-channel top-gate poly-Si TFTs of various sizes were fabricated in these films, using a scanning ELC procedure that involved a small number of exposures, was relatively insensitive to the laser fluence and did not require any patterning of the laser beam or the film. The electrical properties of the devices were correlated to the structural properties of the poly-Si films, determined by atomic force microscopy (AFM) and scanning electron microscopy (SEM).

EXPERIMENTAL DETAILS

The TFTs investigated in this work were fabricated on 4" Corning 1737 glass substrates, coated with a buffer layer consisting of a 100 nm thick SiN_x film and a 500 nm SiO_2 film, both grown by PECVD. This buffer layer prevents the diffusion of impurities from the glass into the Si and limits substrate heating during ELC. On top of the buffer layer, we deposited a 42 nm thick film of a-Si:H by PECVD of SiH_4, at a substrate temperature of 350 °C. PECVD-grown a-Si:H usually contains large amounts of hydrogen, typically 10 at.%. Since hydrogen tends to effuse explosively from the film upon laser irradiation, inducing damage to its surface, dehydrogenation of these films is normally required prior to ELC. However, the particular set of process conditions used during this deposition allowed us to obtain hydrogen concentrations below 3 at.%, as determined from infrared (IR) absorption spectroscopy. At these levels, films can be laser-crystallized without prior dehydrogenation, as we have shown in previous work [6].

The laser crystallization was performed using a XeCl excimer laser, providing pulses with a width of 35 ns and a wavelength of 308 nm (details about the laser process are given in the next section). An optical homogenizer was used to transform the incident beam into a 5×5 mm^2 flat-top profile with a fluence variation of ± 5%. The samples are placed in a vacuum cell evacuated to 10 mTorr. The crystallization process was monitored by measuring the time-resolved transmission of an IR laser ($\lambda = 1.5$ µm), as well as the time-resolved reflection and transmission of a HeNe laser ($\lambda = 632.8$ nm), through the center of the spot irradiated by the XeCl laser.

Seventy-five sets of 64 top-gate, self aligned n-channel TFTs with different sizes (W/L) were fabricated, on each wafer, in the laser-crystallized Si films, using a 4-step photolithography process. First, a 130 nm thick gate oxide was deposited on the Si by low temperature, low pressure CVD at 300 °C. Following this, 150 nm of metal was deposited onto the wafer by DC-sputtering at room temperature, and patterned to form the gate electrodes. After removal of the superfluous gate oxide, the Si was implanted to a dose of $2 \cdot 10^{15}$ arsenic atoms/cm^2 at 25 keV. These dopants were activated by scanning the XeCl laser over the Si at fluence of 310 mJ/cm^2, with 66% overlap between adjacent pulses. The laser pulses induce ultrafast melting and solidification of the Si, during which the As can diffuse to the lattice sites. Note that the implantation/activation process does not affect the Si in the channel, since it is masked by the metal gate. The device active areas were then patterned by photolithography and etched in a SF$_6$ plasma. Next, 450 nm of isolation oxide was deposited by PECVD at 250 °C. Using photolithography, contact holes were defined in this layer, following which a 800 nm thick Al film was sputter-deposited onto the wafer and patterned to define the device interconnects. Finally, the samples were annealed for 3 hrs. at 250 °C to improve the contact properties.

EXPERIMENTAL RESULTS AND DISCUSSION

Influence of the pulse energy on the microstructure of the laser crystallized films

Figure 1 a: *SEM micrograph of a 42 nm a-Si:H film a laser pulse with a fluence of 350 mJ/cm², well above FMT.*

b: *SEM micrograph of the same initial a-Si:H film after exposure to a pulse with a fluence of 310 mJ/cm², just below FMT*

The size of the grains in laser-crystallized films (and thus the device performance) is extremely sensitive to the energy fluence of the laser beam [4]. The critical fluence in the ELC process is the full-melt threshold (FMT), at which the film just melts throughout its entire thickness. For the 42 nm thick films used in this work, the FMT is about 310 mJ/cm². At fluences well below the FMT, the solidification process is triggered by explosive crystallization [7] and typically results in grains with sizes that are smaller than the film thickness. At fluences above FMT, the film is melted completely down to the substrate, removing the template for solidification. The liquid cools below its equilibrium melting point, until homogeneous nucleation [8] initiates crystallization, also leading to small grains. This is illustrated in figure 1a, which shows a SEM image of a 42 nm film annealed at 350 mJ/cm². This film consists of a uniform distribution of small grains with an average size of 39 nm. On the other hand, irradiation of an a-Si film near FMT may result in crystallites that are much larger than the film thickness. This is illustrated in figure 1b, which shows a film irradiated at 310 mJ/cm². The film displayed in this SEM image was obtained from the same starting material as in figure 1a, but consists of grains reaching sizes of 0.5 µm. This effect, referred to as super lateral growth (SLG), is a consequence of the molten material solidifying by lateral growth off discrete solid islands that were not melted by the laser pulse. The density of these islands, which are located at the film-substrate interface, is very low near FMT, and thus the resulting lateral growth may occur over large distances [4].

From the above discussion, it would seem natural to take advantage of the SLG mechanism to prepare films for poly-Si TFTs, since larger grains generally imply better device performance. However, the grain size in the SLG regime (near FMT) is very sensitive to the incident laser fluence, as shown in figure 2, which shows the fluence dependence of the grain size (as well as that of the RMS roughness) of the laser-crystallized films. Therefore, when operating in this regime, inhomogeneities in the laser profile and in the initial films, as well as pulse-to-pulse variations in the laser fluence, can lead to a wide distribution of grain sizes (see figure 1b). These non-uniformities can lead to unacceptable variations in device characteristics. Another important drawback of the SLG regime is that it yields very rough films, as illustrated in figure 2. At FMT, the RMS roughness of the films reaches 12 nm. This roughness induces a high density of trap states at the

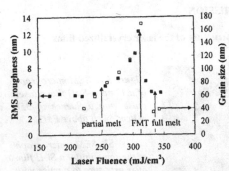

Figure 2: *Dependence of the RMS roughness (closed symbols) and the average grain size (open symbols) of laser-crystallized poly-Si films on the laser fluence. The data plotted in this graph corresponds to films that were initially irradiated above FMT, using a 350 mJ/cm² pulse.*

Si/gate oxide interface. Thermally activated conduction can occur through these states, implying high OFF currents. Laser fluences above FMT, on the other hand, yield much smoother films (RMS roughness: 5 nm). Since the grain size distribution in this regime is also more uniform than in the SLG regime (the grain size is basically insensitive to the fluence), we chose to fabricate TFTs in films annealed above FMT. This was done by exposing the films to a scanning laser beam at a fluence of 350 mJ/cm², using a 67% overlap between adjacent pulses (referred to as "full melt ELC"). One of the films was then partially re-melted using a low fluence laser scan (250 mJ/cm², 50% overlap - referred to as "full+partial melt ELC"). As can be seen in figure 2, this results in a doubling of the grain size with respect to the full melt case, while the roughness is slightly greater. Moreover, a more uniform grain size distribution is expected than at FMT.

Electrical properties of the laser crystallized poly-TFTs

We measured the electrical properties of a selected set of devices in 60 of the 75 die on both wafers using a computer-controlled probe station. Figure 3 compares typical transfer characteristics of TFTs with a gate width W and gate length L of 10 μm (W/L=10/10) obtained using both ELC conditions (i.e., full melt and full+partial melt). From these curves, device parameters including mobility, threshold voltage, subthreshold slope and OFF current were determined. The distribution of these parameters over the wafer is shown in figure 4 for W/L=20/40 TFTs. Figure 4 reveals that TFTs made by full melt ELC have OFF currents below 0.5 pA per μm of gate width, while the other aspects of their performance are not as remarkable (median mobility: 48 cm²/V·s, threshold voltage: 3.5 V and subthreshold slope: 0.87 V/dec.). The partial melt process, on the other hand, leads to a drastic improvement in the performance of the devices: the median mobility increases to 141 cm²/V·s,

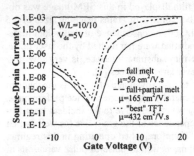

Figure 3: *Transfer characteristics of W/L= 10/10 TFTs made by full melt (solid line) and full+ partial melt ELC (dashed line). The dotted line shows the highest mobility TFT.*

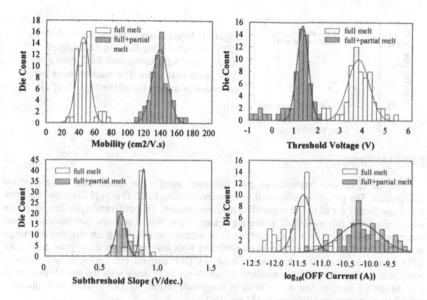

Figure 4: *Histograms showing the distribution of the mobility, threshold voltage, sub-threshold slope and OFF current for W/L=20/40 TFTs, fabricated using full melt ELC (white bars) and full+partial melt ELC (gray bars). The solid lines are Gaussian fits to the distributions. The measurements were performed at a source-drain voltage V_{ds} of 5V.*

while the median threshold voltage and subthreshold slope decreases to 1.1 V and 0.7 V/dec, respectively. Furthermore, the ON-current (not shown) increases from 0.1 mA to 0.5 mA upon partial melting. These changes are attributed to the increase in grain size caused by the partial melt ELC. The increase in the median OFF-current to 5 pA/μm is believed to be related to the slight increase in surface roughness caused by this process.

The properties of the devices fabricated with the two ELC processes are quite uniform, as shown in figure 4. The spread (i.e., the full width at half maximum of a Gaussian fit to the distribution) in the mobility values is about 18 cm²/V·s in the full melt case and 24 cm²/V·s in the partial melt case, for the threshold voltage, the spreads are 1.0 V (full melt) and 0.6 V (full+partial melt), and for the subthreshold slope, the spreads are 0.06 V/dec. (full melt) and 0.11 V/dec. (full+partial melt). These results are in agreement with our expectations based on the fluence dependence of the grain size and are comparable to data reported by other groups [3,9].

The dependence of the median mobility of the TFTs on the gate width is shown in figure 5. For TFTs made using full melt ELC, the median mobility is roughly independent of the device size and has a value of about 45 cm²/V·s. For those made using partial melt ELC, on the other hand, the median mobility decreases from 185 cm²/V·s for W/L=10/10 to 110 cm²/V·s for W/L=160/40. This dependence may be related to the increase of the amount of grain boundary material in the channel. The other device parameters did not depend strongly on the device size.

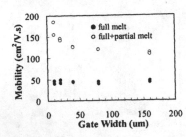

Figure 5: Dependence of the median mobility of TFTs fabricated using full melt ELC (closed symbols) and full+partial melt ELC (open symbols) on the gate width. The measurements were performed at a source-drain voltage V_{ds} of 5V.

CONCLUSIONS

Poly-Si TFTs were fabricated on glass substrates using a low temperature self-aligned top gate process. The poly-Si was obtained by laser-crystallizing PECVD a-Si:H films that were prepared specifically for ELC and required no dehydrogenation. Two different crystallization procedures were used: full melt ELC, producing smooth films with a uniform distribution of small grains, and full+partial melt ELC, resulting in an increase in both the grain size and the roughness. The electrical properties of devices fabricated with both processes show good uniformity. TFTs made using the full melt procedure have low OFF currents (<0.5 pA/μm) and otherwise acceptable performance (median mobility: 45 cm²/V·s). TFTs made with the partial melt procedure have OFF currents that are an order of magnitude larger, but their overall performance is much better (median mobilities up to 185 cm²/V·s).

ACKNOWLEDGEMENTS

This work was performed under the auspices of the U.S. Department of Energy by University of California Lawrence Livermore National Laboratory under contract No. W-7405-Eng-48.

REFERENCES

[1] T. Serikawa, S. Shirai, A. Okamoto and S. Suyama, IEEE Trans. Electron Devices **36**, 1929 (1989).
[2] T. Sameshima, S. Usui, and M. Sekiya, IEEE Electron Device Lett. **7**, 176 (1986).
[3] S. D. Brotherton, D. J. McCulloch, J. P. Gowers, J. R. Ayres and M. J. Trainor, J Appl. Phys. **82**, 4086 (1997).
[4] J. S . Im, H. J. Kim, and M. O. Thompson, Appl. Phys. Lett. **63**, 1969 (1993).
[5] T. Voutsas and M. Hatalis, J. Appl. Phys. **76**, 777 (1994).
[6] D. Toet, P.M. Smith, T. W. Sigmon, T. Takehara, C. C. Tsai, W. R. Harshbarger, and M. O. Thompson, J. Appl. Phys. **85**, 7914 (1999).
[7] M. O. Thompson, G. J. Galvin, J. W. Mayer, P. S. Peercy, J. M. Poate, D. C. Jacobson, A. G. Cullis, and N. G. Chew, Phys. Rev. Lett. **52**, 2360 (1994).
[8] S. R. Stiffler and M. O. Thompson, Phys. Rev. Lett. **60**, 2519 (1988).
[9] R. T. Fulks, J. B. Boyce, J. Ho, G. A. Davis, and V. Aebi, Mat. Res. Symp. Proc. **557**, 623 (1999).

Mat. Res. Soc. Symp. Proc. Vol. 621 © 2000 Materials Research Society

Low Temperature Polycrystalline Si TFTs Fabricated with Directionally Crystallized Si Film

Y.H. Jung, J.M. Yoon, M.S. Yang, W.K. Park, H.S. Soh, Anyang Laboratory, LG-Philips LCD Co. Ltd, Dongan-gu, Anyang-shi, Kyungki-do, KOREA; H.S. Cho, A.B. Limanov, and J.S. Im, Program in Materials Science, Columbia University, New York, NY 10027

ABSTRACT

The comparison of TFTs fabricated on films processed by conventional excimer laser annealing (ELA) and sequential lateral solidification (SLS) demonstrates the dependence of the device characteristics on the microstructure of the device channel region. We report the performance characteristics of non-self-aligned coplanar n- and p-channel low temperature TFTs fabricated on 1000-Å-thick films on Corning 1737 glass substrates that were directionally solidified using SLS. The devices were aligned so that the grain boundaries were parallel to the direction of the source-drain current flow. These results were compared with those obtained from devices fabricated on conventional ELA-processed polycrystalline Si films (with average grain size of ~3000 Å) with identical methods. The values for channel mobility obtained from the SLS TFTs are ~370 cm^2/Vsec for n-channel and ~140 cm^2/Vsec for p-channel devices, compared to ~100 and ~60 respectively for ELA TFTs. Other device characteristics of SLS TFTs were $I_{on}/I_{off} > 10^7$ at Vd=0.1V, and subthreshold slopes less than 0.5V/dec. We further discuss the physical implications of the results and present additional details of the devices.

INTRODUCTION

In order to manufacture TFT-LCD panels incorporating both gate drivers and data drivers, TFT performance is required to exceed that obtained by conventional excimer laser annealing (ELA) methods. The channel mobilities obtained for TFTs fabricated on conventionally crystallized polycrystalline Si films are in the range of ~140 cm^2/Vsec for n-channel TFTs and ~60 cm^2/Vsec for p-channel TFTs; these are sufficient for application in simple circuits such as gate drivers, but insufficient for more complicated circuits such as data drivers. For this reason, it is required that separate modules for data driver IC's - which are fabricated using conventional monolithic semiconductor processes - be attached to most polycrystalline Si TFT-LCD panels developed to date, requiring an additional step in the manufacturing process. It is therefore desirable to incorporate all circuits involved in the operation of the panel on the panel, and furthermore realize incorporation of an entire integrated system on the same panel. For this, TFTs with better electrical properties must be developed.

Currently, one of the most significant frontiers in the development of electrically superior TFTs is the improvement of the microstructure in the active channel region. It is well known that the presence of grain boundaries detrimentally affects the carrier mobility in a semiconductor film, due to carrier trapping at the dangling bonds and formation of a potential barrier at the grain boundary [1]. The most ideal material for the active channel region would therefore be a single crystal, and reduction of grain boundaries in the channel region is desired in general [2].

In this study, n-and p-type TFTs were fabricated on thin Si films that were directionally solidified using the sequential lateral solidification (SLS) method [3-5], and their electrical characteristics were analyzed. These were compared with results obtained from TFTs fabricated on

Figure 1. SEM pictures of typical micro-structures obtained from (a) SLS (directionally solidified) and (b) conventional ELA crystallization (equiaxed polycrystalline structure). Pictures are not shown to scale.

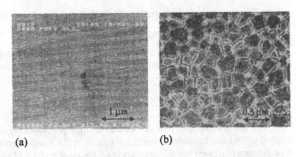

(a)   (b)

conventional ELA-crystallized films. Also, pre-SLS doping was conducted to study its effects on the SLS process and to investigate the role of SLS in activation of dopants.

EXPERIMENT

Coplanar n- and p-channel TFTs were fabricated on Corning 1737 glass panels; all processes, including dehydrogenation annealing, were conducted at temperatures below 400°C. The devices were fabricated by the following sequence of processes. First, an SiO₂ buffer layer 3000-Å-thick was deposited by PECVD on the glass substrate. Then, the 1000 Å precursor a-Si film - the active layer - was deposited and the active channels were patterned as islands of a-Si. Selected source and drain regions were doped with n- or p-type doses before the crystallization step. SLS crystallization served to activate dopants in the pre-doped samples. Comparison was made with devices of which the source and drain regions were doped and activated with a conventional ELA method after SLS crystallization.

The active channel islands were crystallized by the SLS method using a linearly-shaped pulsed excimer laser beam [4] (Lambda Physik LPX 315i, 308 nm XeCl). In this method, each pulse melted a linear region ~2 μm wide and several cm long, and was translated ~0.5 μm relative to the previous pulse in the direction perpendicular to the line of the beam. The beam was shaped by passing it through a long slit and then projecting it onto the target film using a cylindrical lens. Figure 1(a) is an example of the microstructure that results from SLS with such a linear beamlet. Long grains form parallel to the direction of translation of the beamlet, which is typical of the SLS process.

The orientation of the TFT channels was arranged so that it would present the least number of grain boundaries along the path of current flow - that is, so that the direction of current flow would be parallel to the elongated grains (and the grain boundaries). For comparison, devices were fabricated on films crystallized using the conventional ELA method as well, under identical conditions as the SLS processed films, except for the doping and activation steps. Figure 1(b) shows the typical equiaxed polycrystalline microstructure obtained from conventional ELA processes.

Afterwards, the gate insulator, a 1000-Å-thick SiO₂ layer, was deposited, and metallic gate contacts were formed by deposition and patterning of Al and Mo. The patterning was performed so that there would be no overlap of the gate electrode with the source or the drain regions. The alignment was confirmed by reversing the roles of the source and drain electrodes during the testing of the finished TFTs. The results of the reversed tests were identical to those of the normal tests, showing that there was no overlap of gate and source/drain regions - i.e., the devices were symmetrically aligned. Doping and conventional ELA activation of source and drain regions were conducted at this stage for comparison with the pre-doped, SLS-activated, devices. An SiO_2 film and a SiN_x film were deposited in succession to form a passivation layer. After patterning of a contact hole, ITO film was deposited and patterned to form the gate, source, and drain electrodes, completing the TFT fabrication.

Using a 4-point probe method, we measured the sheet resistances of source and drain regions formed depending on the sequence of crystallization and activation, and studied their effects on TFT performance. A sheet resistance of 160 ohm/cm^2 was obtained for samples that were pre-doped with a p-type dose then directionally solidified by the SLS process. For samples SLS-processed then doped and activated using conventional ELA, values of 200 - 450 ohm/cm^2 were obtained depending on the laser energy density used. TFT performance characteristics were found to depend on the sheet resistances measured.

Device performance parameters were measured for SLS-crystallized devices of various dimensions, both n and p-type, and compared to those of analogous ELA-crystallized devices. In addition to the basic device characteristics (I-V curves, mobility, leakage current etc.), measurements on the effects of hot carrier stress were conducted. In the subsequent sections, Vd, Vg, and Id refer to drain voltage, gate voltage, and drain current, respectively.

RESULTS

Figure 2 shows TFT output curves and performance characteristics for (a) n and (b) p-channel devices fabricated on directionally solidified crystalline Si films, compared with those built on conventional ELA-crystallized poly-Si films (dotted line). The channel width and length were both 100 μm. The mobilities obtained were 370 cm^2/Vsec for n-type devices, and 130cm^2/Vsec for p-type devices. These values are roughly 2.5 times greater than those obtained from TFTs on conventional ELA-processed films, and sufficient for the operation of data circuits and other high-performance devices.

As the dopant implantation and ELA activation processes are conducted after self-aligned gate oxide and metal deposition, and affect only the source and drain regions, the directionally solidified microstructure in the channel region (Fig.1a) remained unchanged under post-crystallization doping and ELA activation.

However, the microstructure of the exposed source and drain regions were strongly affected with post-crystallization doping and activation. Fig. 3(a) and (b) show the changes undergone by directionally solidified films that were subsequently doped, then activated by a conventional ELA method, at laser energy densities of 260 mJ/cm^2 and 360 mJ/cm^2, respectively. The change from directional crystalline microstructure to small-grained and equiaxed microstructure is due to activation laser energies above the complete melting threshold. The sheet resistances obtained from films (a) and (b) were 160 ohm/cm^2, and 200 ohm/cm^2 respectively, indicating that the activation of dopants in film (a) was not sufficient, though the directional crystallinity was maintained, while dopants in film (b) were completely activated.

Figure 4 shows a comparison of the device output characteristics for TFTs fabricated on the post-SLS doped and ELA-activated (with a laser energy density of 260 mJ/cm^2) film and on the pre-doped and SLS-activated devices. The performance is similar despite the slight difference in the microstructures of the source and drain regions, which result from ELA activation at energy densities below the complete melting threshold.

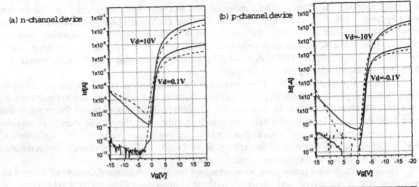

(a)	Solid Line	Dashed Line
Solidification Method	SLS Directional Solidification	Conventional ELA
Mobility	290cm²/V.sec	120cm²/V.sec
Vth	1.2V	2.6V
S-factor	0.41V/dec	0.44V/dec

(b)	Solid Line	Dashed Line
Solidification Method	SLS Directional Solidification	Conventional ELA
Mobility	140cm²/V.sec	70cm²/V.sec
Vth	-3.4V	-2.8V
S-factor	0.46V/dec	0.41V/dec

Figure 2. Id-Vg curves and chart of device performance characteristics from (a) n-channel and (b) p-channel TFTs (width/length = 10µm/20µm) from SLS-processed and conventional ELA-processed Si films.

Figure 5 is a comparison of hot carrier stress characteristics of n-channel TFTs fabricated on SLS-processed material (5-a) and on conventional ELA-processed material (5-b). The TFTs on SLS-processed material were found to have less decrease in channel mobility due to hot carrier stress. The post-hot carrier stress reduction in mobility in SLS-processed n-channel TFTs was from 280cm²/Vsec to 176cm²/Vsec, while the reduction in conventional ELA-processed TFTs was 120cm²/Vsec to 25cm²/Vsec.

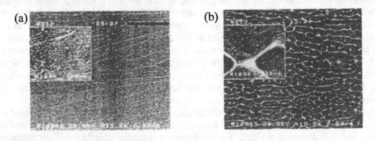

Figure 3. SEM pictures of SLS-processed films after doping and conventional ELA activation at laser energy densities of (a) 260mJ/cm² and (b) 360mJ/cm².

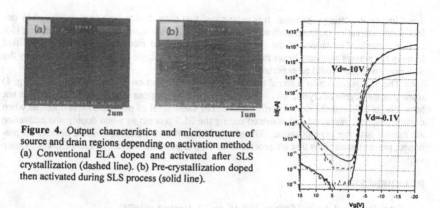

Figure 4. Output characteristics and microstructure of source and drain regions depending on activation method. (a) Conventional ELA doped and activated after SLS crystallization (dashed line). (b) Pre-crystallization doped then activated during SLS process (solid line).

DISCUSSION AND SUMMARY

As a result of this study, we were able to compare the electrical properties of SLS-processed directionally crystallized Si film with those of conventional ELA-processed poly-Si film, by building TFTs on the films and obtaining the performance characteristics. The devices built on SLS-processed films show clear advantages in terms of channel mobilities and hot carrier stress-related reliability. The improvement in channel mobility is most likely due to the reduction in the number of grain boundaries in the channel region. The greater tolerance to hot carrier stress in devices using SLS directionally solidified material can be considered a result of decreased grain-boundary roughness in the material.

The SLS-crystallized devices showed channel mobilities that allow operation of high-performance devices such as those used in data driver circuits, enabling their incorporation onto

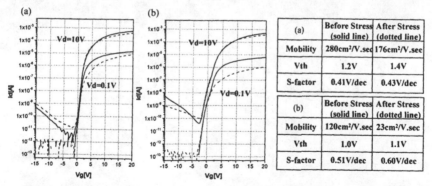

(a)	Before Stress (solid line)	After Stress (dotted line)
Mobility	280cm²/V.sec	176cm²/V.sec
Vth	1.2V	1.4V
S-factor	0.41V/dec	0.43V/dec

(b)	Before Stress (solid line)	After Stress (dotted line)
Mobility	120cm²/V.sec	23cm²/V.sec
Vth	1.0V	1.1V
S-factor	0.51V/dec	0.60V/dec

Figure 5. Hot-carrier stress performance of (a) SLS-crystallized and (b) conventional ELA-crystallized TFTs. Stress conditions: Vg=Vth, Vd=20V, Stress time = 1 hour

thin-films on insulating substrates. It is to be noted that these devices were not optimized by defect-state passivation methods such as hydrogenation (for bulk states) or N_2/H_2 gas forming (for surface states). The values of the channel mobility obtained from the directionally solidified films in the present study are lower than those obtained from single crystal regions formed by the SLS process [6], as would be expected.

The SLS process also is found to have an activation effect on dopants. The results (Fig. 4) show that the degree of crystallinity in the source and drain regions after ELA activation does not have a large effect on the device performance. The same level of performance is obtained when dopants are pre- implanted and activated during the SLS process as when doping and activation are carried out after crystallization. Therefore, fabrication of high-performance devices on pre-doped, pre-patterned films that are processed by SLS is possible without an extra ELA activation step.

REFERENCES

1. J. Seto, *Journal of Applied Physics*, Vol 46, no.12, December 1975
2. J.S. Im and R.S. Sposili, *Crystalline Si Films for Integrated Active Matrix Liquid Crystal Displays (review article), MRS Bulletin,* March 1996
3. R.S. Sposili and J.S. Im, *Applied Physics Letters*, Vol. 69, no.19, (2864) 1996
4. A.B. Limanov, V.M. Borisov, A.Yu.Vinokhodov, A.I. Demin, A.I. El tsov, Yu.B. Kiirukhin, O.B. Khristoforov, Development of Linear Sequential Lateral Solidification Technique to Fabricate Quasi-Single-Crystal Super-Thin Si Films for High-Performance Thin Film Transistor Devices, *Perspectives, Science, and Technologies for Novel Silicon on Insulator Devices, 55-61, P.L.F. Hemment et al. © 2000 Kluwer Academic Publishers*
5. R.B. Bergmann, J. Koehler, R. Dassow, C. Zaczek, J.H. Werner, *Physica Status Solidi A*, vol. 166, no.2, pp. 587-602, April 1998
6. M.A. Crowder, P.G. Carey, P.M. Smith, R.S. Sposili, H.S. Cho, and J.S. Im, *IEEE Electron Device Letters*, Vol. 19, no. 8, (306), August 1998

Mat. Res. Soc. Symp. Proc. Vol. 621 © 2000 Materials Research Society

Low Temperature Laser-Doping Process Using PSG and BSG Films for Poly-Si TFTs

Cheon-Hong Kim, Sang-Hoon Jung, Jae-Hong Jeon and Min-Koo Han
School of Electrical Engineering, Seoul National University,
San 56-1 Shinlim-dong, Kwanak-gu, Seoul 151-742, Korea

ABSTRACT

A simple low-temperature excimer-laser doping process employing phosphosilicate glass (PSG) and borosilicate glass (BSG) films as dopant sources is proposed in order to form source and drain regions for polycrystalline silicon thin film transistors (poly-Si TFTs). We have successfully controlled sheet resistance and dopant depth profile of doped poly-Si films by varying PH_3/SiH_4 flow ratio, laser energy density and the number of laser pulses. The penetration depth and the surface concentration of dopants were increased with increasing laser energy density and the number of laser pulses. The minimum sheet resistance of $450\Omega/\square$ for phosphorus (P) doping and $1100\Omega/\square$ for boron (B) doping were successfully obtained. Our experimental results show that the proposed laser-doping process is suitable for source/drain formation of poly-Si TFTs.

INTRODUCTION

Polycrystalline silicon thin film transistors (poly-Si TFTs) employing the excimer laser annealing have attracted a considerable attention for high quality active matrix liquid crystal displays (AMLCDs) [1]. Although excimer laser annealing ensures high-quality poly-Si films, but the low-temperature doping method for source and drain regions is not well established yet. The ion implantation and the ion shower doping method require high-cost complex equipments and the post-annealing is also required in order to activate the dopants and to cure implant damages.

The excimer-laser doping using doping gas sources or dopant-containing films is a useful method because it is a simple and low-temperature doping process [2-5]. In such processes, excimer laser irradiation melts an a-Si layer and results in liquid-phase incorporation of dopants. In case of laser-doping using doping gas, sophisticated gas chamber and control system are required [4,5]. Furthermore, it is difficult to accurately control the sheet resistance and junction depth because dopants are supplied mainly from the absorbed layers formed either prior to the laser irradiation or during the irradiation [5].

The purpose of our work is to report a simple low-temperature laser-doping method employing phosphosilicate glass (PSG) and borosilicate glass (BSG) films as dopant sources. PSG and BSG films have been deposited by chemical vapor deposition (CVD) method, which is compatible to a conventional poly-Si TFTs process, easy to achieve uniformity and reproducible. We have controlled sheet resistance and depth profile of doped poly-Si films by varying PH_3/SiH_4 flow ratio, as well as laser energy density and the number of laser pulses. Sheet resistance was measured and dopant depth profile was investigated by spreading resistance profiling (SRP) and Hall measurement.

EXPERIMENTS AND RESULTS

A 100nm-thick hydrogenated amorphous silicon (a-Si:H) film was deposited by plasma-enhanced CVD (PECVD) at 280°C on the oxidized Si substrate. To prevent the rapid evolution of hydrogen atoms during the laser irradiation, the a-Si:H film was dehydrogenated at 450°C for 3 hours prior to laser annealing. As a diffusion source of phosphorus atoms, a 50nm-thick PSG film was deposited on it by atmospheric pressure CVD (APCVD) using SiH_4, PH_3 and O_2 gas at 375°C. The dopant concentration in the PSG film was controlled by varying PH_3/SiH_4 flow ratio at the fixed flow rate of O_2 gas. On some samples, a BSG film was also deposited as a diffusion source of boron atoms. Then, the samples were irradiated by 20ns-pulse XeCl excimer laser (λ=308nm). We have varied laser energy density and the number of laser pulses in order to controll sheet resistance and dopant depth profile of doped poly-Si films.

During the laser irradiation, the most of laser energy is absorbed in the underlying a-Si film because the upper PSG and BSG films have little absorption capability at around 300nm [2]. The a-Si surface is melted and the dopants diffuse from the dopant-containing film into the molten a-Si so that the doped poly-Si film is successfully formed.

Four-point-probe technique was used to investigate sheet resistance for the doped poly-Si films. Figure 1 shows the sheet resistance for phosphorus doping as a function of laser energy density. A PSG film was prepared with PH_3/SiH_4 flow ratio of 0.38 and O_2 flow rate of 3000sccm. It is noted that the sheet resistance decreases with increasing laser energy density and the number of laser pulses. The sheet resistance of less than $1k\Omega/\square$ was obtained when laser energy density was larger than $250mJ/cm^2$ with 10pulses.

The sheet resistance of phosphorus-doped poly-Si films for various PH_3/SiH_4 flow ratios and boron-doped poly-Si films versus laser energy density are shown in figure 2. The number of laser pulses was 10. The sheet resistance is controlled by PH_3/SiH_4 flow ratio and both phosphorus (P) doping and boron (B) doping are possible with this technology. The difference between P and B doping is due to the difference of mobility and dopant concentration in dopant-containing films. The minimum sheet resistance of $450\Omega/\square$ for phosphorus (P) doping and $1100\Omega/\square$ for boron (B) doping were successfully obtained. These values are low enough for source/drain formation of poly-Si TFTs.

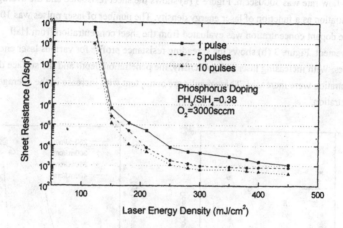

Figure 1. Sheet resistance for phosphorus doping as a function of laser energy density.

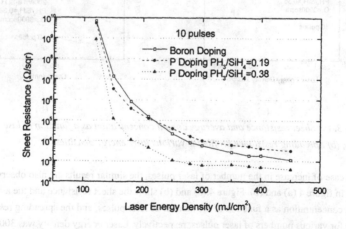

Figure 2. Sheet resistance for phosphorus doping with various PH_3/SiH_4 flow ratios and sheet resistance for boron doping as a function of laser energy density.

Dopant depth profile of phosphorus dopants was investigated by spreading resistance profiling (SRP). The thickness of a-Si:H films was 200nm and PH_3/SiH_4 flow ratio was 0.38 and O_2 flow rate was 3000sccm. Figure 3 (a) shows the sheet resistance and the average dopant concentration as a function of laser energy density. The number of laser pulses was 10. The average dopant concentration was evaluated from the sheet concentration from Hall measurement. Figure 3 (b) shows the spreading resistance profile for various laser energy densities. With increasing laser energy density, the penetration depth and the surface dopant concentration were increased. These results are consistent with increase in the average dopant concentration.

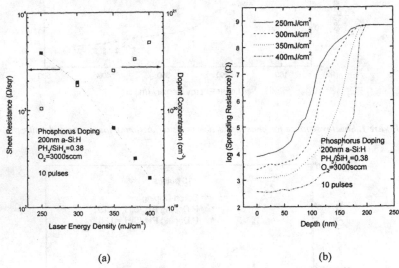

(a)   (b)

Figure 3. (a) Sheet resistance and average dopant concentration as a function of laser energy density, (b) Spreading resistance profile for various laser energy densities.

In case of increase in the number of laser pulses, the similar results are also observed, as shown in figure 4 (a) and (b). Figure 4 (a) and (b) show the sheet resistance and the average dopant concentration as a function of the number of laser pulses, and the spreading resistance profile for various numbers of laser pulses, respectively. Laser energy density was 300mJ/cm². It should be noted that the penetration depth and the surface concentration are less dependent on the number of laser pulses than the laser energy density.

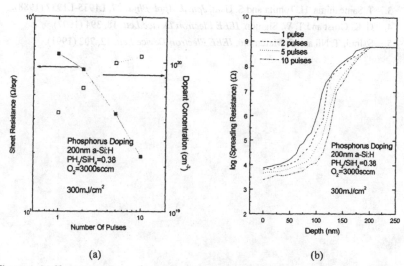

Figure 4. (a) Sheet resistance and average dopant concentration versus the number of laser pulses, (b) Spreading resistance profile for various numbers of laser pulses.

CONCLUSIONS

A simple low-temperature excimer-laser doping process employing phosphosilicate glass (PSG) and borosilicate glass (BSG) films as dopant sources is proposed. We have successfully controlled sheet resistance and dopant depth profile of doped poly-Si films by varying PH_3/SiH_4 flow ratio, laser energy density and the number of laser pulses. The penetration depth and the surface concentration of dopants were increased with increasing laser energy density and the number of laser pulses. The minimum sheet resistance of $450\Omega/\square$ for phosphorus (P) doping and $1100\Omega/\square$ for boron (B) doping were successfully obtained. The proposed laser-doping technique using PSG and BSG films is promising for low temperature poly-Si TFT process.

REFERENCES

1. M. Hack, P. Mei, R. Lujan and A. G. Lewis, *JNCS* 164-166, 727-730 (1993).
2. K. Sera, F. Okumura, S. Kaneko, S. Itoh, K. Hotta and H. Hoshino, *J. Appl. Phys.* 67, 2359, (1990).

3. T. Sameshima, H. Tomita and S. Usui, *Jpn. J. Appl. Phys.* 27, L1935-L1937 (1988).
4. G. K. Guist and T. W. Sigmon, *IEEE Electron Device Lett.* 18, 394 (1997).
5. S. Inui, T. Nii and S. Matumoto, *IEEE Electron Device Lett.* 12, 702 (1991).

Mat. Res. Soc.Symp. Proc. Vol. 621 © 2000 Materials Research Society

High Voltage Effects In Top Gate Amorphous Silicon Thin Film Transistors

N. Tosic[1], F. G. Kuper[1,2] and T. Mouthaan[1]
[1] University of Twente, MESA+ Institute,
P.O. Box 217, 7500 AE Enschede, The Netherlands
[2] Philips Semiconductors, MOS4YOU, Nijmegen, The Netherlands

ABSTRACT

In this paper, an analysis of the high voltage induced degradation in top gate amorphous silicon Thin Film Transistors (TFT) will be shown, including the aspect of self-heating. It will be shown through experimental results that the degradation level under high voltages on drain and gate is different for TFT's with different channel lengths. In addition, the temperature distribution over the TFT area for devices with different channel length is simulated. Simulation shows that the peak of temperature distribution is located at the drain/channel edge and that level of thermal heating depends on the channel length.

INTRODUCTION

The degradation induced by high voltages and electrostatic discharge (ESD) is one of the most critical reasons for lower production yield in display manufacturing.

In our previous work [1], considering the effect of the electrostatic discharge on a TFT, we have shown that even a low ESD zap degrades TFT's and a high ESD voltage induces electrostatic breakdown. The degradation was shown through the threshold voltage and sub - threshold slope change, and it was attributed to creation of defect-states at amorphous Si /gate dielectric interface. The creation of defects was observed in the region where electrical field was higher than a critical value. It was also shown that hard breakdown is not due to this pre - breakdown degradation. In addition, it was noticed that breakdown voltage depends on the channel length. However, the physical origin of this dependence between breakdown voltage and channel length was not clear and it will be considered in this paper.

According to Tada et al. [2], under weak ESD whose voltage is not so high as breakdown voltage, a "soft" failure of a-Si TFT is noticed and is caused by non-adiabatic heating of the channel. In this paper, we will show that level of this thermal heating is correlated with the size of the channel length.

EXPERIMENTAL

Tested devices were top-gate amorphous silicon thin film transistors. A number of devices with different channel dimensions was used in the experimental. The channel width (W=100 μm) was constant, but the channel length was varied (L=4, 6, 10 and 100 μm).

Series of high voltage stresses were applied in the following way: during each stress serial, the gate voltage V_G was kept constant, and the drain voltage was increased from 0 to 90V with step 10V. For each drain voltage measurement was repeated 10 times, with time interval 1 s. The measurement is repeated 10 times in order to show if there is thermal degradation. Stress time was short (~ 4μs). Four serials of measurements were carried-out. In each series the gate voltage was increased.

In Fig. 1. are shown measurements of I_D over time in the third series of measurements for a TFT with channel length L=10μm. For higher drain voltages (>50V), I_D firstly increases, and later decreases. The first point represents a current burst due to joule heating, but later a process of "self-cooling" sets-in. This lowering of the current is due to threshold voltage degradation. Therefore, the degradation of current in every next time steps is much less steep then in the first second.

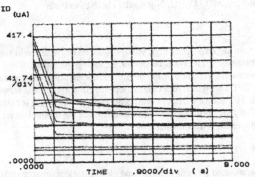

Figure 1. *Thermal heating effect on I_D monitored on a TFT with W/L=100μm/10μm. Biasing: V_G=80V, V_D=0:90V (step=10V)*

In Fig. 2. is shown I_D over time for the forth series of measurements for a TFT with channel length L=100μm. In contrast to Fig. 1, even for higher drain voltages, I_D current stays constant in time. It means that joule heating is very low in devices with a longer channel so the process of thermal induced degradation in not started.

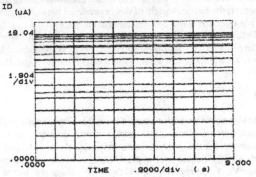

Figure 2. *Characteristics of a TFT with W/L=100μm/100μm showing an absence of thermal heating. Biasing: V_G=100V, V_D=0:90V (step=10V)*

Before the catastrophic failure occurs, the devices are heavily degraded, showing threshold voltage increase. In order to follow this degradation, after each stress serial the transfer characteristics $I_D(V_G)$ were measured. In Fig. 3 and 4 are shown transfer characteristics for TFT's with L=10 and 100 µm, respectively. It is shown that in a TFT with L=100µm, in the second stress serial, a shift of the transfer characteristics, although much smaller than in shorter devices, is measured. As this shift could not be recognized in the time measurements (Fig. 2), it can be concluded that this shift is induced by other reason than joule heating. As it was not be detected in the time measurements, it is possible that this process is very fast so it happens already in the first measurements in a $I_D(t)$ curve for a certain drain voltage. It can be concluded that, in principle, two degradation mechanisms contribute to deterioration of a-Si TFT's under high biasing conditions. The one found only in the shorter devices is thermally activated and another one, which is also recognized in the long channel device, is not.

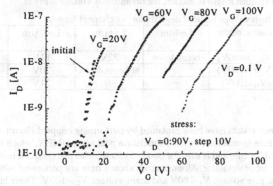

Figure 3. *Pre-breakdown degradation of transfer characteristic in TFT with W/L=100µm/10µm*

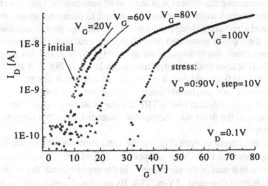

Figure 4. *Pre-breakdown degradation of transfer characteristic in a TFT with W/L=100µm/100µm*

Further, the sub-threshold slope shown in the Fig. 3 and 4 gives an additional indication that two degradation mechanisms exist. In long channel device (Fig. 4) the sub-threshold slope is constant. It implies that shift of the transfer characteristic is only due to charge trapping in the gate insulator. It can be concluded from the stability of the drain current shown in Fig. 2 that it is not thermally activated process. In shorter devices (Fig. 3), the sub-threshold slope changes after each stress serial, implying that both degradation processes are present: charge trapping in the gate dielectric and creation of defect states in amorphous silicon/gate dielectric interface. This process is due to thermal heating of the short-channel devices.

An overview of the pre-breakdown degradation for the devices with four different channel lengths and with a constant channel width (W=100μm) is given in the Table 1. It is shown that in TFT's with a shorter channel length (<100μm) strong thermal heating exists under high voltages on drain and gate. This thermal effects at the end lead into electrical breakdown of devices. So, if the channel length is shorter, the breakdown voltage is lower.

Table I. Level of high voltage induced degradation vs channel length

shift of transfer characteristic after:	L=4μm	L=6μm	L=10μm	L=100μm
2. stress series (V_G=50/60V)	breakdown	32V	11V	2V
3. stress series (V_G=70/80V)		at breakdown	20V	8V
4. stress series (V_G=100V)			21V	16V

SIMULATIONS

Experimental results have been explained by performing coupled electro-thermal simulations. For these simulations we used simulation package LUMEX, which was developed at the University of Twente and is used for coupled electro-thermal simulations as a pre-processor for the circuit simulator. All simulations shown here are performed with the same biasing conditions: gate voltage V_G= 60V and drain voltage V_D= 100V. These biasing conditions are similar to the conditions applied in the 2. stress series in the experimental.

The area surrounding the devices was same in all simulations. Only the channel length was varied, while the channel width was constant, as well as drain and source metal contacts widths. The simulated structure is designed according to the top-gate device built on a glass substrate and from the top side passivated over all device area, used in the experimental procedure. The temperature distribution shown in the Figs. 5, 6, 7, and 8 is temperature of the glass substrate in the first layer under amorphous silicon layer. In Fig. 5, 6 and 7 is in the channel length area on x-axis used finer grid than it is for the surrounding substrate. In Fig. 8 the grid was uniform almost all over x-axis, as channel length was too long (L=100μm).

The temperature distribution over a TFT is same in all four cases. The temperature peak is located always close to the drain side. Across the channel length, the temperature decreases from drain to source.

For the channel length of 4μm (Fig. 5), the simulation shows that the TFT is overheated. The simulated temperature so high that is not possible (ΔT=2400°C), so it implies that device is already broken down, as it was in the experiment. In the experimental, the device with L=6 μm is in the second serial heavily degraded (ΔV_T=32V). Its simulated temperature is relatively high (ΔT=120°C), as shown in Fig. 6.. The device with L= 10μm (Fig. 7) shows lower heating (ΔT=28°C). Finally, the device with the length 100μm as shown in Fig. 8 is not heated (<1 °C).

CONCLUSIONS

In this paper is shown that electrical properties are strongly influenced by high voltages applied on drain and gate. This influence is enhanced by thermal heating. The experimental suggested that two degradation mechanisms play a role in the pre-breakdown deterioration: charge trapping and defect creation. The former mechanism is not, but the later mechanism is thermally activated. Therefore, in long channel devices, which have much less thermal heating under high voltages, only charge trapping mechanism sets-in. In shorter devices, where thermal heating is harder, defect creation has a dominant role in deterioration mechanism. For that reason, breakdown voltage is directly proportional with the channel length.

REFERENCES

1. N. Tosic, F.G. Kuper, T. Mouthaan, "Transmission line model testing of top-gate amorphous silicon thin-film transistors", *Proc. of IRPS 2000 Conference.*
2. M. Tada, S. Uchikoga, M. Ikeda, "Power-density-dependent failure of amorphous Si TFT", Proc. of AM-LCD '96, pp. 269-272.

ACKNOWLEDGEMENT

This work is supported by the Technology Foundation STW under the project number TEL.3932. The authors wish to thank Jeremy Sandoe, John Hughes and Steve Deane, from Philips Research Laboratories, Redhill, U.K. for help with entering the theory behind the electrothermal simulations. Also, we are very thankful to Benno Krabenborg from Philips Semiconductors, Nijmegen, The Netherlands for his help with the LUMEX program.

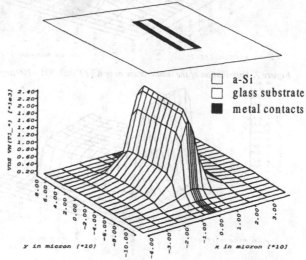

Figure 5. Distribution of the temperature over a TFT with W/L=100µm/4 µm

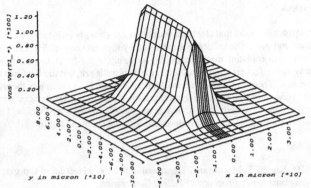

Figure 6. Distribution of the temperature over a TFT with W/L=100 μm/6 μm

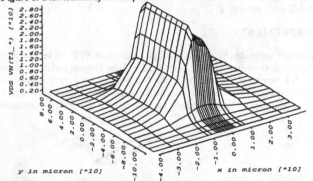

Figure 7. Distribution of the temperature over a TFT with W/L=100 μm/10μm

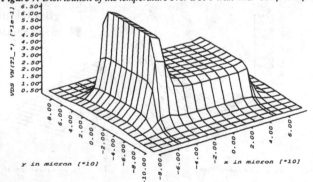

Figure 8. Distribution of the temperature over a TFT with W/L=100 μm/100μm

Mat. Res. Soc. Symp. Proc. Vol. 621 © 2000 Materials Research Society

Plastic Deformation of Thin Foil Substrates with Amorphous Silicon Islands into Spherical Shapes

Pai-hui I. Hsu, Min Huang, Sigurd Wagner, Zhigang Suo, and J. C. Sturm
Center for Photonics and Optoelectronic Materials (POEM)
Princeton University
Princeton, NJ 08544, U.S.A.

ABSTRACT

There is a growing interest in the application of large area electronics on curved surfaces. One approach towards realizing this goal is to fabricate circuits on planar substrates of thin plastic or metal foil, which are subsequently deformed into arbitrary shapes. The problem that we consider here is the deformation of substrates into a spherical shape, where the strain is determined by geometry and cannot be reduced by simply using a thinner substrate. The goal is to achieve permanent, plastic deformation in the substrates, without exceeding fracture or buckling limits in the device materials.

Our experiments consist of the planar fabrication of amorphous silicon device structures onto stainless steel or Kapton® polyimide substrates, followed by permanent deformation into a spherical shape. We will present empirical experiments showing the dependence of the results on the island/line size of the device materials and the deformation temperature. We have successfully deformed Kapton® polyimide substrates with 100 μm wide amorphous silicon islands into a one steradian spherical cap, which subtends 66 degrees, without degradation of the silicon. This work demonstrates the feasibility of building semiconductor devices on plastically deformed substrates despite a 5% average biaxial strain in the substrate after deformation.

INTRODUCTION

The applications of traditional large-area electronics, such as displays, are limited by the fact that glass substrates are rigid and easily breakable. Previous work [1,2] has demonstrated that transistors on thin metal or plastic foil substrates can be rolled around a cylinder down to 0.5mm radius of curvature with no adverse effects. In these cases, the transistor is compressed when bending inwards and elongated when bending outwards. The strain due to such cylindrical deformation can be decreased by reducing the thickness of the substrates. In this work we permanently deform the substrates into the final shape of a spherical dome. To deform a flat sheet into a spherical structure, the initial substrate is stretched so that the surface area increases. The average strain in this case is determined by geometry and is independent of the substrate thickness. Another important difference is that while the device can be either compressed or stretched when rolling the substrate, spherically deforming the substrate leaves the devices in tension in all cases.

EXPERIMENTS AND RESULTS

The thin film structures are first processed on planar thin foil substrates of stainless steel or Kapton® polyimide, with a typical thickness of 50 μm. The foil is placed over a circular hole and clamped by a circular ring (Figure 1). Pressurized gas is then used to deform the material inside the clamped ring into the shape of a spherical dome. A straightforward calculation can estimate the average radial strain, $\varepsilon_{avg,rad}$, which is necessary to expand the foil to a spherical shape that subtends a given half-angle (θ).

$$\varepsilon_{avg,rad.} = \frac{\theta - \sin\theta}{\sin\theta}, \ \theta \text{ in radians} \tag{1}$$

Note that this strain is determined solely by the higher surface area of the spherically deformed cap compared to that of the original foil. Unlike the case of deforming the foil into a cylindrical shape, the strain cannot be reduced by making the foil thinner for any given subtended angle. This relationship is shown in Figure 2.

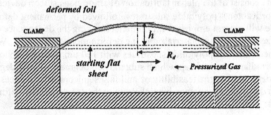

Figure 1. Design of the apparatus used for deformation.

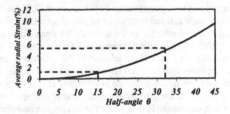

Figure 2. Average radial strain vs. half-angle. For a subtended half-angle of 15°, the average strain is 1%, and for a subtended half-angle of 33° (corresponding to one steradian of solid angle), the average strain is 5%.

When plastic or ductile metals are subjected to large strains, they deform plastically as seen in Figure 3. Therefore the deformation largely persists after the pressure is removed. For example, Figure 4 shows the height vs. radius of a 25 μm stainless steel sheet after the pressuring gas has been released. The foil was deformed into the shape of a spherical dome with a half-angle of 33°. On the other hand, inorganic semiconductor materials are brittle and can only be elastically deformed. For the 66° field of view spherical cap (Figure 4), the average strain is 5%, which far exceeds the breaking limit of the semiconductor materials. The brittle materials we examined include a sandwich of 100 nm amorphous silicon on top of 400 nm Si_3N_4 (Figure 5(a)). A uniform layer of such brittle materials deposited on the substrate cracked after it was deformed into a 66° dome.

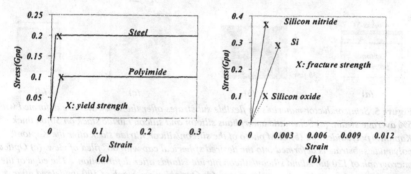

Figure 3(a). *Ideal stress-strain relationship for substrate materials. For compliant materials, the deformation is first elastic (linear region in the above diagram). After the deformation reaches the yield strength, the deformation becomes permanent (plastic deformation). (b) Ideal stress-strain relationship for brittle materials. The typical breaking strain limit is smaller than 1% for such ceramic materials [3]. The curves in both figures are an approximation of the stress-strain relationship for perfectly elastic/plastic materials. In practice, materials do no exhibit perfectly linear correlation between stress and strain.*

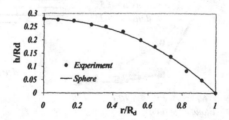

Figure 4. *Height of deformed 25 μm stainless steel sheet as a function of radius, with the height and radius normalized by the radius of the deformed region R_d. The solid line represents a spherical shape subtending a half angle of 33°.*

To overcome this limitation, we then examined the concept of fabricating islands of hard material (for eventual devices) on soft substrates. The qualitative concept was that the soft substrate could flow beneath the island so that the island itself might not be excessively strained. This was tested using a sandwich of 100 nm of amorphous silicon and 400 nm of silicon nitride deposited on 50 μm of Kapton® polyimide. Both layers were deposited by PECVD at 200°C and then patterned into square islands by dry etching. Using such a process, and performing deformation at 150°C, islands up to 100 μm without cracks could be obtained.

The likelihood of obtaining crack-free islands depends primarily on the size of the island (with smaller islands easier to obtain) and secondarily on the island density. The results are presented in Figure 6.

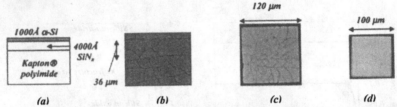

Figure 5. *Semiconductor materials on flexible substrates after the substrates are deformed with 5% average strain. (a) Schematic amorphous silicon and silicon nitride stack on 50μm thick Kapton® polyimide film. (b) SEM photo of the silicon/silicon nitride stack after the Kapton® polyimide substrate is deformed into the desired spherical cap with 66° filed of view. (c) Optical micrograph of 120 μm island silicon/silicon nitride islands after deformation. The edge of the picture is out of focus due to curved surface. (d) Optical micrograph of 100 μm island after deformation at 150°C.*

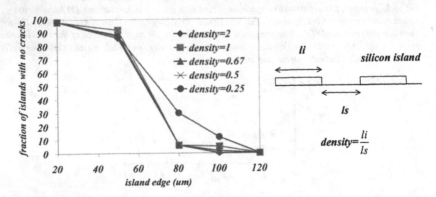

Figure 6. *Fraction of square α-Si/Si₃N₄ islands on polyimide with no cracks after the substrate is deformed with 5% average strain as a function of the island edge, for different island densities. The island density is defined as the island edge size over the island spacing.*

MODELING AND DISCUSSION

To further understand the deformation of the islands, we used ABAQUS [4], a finite element analysis program, to model the strain distribution in the substrate. The test structure consisted of 0.5 μm thick square silicon islands on a 50 μm thick polyimide substrate in the cylindrical coordinate with the z-axis perpendicular to the center of the silicon island. The whole structure was then deformed under the same boundary conditions as in the experiment. The results of this analysis for the island on the top center of the dome are shown in Figure 7.

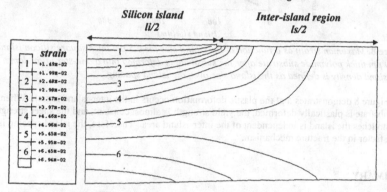

Figure 7. *Contour plot of circumferential strain distribution in the 50 μm thick substrate after deformation into a one-steradian dome from finite element analysis. The island is 0.5 μm thick with length of li. The island to island spacing is ls. Contour lines labeled 1 to 6 indicate strain linearly increasing from 1.5×10^{-2} to 6.5×10^{-2}.*

The simulation of the strain distribution (Figure 7) highlights two important facts:

1. The stiff island confines the substrate underneath itself. The strain in the top substrate, which is just below the island, is small and quite uniform.
2. The strain increases as one moves radially away from the island. In the inter-island region, the strain in the substrate is so high that the substrate is plastically deformed.

By patterning the uniform silicon layer into isolated islands, the plastic deformation takes place in the inter-island region. This is the main reason why the silicon islands are intact despite a 5% average strain in the substrate after deformation. For an island without cracks, the strain in the island must be less than the maximum strain at its fracture strength. Our experimental results show that few islands larger than 100 μm are intact after deformation. This indicates that the strain in the island increases with its size. Finite element analysis demonstrates that the strain in the island indeed correlates to the island length, and the strain is not strongly dependent on the island spacing when the island is small (Figure 8).

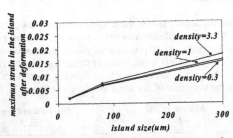

Figure 8. *Maximum strain as a function of the island size for 0.5 μm thick square silicon islands on 50 μm thick polyimide substrate after the substrate is deformed into a one-steradian dome. The island density is defined as the island size over the island spacing.*

Figure 8 demonstrates that the plastic deformation occurs in the inter-island region. When the substrate is plastically deformed, the yield strength is almost constant, and the sheer strength that stretches the island is independent of the inter-island area. Therefore, the island spacing is a weak factor in the fracture mechanism.

SUMMARY

The concept of developing spherically-shaped electronics by deforming thin foil substrates with patterned device islands on their surface has been investigated. We find that the strain in the island increases with the island size, but is only weakly dependent on island density. 100 μm silicon islands on a plastically deformed substrate in the shape of a spherical dome subtending 66° have been fabricated so it appears feasible to build circuits in such structures.

ACKNOWLEDGEMENT

The authors gratefully acknowledge the support from DARPA/ONR and the assistance of H. Gleskova with PECVD deposition.

REFERENCE
1. E. Y. Ma, S. D. Theiss, M. H. Lu, C. C. Wu, J. C. Sturm, and S. Wagner, Thin Film Transistors For Foldable Displays, *Tech. Dig. Int. Electron Devices Meet.*, p. 535-538 (1997).
2. H. Gleskova and S. Wagner, Failure Resistance of Amorphous Silicon Transistors Under Extreme In-plane Strain, *Appl. Phys. Lett.*, **75**, 3011 (1999).
3. W. D. Callister, JR., *Materials Science and Engineering: An introduction*, 4th ed.(John Wiley & Sons, Inc., New York, 1997), p. 401.
4. ABAQUS, Version 5.8, Hibbitt, Karlsson & Sorensen, Inc., Rhode Island, 1999.

Poster Session:
TFT IV: Processes
and Devices

Mat. Res. Soc. Symp. Proc. Vol. 621 © 2000 Materials Research Society

CRYSTALLIZATION OF AMORPHOUS Si THIN FILMS USING A VISCOUS Ni SOLUTION

Jin Hyung Ahn, Sung Chul Kim, and Byung Tae Ahn
Department of Materials Science and Engineering, Korea Advanced Institute of Science and Technology, Taejon, 305-701, Korea

ABSTRACT

We prepared a viscous Ni solution by dissolving $NiCl_2$ in 1N HCl and mixing it with propylene glycol to control the amount of Ni on Si surface. A uniform film was formed after spin coating and oven dry. The a-Si films deposited by LPCVD with Si_2H_6 gas were crystallized more uniformly and more reproducibly. And the crystallization was enhanced from 600°C, 30h to 500°C, 10h. The surface roughness of poly-Si film crystallized with the viscous solution was much smaller than that of poly-Si film crystallized from Ni/Si direct contact. The TFT mobility was improved even though the crystallization temperature was much lower.

INTRODUCTION

It generally takes tens of hours to crystallize a-Si films at 600°C. So, there have been many studies to enhance the solid-phase crystallization. One of the methods is to deposit metals on Si surface. Metals can be deposited using sputtering or evaporation. Lee fabricated high-performance thin film transistors by laterally crystallizing the channel area from the source/drain area on which 5Å thick Ni was deposited by sputtering [1].

But metals also can be deposited using metal solutions. The amount of metal deposited on Si surface can be controlled by the concentration of the solution. Sohn reported that SPC was enhanced using various metal solutions [2]. Yoon reported that an a-Si film could be crystallized at 500°C in 20h using a Ni solution [3]. Kalkan reported that crystallization time could be reduced from 18h to 10min at 600°C using a Ni solution [4]. Sohn and Yoon spin-coated metal solutions and Kalkan dropped the solution onto a 200°C a-Si film and dried it. They all used a diluted acid like HCl and HNO_3 as the solvent. But the solutions using diluted acid as the solvent have some limitations. Metals such as Ni and Al of which electronegativity is not larger than Si cannot be deposited on the Si surface [2]. And the solution exists on a-Si films in uncontrollable forms of droplet because Si surface is hydrophobic, resulting in non-uniform and irreproducible deposition.

In this study we made a viscous Ni solution dissolving $NiCl_2$ in a 1N HCl+propylene glycol solvent. The viscous solution formed a uniform film when spin coated, and enough amount of Ni could be uniformly deposited after oven dry. As a result, crystallization temperature could be lowered and the crystallization process could be more uniform and reproducible.

EXPERIMENT

100nm thick a-Si films were deposited by low-pressure chemical vapor deposition at 530°C using Si_2H_6 on oxidized Si wafers. The films were cleaned in boiling $H_2SO_4+H_2O_2$ solution and dipped in diluted HF solution to remove oxide on the surface before solution coating. Ni solution was prepared by dissolving NiCl$_2$ in a 1N HCl+propylen glycol solvent. Propylene glycol is an organic solvent that makes the solution viscous. The volumetric ratio between 1N HCl and propylene glycol was fixed to 4:1. A uniform film was formed on Si surface by spin coating the Ni solution and the film was dried at 230°C. The Ni-adsorbed a-Si films were annealed in Ar atmosphere with a conventional type of a tube furnace. Intrinsic a-Si film and Ni-deposited a-Si film from the Ni solution with only 1N HCl solvent were also annealed for the comparison.

Surface morphology of Si films was observed by optical microscope and scanning electron microscope. The amount of Ni deposited on Si films was measured by dissolving the adsorbate on the Si film into 1% HNO$_3$ solution and analyzing it with inductively coupled plasma mass spectroscopy (ICP-MS). The crystalline fraction of the annealed Si films was evaluated with Si (111) peak intensity of X-ray diffraction.

TFTs were fabricated with the poly-Si films, crystallized at 530°C for 5h from a-Si with Ni coating from 0.1M Ni (1N HCl+PG) solution. TFTs were also fabricated with the poly-Si films, crystallized at 600°C for 48h without Ni for the comparison. A 130nm thick SiO$_2$ was deposited by LPCVD as the gate insulator. A 100nm thick a-Si was deposited as the gate. Source, drain and gate were doped with phosphorous in PH$_3$/H$_2$ plasma. The specimens were annealed at 600°C for 48h to crystallize the gate and activate the dopant. After depositing 300nm thick SiO$_2$ by LPCVD as the interlayer insulator and opening contacts, 500nm of Al was deposited by sputtering. Alloying was done in 10.4%H$_2$/N$_2$ at 450°C for 30min.

RESULTS

Figure 1 shows the optical microscopic surfaces of LPCVD Si films after annealing at 500°C for 10h. Ni was deposited on a-Si films from 1,000ppm Ni in 1N HCl solution (Fig. 1a) and 0.1N Ni in 1N HCl+propylene glycol solution (Fig. 1b). Lumps of adsorbate that are originated from the droplets of the solution are seen in Fig. 1a. And the film around the lumps was crystallized while the other area remained as amorphous. The lumps maybe originated from the agglomeration of Ni droplets. But the lumps exist in uncontrollable forms because Si surface is hydrophobic, resulting in the nonunifom and irreproducible crystallization. Thus, it is inevitable to use a physical method to deposit uniformly on Si surface.

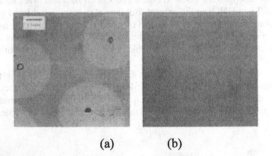

(a)   (b)

Figure 1. SEM surface micrographs of Si films after annealing at 500°C for 10h. (a) : Ni was adsorbed from 1,000ppm Ni 1N HCl solution; (b) : Ni was adsorbed from 0.1M Ni (1N HCl+PG) solution.

By employing Ni solution with 1N HCl+PG solvent, a uniform film was coated because of its viscosity. And no lumps of adsorbate were detected after drying the film. And a-Si film was uniformly crystallized as seen in Fig. 1b. And no lumps were detected even in a x100,000 SEM magnification after crystallization.

Figure 2 shows the annealing-time dependence of crystalline fraction of three types of Si films : a-Si films without Ni adsorption, a-Si films with Ni deposition from 1000ppm Ni 1N HCl solution, and a-Si films with Ni deposition from 0.1M Ni (1N HCl+PG) solution. First two types of a-Si films were annealed at 600°C and a-Si films with Ni deposition from 0.1M Ni (1N HCl+ PG) solution were annealed at 500°C. It took about 30h to fully crystallize for intrinsic a-Si films. And the crystallization was not enhanced for a-Si films with Ni deposition from 1000ppm 1N HCl solution. But a-Si films with Ni deposition from 0.1M Ni (1N HCl+ PG) solution crystallized in 10h even though the annealing temperature was 100°C lower. And the crystallization was further enhanced with microwave annealing that a-Si films with Ni deposition from 0.005M Ni (1N HCl+PG) solution was fully crystallized in 8h at 480°C. We can see that Ni deposition from (1N HCl+PG) viscous solution improves the uniformity of the crystallized film and enhances crystallization of a-Si film.

Figure 3 shows the XPS spectra of Ni and Cl in the adsorbate on Si films before and after an annealing (530°C, 5h). Cl existed as $NiCl_2$ before the annealing. But Cl was not detected after annealing. The spectra of Ni before and after annealing were similar to each other except that the peak in the vicinity of 855eV broadened. It can be considered that Ni exists as $NiCl_2$ and the $NiCl_2$ reacts with Si to form Ni in Si and Cl_2 during annealing.

Remaining adsorbate after annealing could be removed simply with $H_2SO_4+H_2O_2$ solution. The quantity of Ni and Cl incorporated into the Si film was below the detection limit of AES. In SIMS depth profile of Ni and Cl in the poly-Si films, Cl was uniformly distributed in the Si film. Ni content in the Si films was higher at Si/SiO_2 interface than in the bulk of the film. This is the same result with Ni-silicide induced crystallization of amorphous Si films [6].

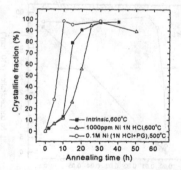

Figure 2. Crystalline fraction of three types of Si films : a-Si films without Ni adsorption, a-Si films with Ni deposited from 1000ppm Ni 1N HCl solution, and a-Si films with Ni deposited from 0.1M Ni (1N HCl+PG) solution. First two types of a-Si films were annealed at 600°C the third type of a-Si films was annealed at 500°C.

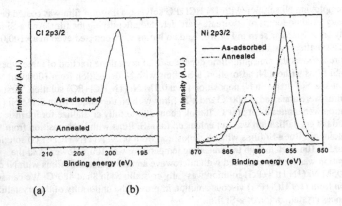

(a)   (b)

Figure 3. XPS spectra of Ni and Cl in the adsorbate on Si films before and after an annealing (530°C, 5h).

Figure 4 shows the concentration of Ni deposited on Si films and Si crystalline fraction of Si films annealed at 500°C for 10h as a function of the concentration of Ni (1N HCl+PG) solution. The concentration of Ni deposited was determined by dissolving Ni into 1% HNO_3 and analyzing it with ICP-MS. The concentration of Ni deposited linearly increased from $2 \times 10^{14} cm^{-2}$ to $5.7 \times 10^{15} cm^{-2}$ as the concentration of the solution increased from 0.001M to 0.1M.

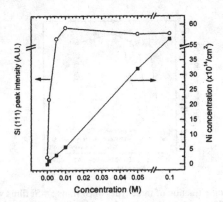

Figure 4. Concentration of Ni adsorbed on Si films and Si crystalline fraction of Si films annealed at 500°C for 10h as a function of the concentration of Ni (1N HCl+PG) solution.

For metal solutions using a diluted acid only, metal deposition depends on the pH as well as the concentration of the solution [5]. But for the viscous Ni solution the adsorption is dominated by a physical process. So the concentration of Ni deposited is proportional to the concentration of the solution. Unlike the concentration of Ni deposited, the crystallization saturated at some concentration of the solution. More than 90% of the film was crystallized at 0.005M and the crystalline fraction did not increase much.

Crystallization of a-Si films with Ni deposition from Ni (1N HCl+PG) solution was slower than that of a-Si films with Ni deposited by sputtering. A-Si films with 5Å Ni deposition fully crystallized even in 3h at 500°C, while it took about 10h for a-Si films adsorbed from 0.005M Ni (1N HCl+PG) solution. The concentration of Ni with the 5Å Ni thickness is similar with that of Ni deposited from 0.005M solution. The faster crystallization in the Si films with pure Ni metal layer is due to higher reactivity between Ni and Si to form Ni-silicide. In the Si film with Ni deposited from Ni solution, the surface is coated with $NiCl_2$ and the reactivity between $NiCl_2$ and Si seems lower than that between Ni and Si.

Figure 5 shows the AFM images of the surface of the poly-Si film with Ni deposited from 0.1M Ni (1N HCl+PG) solution and that of the poly-Si film with 20Å thick Ni metal film. The surface roughness of the LPCVD a-Si film was 19Å in rms value. The surface of the poly-Si film with Ni deposited from Ni (1N HCl+PG) solution was not roughened after annealing (18Å in rms value). But the surface of the poly-Si film with Ni metal film was roughened after annealing (61Å in rms value).

Figure 6 shows the $logI_D$-V_G curves of TFTs fabricated with poly-Si films with 0.1M Ni (1N HCl+PG) solution and intrinsic poly-Si films (no Ni deposition). Drain current as a function of gate voltage was depicted in logarithmic scale. The width and length of the TFTs were 15 μm and 20 μm, respectively, and the voltage between the source and the drain was 1V. The characteristics of the TFTs were similar to each other. The field effect mobility of the TFT with Ni adsorption was 27.1 cm^2/Vsec and that of the TFT of intrinsic Si was 23.3 cm^2/Vsec. The threshold voltage of the TFT with Ni deposition was 8.4V and that of the TFT of intrinsic Si was 7.3V. The on/off ratio earned from the maximun and minimum current in the $logI_d$ vs. V_g plot (Figure 6) was about 10^5 for both TFTs.

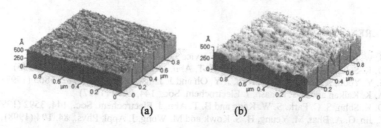

(a)   (b)

Figure 5. AFM images of (a) the surface of the poly-Si film with Ni deposited from 0.1M Ni (1N HCl+PG) solution and (b) that of the poly-Si film with 20Å thick Ni metal film.

CONCLUSIONS

We enhanced the crystallization of amorphous Si films employing a viscous metal solution. A-Si films with Ni deposited from Ni (1N HCl+propylene glycol) solution enhanced the crystallization. And the crystallization was uniform and reproducible. The crystallization was slower than that by Ni sputtering. But the surface of the poly-Si film using the solution was smoother than that using sputtering. And the characteristics of the TFT using Ni solution were similar to that without Ni adsorption, even the crystallization temperature and time were reduced.

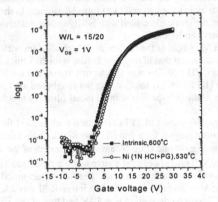

Figure 6. I_D-V_G characteristics of TFTs fabricated with poly-Si film with Ni deposited from 0.1M (1N HCl+PG) solution and intrinsic poly-Si film (no Ni deposition).

REFERENCES

[1] S. W. Lee and S. K. Joo, IEEE Electron Device Lett., 17, 160 (1996)
[2] D. K. Sohn, J. N. Lee, S. W. Kang and B. T. Ahn, Jpn. J. App. Phys., 35, 1005 (1996)
[3] S. Y. Yoon, K. H. Kim, C. O. Kim, J. Y. Oh and J. Jang, J. Appl. Phys., 82, 5865 (1997)
[4] A. K. Kalkan and S. J. Fonash, J. Electrochem. Soc., 144, L297 (1997)
[5] D. K. Sohn, S. C. Park, S. W. Kang and B. T. Ahn, J. Electrochem. Soc., 144, 3592 (1997)
[6] Z. Jin, G. A. Bhat, M. Yeung, H. S. Kowk and M. Wong, J. Appl. Phys., 84, 194 (1998)

Mat. Res. Soc. Symp. Proc. Vol. 621 © 2000 Materials Research Society

Top Gate Self-Aligned µc-Si TFTs using Layer-by-Layer Method and Ion-doping Method

Yukihiko Nakata, Yasuaki Murata* and Masaya Hijikigawa*
Sharp Laboratories of America, LCD Process Tech. Department
5700 NW Pacific Rim Boulevard, Camas, WA 98607, U.S.A.
*Sharp Corporation, LCD Development Group
2613-1, Ichinomoto-cho, Tenri, Nara 632-8567, Japan

ABSTRACT

We investigated the Layer-by-Layer method to obtain thin µc-Si films. Good quality µc-Si films can be obtained at a thickness of 600 Å. Ion-doping of phosphor to i-type µc-Si was investigated to realize the top-gate, self-aligned µc-Si TFT. High conductivity, more than 1 ohm-cm, can be obtained. Using the combination of Layer-by-Layer deposition method and ion-doping method, we fabricated top gate self-aligned µc-Si TFTs. The mobility of TFTs with 600Å-thick µc-Si channel layer is 1.5 cm^2/Vs, which is about three times that of a-Si TFT's mobility. The top gate self-aligned µc-Si TFT is very promising technology for large LCD with high mobility.

INTRODUCTION

Amorphous silicon (a-Si) TFTs are usually used for large size LCD panels, but their mobility is small (i.e. 0.5 cm²/Vs). On the other hand, poly-Si TFTs typically demonstrate higher mobility (i.e. more than 100 cm²/Vs) and strong effort has been placed in the improvement of the crystallization process to enhance this performance even more. However, the reproducibility of poly-Si TFTs is still issue in order to apply to large size LCD panels.

It is expected for new materials to easily apply to large size LCD with high mobility and good reproducibility. One of the candidates is microcrystalline Si (µc-Si). The formation kinetics of µc-Si has been investigated using conventional Plasma Enhanced Chemical Vapor Deposition (PE-CVD) [1], because of their promising characteristics.

The formation of µc-Si has been investigated using the Layer-by-Layer method [2]. Also, µc-Si films were made by VHF-CVD and Layer-by-Layer method [3]. However, these films have film thickness of more than 2000-5000Å and they do not have the same thin film characteristics as films of thickness of less than 1000Å. Top gate µc-Si TFTs deposited by conventional PE-CVD have been investigated [4]. However, the film thickness was 3000Å and they were not the self-aligned TFT structure. On the other hand, n-type, µc-Si that deposited by conventional PE-CVD was used to improve the TFT characteristic [5]. However, there is no report on the combination of ion-doping and µc-Si films.

In case of µc-Si deposition, there is the incubation layer and the crystallinity is improved as the film thickness increases. In other words, thicker film is neccesary to get good film quality, in the case of conventional PE-CVD. However, the film thickness of the TFT channel have to be reduced to less than 1000Å, to suppress the reverse current and the photo-excited carrier generation. Therefore, in this paper, the Layer-by-Layer method was investigated to obtain the high quality µc-Si films with less than 1000Å thickness. Also, the ion-doping method was investigated to make the top gate self-aligned µc-Si TFTs with the channel at the top of the high quality i-type μc-Si. As our conclusion, we propose the top gate self-Aligned μc-Si TFTs using Layer-by-Layer method and ion-doping method.

EXPERIMENTAL DETAILS

The PE-CVD equipment for the Layer-by-Layer method is shown in Fig 1. Glass substrate size is 320mm x 400 mm and films can be deposited on two substrates, which are located on the both side of the heater. The electrodes are ladder-type and the gases are supplied from the back side of the ladder electrodes.

The timing chart of the Layer-by-Layer method using PE-CVD is shown in Fig. 2. H_2 gas is flows at 3000sccm continuously, SiH_4 gas is flows at 130sccm and is turned on and off periodically for a-Si deposition and hydrogen plasma treatment. Total gas pressure is kept at 93.1Pa and RF power is 25 W. The deposition time, the hydrogen plasma treatment time and the total film thickness were varied as parameters.

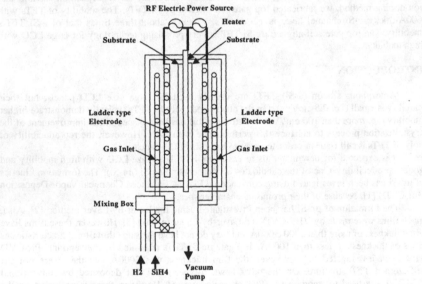

Figure 1. *Plasma Enhanced CVD Equipment for Layer-by-Layer Method.*

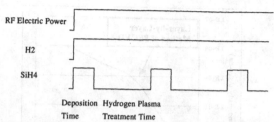

RF Electric Power

H2

SiH4

Deposition   Hydrogen Plasma
Time   Treatment Time

Figure 2. Timing Chart of Layer-by-Layer Method using Plasma Enhanced CVD.

DATA and DISCUSSIONS

(1) μc-Si Film Deposition using Layer-by-Layer Method

The dependence of the conductivity on the Si film thickness per cycle using the Layer-by-Layer method is shown in Fig. 3. The hydrogen treatment time per cycle is 75 sec. The total film thickness is 600 Å. The high conductivity indicates that the films are μc-Si. The μc-Si films can be obtained using a deposition rate of less than 15Å/cycle.

Figure 4 shows the dependence of the Si film conductivity by the Layer-by-Layer method and the conventional PE-CVD on the total film thickness. The conductivity of Si films by the Layer-by-Layer method is higher than that of conventional PE-CVD a-Si films. This indicates the Layer-by-Layer method is very useful to get μc-Si material at a thickness of less than 1000Å. As a result, good quality μc-Si films can be obtained at a thickness of 600Å, using Layer-by-Layer method, compared with the Si films by the conventional PE-CVD without hydrogen plasma treatments.

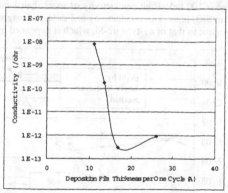

Figure 3. Conductivity of Si films vs. deposition film thickness per cycle by Layer-by-Layer method.

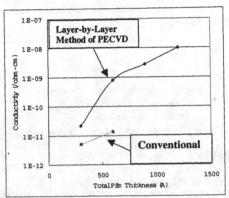

Figure 4. Conductivity of Si films by Layer-by-Layer method and conventional PE-CVD vs. total film thickness.

(2) Ion-doping to μc-Si Films

Ion-doping of phosphor to i-type μc-Si was investigated to realize the top gate self-aligned μc-Si TFT. The ion doping conditions (acceleration voltage and SiO₂-cap-layer thickness) were optimized because a 600Å-thick μc-Si film is too thin to dope and activate the phosphor efficiently. The dose amount of phosphor was 2E16 ions/cm². We found that the conductivity of μc-Si could not be increased to higher than 1E-3 /ohm-cm, in the case of 1000Å-thick cap-SiO₂ and acceleration voltage of 10 keV –100 keV.

The dependence of the μc-Si conductivity on the cap SiO₂ thickness is shown in Fig. 5. The accelerated voltage is 100 keV. High conductivity of more than 1 /ohm-cm can be obtained using a 2000Å-thick cap SiO₂ layer. This conductivity is 3 orders of magnitude higher than that of n+-type a-Si and is the same to that of n+-type μc-Si, which is deposited by PE-CVD.

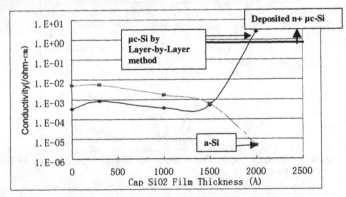

Figure 5. Conductivity of μc-Si by Layer-by-Layer method and a-Si vs. Cap SiO₂ film thickness after ion-doping. Accelerated voltage : 100 keV, Dose amount: 2E16 ions/cm²

(3) Fabrications of Top Gate Self-Aligned µc-Si TFTs

Using the combination of Layer-by-Layer µc-Si deposition method and ion-doping method, we fabricated top-gate, self-aligned µc-Si TFTs were made. The process flow is shown in Fig. 6. At first, µc-Si film of 600Å was deposited on the glass substrate. After that, the µc-Si film was patterned and 2000 Å SiO₂ film was deposited. After Al gate formation, phosphor was ion-doped using Al gate as the mask to make the n-type µc-Si region. After that, SiO₂ inter layer, Al source and drain were made.

The mobility of the µc-Si TFT with 600Å-thick channel is 1.5 cm²/Vs, which is about three times of a-Si TFT's mobility. Based on our results, we believe that top gate self-aligned µc-Si TFT using the Layer-by-Layer method and the ion-doping method is very promising technology for large LCD with high mobility.

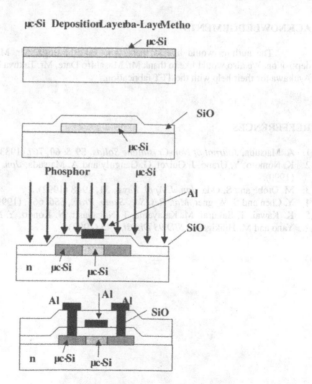

Figure 6. *Process Flow of Top Gate Self-Aligned µc-Si TFT using Layer-by-Layer method and Ion-doping method.*

CONCLUSIONS

1) We investigated the Layer-by- Layer method to obtain the thin μc-Si films. We found that good quality μc-Si films can be obtained at a thickness as thin as 600Å.
2) The ion-doping of phosphor to i-type μc-Si was investigated to realize the top gate self-aligned μc-Si TFT. High conductivity, more than 1 /ohm-cm, can be obtained.
3) By the combination of the Layer-by-Layer method and the ion-doping method, top gate self-aligned μc-Si TFTs were made.
4) The mobility of the μc-Si TFT with 600Å-thick channel is 1.5 cm^2/Vs, which is about three times that of a-Si TFTs.
5) The top gate self-aligned μc-Si TFT is very promising technology for large LCD with high mobility.

ACKNOWLEDGEMENTS

The authors would like to thank Mr. Takeshi Itoga, Mr. Masaki Fujiwara for μc-Si deposition. We also would like to thank Mr. Masahiro Date, Mr. Takuya Matsuo and Mr. Michiteru Ayukawa for their help with the TFT fabrication.

REFFERENCES

1. A. Matsuda, *Journal of Non-Crystalline Solids*, **59 & 60**, 767 (1983).
2. K. Nomoto, Y. Urano, J. Guizot, G. Ganguly and A. Matsuda, *Jpn. J. Appl. Phys.*, **29**, L1372 (1990).
3. M. Otobe and S. Oda, *Jpn. J. Appl. Phys.*, **31**, 1948 (1992).
4. Y. Chen and S. Wagner, *Mat. Res. Soc. Symp. Proc.*, **556**, 665, (1999)
5. K. Kawai, T. Sakurai, M. Katayama, T. Nagayasu, N. Kondo, Y. Nakata, S. Mizushima, K. Yano and M. Hijikigawa, *SID 93 Digest*, **743** (1993).

Mat. Res. Soc. Symp. Proc. Vol. 621 © 2000 Materials Research Society

Nd:YVO₄ Laser Crystallization for Thin Film Transistors with a High Mobility

Ralf Dassow[1], Jürgen R. Köhler[1], Melanie Nerding[2], Horst P. Strunk[2], Youri Helen[3], Karine Mourgues[3], Olivier Bonnaud[3], Tayeb Mohammed-Brahim[3], and Jürgen H. Werner[1]
[1]Universität Stuttgart, Institut für Physikalische Elektronik, Pfaffenwaldring 47, D-70569 Stuttgart, Germany, r.dassow@ipe.uni-stuttgart.de
[2]Universität Erlangen-Nürnberg, Institut für Werkstoffwissenschaften, Lehrstuhl für Mikrocharakterisierung, Cauerstr. 6, D-91058 Erlangen, Germany
[3]Groupe de Microélectronique et Visualisation, Université de Rennes 1, Campus de Beaulieu, F-35042 Rennes Cedex, France

ABSTRACT

We crystallize amorphous silicon films with a frequency doubled Nd:YVO₄ laser operating at a repetition frequency of up to 50 kHz. A sequential lateral solidification process yields polycrystalline silicon with grains longer than 100 μm and a width between 0.27 and 1.7 μm depending on film thickness and laser repetition frequency. The average grain size is constant over the whole crystallized area of 25 cm². Thin film transistors with n- type and p-type channels fabricated from the polycrystalline films have average field effect mobilities of μ_n = 467 cm²/Vs and μ_p = 217 cm²/Vs respectively. As a result of the homogeneous grain size distribution, the standard deviation of the mobility is only 5 %.

INTRODUCTION

Polycrystalline silicon fabricated at low temperatures is of great interest for future electrical devices on cheap and /or transparent substrates (e. g. glass). Laser crystallization of amorphous silicon deposited at low temperatures is the most promising technique to produce the desired material with excellent electrical properties. The established process is based on the irradiation of each part of the silicon film with a number of n = 10 ... 50 pulses of an excimer laser [1]. To improve the electrical properties of the devices, crystallization of the active layer is often performed in vacuum [2]. Thin film transistors (TFTs) with the highest reported electron mobility μ = 640 cm²/Vs were deposited and crystallized completely in ultra high vacuum [3]. Unfortunately, with the established processes the grain sizes and therefore also (due to grain boundaries [4]) the electrical properties of the TFTs depend strongly on the laser's local energy density D. For example, an increase of D from D = 406 mJ/cm² to D = 420 mJ/cm² results in a decrease of the grain size d from d = 500 nm to d = 30 nm [5]. Due to this sensitivity of d on D, temporal and spatial fluctuations of the laser energy lead to considerable scatter in TFT properties. Sequential lateral solidification (SLS) increases the grain size [6] and hence reduces the number of grain boundaries within the devices. This reduction should significantly improve the electrical properties. Simultaneously the SLS process reduces the sensitivity of the grain size on the energy density D. Unfortunately, up to now only excimer lasers were used in combination with an SLS process. The repetition rate of this type of laser has an upper limit of f ≈ 300 Hz. As a consequence the crystallization process is slow.

This work presents the results of non-vacuum laser crystallization utilizing an SLS process in combination with a Nd:YVO$_4$ laser operating at a repetition rate of up to 50 kHz. We produce n- and p-channel thin film transistors with average field effect mobilities of μ_n = 467 cm^2/Vs and μ_p = 217 cm^2/Vs, for electrons and holes respectively. The structural homogeneity of our film results in a standard deviation for the field effect mobility of only 5%.

EXPERIMENTAL DETAILS

Silicon Deposition

We use two different types of samples for our crystallization experiments. Sputtering deposits silicon films with different film thicknesses d = 50 ... 300 μm on 5 x 5 cm^2 Corning 1737F glass. With these films, we investigate the influence of d and of the crystallization parameters on the crystalline structure. Further, we deposit an a-Si layer with a thickness of d = 150 nm by low pressure chemical vapor deposition (LPCVD) on a SiO$_2$ buffer layer. The SiO$_2$ serves as a diffusion barrier and is deposited on a 5 x 5 cm^2 HOYA NA 40 glass substrate by atmospheric pressure chemical vapor deposition (APCVD). Further details of the deposition technique were given in Ref. [7].

Laser Crystallization

Figure 1 shows the experimental setup for the crystallization of the amorphous silicon film with a pulsed Nd:YVO$_4$ laser. The laser emits light with a wavelength λ = 532 nm at a tunable repetition frequency f < 100 kHz. Beam shaping of the TEM$_{00}$ beam with one cylindrical and one spherical lens forms an elliptical beam cross section at the surface of the substrate. The energy density has a Gaussian height h = 12 μm in x-direction and a width w = 460 μm in the perpendicular y-direction. A translation stage moves the sample with a scanning speed v in x-direction. The stage limits v to the maximum speed v_{max} = 25 cm/s. Between each pulse of the laser the sample is translated by a distance Δx = v / f. We tune the repetition frequency f to the scanning speed v to obtain a translation Δx = 0.5 μm. In our experiments we use the scanning velocities v = 10 mm/s and v = 25 mm/s with the corresponding repetition frequencies f = 20 kHz and f = 50 kHz respectively.

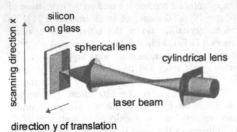

scanning direction x

silicon on glass

spherical lens

cylindrical lens

laser beam

direction y of translation between scans

Figure 1. Experimental setup. A cylindrical and a spherical lens shape the TEM$_{00}$ laser beam to form an elliptical beam cross section at the surface of the substrate.

The laser pulse energy E is adjusted to completely melt an elliptical area with a width w_m = 200 μm and a height h_m = 5 μm. Table I shows the required energy E for films with a different thickness d. The maximum pulse energy of our Nd:YVO$_4$ laser decreases from E_{max} ≈ 100 μJ to at E_{max} ≈ 10 μJ as the frequency f increases from f = 10 kHz to f = 100 kHz. Therefore, we reduced the film thickness to crystallize samples at high repetition frequencies. With the maximum scanning speed v_{max} = 25 cm/s and the corresponding f = 50 kHz the pulse

Table I. Pulse energy E required to completely melt an elliptical area with the width $w_m = 200$ μm and the height $h_m = 5$ μm for silicon films on glass with a different thickness d.

Film Thickness d (nm)	50	75	150	300
Pulse Energy E (μJ)	12.4	14.6	20.4	27.9

energy is sufficient to crystallize a silicon film with a maximum thickness d = 50 nm.

As the film cools after the laser pulse, lateral growth proceeds from the edges of the completely molten area [6]. The next laser pulse again melts an elliptical area, which is translated by the distance $\Delta x = 0.5$ μm with respect to the previous one. Now the edge of the completely molten area is located within the laterally grown grains produced by the previous step. Hence these grains act as seed and are elongated as the film cools. Repeating the pulsed laser irradiation during the translation of the sample results in long elongated grains with a length l exceeding by far the single pulse lateral growth length $\Delta l \approx 1$ μm [8].

After a completed scan in x-direction, we translate the sample by a distance $\Delta y = 130$ μm in y-direction. The distance y is smaller than the width w_m of the complete molten area to have an overlap between to subsequent scans.

RESULTS

Transmission Electron Microscopy

We investigate the influence of the film thickness d and the laser repetition frequency f on the crystalline structure of the films with transmission electron microscopy (TEM). Figures 2a to 2d show grains which are elongated in the scanning direction of the SLS process. The grain width increases with increasing d (see Figs. 2a to 2c) and also with a increasing f (see Figs 2a, 2d). We calculate the average grain width w_G of the different samples by evaluating a large number of TEM pictures. Figure 2e shows an increase from $w_G = 0.27$ μm to $w_G = 1.7$ μm as the film thickness d increases from d = 50 nm to d = 300 nm.

The almost linear increase of the grain width w_G correlates with the influence of the film thickness d on the melt duration t_m. Our numerical analysis using a two-dimensional thermodynamic computer simulation [9] shows that an increase from d = 50 nm to d = 300 nm results in an almost linear increase of the melt duration from $t_m = 43$ ns to $t_m = 208$ ns. Therefore, we attribute the increase in the grain width w_G to the increase of t_m, which means that the grains have more time to grow and therefore w_G increases. The grain width parameter

$$w_t = \frac{dw_G}{dt_m} \qquad (1)$$

represents the increase of w_G per increase of t_m. Using our experimental and simulation results we calculate $w_t \approx \Delta w_G / \Delta t_m = (300 - 50)$ nm / $(208 - 43)$ nm = 8.7 nm/ns.

Figure 2e shows an increase from $w_G = 0.27$ μm to $w_G = 0.45$ μm as the laser frequency increases from f = 20 kHz to f = 50 kHz using 50 nm thick films. At the same time the pulse energy density D_m required for completely melting the silicon decreases by about 12 %. Again an increase of the melt duration t_m, now caused by an increasing temperature of the silicon film, explains the raising of w_G as well as the decrease of D_m: i) Our numerical simulations indicate that a reduction of D_m by 12 % is a result of an increase of the silicon film temperature T prior to

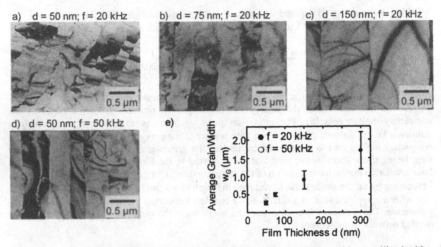

Figure 2: a) - c) TEM pictures of samples with a film thickness d = 50 ... 150 crystallized with a repetition frequency f = 20 kHz. d) TEM picture of a sample with d = 50 nm crystallized at f = 50 kHz. e) The average grain width w_G increases with increasing film thickness d and increasing repetition frequency f.

the onset of the laser pulse from T = 300 K to T = 660 K. ii) By inserting the enhanced silicon temperature T = 660 K into our simulation, we obtain an increase of this melt duration from t_m = 43 ns to t_m = 65 ns. The increasing temperature t_m of the silicon film can be explained by a rise of the average laser beam power as the frequency f increases at a constant pulse energy E. Now we calculate the grain width parameter $w_t \approx (0.45 - 0.27)$ nm / (65 - 43) ns = 8.2 nm/ns.

The grain width parameters w_t are approximately the same in the two considered cases of film thickness variation and temperature variation due to different repetition frequencies f.

Homogeneity

We investigate the homogeneity and the crystalline structure of the laser crystallized samples with optical and scanning electron microscopy (SEM). The bottom part of Fig. 3a shows a sample after Secco etching. A single laser scan with an energy density distribution shown in the upper part of Fig. 3a crystallized the silicon. Secco etching not only removes the silicon at the defects of the crystalline structure, but also affects the glass due to the HF content of the solution. The latter effect results in a lift-off of small grained silicon. Only large grains produced by the SLS process remain on the glass and appear as bright areas in the optical microscope. In the center of the laser beam, where the energy density D is higher than the energy density D_m, which is required for complete melting, the SLS process results in long elongated grains. D_m is the energy density required to completely melt crystalline silicon. The region of the large grained silicon has a width $w_{LG} \approx 200$ µm. In the outer part of the laser beam with an energy density D < D_m the silicon film is not melted completely. The resulting grain size depends on the local energy density, but is smaller than the film thickness d [10]. These regions appear as dark areas in Fig.

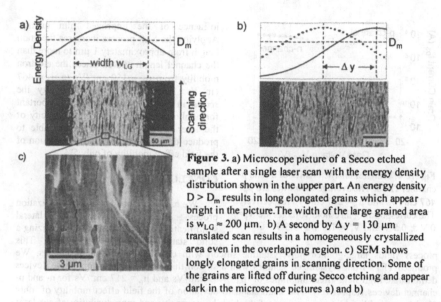

Figure 3. a) Microscope picture of a Secco etched sample after a single laser scan with the energy density distribution shown in the upper part. An energy density $D > D_m$ results in long elongated grains which appear bright in the picture. The width of the large grained area is $w_{LG} \approx 200\ \mu m$. b) A second by $\Delta y = 130\ \mu m$ translated scan results in a homogeneously crystallized area even in the overlapping region. c) SEM shows longly elongated grains in scanning direction. Some of the grains are lifted off during Secco etching and appear dark in the microscope pictures a) and b)

3a. Figure 3b shows the energy density distribution of two subsequent scans with the translation $\Delta y = 130\ \mu m$ between them and the resulting microscope picture after Secco etching. The crystalline structure is the same throughout the whole crystallized area, even in the overlap region. The homogeneity in this region a result of two different crystalline growth mechanisms: i) In the region with $D > D_m$, the laser pulse completely remelts the large grained silicon produced by the previous laser scan. During the cooling, lateral growth again results in long grained material. ii) In the outer part of the beam the energy density $D < D_m$ is not high enough to completely melt the large grained silicon produced by the previous scan. The result is a liquid film on top of the remaining crystallites. Now these crystallites act as seed for columnar growth of the grains which rebuilds the previous grain structure.

Scanning electron microscopy validates that the bright areas of Figs. 3a, 3b correspond to the laterally elongated grains. The picture height is equal to the channel length $l_{CH} = 15\ \mu m$ of the TFTs we produce from the laser crystallized silicon. The grain length exceeds the channel length of the devices.

TFT Results

We fabricate TFTs with a channel length $w_{CH} = 15\ \mu m$ parallel to the scanning direction and a width $w_{CH} = 30\mu m$ (for details see Ref. [7]). Figure 4 shows typical drain current (I_D) versus gate voltage (U_G) characteristics of p- and n-channel TFTs. The average electron and hole field effect mobilities of the n-channel and p-channel devices are $\mu_n = 467\ cm^2/Vs$ and $\mu_p = 217\ cm^2/Vs$ respectively. A comparison of these results with transistors produced by the same TFT process but with silicon from conventional excimer laser crystallization [7] shows the distinct

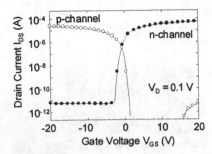

influence of the improved film quality: Applying the SLS process increases the grain length from approximately 1 μm to more than the channel length and enhances the electron mobility from $\mu_n \approx 150$ cm^2/Vs to $\mu_n \approx 467$ cm^2/Vs. Apart from a high mobility the *scatter* in the electrical properties is important for applications. Due to the homogeneity of the laser crystallized films we are able to produce devices with a standard deviation of the field effect mobility of only 5%.

Figure 4. Transfer characteristics of n- and p-channel TFTs. The average mobilities are $\mu_n = 467$ cm^2/Vs and $\mu_p = 217$ cm^2/Vs respectively. The corresponding threshold voltages are $V_t = 0$ V $\pm$ 0.5 V and $V_t = -1.8$ V $\pm$ 0.5 V.

CONCLUSIONS

We have developed a laser crystallization process based on sequential lateral solidification of amorphous silicon utilizing a solid state Nd:YVO$_4$ laser system. This process yields large-grained polycrystalline silicon with excellent electronic properties. We produced TFTs from crystallized samples with a silicon thickness d = 150 nm. These devices demonstrate high field effect mobilities of $\mu_n = 467$ cm^2/Vs and $\mu_p = 217$ cm^2/Vs for n- and p-channel devices respectively. The small standard deviation of the field effect mobility of only 5% is remarkable and a consequence of the high homogeneity and reproducibility of our laser crystallization process. An increasing repetition rate, going hand-in-hand with an increasing scanning speed, not only yields a higher throughput but also results in larger grains. Increasing the repetition rate should therefore further improve the electronic properties of the devices produced from crystallized films.

REFERENCES

1 M. Stehle, J. Non-Cryst. Solids **218**, 218 (1997).
2 A. T. Voutsas, A. M. Marmorstein, and R. Solanki, J. Electrochem. Soc. **146**, 3500 (1999).
3 A. Kohno, T. Sameshima, N. Sano, M. Sekiya, and M. Hara, IEEE Trans. Electron. Devices **42**, 251 (1995).
4 J. Levinson, F. R. Sheperd, P. J. Scanlon, W. D. Westwood, G. Este, and M. Ryder, J. Appl. Phys. **53**, 1193 (1982).
5 G. K. Giust, and T. W. Sigmon, Appl. Phys. Lett. **70**, 767 (1997).
6 R. S. Sposili, and J. S. Im, Appl. Phys. Lett. **69**, 2864 (1996).
7 Y. Helen, K. Mourgues, F. Raoult, T. Mohammed-Brahim, O. Bonnaud, R. Rogel, S. Prochasson, P. Boher, and D. Zahorski, Thin Solid Films **337**, 133 (1999).
8 R. Dassow, J. R. Köhler, M. Grauvogl, R. B. Bergmann, and J. H. Werner, Solid State Phenom. **67-68**, 193 (1999).
9 unpublished.
10 J. S. Im, and H. J. Kim, Appl. Phys. Lett. **63**, 1969 (1993).

Mat. Res. Soc. Symp. Proc. Vol. 621 © 2000 Materials Research Society

Location Control of Laterally Columnar Si Grains
by Dual-Beam Excimer-Laser Melting of Si Thin-Film

Ryoichi Ishihara
Laboratory of Electronic Components, Technology and Materials (ECTM)
Delft Institute of Microelectronics and Submicrontechnology (DIMES)
Delft University of Technology
Feldmannweg 17, POBox 5053, 2600 GB Delft, The Netherlands

ABSTRACT

The offset of the underlying TiW is introduced in the island of Si, SiO_2 and TiW on glass. During the dual-beam excimer-irradiation to the Si and the TiW, the offset in TiW acts as an extra heat source, which melts completely the Si film near the edge, whereas the Si inside is partially melted. The laterally columnar Si grains with a length of 3.2 μm were grown from the inside of the island towards the edge. By changing the shape of the edge, the direction of the solidification of the grain was successfully controlled in such a way that the all grain-boundaries are directed towards the edge and a single grain expands. The grain-boundary-free area as large as 4 μm × 3 μm was obtained at a predetermined position of glass.

INTRODUCTION

TFTs inside a large Si grain, i.e., crystal-Si (c-Si) TFTs [1] will solve the problems of the excimer-laser crystallized polycrystalline-Si (poly-Si) TFTs [2], such as the high leakage current, the relatively low field-effect mobility and the large deviation of those characteristics [3]. In order to produce the c-Si TFTs on a large glass substrate, it is essential to control the position of the large Si grain at the predetermined position on a glass. The position of surviving crystal seeds, from which crystal grains are grown, can be controlled either by shaping the irradiation beam [4,5], by shadowing the beam with a cap layer [6] or by local structural variation of the underlying material [7,8] which locally modifies the heat extraction rate towards the substrate. The number of the seeds can be controlled to be the unity by making the size of the low-temperature area small enough. In the last method, a single crystal grain as large as 6.5 μm can successfully be grown at the predetermined position on glass [9]. Another way to have a single crystal area is to select one crystal out of many crystal seeds using temperature gradient so that the grain boundary is swept away [10]. The method, however, has a limitation of the size of single grain since the width of Si film has to be gradually narrowed along the growing direction.
In this study, a novel location-control method of the single crystal area is proposed. This method is based on the local heat supply that melts Si completely beside the surviving seeds and selects one single crystal during the growth using the lateral temperature gradient. The method allows one to have a large crystal Si area thanks to (1) slow cooling rate of the melt and (2) a widening Si region rather than narrowing. The method would also allow one to have a preferred crystallographic orientation, which is an important parameter for the c-Si TFTs.

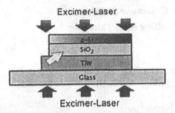

Excimer-Laser

a-Si
SiO₂
TiW
Glass

Excimer-Laser

Figure 1: Schematic cross sectional view of the structure proposed in this study.

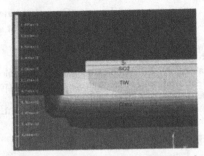

Figure 2: Simulated temperature profile right after laser irradiation

CONSEPT OF THE NEW METHOD

The method proposed in this study uses the structure having an offset of the TiW at the edge of the tri-layer of Si, SiO₂ and TiW on a glass, as depicted in Fig. 1. The two excimer-laser lights irradiate the front (Si) and back (TiW) side, respectively. The metal absorbing the light suppresses fast heat flow towards the glass substrate and results in the large grain [11]. The offset of TiW at the edge absorbs two laser-lights from front and back side and, as a result of that, the temperature becomes higher compared with the one covered with SiO₂ and a-Si. To predict this effect, 2D transient temperature profile has been simulated by the finite element method. The details of the simulation can be found in another article [12]. Figure 2 shows a temperature profile of a cross section of the structure right after the laser irradiation (at a time of 66 ns). It can been seen that the additional heat in the offset of TiW diffuses into the TiW under the tri-layer structure and eventually heats up the Si film laying on top. The horizontal temperature gradient is formed in the Si over a length of about 2.5 µm. It was found that the temperature difference in the Si between the edge and 2.5 µm inside the island is 130 K. The 2.5 µm of the Si from the edge is completely melted, whereas the Si far inside the tri-layer structure is not melted completely. It is thus expected that the partially melted Si solidifies as the small grains and the seeds that are situated near the completely melted zone will grow further into the molten-Si.

During the solidification, undercooling of Si will take place at the growing solid / liquid interface. The solidification is controlled by a rate at which latent heat is conducted away from solid / liquid interface as follows:

$$-\kappa_S \frac{\partial T_S}{\partial x} = -\kappa_L \frac{\partial T_L}{\partial x} - vL, \qquad \text{Eq. (1)}$$

where κ is the thermal conductivity, T the temperature, the subscripts S and L stand for solid and liquid, v the velocity of solid / liquid interface and L the latent heat of fusion per unit volume. It

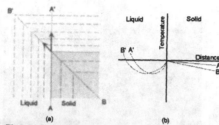

Figure 1: Schematic diagram of isotherms (a) and cross-sectional temperature profile (b).

is clear from Eq. (1) that the steeper temperature gradient in the liquid gives the faster solidification velocity, since it releases more latent heat towards the liquid. In other words, the solidification follows a direction perpendicular to the isotherm. If the edge of the island is formed in a triangle shape, the isotherms can be schematically shown in Fig. 3 (a). Since the direction A-A' has the steeper temperature gradient in the liquid, as shown in Fig. 3 (b), the crystal growth in the direction A-A' will protrude than that along the direction B-B'. In this way, the crystal growth can be controlled so that the grain boundary is swept away and the one single crystal expands.

EXPERIMENTAL

A 550-nm-thick TiW layer is sputtered on the 4-inch glass substrate (Hoya NA 35), which has a transmittance of 83 % for a wavelength of 308 nm. Subsequently, an intermediate SiO_2 layer was deposited by LPCVD using SiH_4 and O_2 at 425°C with a thickness of 400 nm. A 100-nm-thick a-Si was then deposited by LPCVD using SiH_4 at 545°C. In order to transmit the light through a part of the substrate, the tri-layer was patterned as islands with a width and a length of 50 μm and 8 cm, respectively. At the edge of the island, the offset of TiW was formed with a length of 0.5 μm by over-etching of both the a-Si and SiO_2 layers. The edge of the tri-layer structure has been designed with a triangular-shape for various angles θ on the top. The length at the base of the triangle was fixed at about 3.0 μm. The θ value was varied from 180° (linear line) to -90° (fan-shaped). The XeCl excimer-laser light with a wavelength of 308 nm and a pulse duration of 66 ns (XMR 7100 system) irradiated the back of the glass substrate. Part of the light was transmitted through the substrate and was reflected by a mirror composed of Al having a reflectance of 90 % and irradiated onto the sample from the front. Irradiating energy density E_B to the back was varied from 415 mJ/cm^2 to 470 mJ/cm^2. The sample was treated with diluted Schimmel etchant for defect-delineation. Then the surface morphology of the sample was inspected by SEM. Combined with the SEM, the crystallographic orientation was determined on a "grain-by-grain" basis, using the electron backscattering pattern technique (EBSP) [12, 13].

RESULTS AND DISCUSSION

Figure 4 shows a planer-view SEM image around the edge of the tri-layer structure with θ = 180° (linear line) after the dual-beam laser irradiation. One can see that the inside of the island (right) has grains as small as 300 nm and the lateral growth takes place from the inside towards the edge (left) that has the TiW offset. The average width and length of the columnar grains were 0.8 μm

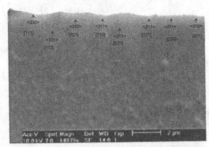

Figure 5: Crystallographic orientation perpendicular to the surface and along the growing directions measured by EBSP

Figure 4: SEM image of Si film near the offset of TiW with θ of 180°

and 3 µm, respectively. The EBSP measurement, as the one example shown in Fig. 5, indicates that the crystallographic axes near <110> are predominant in the direction perpendicular to the surface of the large columnar. This results from the fact that the primary grain selection occurred just before the large columnar grain growth and only large grains having low surface free-energies survived. It was also found that the crystallographic axis along the growth direction (from bottom to top in the Figure) has two preferred orientations near <100> and <211>. If the liquid / solid interface is perpendicular to the surface and flat, the crystallographic orientation of the growing interface has those indexes. It was reported that the growing interface has a crystallographic axis of <100> in the zone-melting (ZMR) of Si film [14].

Figure 6 shows a SEM image of the sample with θ of 60°. One can see that the grain boundaries

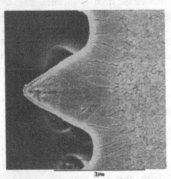

Figure 6: SEM image of Si near the offset of TiW with θ of 60° (triangular-shape)

Figure 7: SEM image of Si near the offset of TiW with θ of 0 (square-shape)

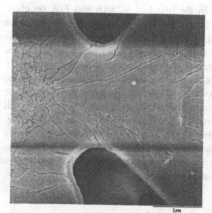

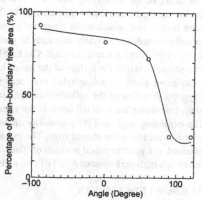

Figure 8: SEM image of Si near the offset of TiW with –90° (fan-shape)

Figure 9: Percentage of GB free area of 1 μm × 1μm in the center of triangle edge

are directed towards the edge of the tri-layer and a single grain expands in the triangular edges. As a result of that, the grain boundary free area was successfully formed in the triangular edge. The length of the grain was 1.8 μm. The further grain growth was stopped by another formation of "petal-shaped" grains ahead, which has a defective core in the center and would be nucleated spontaneously from the melt.

Figure 7 shows SEM image for θ of 0 with which the edge has eventually the square-shape. All grain boundaries are swept towards the offset in TiW, and the grain-boundary-free area with a length of 4 μm has been successfully formed in the rectangle. The lateral length is longer than that for the previous condition because of the fact that the heat is stored more in the tri-layer because of more contact area of the TiW offset at the end of the square, resulting in the delay in the trigger of the spontaneous nucleation from the melt. Figure 8 shows a SEM image for θ of -90° with which the edge has eventually a fan-shape. In this case, the one crystal could successfully expand more than the width of the bottom of the shaped-edge. Figure 9 summarizes the percentage of a boundary-free area in the area of 1 μm × 1 μm at the center of the shaped edge. It should be noted that 30 positions were investigated for each angle. It was found that decreasing θ from 120° increases dramatically the frequency of a single crystal in the triangle. This is because the angle change in the isothermals becomes strong with the smaller θ values. The percentage reached 92% at θ of –90°. The percentage will reach 100% either by decreasing θ more, or by decreasing the width of the bottom of the shaped-edge, which will physically select the expanding grain (the necking).

CONCLUSIONS

The offset of the underlying TiW is introduced in the island of Si, SiO_2 and TiW on glass. During the dual-beam excimer-irradiation to the Si and the TiW, the offset in TiW acts as an extra heat source, which melts completely the Si film near the edge, whereas the Si inside is partially melted. The laterally columnar Si grains with a length of 3.2 μm were grown from the inside of the island towards the edge. The EBSP measurement indicated that the crystallographic axis perpendicular to the surface of the grains have low index numbers. It was also found that the grain grows along crystallographic axes near <100> or <211>. By modifying the shape of the edge, the direction of the solidification was successfully controlled in such a way that the all grain-boundaries are directed towards the edge and a single grain expands in the shaped edge. The expanding angle and the probability of the single grain increases with decreasing the angle on top of the triangular-shaped edge. The grain-boundary-free area as large as 4 μm × 3 μm was obtained at a predetermined position of glass. These features will make the method effective in realizing a high-performance c-Si TFTs on a glass substrate.

ACKNOWLEDGEMENTS

The author would like to acknowledge Professor dr. C. I. M. Beenakker and Dr. J. W. Metselaar for encouragement and support. The author is grateful to Dr. P. F. A. Alkemade for discussion and technical assistance of EBSP. This work was partially supported by the Dutch Technology Foundation STW Project No. DEL66.4542 and the Dutch Fundamental Research for Matters (FOM).

REFERENCES

1. R. Ishihara and M. Matsumura: Jpn. J. Appl. Phys. 36 **10** (1997) 6167.
2. T. Sameshima, S. Usui and M. Sekiya: IEEE Electron Device Lett. EDL-7 No.5 (1986) 276.
3. F. V. Farmakis, J. Brini, G. Kamarinos, C. T. Angelis, C. A. Dimitriadis and M. Miyasaka: Proceedings of the 29th Euro. Solid State Device Res. Conf., (1999) 696.
4. R. S. Sposili and J. S. Im: Appl. Phys. Lett. 69 (1996) 2864.
5. C. Oh, M. Ozawa and M. Matsumura, Jpn. J. Appl. Phys. 37 (1998) L492.
6. L. Mariucci, R. Carluccio, A. Pecora, V. Foglietti, G. Fortunato, P. Legagneux, D. Pribat, D. Della Sala, J. Stoemanos: Thin Solid Films 337 (1999) 137.
7. R. Ishihara and A. Burtsev: Jpn. J. Appl. Phys. 37 3B (1998) 1071.
8. P. Ch. van der Wilt and R. Ishihara: Physica Status Solidi (a), 166 2 (1998) 619.
9. R. Ishihara: Digest of Technical Papers of AMLCD'98 (1998) 153.
10. A. Hara and N. Sasaki: Digest of Technical Papers of AMLCD'99 (1999) 275.
11. K. Shimizu, O. Sugiura, and M. Matsumura: IEEE Trans. Electron Devices Vol. 40 (1993) 112.
12. R. Ishihara, P. Alkemade and A. Burtsev: to be published in Jpn. J. Appl. Phys.
13. J. A. Venables and C. J. Harland: Phil. Mag., 27 (1973) 1193.
14. M. W. Geis, H. I. Smith, D. J. Silversmith, R. W. Mountain and C. V. Thompson: J. Electrochem. Soc. 130 (1983) 1178.

Mat. Res. Soc. Symp. Proc. Vol. 621 © 2000 Materials Research Society

Crystal Growth of Laser Annealed Polycrystalline Silicon As a Function of Hydrogen Content of Precursors

Takuo Tamura, Kiyoshi Ogata, Michiko Takahashi[1], Kenkichi Suzuki[1], Hironaru Yamaguchi, and Satoru Todoroki
Production Engineering Research Laboratory, Hitachi Ltd.
292 Yoshida-cho, Totsuka-ku, Yokohama 244-0817, Japan
[1]Image-related Device Development Center, Displays, Hitachi Ltd.
3300 Hayano, Mobara, Chiba 297-8622, Japan

ABSTRACT

The influence of hydrogen in a precursor on excimer laser crystallization behavior was investigated. The crystal orientation and lattice constant of polycrystalline silicon films were analyzed by X-ray diffraction measurement. An intensity ratio of 111 to 220 was used as an index of the (111) preferred orientation. A randomly oriented film with an index of 2 at lower energy density changed to the highly (111) oriented phase with an index of more than 15 at higher energy density. It was observed in the PE-CVD films that increasing the hydrogen in the precursor films decreased the orientation index. The lattice constant of the laser-annealed PE-CVD silicon was found to be larger than that of the PVD silicon and to decrease with an increase in laser energy density. The network structure of as-deposited PVD film with less hydrogen content was denser than that of as-deposited PE-CVD. The network structure of the precursor strongly affected the crystal growth, and the structure of the ELC poly-Si was still affected by the precursors, even though the hydrogen content decreased after laser annealing.

INTRODUCTION

Low-temperature polycrystalline silicon (LTPS) is attractive for many manufacturers of flat panel displays because it has higher electrical performance and can be cheaply produced[1]. Display technologies have been based on a-Si thin film transistor (TFT), however higher performance of the LTPS TFTs is necessary for next-generation LCD applications.

In the LTPS process, an excimer laser crystallization (ELC) technique is used to crystallize precursor amorphous silicon films. The crystallographic features of the LTPS films strongly correspond to electrical characteristics, such as the field-effect mobility and threshold voltage of the TFT[2-3]. Precursors and ELC conditions can vary the crystallinity of the LTPS films, however the mechanism of the ELC had not been thoroughly revealed. Kuriyama et al. reported that the grain growth phenomenon caused by excimer laser crystallization was strongly affected by precursors[4]. The authors reported that higher hydrogen concentration in precursor film caused an ablation or pinhole in the ELC film[5]. Detailed investigations of such crystallization mechanisms and the influences of hydrogen concentration are necessary for the industrialization of LTPS technologies.

EXPERIMENTAL

Two kinds of silicon films were prepared to investigate the influence of precursors (see Table I). Amorphous silicon films were deposited using a plasma-enhanced chemical vapor deposition (PE-CVD) method under the condition of Ts=380°C with a SiH_4 source. Amorphous silicon films deposited by physical vapor deposition (PVD) methods were also prepared in high vacuum[6]. The 50-nm-thick precursor films were deposited on silica glass. The PE-CVD films were furnace-annealed in N_2 ambient at 450°C for 5, 30, and 60 minutes to eliminate hydrogen in the films. Hydrogen contents were measured using secondary ion mass spectrometry (SIMS) method.

Excimer laser crystallization was carried out using a XeCl laser (λ=308 nm) at room temperature in a vacuum condition. Irradiated energy densities ranged from 340 to 520 mJ/cm^2 for PE-CVD and 360 to 500 mJ/cm^2 for PVD, respectively. The irradiated area was 0.3 x 150 mm^2, and the beam was scanned along the short axis of the slit pattern. The illumination overlap was about 97%.

Crystallinity analysis of the polycrystalline silicon film was performed using μ-Raman spectroscopy. The Raman spectra were measured at room temperature in a backscattering geometry with a 514.5 nm line of an Ar ion laser. A 1mW focused laser beam with a 1μm diameter was used in this measurement. The scattered light was analyzed by using a Jovin-Ybon T64000 triple monochromator in a subtractive configuration. The Raman peak frequency was determined to an accuracy of 0.5 cm^{-1}.

The Surface morphology of the polycrystalline silicon films was observed by using scanning electron microscopy (SEM). Grain boundaries of the polycrystalline films were clearly observed after Secco etching (HF and $K_2Cr_2O_7$). Atomic force microscopy (AFM) measurements were also performed using the tapping mode of the Nanoscope IIIa from Digital Instruments.

The X-ray diffraction spectra of the films were measured by using a θ-2θ type diffractometer with an X-ray source of Cu-kα with 50kV voltage and a 300 mA current. All the patterns were measured at 0.1°/min. by a step of 0.01° and in the conventional θ-2θ geometry. Peak widths (FWHM) of 111 were used to estimate a grain size, and 2θ angle of the 111 peak was used to calculate the lattice constant. The intensity ratio of 111 to 220 was used as an index of the (111) preferred orientation.

X-ray reflectivity measurements were performed in order to estimate a density of the precursors. The measurements were carried out using Rigaku ATX-G under a total reflection condition.

Table I. List of prepared samples

Sample ID	a	b	c	d	e
Silicon deposition	PE-CVD				PVD
Excimer laser annealing (mJ/cm^2)	-	340~500	340~500	340~500	360~500
Hydrogen contents before ELA (atom%)	4.4	1.8	1.2	0.8	0.05
Hydrogen contents after ELA (atom%)	-	0.02	0.04	0.02	-

RESULTS

The hydrogen contents of the as-deposited PVD film (#e) were 0.05 atom%, and that of as deposited PE-CVD (#b, c, d) ranged from 0.8 to 1.8 atom% decreased to 0.02~0.04 atom% after ELA of 520mJ/cm^2 (see Table I).

In the Raman spectra of the ELC silicon films, a sharp peak corresponding to the first order optical phonon was observed at a frequency of about 515 cm^{-1}. The Raman intensity and FWHM as a function of laser energy density are shown in Figure 1 and 2. The Raman intensity was increased with an increase of laser energy density at less than 440 mJ/cm^2 and then decreased with an increase in the energy. On the other hand, the FWHM value decreased with increasing laser energy density up to 440 mJ/cm^2 and the value was nearly constant at an energy density of more than 440 mJ/cm^2. No obvious differences were observed for different precursors in this Raman measurement.

The grain size of samples #b, #c, #d and #e observed by SEM are shown in Figure 3. Grain size and surface roughness, Rmax and Rms of sample #d from AFM observations are shown in Figure 4.

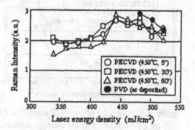

Figure 1. Raman intensity of poly-Si as a function of laser energy density.

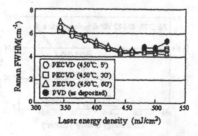

Figure 2. Raman FWHM of poly-Si as a function of laser energy density.

Increasing the laser energy density increased the grain size slightly at less than 400 mJ/cm^2, however the grain size increased rapidly at more than 400 mJ/cm^2. The grain size observed by SEM and AFM showed good agreement. The sample #d showed a larger roughness value, Rmax at energy of 440 and 480 mJ/cm^2 and this tendency agreed well with Raman intensity data.

Two major peaks overlap in the X-ray scattering of an amorphous silica substrate at about $2\theta=28.5°$ and $47.5°$. They correspond to silicon 111 and 220. Figure 5 shows the peak widths (FWHM) of the 111 as a function of the laser energy density. It decreased with an increase of laser energy density and saturated at a FWHM value of 0.2°. The theoretical grain size was calculated using Scherer's formula. These calculated values to the vertical grain size because the measured crystal plane was parallel to the sample surface. From the formula, the vertical grain size was estimated to become saturate at about 50 nm. This saturation is due to the crystal growth being limited by the 50-nm film thickness. In this measurement, no FWHM differences were observed for the different precursors.

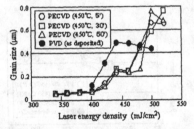

Figure 3. *Grain size of poly-Si analyzed by SEM as a function of laser energy density.*

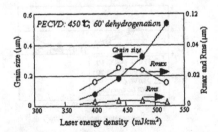

Figure 4. *Grain size, Rmax, and Rms of poly-Si analyzed by AFM as a function of laser energy density.*

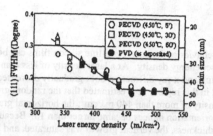

Figure 5. *(111) FWHM of poly-Si measured by XRD as a function of laser energy density.*

Figure 6 shows the (111) orientation index of the PE-CVD films after ELC of 520 mJ/cm². Increasing hydrogen in the precursor films decreased the orientation index. The silicon film with hydrogen contents of 1.8, 1.2 and 0.8% showed indices of 16, 20 and 27, respectively.

Figure 7 shows the relationship between lattice constant and laser energy density. The values of PVD films were nearly constant and smaller than that of the PE-CVD films. The index gradually decreased with an increase in laser energy density.

The density of precursor films was determined by a critical angle of x-ray reflectivity measurement. The analyzed densities of #a (PE-CVD, as deposited), #d (PE-CVD, annealed at 450°C for 60 min.) and #e (PVD) were 2.23, 2.28 and 2.30 g/cm³, respectively.

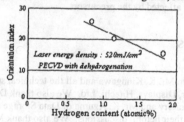

Figure 6. *Orientation index of poly-Si as a function of hydrogen content.*

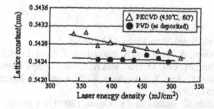

Figure 7. *Lattice constants of different precursors as a function of laser energy density.*

DISCUSSION

The grain growth behavior of the polycrystalline silicon films can be grouped into two categories due to the laser energy density. At a laser density of less than 440mJ/cm^2, the peak width of Raman scattering and x-ray diffraction both gradually decreased and 111 orientation indices were less than 2 in this region. It is estimated that the random growth of small grains was dominant. At a laser density of more than 440 mJ/cm^2, the horizontal grain size increased drastically and the 111 orientation indices became larger than 15. Because the vertical grain size was limited to the film thickness, the vertical growth was saturated, and lateral growth parallel to the (111) plane became dominant. Therefore, (111) orientation can be used as an index of crystal growth. As shown in figure 6, the (111) preferred orientation increased when hydrogen in the precursor was decreased by PE-CVD, that is, higher crystal growth was attained with less hydrogen precursor. Since it is impossible for complete dehydrogenation to be caused only by one-shot laser irradiation, it was concluded that hydrogen degraded the crystal growth.

Another interesting feature was the difference between the PVD and PE-CVD films. The lattice constant of PE-CVD (#d) was larger than that of PVD (#e) before ELA and decreased slightly with an increase in laser energy density. It became close to that of PVD after 500 mJ/cm^2 irradiation. The density analyzed by X-ray reflectivity measurement showed a consistent result. The network structure of as-deposited PVD film with less hydrogen content was denser than that of as-deposited PE-CVD. And the network structure of the precursor strongly affected the crystal growth. Although the hydrogen content decreased after laser annealing, the structure of the ELC poly-Si was still affected by the precursors.

ACKNOWLEDGEMENTS

Special thanks are due to Mr. K. Kinugawa and all the technical staff of Image-related Device Development Center, Displays, Hitachi, Ltd. We also thank Dr. Kataoka of Production Engineering Research Laboratory and Mr. K. Azuma of Data Storage and Retrieval Systems Division of Hitachi Ltd. for their helpful discussions. We also thank Ms. A. Takase of Rigaku Corp. for her help to x-ray total reflection analyses.

REFERENCES

1. S. Chen and I. C. Hsieh, Solid State Tech., **113** (1996).
2. K. Kitahara, S. Murakami, A. Hara, and K. Nakajima, J. Appl. Phys. Lett., **72**, 2436 (1998).
3. K. Park, J. Jeon, C. Park, M. Lee, M. Han, and K. Choi, J. Appl. Phys. Lett., **75**, 460 (1999).
4. H. Kuriyama, T. Nohda, Y. Aya, T. Kuwahara, K. Wakisaka, S. Kiyama, and S. Tsuda, Jpn. J. Appl. Phys., **33**, 5657 (1994).
5. M. Takahashi, M. Saitoh, K. Suzuki, and K. Ogata, (2000) (in press).
6. E. Demaray, SID 99 Digest p.853 (1999).

Mat. Res. Soc. Symp. Proc. Vol 621 © 2000 Materials Research Society

Sub-grain Boundary Spacing in Directionally Crystallized Si Films Obtained via Sequential Lateral Solidification

M. A. Crowder, A. B. Limanov, and James S. Im
Division of Materials Science and Engineering, Department of Applied Physics and Applied Mathematics, School of Engineering and Applied Science, Columbia University, New York, New York 10027

Abstract

In this paper, we report on the average linear density of sub-grain boundaries that are found in directionally solidified microstructures obtained via sequential lateral solidification of Si thin films. Specifically, we have characterized the dependence of the sub-grain boundary density on the film thickness, incident energy density, and per-pulse translation distance. The investigation was confined to analyzing directionally solidified microstructures obtained using straight-line beamlets. It is found that the average spacing of the sub-grain boundaries depended approximately linearly on the film thickness, where it varied from ~0.28 μm at a thickness of 550 Å to ~0.75 μm at 2,000 Å. In contrast, variations in either the energy density or the per-pulse translation distance within the investigated SLS process parameter domain were found to have a negligible effect on the spacing. Discussion is provided on a preliminary model that invokes polygonization of thermal-stress generated dislocations, and on implications of the dependence of device performance on the film thickness.

Introduction

Sequential lateral solidification (SLS) is a pulsed-beam based crystallization method [1, 2] that permits flexible control of grain boundaries. Such a capability during crystal growth is attained by controlling the locations, shapes, and extent of melting induced by the incident laser pulses and iteratively irradiating and sequentially translating the films [3]. This ability to manipulate the locations of high-angle grain-boundaries in the crystallized films can be used to realize low defect-density crystalline Si films with various microstructures, including (1) large-grained and grain-boundary-location controlled polycrystalline films, (2) directionally solidified microstructures, or (3) location-controlled single-crystal regions [4].

In general, there can be a variety of intragranular structural defects that may form during the growth in a crystallization process [5]. Previously, we have identified that sub-grain boundaries constitute the most prevalent intragrain defect found in SLS processed Si films [1, 2]. (Twins constitute additional intragrain defects observed in SLS processed Si films [6]). Since crystalline defects within the active portion of the semiconductor can only degrade the performance of microelectronic devices, it is important to characterize and analyze the extent to which the defects appear, particularly as a function of primary external parameters. Such quantitative analysis of the sub-grain boundaries has not been previously conducted, and to accomplish this constitutes the primary goal of this paper. The results obtained in this investigation reveal that the average sub-grain boundary spacing in directionally solidified SLS processed films depends sensitively on the film thickness and is relatively unaffected by either the SLS processing energy density or the per-pulse translation distance. We conclude from these findings that the performance of thin-film transistors (TFTs) fabricated on such a microstructure would likely increase with increasing film thickness.

Mat. Res. Soc. Symp. Proc. Vol 621 © 2000 Materials Research Society

Experimental

Thin amorphous Si films of varying nominal thicknesses (550, 1,000, and 1,500 Å-thick) were deposited at 250 °C using a PECVD process on Corning 1737 glass plates coated with a 3,000 Å-thick SiO_2 buffer layer. These samples were subsequently dehydrogenated at 450 °C.

The samples were irradiated using a 308 nm XeCl excimer laser with a 28 ns pulse duration at energy densities ranging from below the complete melting threshold to well above the agglomeration threshold. The incident laser beam was spatially homogenized along both axes and then patterned into straight-slit beamlets using a reticle mask. These beamlets were then demagnified to increase the energy density and projected onto the films. The size of the beamlets at the sample was 5 μm by 50 μm in the shape of straight slits. The beam energy was maintained using a computer controlled two-plate energy attenuator and was monitored using a pyroelectric energy meter. Finally, the samples were translated between pulses on a three-axis linear air-bearing stage having sub-micron resolution. More specific details of the SLS system and procedures can be found in reference [7]. Following SLS processing, the samples were Secco-etched and examined using optical and scanning electron microscopes.

The average linear density of sub-grain boundaries was calculated within the SLS processed regions of the Secco-etched samples using a simple grain-boundary counting method. The total number of boundaries—including sub-grain and grain boundaries, but not closely spaced twins and other intragrain defects that may appear in some grains—encountered along a 50 μm line near the end of the scan extending perpendicularly to the scan direction was recorded and averaged for several samples processed under similar conditions. This procedure was repeated for several experimental conditions where the per-pulse translation distance and the energy density were systematically and incrementally varied.

Results

Sub-grain boundaries appear within the SLS crystallized region as linear defects that run approximately parallel to the motion of the solidification interface [1, 2]. The general pattern consists of a river-like or wishbone structure where sub-grains can be either eliminated by the sub-grains on either side of them, or bifurcated by the appearance of a sub-grain boundary in the central region of the sub-grain; such convergence and generation of sub-grain boundaries are shown in Fig 1. Sub-grain boundaries can be identified as *grain boundaries that must—at least initially—be low-angle in nature* as they originate within the interior region of a single grain. Conversely, we define grain boundaries as the boundaries that are formed via grains solidifying from different seeds, and therefore would typically have high-angle misorientations. This definition follows the convention that was previously adapted to describe the lineage defects observed in zone-melting recrystallized Si films [8]. It is found that the average spacing between the boundaries remains approximately constant after a short transition region (a few microns in length along the direction of translation) within which the majority of the large number of grains that are initiated at the onset of lateral growth are rapidly occluded [9]. The average spacing values we report in the present work have been determined well within the area where such steady-state conditions are established.

Fig. 2a shows the average spacing of the sub-grain boundaries within SLS processed Si films versus the film thickness, where it can bee seen that the spacing increases linearly with increasing film thickness and has a slope of ~3.46 μm/μm. The effects of varying the processing parame-

Mat. Res. Soc. Symp. Proc. Vol 621 © 2000 Materials Research Society

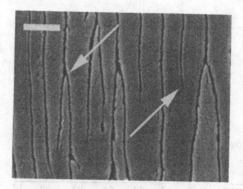

Figure 1: SEM micrograph of Secco-etched SLS-processed 550 Å-thick PECVD Si film showing convergence and generation of sub-grain boundaries (as indicated by the arrows). A 500 nm scale bar is shown. The scan direction extends from the bottom to the top of the image.

ters (i.e, per-pulse translation distance and incident energy density) with respect to the sub-grain boundary spacing can bee seen in Figs. 2b and 2c. Variations in these parameters within the investigated range were seen to have no notable trends on the spacing of the sub-grain boundaries within the crystallized regions.

Discussion

The experimental results presented in this paper unambiguously identifies the film thickness as being the primary factor—of the three external parameters investigated—that affects the average spacing between the sub-grain boundaries. It is also apparent that variations in the incident energy density and the per-pulse translation distance have a negligible effect on the spacing for the conditions that were explored within the present experiments.

In general, lineage patterns of varying nature have been frequently observed in directionally solidified materials, where there can apparently be a number of causes [5]. As far as the present defects are concerned, we can note—as has been previously pointed out [1]—they most closely resemble the sub-grain boundaries that were observed in zone-melting recrystallization (ZMR) of Si films [10]. Nevertheless, compared to the sub-grain boundaries observed in ZMR, the sub-grain boundaries seen in SLS are more closely spaced ($\sim$ tens of μm in ZMR vs. < 1 μm in SLS [8]) and appear to have, as indicated by preliminary results, a higher degree of misorientation ($\sim 1°$ in ZMR vs. several degrees in SLS).

Although there were several models (ranging from plastic deformation due to mechanical buckling of the films to thermal-conductivity-difference-induced protrusions of crystal) [8, 11–13] that were proposed to account for the formation of sub-grain boundaries encountered in ZMR, the situation can be summarized as being that no mechanism was definitively established as being responsible.

While it is likely to be the case that the fundamental underlying causes of sub-grain boundaries for both SLS and ZMR processes are at least related and probably identical, there exists a number of important differences that can be recognized between the two processes. For example, ZMR is a near-equilibrium transformation process involving relatively thick Si films where the interfacial

Mat. Res. Soc. Symp. Proc. Vol 621 © 2000 Materials Research Society

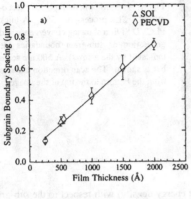

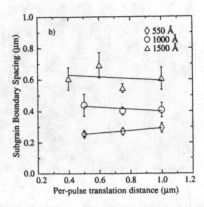

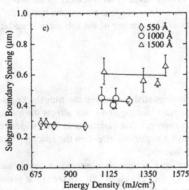

Figure 2: a) Average spacing between sub-grain boundaries in SLS processed PECVD films of varying thicknesses. Average sub-grain boundary spacing in an SLS processed 500 Å-thick SIMOX film is shown for comparison. Average sub-grain boundary spacings for various thickness PECVD films crystallized via the SLS process as a function of b) per-pulse translation distance (and fixed nominal energy density), and c) energy density (and fixed per-pulse translation distance).

undercooling is typically less than 0.01 °C, while in the SLS process, the extent of undercooling experienced during lateral growth of substantially thinner films is on the order of 100–200 °C. As well, the thermal environment under which solidification is conducted is fundamentally different in character for the ZMR and SLS processes: whereas ZMR is a thin-film version of traditional/conventional directional solidification carried out under steady-state conditions, SLS involves a series of melting and solidification iterations that are extremely transient. For instance, the total average time interval during which a portion of the material remains molten (i.e., the dwell time) in ZMR is on the order of seconds, while in SLS it is less than one microsecond. In view of these and additional [1] differences, one must be cautious in applying a defect formation model to these processes.

Of the various mechanisms that can be considered, we presently feel that the general concepts introduced by Baumgart and Phillipp [13] for explaining the formation of sub-grain boundaries in

Mat. Res. Soc. Symp. Proc. Vol 621 © 2000 Materials Research Society

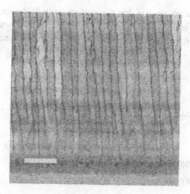

Figure 3: SEM micrograph of 500 Å-thick SLS processed silicon-on-insulator (SOI) film showing beginning of scan and initiation of sub-grain boundaries. The film was Secco-etched to reveal defects. A 1 μm scale bar is shown. The scan direction extends from the bottom to the top of the image.

ZMR processed thin films may also be applicable to the present situation. The crux of this model corresponds to plastic deformation of the solidified film occurring as a result of mechanical stresses of thermal origin that exceed the yield stress (i.e., generation and motion of dislocations). Subsequent polygonization of dislocations that takes place to reduce the interal strain energy leads to the appearance of low-angle grain boundaries. When interpreted within the framework of this plastic deformation model, the higher density of dislocations in SLS processed films as compared to ZMR can be attributed to the higher thermal stress arising from the more severe thermal environment and to the reduced thickness of the films.

Finally, we note two points regarding the presence of sub-grain boundaries in SLS: (1) the performance of thin-film transistor devices fabricated on regions containing sub-grain boundaries aligned parallel to the source-drain region have been shown to have very high-performance characteristics in spite of their presence within the active channel portion of the devices [14]; and furthermore, (2) it has been demonstrated that the presence of sub-grain boundaries can be eliminated by tailoring the shape of the beamlet to cause the sub-grain boundaries to diverge during the SLS process. This particular approach has been shown to be capable of producing location-controlled dislocation- and sub-grain boundary-free regions that are several microns in width and unlimited in length [3,4]. As well, due to the flexible nature of the SLS process, we are confidant that additional/alternative SLS schemes can be further developed to reduce or eliminate sub-grain boundaries in SLS crystallized films.

Clearly, more experimental and theoretical analyses consisting of detailed microstructural characterization and thermo-mechanical profile analysis must be carried out in order to definitively identify and elaborate on the details of the sub-grain boundary formation mechanism in SLS. To this end, one potentially effective approach in deciphering the phenomenon may correspond to conducting SLS experiments on silicon-on-insulator (SOI) films. Figure 3 shows preliminary results obtained from SLS of a (100)-oriented 500 Å-thick SOI film that was prepared using a SIMOX approach. The figure shows, among other things, that the formation of sub-grain boundaries is intrinsic to the SLS process, and further that the sub-grain boundary spacing in (100)-oriented SOI films is comparable to that observed in similar thicknesses of PECVD films (Fig. 2a). In addition to SOI SLS experiments, a recently developed numerical simulation should be of benefit in analyzing the thermal profiles and stresses present in the SLS process during solidification [15].

Mat. Res. Soc. Symp. Proc. Vol 621 © 2000 Materials Research Society

Finally, we note that the findings in the present work predict that the performance of devices fabricated on directionally solidified SLS-processed materials may improve as the film thickness increases, and should not depend sensitively on the processing energy density and per-pulse translation distance.

Acknowledgments

This work was supported by DARPA under project N66001-98-1-8913.

References

[1] J. S. Im and R. S. Sposili, MRS Bulletin **xxi**, 39 (1996).

[2] R. S. Sposili and J. S. Im, Appl. Phys. Lett. **69**, 2864 (1996).

[3] J. S. Im, R. S. Sposili, and M. A. Crowder, Appl. Phys. Lett. **70**, 3434 (1997).

[4] J. S. Im, M. A. Crowder, R. S. Sposili, J. P. Leonard, H. J. Kim, J. H. Yoon, V. V. Gupta, H. J. Song, and H. S. Cho, Physica Status Solidi **166**, 603 (1998).

[5] B. Chalmers, Principles of Solidification (John Wiley & Sons, Inc., New York, 1964).

[6] A. B. Limanov, V. M. Borisov, A. Y. Vinokhodov, A. I. Demin, A. I. El'tsov, Y. B. Kirukhin, and O. B. Khristoforov, in Perspectives, Science and Technologies for Novel Silicon on Insulator Devices, edited by P. L. F. Hemment (Kluwer Academic Publishers, New York, 2000), pp. 55–61.

[7] R. S. Sposili, M. A. Crowder, and J. S. Im, New excimer laser crystallization system for conducting the sequential lateral solidification (SLS) process, to be published in these proceedings.

[8] M. W. Geis, H. I. Smith, and C. K. Chen, J. Appl. Phys. **60**, 1152 (1986).

[9] H. J. Kim and J. S. Im, Appl. Phys. Lett. **68**, 1513 (1996).

[10] E. I. Givargizov, Oriented crystallization on amorphous substrates (Plenum Press, New York, 1991).

[11] J. M. Gibson, L. N. Pfeiffer, K. W. West, and D. C. Joy, in Silicon-On-Insulator and Thin Film Transistor Technology, edited by A. Chiang, M. W. Geis, and L. N. Pfeiffer (Mater. Res. Soc. Proc., Pittsburg, Penn., 1986), Vol. 53, pp. 289–299.

[12] A. Kamgar, S. Nakahara, and R. V. Knoell, in Beam-Solid Interactions and Transient Processes, edited by M. O. Thompson, S. T. Picraux, and J. S. Williams (Mater. Res. Soc. Proc., Pittsburg, Penn., 1987), Vol. 74, pp. 571–576.

[13] H. Baumgart and F. Phillipp, in Energy Beam-Solid Interactions and Transient Thermal Processing, edited by D. K. Biegelsen, G. A. Rozgonyi, and C. V. Shank (Mater. Res. Soc. Proc., Pittsburg, Penn., 1985), Vol. 35, pp. 593–598.

[14] Y. H. Jung, J. M. Yoon, M. S. Yang, W. K. Park, H. S. Soh, H. Cho, A. B. Limanov, and J. S. Im, Low temperature poly-Si TFTs fabricated with directionally crystallized Si films, to be published in these proceedings.

[15] J. P. Leonard and J. S. Im, in Nucleation and Growth Processes in Materials, edited by A. Gonis, P. E. A. Turchi, and A. J. Ardell (Mater. Res. Soc. Proc., Pittsburg, Penn., 2000), Vol. 580, to be published.

Mat. Res. Soc. Symp. Proc. Vol 621 © 2000 Materials Research Society

Sequential Lateral Solidification of PECVD and Sputter Deposited a-Si Films

M. A. Crowder, Robert S. Sposili, A. B. Limanov, and James S. Im
Division of Materials Science and Engineering, Department of Applied Physics and Applied Mathematics, School of Engineering and Applied Science, Columbia University, New York, New York 10027

Abstract

We have investigated sequential lateral solidification (SLS) of amorphous Si films that have been prepared via PECVD and sputter deposition methods. The focus of the work was on identifying and analyzing the energy density and per-pulse translation distance parameter space that permits SLS of these films. Experimental details include the use of a two-axis projection irradiation system to image a straight-slit beamlet pattern onto the sample, and analyzing the resulting microstructures by SEM and optical microscopy of Secco-etched samples. High-temperature-deposited LPCVD films were also examined to enable further comparative analysis. We conclude from these results that there are no major differences in both the SLS process characteristics and the resulting microstructure among the investigated films (provided that the films are dehydrogenated in the case of PECVD a-Si). Based on the controlled super-lateral growth (C-SLG) model of the SLS process, we attribute these findings to the fact that the SLS method involves—as one of its essential features—complete melting of the Si film at fluences that are sufficient to thoroughly melt crystalline Si films, during which all pre-irradiation phase and microstructural details are erased.

Background and Motivation

Sequential lateral solidification (SLS) is a controlled super-lateral growth (C-SLG) [1] process that can be implemented in order to generate a variety of crystalline microstructures. Depending on the details of the process, such as the shape of the beamlets that are incident on the sample and the sequence of translation between the pulses, the method can produce crystalline Si films with grain-boundary controlled polycrystalline or directionally solidified microstructures, as well as location-controlled single-crystal Si regions [1–5].

An important and enabling characteristic of the SLS process is that the method is compatible with low-cost, low-temperature substrates such as glass or plastic. As such, it is of relevance to systematically evaluate the applicability of the SLS method to those films that are deposited using low-temperature techniques (e.g., PECVD or sputter-deposited films) that are compatible with such substrates. In general, as-deposited films prepared using different deposition methods can contain different concentrations of impurities, varying degrees of structural relaxation and medium range order, and different distributions of embedded microcrystals [6]. Given that the far-from-equilibrium first-order phase transformation sequence can often be critically affected by such details in general, it is conceivable to expect differences in the crystallization characteristics and resulting microstructures of the films.

In this paper, we focus on analyzing one of the important factors in implementing the SLS process on PECVD and sputter deposited films, namely that of identifying the experimental parameter space within which the method can be properly executed. For a given per-pulse translation distance, the energy density processing window for SLS is bounded at an upper end due to the onset of agglomeration, and the lower limit due to the failure to grow continuous grains [2]. By

Mat. Res. Soc. Symp. Proc. Vol 621 © 2000 Materials Research Society

varying these parameters, we can quantitatively identify the energy density/per-pulse translation distance process window in which SLS produces continuous lateral growth of the crystals.

Experimental

The samples used for the parametric study consisted of ~1,000 Å-thick films prepared via PECVD and sputter deposition methods. It is noted that the thickness given is the nominal thickness of the film. The sputter deposited films were prepared via DC sputtering on thermally oxidized Si wafers with a 2.0 μm-thick oxide layer. The first set of PECVD films were deposited at 250 °C on Corning 1737 glass substrates that were coated with a 3,000 Å-thick SiO_2 buffer layer. Following deposition of the a-Si films, they were dehydrogenated at 450 °C. The second set of PECVD films were deposited at 250 °C on Corning 1737 glass substrates coated with a 4,000 Å-thick SiO_2 layer on top of a 1,000 Å-thick SiN buffer layer. These films were then dehydrogenated at 450 °C for one hour. LPCVD a-Si films were also prepared on Si substrates having a 2.0 μm-thick thermal oxide layers using a higher temperature process (i.e., decomposition of silane at 580 °C) and analyzed for comparison.

The samples were irradiated using a 308 nm XeCl excimer laser with a 28 ns pulse duration at energy densities ranging from below the complete melting threshold to well above the agglomeration threshold. The incident laser beams were spatially homogenized along both axes and then patterned into straight-slit beamlets using a reticle mask. These beamlets were then demagnified to increase the energy density and projected onto the films. The size of the beamlets at the sample was 5 μm by 50 μm in the shape of straight slits. The beam energy was maintained using a computer controlled two-plate energy attenuator and was monitored using a pyroelectric energy meter. (The energy density at the sample plane was estimated by taking into account the transmission of all the optical components in the SLS processing system and the demagnification of the imaged beamlets onto the sample itself). Finally, the samples were translated between pulses on a three-axis linear air-bearing stage having sub-micron resolution. More specific details of the SLS system utilized in the current experiment can be found in reference [7]. Following SLS processing, the samples were Secco-etched and examined using optical and scanning electron microscopes.

Results

Fig. 1 shows planar-view SEM micrographs for the second set of 1,000 Å-thick PECVD films at varying energy densities and a fixed per-pulse translation distance of 0.4 μm. Fig. 1a shows the effect of processing at energy densities that are greater than the upper bound for the processing window, where agglomeration of the films occurs due to dewetting of the molten Si on the underlying oxide layer. At the other extreme, Fig. 1i shows the effects of irradiating with insufficient energy densities to induce complete melting of the Si film over sufficiently wide regions (i.e., below the threshold for complete melting). The resulting microstructure consists of nearly equiaxed grains with occasional grains that are elongated in the direction of the scan. As the fluence increases slightly, the fraction of continuously grown grains in the directionally solidified material increases dramatically until no discontinuities are observed. Between these low and the high ends of the window, the SLS method produces directionally solidified microstructures that are essentially identical in appearance (Figs 1b to h).

Mat. Res. Soc. Symp. Proc. Vol 621 © 2000 Materials Research Society

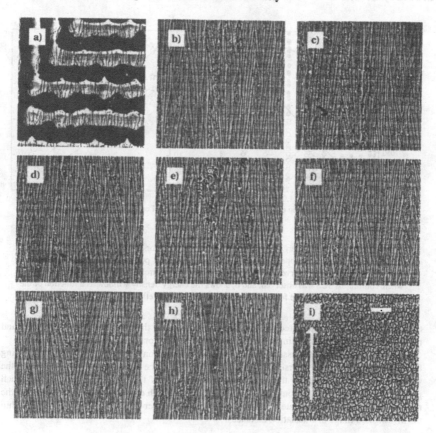

Figure 1: SEM micrographs of Secco-etched 1,000 Å-thick PECVD films irradiated at a) 1.52, b) 1.35, c) 1.30, d) 1.26, e) 1.24, f) 1.15, g) 1.11, h) 1.04, and i) 0.92 J/cm² with a per-pulse translation distance of 0.4 μm. A 5 μm scale is shown in i).

Fig. 2a reveals the dependence of the energy density on the per-pulse translation distance, showing the entire parameter space for the second set of 1,000 Å-thick PECVD films (a subset of which is shown in Fig. 1). The filled circles indicate the points within the parameter space where continuous SLS is observed (e.g., Figs. 1b-h), while the diamonds and circles indicate the regions where agglomeration (Fig. 1a) or discontinuous SLS (Fig. 1i) occurred, respectively. The SLS processed regions were identified as being discontinuous (i.e., processed outside the window) when there existed a single discontinuity extending across the breadth of the scan. The lines demarcating the boundaries are linear regression fits of the averages of the maximum and minimum points of

Mat. Res. Soc. Symp. Proc. Vol 621 © 2000 Materials Research Society

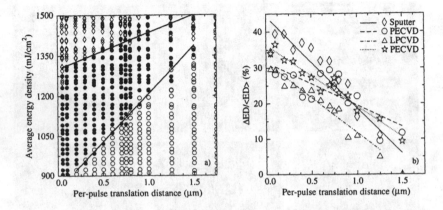

Figure 2: Plots showing energy density processing windows for a) energy density vs. data for PECVD on SiN where the open diamonds and circles indicate points where agglomeration and discontinuous SLS was observed, respectively, while the filled circles indicate the parameter space where continuous SLS was observed; and b) shows the energy density window as a function of the per-pulse translation distance for the four different deposition methods.

the continuous and agglomerated regions, respectively, and of the averages of the maximum and minimum points of the discontinuous and continuous SLS regions, respectively.

Fig. 2b captures the processing window-related information shown in Fig. 2a by projecting the energy density window versus per-pulse translation distance as $\Delta ED / < ED >$, where the window, ΔED, is divided by the mean energy, $< ED >$ of the processing window in which proper SLS takes place (i.e., continuous lateral epitaxial growth of grains). It can be seen that the resulting quantity is approximately an inverse function of the per-pulse translation distance within the investigated range of translation distances.

Discussion

The processing parameter domain within which a materials process can be properly implemented is, in general, an important factor. This can be a particularly significant issue in pulsed-laser processing of materials since there exist intrinsic pulse-to-pulse energy variations and spatial non-uniformities in intensity that can adversely affect the microstructural uniformity of the resulting material. In fact, such a situation is encountered in "conventional" excimer laser annealing (ELA) of amorphous Si films within the process conditions that have been known to produce the highest performance TFT devices [8]. Systematic investigations that correlate laser-crystallized microstructures to device results have definitively shown that these conditions lie within the super-lateral growth (SLG) regime [8, 9]. Based on the analysis of the SLG phenomenon, it was noted previously [10–12] that conventional ELA of Si films within the SLG regime will encounter

Mat. Res. Soc. Symp. Proc. Vol 621 © 2000 Materials Research Society

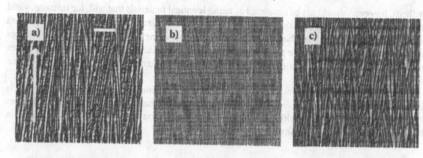

Figure 3: SEM micrographs of Secco-etched 1,000 Å-thick a) sputter deposited, b) PECVD, and c) LPCVD films irradiated at average energy densities of 1.356, 1.149, and 1.515 J/cm^2, respectively, and with per-pulse translation distances of 0.75 μm. A 5 μm scale bar and an arrow indicating the scan direction are shown in a).

processing- and materials-related difficulties and limitations.

As is evident in Fig. 2b, the present results pertaining to the SLS process window clearly indicate that such problems are not encountered in SLS processing of PECVD and sputter deposited films. This important processing characteristic stems from the relative ease with which the requirements for the SLS process can be satisfied: complete melting of the film with the solid–liquid interface boundary falling within the previously regrown region so as to subsequently permit continuous epitaxial regrowth of crystals during iterative processing steps. The overall shape of the SLS processing window, as seen in Fig. 2a, is a reflection of 1) the stochastic aspect of the process in meeting the above requirements, and 2) the tendency of molten Si to agglomerate on an SiO$_2$ underlayer. The detailed consideration of the manner in which these factors are manifested in SLS also dictate that the process windows will be affected by the width of the molten zone, and the results shown in these experiments used a 5 μm wide molten zone. {Additional experimental results verify such a view as they show that the use of 3 μm (10 μm) molten zones can lead to wider (narrower) processing windows for these films.}

As far as the dependence of the SLS process window on the precursor material is concerned, Figs. 2b and 3 clearly show that the effect is, at best, minimal. The reason for the invariance in the SLS process window can be attributed to the fact that the SLS process involves complete melting of the crystalline film, which effectively erases any solid-phase microstructural history within the film. More specifically, the combination of melt duration and thermal history encountered during a single pulse should lead to liquids with an equilibrium density distribution of sub- and near-critical clusters. The type of complex melting scenarios [13] that may be manifested in as-deposited a-Si films at lower energy densities should not be encountered. (We note that small variations in the SLS process window are not unexpected as they can stem from slight differences in hydrogen content and actual film thickness of the samples.)

An important point to note regarding the microstructural uniformity of the SLS processed material is the apparent similarity of the samples that are processed within the entire range of the energy density processing window, as is evident in Fig. 1. This apparent microstructural semblance between the resulting SLS processed materials that are prepared using different deposition

Mat. Res. Soc. Symp. Proc. Vol 621 © 2000 Materials Research Society

methods, however, should not be construed as being identical materials that will, for instance, yield similar device results. We note that the details of the microstructural defects were not scrutinized in this study, other than those that are revealed by Secco-etching and observed in either SEM or optical microscopy. As well, the possible presence and varying amount of non-volatile impurities in the original film will persist after the SLS process, and can potentially affect the resulting device characteristics.

On the other hand, for a given deposition method we may conclude from the obtained results and established relationship between the microstructure and performance of semiconductor devices that the performance of the resulting devices on the material should be insensitive to variations in the SLS processing energy density.

Acknowledgments

This work was supported by DARPA under project N66001-98-1-8913.

References

[1] J. S. Im, M. A. Crowder, R. S. Sposili, J. P. Leonard, H. J. Kim, J. H. Yoon, V. V. Gupta, H. J. Song, and H. S. Cho, Physica Status Solidi **166**, 603 (1998).

[2] R. S. Sposili and J. S. Im, Appl. Phys. Lett. **69**, 2864 (1996).

[3] J. S. Im, R. S. Sposili, and M. A. Crowder, Appl. Phys. Lett. **70**, 3434 (1997).

[4] A. B. Limanov, V. M. Borisov, A. Y. Vinokhodov, A. I. Demin, A. I. El'tsov, Y. B. Kirukhin, and O. B. Khristoforov, in Perspectives, Science and Technologies for Novel Silicon on Insulator Devices, edited by P. L. F. Hemment (Kluwer Academic Publishers, New York, 2000), pp. 55–61.

[5] J. R. Kohler, R. Dassow, R. Bergmann, J. Krinke, H. P. Strunk, and J. H. Werner, Thin Solid Films **337**, 129 (1999).

[6] T. Kamins, Polycrystalline silicon for integrated circuit applications (Kluwer Academic Publishers, New York, 1988).

[7] R. S. Sposili, M. A. Crowder, and J. S. Im, New excimer laser crystallization system for conducting the sequential lateral solidification (SLS) process, to be published in these proceedings.

[8] D. Pribat, P. Legagneux, F. Plais, C. Reita, F. Petinot, and O. Huet, in Flat Panel Display Materials II, edited by M. K. Hatalis, J. Kanicki, C. J. Summers, and F. Funada (Mater. Res. Soc. Proc., Pittsburg, Penn., 1997), Vol. 424, pp. 127–140.

[9] S. D. Brotherton, D. J. McCulloch, J. P. Gowers, J. R. Ayres, and M. J. Trainor, J. Appl. Phys. **82**, 4086 (1997).

[10] H. J. Kim, J. S. Im, and M. O. Thompson, in Microcrystalline semiconductors: materials science and devices, edited by P. M. Fauchet, C. C. Tsai, L. T. Canham, I. Shimizu, and Y. Aoyagi (Mater. Res. Soc. Proc., Pittsburg, Penn., 1994), Vol. 283, p. 703.

[11] J. S. Im and H. J. Kim, Appl. Phys. Lett. **64**, 2303 (1994).

[12] J. S. Im, H. J. Kim, and M. O. Thompson, Appl. Phys. Lett. **63**, 1969 (1993).

[13] J. H. Yoon and J. S. Im, Metals and Materials **5**, 525 (1999).

Mat. Res. Soc. Symp. Proc. Vol. 621 © 2000 Materials Research Society

Sequential Lateral Solidification of Ultra-Thin a-Si Films

Hans S. Cho, Dongbyum Kim, Alexander B. Limanov, Mark A. Crowder, and James S. Im, Program in Materials Science and Engineering, Department of Applied Physics and Applied Mathematics, School of Engineering and Applied Science, Columbia University, New York, NY

ABSTRACT

This paper demonstrates that Sequential Lateral Solidification (SLS) of Si can be carried out on films as thin as – and potentially much thinner than – 250 Å. When compared to thicker Si films, however, the SLS-processed ultra-thin films contain more twins, and successful processing requires irradiation within a narrower laser energy density range and a smaller per-pulse translation distance. The physical interpretation of these findings is formulated by analyzing the details of the microstructures observed in single-pulse-irradiation-induced Controlled Super-Lateral Growth (C-SLG) experiments. SEM and TEM analyses reveal complicated microstructural details that we interpret as originating from breakdown of epitaxial growth during lateral solidification, an effect that is detrimental to the SLS process. Based on considerations of far-from-equilibrium solidification of Si, it is argued that undercooling of the solidification interface below a threshold value at which solidification no longer proceeds epitaxially – arising from reduction in interfacial recalescence during lateral solidification of ultra-thin Si films, relative to that of thicker films – is responsible for the breakdown. Based on this model, we discuss how external parameters may be adjusted so as to permit optimal crystallization of ultra-thin Si films using SLS.

INTRODUCTION

Pulsed-laser annealing (PLA) of thin amorphous silicon (a-Si) films deposited on glass substrates has been the focus of much interest with the growing demand for high-performance thin film transistors (TFTs). As various applications for thin Si film-based devices are developed, and as the design rule for devices decreases, it is anticipated that more interest will be directed towards films of increasingly small thickness. Deposited as a-Si or small-grained polycrystalline Si on SiO_2 surfaces, these films require additional processing to improve the electronic properties of devices that would eventually be fabricated on them.

Sequential Lateral Solidification (SLS) represents an effective low-temperature PLA method that can produce high quality crystalline films from amorphous and polycrystalline precursor materials [1,2]. The SLS process utilizes the Controlled Super-Lateral Growth (C-SLG) technique [3] that relies on inducing complete melting of localized regions and subsequent lateral growth from adjacent unmelted regions. Using repeated C-SLG-inducing laser irradiations, combined with systematic and coordinated spatial translations of the film relative to the beamlets, SLS makes possible the realization of various microstructures, from grain-boundary-location-controlled polycrystalline Si films to location-controlled single-crystal regions, on amorphous substrates.

In this study, SLS processing of ultra-thin films is conducted and analyzed. The details of SLS are analyzed by studying the features found in C-SLG microstructures of the films. Previous C-SLG and SLS investigations to date have dealt with films of thickness in the range of 500 Å to 2500 Å. There are reasons to anticipate, based on a number of previous investigations

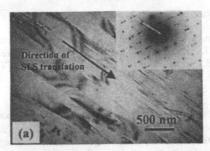

Figure 1. TEM images of directionally solidified microstructure resulting from SLS processing. (b): area of (a) taken at twice the magnification. Inset in (a) is a selected area diffraction pattern of a twinned sub-grain region.

[4-9], that one could encounter more complicated transformation behavior in ultra-thin Si films, which in turn can lead to more varied microstructures than have been observed in thicker films. The experimental results we report and analyze in this investigation reveal that this is indeed the case.

EXPERIMENT

The test samples consisted of 250-Å-thick PECVD a-Si films, deposited on SiO_2 buffer-layer (3000Å PECVD) coated Corning 1737 glass substrates. SLS crystallization was conducted using a XeCl (308 nm wavelength) excimer laser (Lambda Physik LPX 315i) with FWHM pulse duration of 28 nsec. The laser pulses were projected through a cylindrical lens system to produce beamlets ~2 μm wide incident on the a-Si film surface. The substrate was spatially translated perpendicular to the slits between pulses at distances of 0.2 - 0.5 μm per pulse. Both energy density and per-pulse translation distance were varied during SLS to produce a parametric plot that gives information about the microstructures produced at these conditions, from which a process window for continuous SLS (i.e., continuous growth of grains) can be obtained. The discontinuous SLS (agglomeration) regime was defined as the set of processing conditions for which discontinuations of the grains (agglomeration) were (was) detected in the processed film.

In order to understand the phenomenological details encountered during the C-SLG process (which constitutes the basis of the SLS process) the microstructures in regions irradiated by individual pulses using the above configuration were examined in detail. The film was irradiated at various laser energy densities. Microstructural analyses were conducted using TEM, SEM, and optical microscopy. Secco etching was performed prior to optical and SEM analysis. TEM samples were obtained by detaching portions of the film from the glass substrate using chemical etching of the buffer layer and substrate.

RESULTS

Figure 1 shows an area of a 250-Å-thick Si film directionally crystallized by the SLS process using a per-pulse translation distance of 0.5 μm. The parametric plot in Fig. 2 reveals a

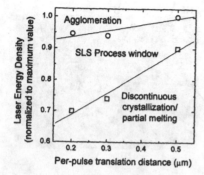

Figure 2. Parametric analysis of the SLS process for 250 Å film, showing continuous SLS process window. Squares and circles indicate the continuous SLS and the agglomeration threshold, respectively.

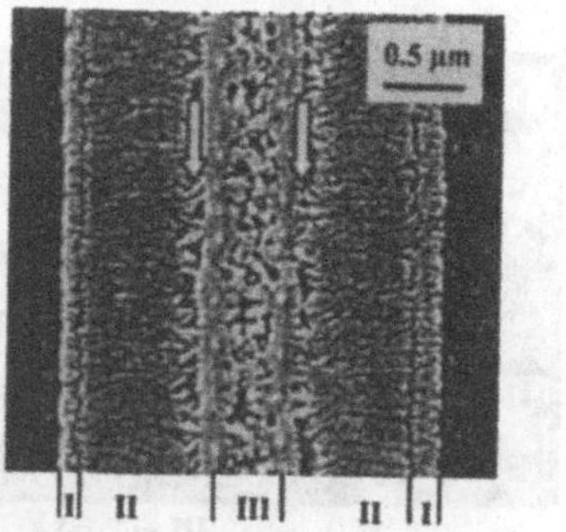

Figure 3. SEM image of microstructure resulting from single-pulse irradiation of 250 Å film. White arrows indicate areas of the film preferentially removed by Secco etching.

parameter domain in which directionally solidified microstructure (i.e., continuous SLS) is obtained. For a given per-pulse translation distance, the energy density window for obtaining continuous SLS was found to be smaller than for thicker films, roughly 20% of the mean energy density at a per-pulse translation distance of 0.3 μm, compared to ~30% for films 500 Å and thicker [10]. As was the case for thicker films [1], sub-grain boundaries, grain boundaries, and twins were found in SLS-crystallized ultra-thin films (Fig.1). The distance between the sub-grain boundaries was roughly 0.15 – 0.2 μm. One of the more prominent microstructural characteristics of the SLS-processed ultra-thin films is that most of the crystallized grains contained a high density of twins, as can be seen in Fig.1 and the inset selected-area diffraction image.

Figures 3 and 4 show SEM and TEM images, respectively, of C-SLG microstructure resulting from single-pulse irradiation that induced complete melting within the irradiated portion of the film. From the SEM image, one can observe that the overall microstructural pattern is roughly similar to that typically obtained from thicker films [3]: two areas of small grains (marked "I" in Fig.3); two columns of large elongated grains (II), and a central region consisting of highly defective areas intermixed with small microcrystalline grains (III). There are also two rows of grains (D and E in Fig.4-b) on the border of Regions II and III, with protrusions appearing as white spots between these rows (Fig.3).

However, TEM analysis clearly reveals features that are not observed in thicker films, in the area of Region II that borders Region III (Fig.4-b). These are labeled as area (B), in which the microstructure is highly defective, and (C), in which there appear to be no identifiable crystalline features. These are identified as the regions preferentially removed by Secco etching, observed in SEM (white arrows in Fig.3). The C-SLG microstructures of thicker films do not contain such defective regions, as the region corresponding to Region II consists entirely of large grains such as those in area (A).

Therefore, the observed microstructures in Region II, labeled (A) - (E) along the direction of the black arrow (Fig.4-b), are as follows: (A) elongated crystalline grains (some containing a

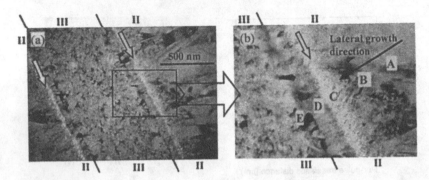

Figure 4. TEM images of typical C-SLG microstructure observed in 250 Å films. (b) is enlarged view of inset area in (a); labeled microstructures are described in text.

high density of twins), (B) highly defective region, (C) an amorphous-like region where no crystalline features are visible, and (D) re-appearance of crystalline material in the form of a row of non-equiaxed grains; in Region III adjacent to (D), there is also a parallel row of crystalline non-equiaxed grains (E).

The lateral lengths of (A) and (II) are plotted versus incident laser energy density in Fig.5. The trend is similar to that of C-SLG distance in thicker films in that with increasing energy density, the total length of (II) and the length of (A) increase, and the width of the central equiaxed region (III) decreases (i.e., the total length of the remaining irradiated cross-section increases) [3]. At relatively high incident energy density, the highly-defective/amorphous regions (B, C) become less pronounced, though the grains identified in (D) increase in length.

DISCUSSION

As can be observed from Fig. 4, C-SLG microstructure obtained in ultra-thin films is largely similar to, but contains certain features that make it distinct from, that obtained in thicker films. The primary microstructural difference we recognize as being harmful for SLS – but fundamentally interesting – pertains to the appearance of extremely defective and microcrystalline/amorphous regions (B, C) within the C-SLG microstructure.

Based on what has been established about the C-SLG process [3] and rapid solidification behavior of Si, we propose that the basic C-SLG melting and solidification scenario for ultra-thin films remains the same as the established C-SLG scenario (i.e., lateral growth commencing from a pre-existing solid-liquid interface until it terminates by coming into contact with another solidification front, which may originate from nucleated grains or the opposing C-SLG growth front). According to this interpretation, the structures (B), (C) and (D) are formed, in that order, during lateral growth (i.e., in the same manner as (A)). It is argued that the difference in the microstructure observable in (B) and (C) with that of (A) originates from the breakdown in epitaxial growth during the lateral solidification. This breakdown, we further argue, stems from the deeper interfacial undercooling – associated with thinner Si films and arising from reduction of interfacial recalescence – that dips below a threshold value, the point below which solidification no longer proceeds epitaxially.

Thus, according to this model, it is possible to account for the entire microstructural pattern observed in Region II by simply keeping track of the interface temperature during lateral solidification and the commensurate sequence of the lateral solidification as follows. After the maximum extent of melting is reached, the lateral epitaxial growth initiates as the interface temperature finds itself below the melting point, resulting in large grains (A). As the interface temperature continuously decreases and eventually drops below a certain threshold value (which must presumably be close to but higher than the full interfacial amorphization threshold), the lateral growth proceeds non-epitaxially (B, C). Concurrently, within the central portion of the completely melted area (region III), deep supercooling of the bulk liquid triggers copious nucleation followed by rapid growth ahead of

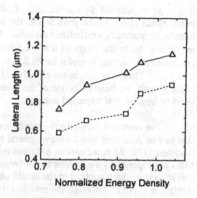

Figure 5. The length of epitaxial lateral growth (A) indicated by squares, and the total lateral growth distance (Region II) indicated by triangles, plotted versus the laser energy density, normalized to the values in Fig.2.

the interface (beyond the characteristic nucleation denuded zone caused by interfacial recalescence [11]). As the C-SLG growth front approaches the growing solids in Region III, the latent heat of transformation released from the growing nuclei combined with that released from the C-SLG interface causes a rise in interface temperature to a value above the threshold, resulting in a return to crystal-growth mode of solidification (D), at which point many grains appear and grow laterally. The grains identified in region (E) are formed when lateral growth proceeds, also with increasing interfacial temperature, from the nucleated solids into the nucleation-denuded zone.

Such experimental results and the interpretation are mostly consistent with previous works that investigated far-from equilibrium solidification behavior of Si. Over the past two decades, it has been predicted [4], observed [5-7], and established [7,9] that, given sufficient undercooling of the interface (and commensurate increase in growth rate), liquid Si transforms to a-Si via interfacial amorphization (i.e., a discontinuous first order phase transition). Although it is not fully certain whether the structure in (C) is fully amorphous, merely extremely defective, or a mixture of amorphous and microcrystalline regions, it is certain that the conditions encountered during lateral growth have produced interface undercooling and interface velocities that are on the order of the values associated with the interfacial amorphization threshold. Indeed, numerical analyses of C-SLG in 1000 Å films have approximated the quasi-steady-state lateral growth velocities to be in the range of 10 m/sec [9], a value approaching those obtained for the amorphization during vertical re-growth (e.g., 15 m/sec for (100) orientation) [8]. As the lateral growth velocities during C-SLG of ultra-thin films are presumably higher (i.e., deeper interfacial undercooling is attained), the appearance of non-epitaxial growth is not surprising, and, in fact, expected. We plan to conduct additional experimental and analytical work in order to clarify the physical details associated with the breakdown of epitaxial growth and the re-emergence of crystalline lateral growth observed in these films.

As regards the SLS processing of ultra-thin films, the above result shows that only the initial epitaxial crystalline growth (i.e., the pre-breakdown portion) is useful for producing continuous directionally crystallized material. Subsequently, even though the total lateral growth length may be in the range of 0.8 – 1.5 µm, the corresponding lengths of the laterally grown crystalline region that is useful for SLS (i.e., the effective C-SLG distance) ranges from 0.6 µm to 0.9 µm. This reduction in the effective SLG distance due to the interfacial breakdown of epitaxial growth, we believe, is one of the main factors responsible for the decrease in the process window beyond that expected according to the thickness-dependent trends observed in thicker films.

The observed average sub-grain boundary spacing of ~0.2 µm is consistent with the value that can be predicted based on the general trend obtained from SLS-processed films of greater thickness [12]. As noted earlier, a distinct microstructural feature of the SLS processed ultra-thin films corresponds to the prevalence of twins within a large fraction of grains (Fig.4). This observation is in sharp contrast to the results obtained in 500Å films, in which twinning defects are rarely found [13]. Although the precise origin of this difference is yet to be revealed in detail, we presently feel that it is also likely to be associated with the interfacial undercooling in the present films approaching the threshold associated with breakdown of epitaxial growth.

The above model and discussions suggest that even deeper interfacial undercooling during lateral solidification that will be experienced, for example, in thinner films, with more conductive substrates, and/or using laser pulses of shorter duration, will result in more prompt breakdown of the crystalline growth and full interfacial amorphization during lateral growth (as well as thorough amorphization within the central region, where the transformation is triggered by nucleation [7]). The model also indicates that it will be possible to prevent such a breakdown in the epitaxial growth (i.e., sustain the interface temperature well above the amorphization threshold), by properly adjusting external parameters. This can be accomplished, for instance, by decreasing the effective heat conduction into the substrate, applying substrate heating, and/or by increasing the laser pulse duration.

This work was funded by DARPA under project N66001-98-1-8913.

REFERENCES

1. R.S. Sposili and J.S. Im, *Appl. Phys. Lett.* **69** (19), 1996
2. J.S. Im, R.S. Sposili, and M.A. Crowder, *Appl. Phys. Lett.* **70** (25), 1997
3. J.S. Im, M.A. Crowder, R.S. Sposili, J.P. Leonard, H.J. Kim, J.H. Yoon, V.V. Gupta, H. J. Song, and H.S. Cho, *phys. stat. sol. (a)* **166** (603) 1998
4. D. Turnbull, *Mat. Res. Soc. Symp. Proc.* **Vol. 13** (131-134), 1983
5. P.L. Liu, R. Yen, and N. Bloembergen, *Appl. Phys. Lett.* **34** (12), 1979
6. T. Sameshima and S. Usui, *Appl. Phys. Lett.* **59** (21), 1991
7. T. Eiumchotchawalit, J.S. Im, *MRS Symposium Proceedings* **Vol. 321**, (725) 1994
8. A.G. Cullis, N.G. Chew, H.C. Webber, and D.J. Smith, *Journal of Crystal Growth* **68** (624-638), 1984
9. H.J. Kim and J.S. Im, *MRS Symposium Proceedings* **Vol. 397**, (401) 1996
10. R.S. Sposili and M.A. Crowder, unpublished
11. J.S. Im, H.J. Kim, and M.O. Thompson, *Appl. Phys. Lett.* **63** (14), 1993
12. M.A. Crowder, A.B. Limanov, and J.S. Im, published in the present volume.
13. M.A. Crowder, A.B. Limanov, and J.S. Im, to be published

Mat. Res. Soc. Symp. Proc. Vol. 621 © 2000 Materials Research Society

High-Performance Poly-Si TFTs With Multiple Selectively Doped Regions In The Active Layer

Min-Cheol Lee, Juhn-Suk Yoo, Kee-Chan Park, Sang-Hoon Jung, Min-Koo Han, and Hyun-Jae Kim*

School of Electrical Engineering, Seoul National University, Seoul, 151-742, KOREA
AMLCD division, Samsung Electronics, Kyung-gi Do, KOREA*

ABSTRACT

We have proposed and fabricated a new poly-Si TFT that employs selectively doped regions between the source and drain in order to reduce leakage current without the sacrifice of the on current. In the proposed poly-Si TFTs, the selectively doped regions where doping concentration is identical to that of source/drain, reduce the effective channel length during the on state. Under the off state, the selectively doped regions may reduce the lateral electric field induced in the depletion region near drain so that the leakage current reduces considerably. The experimental data of the proposed TFT shows that it has the high on-current, low leakage current and low threshold voltage when compared with conventional TFT. The fabrication steps for the proposed TFT are reduced because ion-implantation for source/drain and selectively doped regions is performed simultaneously prior to an excimer laser irradiation. It should be noted that, in the proposed TFT, only one excimer laser annealing is required while two excimer laser annealing steps are required in conventional TFT.

INTRODUCTION

In the active matrix liquid crystal display (AMLCD) application, low temperature poly-Si thin-film transistors (TFTs) fabricated by excimer laser annealing have been widely investigated due to high mobility and integration of pixel and peripheral circuit on the inexpensive glass substrate [1]. However, the leakage current is considerably high due to the trap density of grain boundary [2]. It is well known that leakage current is originated from the electron-hole generation stimulated by electric field via trap-state on the grain boundary. Most previously reported structures, such as LDD (Lightly Doped Drain), undoped offset, successfully reduce the leakage current of poly-Si TFTs by the reduction of the electric field in the depletion region near the drain. However, in those device structures, the on current is also decreased due to high resistive region of LDD or undoped offset so that on/off current ratio is decreased [3, 4].

The purpose of our work is to propose and fabricate a new poly-Si TFT structure with selectively doped region between source and drain in order to improve the on/off current characteristics simultaneously. In the proposed poly-Si TFT, we employ the selectively doped regions in the active layer by an ion implantation as well as ion shower and the doping concentration of implanted ions is identical to that of source/drain.

DEVICE STRUCTURE AND EXPERIMENT

The proposed poly-Si TFT has the selectively doped regions with in the active layer as shown in *Figure.1*. The selectively doped regions are orderly located and their doping concentration is identical to source/drain because ion-implantation steps for the selectively doped regions and the source/drain are achieved by one masking step prior to excimer laser irradiation. *Figure.2* shows the schematic sketch of device operation under the on- and off state. Under the off state, the selectively doped region may redistribute the lateral electric field throughout the active layer. By inserting selectively doped regions into the active layer, a few potential barriers are located between the source and drain, which results in the lateral field reduction in the depletion region near the drain. As shown in *Figure.2 (a)*, the depletion region, where peak electric field is located, exists at the undoped/n+ doped region interface and reduces the electric field induced near the drain. Due to the relaxed electric field, the leakage current of the proposed TFT can be considerably reduced. Under the on state, the effective channel length is equivalent to the summation of undoped region length, $L_{eff} = L_1 + L_2 + L_3$, as shown in *Figure.2 (b)*. As the effective channel length become smaller, the on current characteristics can be increased considerably when compared with conventional or conventional LDD TFT.

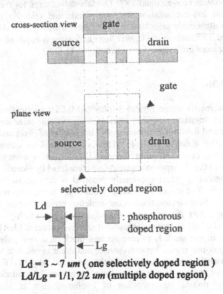

Figure.1 the device structure of the proposed poly-Si TFT

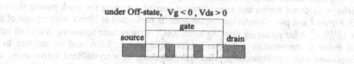

(a) off state operation

(b) on state operation

Figure. 2 the device operation of the proposed TFT under on and off state

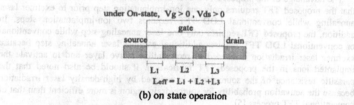

(a) Ion-implantation for source/drain
and selctively doped regions

(B) Excimer Laser Annealing for
recrystallization and activation

(c) Gate Oxide and Gate electrode
formation and passivation

Figure.3 the key process of the proposed TFT

Figure. 3 shows the key fabrication step of the proposed TFT. The mask step for selectively doped region and source/drain is carried out using the mask pattern shown in *Figure. 1* and phosphorous ion-implantation is performed at 30keV with dosage of 5 × 10^{15}/cm^2. After removing masking layer, XeCl excimer laser annealing is carried out in order to recrystallize a-Si film whose thickness is 800Å and to activate the implanted ions simultaneously. After the patterning of recrystallized active layer, a 1000Å thick TEOS oxide and a 4000Å thick aluminum layer is deposited on the active layer. Aluminum layer is patterned using RIE and finally, metallization and passivation are achieved to complete device fabrication.

The advantage of the proposed TFT over conventional one, in device fabrication, is that the proposed TFT requires only one ion-implantation step prior to excimer laser annealing while conventional LDD TFT requires two ion-implantation steps. In addition, the proposed TFT requires only one laser annealing step while conventional or conventional LDD TFT requires additional excimer laser annealing step because excimer laser irradiation is needed to recrystallize active layer and to activate the implanted ions in the proposed TFT fabrication. It should be also noted that the parasitic resistance of the source/drain is reduced by high-density laser irradiation because the activation probability for source/drain region is more efficient than that of conventional TFT process [5].

RESULT AND DISCUSSION

We fabricated two types of poly-Si TFTs with single selectively doped region and that with multiple doped regions. *Figure. 4 (a)* shows the transfer characteristics of poly-Si TFT with one selectively doped region located in the middle of active layer. Under the on state, the on current and threshold characteristics are improved considerably compared with conventional TFT. In case of Ld=7μm, the on current and threshold voltage are 0.3mA and 3.6V respectively while, in conventional one, the on current and threshold voltage are 0.07mA and 7.6V respectively. Under the off state, the leakage current of the proposed TFT is lower than that of conventional one by the magnitude of 1 order due to the lateral electric field relaxation near drain under the off state. High on/off current ratio as high as 10^8 orders is obtained when doped length is 7μm and channel length is 10μm.

In the proposed TFTs, the on current is increased and leakage current is reduced. This is the very unique characteristic of the proposed TFT with low leakage current and high on current. In the most previously reported poly-Si TFT structures, leakage current is decreased but the on current is also decreased due to the addition of high resistive region in device structure. However, in our TFTs, the on current is much higher than that of conventional one due to the short effective channel length.

Figure. 4 (b) shows the transfer characteristics of the TFT with multiple doped regions. We designed the selectively doped stripes with Ld/Lg = 1/1, 2/2[μm] which are repeatedly arranged between source and drain as shown in *Figure. 1*. In case of TFT with multiple doped regions, leakage current is much lower than that of TFT with single selectively doped region. In case of 1/1 TFT in *Figure. 4 (b)*, five potential barriers are formed between the source and drain so that the leakage current is significantly reduced compared with TFT that has one selectively doped region because the number of potential barriers formed by multiple doped regions is increased.

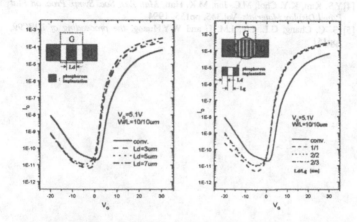

(a) Single selectively doped region   (b) multiple selectively doped regions

Figure. 4 the transfer characteristics of the TFT with selectively doped region

The on current of the proposed TFT is also improved due to short effective channel length. In our experiment, I_{max}/I_{min} of the 1/1 TFT is up to 10^8 while that of the conventional one is 3×10^6.

CONCLUSION

We propose and fabricate a new poly-Si TFT with selectively doped region successfully. In the proposed TFT, there is the unique property of poly-Si TFT with low leakage current and high on current. By inserting selectively doped region into the active layer, the effective channel length is significantly reduced under the on state and leakage current can be considerably reduced under the off state. The number of process steps required in new poly-Si TFT can be reduced compared with conventional LDD TFT because ion-implantation for selectively doped regions and source/drain are performed prior to excimer laser annealing so that only one ELA process step is required while two ELA steps, that is, recrystallization and activation, are required.

REFERENCE

[1] T. Sameshima, M. Hara and S. Usui: *Jpn. J. Appl. Phys.* **28** (1989) L2132
[2] J.G.Fossum, A.Ortiz-Conde, H.shichijo, and S.K.Banerjee, *IEEE Trans. Electron Devices*, **32**, pp1878, 1989
[3] C.M. Park, B.H. Min, K.H. Jang, M.K. Han, *SSDM Extended Abstract*, p.656,

1995

[4] Y.S. Kim, K.Y. Choi, M.C. Jun, M.K. Han, *Mat. Res. Soc. Symp. Proc. on Flat Panel Display Materials*, Vol.345, pp155, 1994

[5] S. C. Chang, G.L. Lin, I.M. L and W.Y.Huang, *the proceeding of IDW'99*, pp151

Mat. Res. Soc. Symp. Proc. Vol. 621 © 2000 Materials Research Society

Oxide Charging Effects by PH₃/He Ion Shower Doping

Cheon-Hong Kim, Juhn-Suk Yoo, Kee-Chan Park and Min-Koo Han

School of Electrical Engineering, Seoul National University,

San 56-1 Shinlim-dong, Kwanak-gu, Seoul 151-742, Korea

ABSTRACT

We report the oxide charging effects on metal oxide semiconductor (MOS) structure caused by PH₃/He ion shower doping. The parallel negative shift of flat-band voltage occurred for the ion-doped PETEOS samples even after thermal annealing. When the ion dose was higher, this shift was larger. These results show that a considerable amount of positive charges were induced inside the oxide films after PH₃/He ion shower doping process. For the same ion dose, the flat-band voltage shift is larger when the thickness of PETEOS is thicker. When the ion dose was 1.5×10^{17}cm^{-2} and the thickness of PETEOS was 80nm, the shift of flat-band voltage was larger than -7V. We can conclude that PH₃/He ion shower doping process induces the positive charges, which result in the flat band voltage shift of MOS capacitors, in the bulk oxide films when oxide films are exposed to ion shower doping.

INTRODUCTION

Polycrystalline silicon (poly-Si) thin film transistor (TFT) employing the excimer laser annealing is a promising device for active matrix liquid crystal displays (AMLCDs) due to its high mobility and large current capability [1]. For source and drain doping in large-area processing, low-energy ion shower doping technique using a broad ion beam without mass separation and beam scanning is widely used instead of conventional ion implantation techniques [2,3].

It is well known that plasma process usually causes electrostatic charging damages to the integrity of gate oxide [4]. Recently, it has been reported that the oxide charging damages during the hydrogen plasma passivation would antagonize the effect of hydrogenation [5]. Also, it has been found that the hydrogen plasma passivation applied on metal oxide semiconductor (MOS) structure induces positive charges in the oxide, causing the flat band voltage shift of device [6].

However, the oxide charging effects by ion shower doping have been reported scarcely. Unlike conventional plasma treatments, the ions diffused from plasma region to acceleration mesh are accelerated by a potential difference between the mesh and a sample stage and are radiated onto samples in the ion shower apparatus [2,3]. Charge accumulation during the ion shower doping is effectively suppressed using the double meshes, that is, acceleration mesh and grounded mesh [2]. While PH_3/He ion shower doping is performed, various positive ions, such as H^+, P^+, PH^+ etc., are radiated on samples and implanted into them. Oxides should be damaged by these ions like the case of ion implantation [7].

The purpose of our work is to report the oxide charging effects on MOS structure caused by PH_3/He ion shower doping. From high frequency (1MHz) capacitance-voltage (C-V) characteristics, we have verified that considerable amount of positive charges were induced inside the oxide films after PH_3/He ion shower doping process.

EXPERIMENTS AND RESULTS

A Plasma-enhanced tetra-ethyl-ortho-silicate (PETEOS) film was deposited at 390°C on p-type silicon (Si) wafer with the resistivity of about 10Ωcm. The thickness of PETEOS films was 40nm and 80nm. The ion shower doping was performed for some samples using RF plasma source (13.56MHz) at room temperature. Doping source was gas mixture of 1% PH_3 diluted in He and pressure was about 10^{-4}Torr. The diameter of ion beam was about 20cm and the ion current density was fixed at 20μA/cm^2. The acceleration voltage was 6kV and the ion doping time was varied from 10min to 20min. The resultant ion dose was 7.5×10^{16}cm^{-2} and 1.5×10^{17}cm^{-2}, respectively. After aluminum (Al) metallization, samples were thermal-annealed at 400°C for 2hours. This post-annealing step is required to activate dopants and to cure damages in poly-Si films exposed to ion shower doping. In order to observe the practical situation, this annealing step was performed. We have investigated high frequency (1MHz) capacitance-voltage (C-V) characteristics.

Figure 1 shows high frequency C-V characteristics of MOS capacitors without and with ion shower doping after thermal annealing. The thickness of PETEOS was 40nm. It is noted that the parallel negative shift of flat-band voltage occurred for the ion-doped PETEOS samples. The flat-band voltage shift of PETEOS exposed to ion shower doping with the ion dose of $7.5 \times 10^{16} cm^{-2}$ was -1.6V. This result shows that a considerable amount of positive charges indeed were induced inside the oxide films after PH_3/He ion shower doping process. Since the penetration depth is less than 15nm even in Si films [2], the ions cannot penetrate into Si substrate. Therefore, it is assumed that the degradation of SiO_2/Si interface is negligible. Therefore, PH_3/He ion shower doping process induces the positive charges, which result in the flat band voltage shift of MOS capacitors, in the bulk oxide films when oxide films are exposed to ion shower doping.

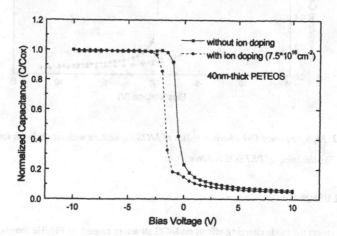

Figure 1. *High frequency (1MHz) C-V characteristics of MOS capacitors without and with ion shower doping after thermal annealing. The thickness of PETEOS is 40nm.*

High frequency C-V characteristics of MOS capacitors for 80nm-thick PETEOS without and with ion shower doping after thermal annealing are shown in figure 2. For PETEOS samples exposed to ion shower doping, the ion dose was $7.5 \times 10^{16} \text{cm}^{-2}$ and $1.5 \times 10^{17} \text{cm}^{-2}$, respectively. The parallel negative shift of flat-band voltage occurred for PETEOS samples exposed to ion shower doping like the case of figure 1. When the ion dose was higher, this shift was larger. When the ion dose was $1.5 \times 10^{17} \text{cm}^{-2}$, the shift of flat-band voltage was larger than $-7V$. For the same ion dose, the flat-band voltage shift is larger in case of 80nm-thick PETEOS.

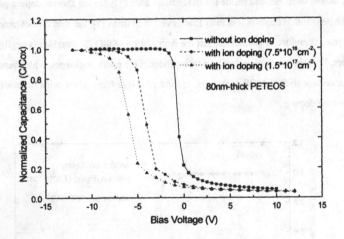

Figure 2. *High frequency C-V characteristics of MOS capacitors without and with ion shower doping. The thickness of PETEOS is 80nm.*

CONCLUSIONS

We report the oxide charging effects on MOS structure caused by PH_3/He ion shower doping. From high frequency (1MHz) capacitance-voltage (C-V) characteristics, we have verified that a considerable amount of positive charges were induced inside the oxide films after PH_3/He ion shower doping process. The parallel negative shift of flat-band voltage occurred for the ion-doped PETEOS samples even after thermal annealing at 400°C for 2 hours.

This shift was larger when the ion dose was higher. For the same ion dose, the flat-band voltage shift is larger when the thickness of PETEOS is thicker. These results show that a considerable amount of positive charges were incorporated inside the oxide films after PH_3/He ion shower doping process. When the ion dose was $1.5 \times 10^{17} cm^{-2}$ and the thickness of PETEOS was 80nm, the shift of flat-band voltage was larger than $-7V$.

REFERENCES

1. M. Hack, P. Mei, R. Lujan and A. G. Lewis, *JNCS* 164-166, 727-730 (1993).
2. J. M. Jun, S. S. Yoo, K. H. Lee, H. K. Kang, K. N. Kim and J. Jang, *Journal of the Korean Physical Society,* 27, S95-S98, (1993).
3. A. Mimura, G. Kawachi, T. Aoyama, T. Suzuki, Y. Nagae, N. Konishi and Y. Mochizuki, *IEEE Trans. Electron Device,* 40, 513-520 (1993).
4. J. P. McVittie, *IEDM Tech. Dig.,* 433-436 (1997).
5. K. Y. Lee, Y. K. Fang, C. W. Chen, K. C. Huang, M. S. Liang and S. G. Wuu, *IEEE Electron Device Lett.,* 18, 187-189 (1997).
6. C. F. Yeh, T. Z. Yang and T. J. Chen, *IEEE Trans. Electron Device,* 42, 307-314 (1995).
7. M. Takase, K. Eriguchi and B. Mizuno, *Jpn. J. Appl. Phys.,* 36, 1618-1621 (1997).

Mat. Res. Soc. Symp. Proc. Vol. 621 © 2000 Materials Research Society

The Dependence of Poly-Si TFT Characteristics on the Relative Misorientation Between Grain Boundaries and the Active Channel

Y.H. Jung, J.M. Yoon, M.S. Yang, W.K. Park, H.S. Soh, Anyang Laboratory, LG-Philips LCD Co. Ltd, Dongan-gu, Anyang-shi, Kyungki-do, KOREA; H.S. Cho, A.B. Limanov, and J.S. Im, Program in Materials Science, Columbia University, New York, NY 10027

ABSTRACT

We present device-related experimental results that quantitatively reveal the effect of varying the active channel/grain-boundary misorientation on the resulting TFT characteristics. Specifically, using low-temperature SLS processes, we have fabricated and analyzed n-channel and p-channel devices (40 μm width x 8 μm length) with three different orientations of the channel with respect to the grain boundaries: parallel, 45° inclined, and perpendicular on Corning 1737 glass substrates.

The results reveal that the TFTs with the best (worst) characteristics were obtained for the devices with parallel (perpendicular) alignment. In general, for both n- and p-channel devices, the most prominent orientation-dependent effects were observed in the values of the field effect mobilities, which were 340, 227, and 141 cm^2/Vsec for n-channel devices and 145, 105, and 80 cm^2/Vsec for p-channel devices, in the order of increasing orientation mismatch. In contrast, no notable effect was manifested in the leakage currents, while small effects were seen for the sub-threshold slopes and threshold voltages. The degradation of device performance under hot-carrier stress was found to decrease with increasing orientation mismatch.

INTRODUCTION

Integration of TFT-LCD panels by fabricating the gate driver and data driver circuits onto the same thin Si film on an insulating substrate demands higher TFT device performance than is available with conventional ELA-processed poly-Si TFTs. The channel mobility for poly-Si TFTs fabricated on films crystallized using conventional ELA methods is ~140 cm^2/Vsec for n-channel devices and ~60 cm^2/Vsec for p-channel devices, values that are insufficient for the operation of driver circuits, restricting the use of TFTs for those devices.

The first step in improving device performance in TFTs is to improve the quality of the Si thin film. For poly-Si TFTs, the electrical properties of the device are limited by the presence of grain boundaries in the channel region. In particular, channel mobility is limited by the formation of potential barriers at the grain boundaries, which are the result of trapping states due to dangling bonds [1]. Therefore, improvement in electrical properties of polycrystalline Si thin films can be achieved by reducing the number of grain-boundaries and by controlling their location within the channel region [2]. The Sequential Lateral Solidification (SLS) method [3-6] has been developed to achieve this goal by controlling the location and orientation of grain boundaries or by eliminating grain boundaries in the channel region altogether. For example, the directional solidification form of SLS produces material with long, roughly parallel grains and grain boundaries [5].

Devices fabricated on Si films directionally solidified using the SLS-process were found to exhibit device characteristics superior to those fabricated on films crystallized using the

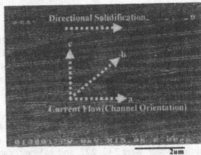

a. Parallel

b. 45 Degree Misoriented

c. 90 Degree Misoriented

Figure 1. Experimental configuration: Schematic representations of TFT devices in various orientations (above), and illustration of the relative orientation mismatch of TFT channels with respect to grain boundaries of directionally solidified SLS-processed film (shown in SEM picture at right).

conventional ELA process [7], the microstructure of which consists of randomly located equi-axed grains.

In this study, the SLS method was used to directionally crystallize a-Si films on glass substrates for fabrication of TFTs. The devices were aligned relative to the long grain-boundaries in three configurations: parallel, perpendicular, and at a 45 degree angle, in order to investigate the effect that the relative orientations have on the electrical properties of the device. We were able to confirm that the optimum device performance was obtained when the channel direction was aligned parallel to the grain boundaries.

EXPERIMENTAL METHODS

As Corning 1737 glass panels were used as the substrate material for the coplanar n- and p-channel TFTs. All processes, including dehydrogenation annealing, were conducted at temperatures under 400°C. First, an SiO_2 buffer layer 3000-Å-thick was deposited by PECVD on the glass substrate. Then, the 1000 Å precursor a-Si film - the active layer - was deposited and the active channels were patterned as islands of a-Si. The active channel islands were oriented parallel, perpendicular, and 45° relative to the direction of SLS crystallization as shown in Figure 1, in order to study the effect the number of grain boundaries in the channel region has on the device performance. The source and drain regions were doped with n- or p-type doses before the SLS crystallization step, which activates the dopants.

A pulsed excimer laser beam (Lambda Physik LPX 315i, with discharge wavelength of 308 nm, using XeCl as discharge gas) shaped into a long linear beamlet was used to directionally crystallize the a-Si film [5]. In this method, each pulse melted a linear region ~2 μm wide and several cm long, and was translated ~0.5 μm relative to the previous pulse in the direction perpendicular to the line of the beam. The beam was shaped by passing the beam through a long slit, then projecting it onto the target film with a cylindrical lens [4]. Long grains form parallel to the direction of translation of the beamlet, which is typical of the SLS process.

Afterwards, the gate insulator, a 1000-Å-thick SiO_2 layer, was deposited, and metallic gate contacts were formed by depositing and patterning Al and Mo. The patterning was performed so that there would be no overlap of the gate electrode with the source or the drain regions. An SiO_2 film and a SiNx film were deposited in succession to form a passivation layer. After patterning of a contact hole, ITO film was deposited and patterned to form the gate, source, and drain electrodes, to complete the fabrication of the TFT. The channel width and length of the devices fabricated for this study were 40 μm and 8 μm, respectively.

The device performance characteristics of TFTs with different degrees of active channel/grain boundary mismatch were measured and compared. In addition to the basic device characteristics (I-V curves, mobility, leakage current etc.) at room temperature, measurements on the effects of hot carrier stress were conducted. In the subsequent sections, Vd, Vg, and Id refer to drain voltage, gate voltage, and drain current, respectively.

RESULTS AND DISCUSSION

Figure 1 shows an SEM image of SLS-processed directionally crystallized film and schematic diagrams illustrating and indicating the misorientation of the fabricated TFT channels relative to the (roughly) unidirectional grain boundaries. The grains are continuous and generally free of grain boundaries in the solidification direction, but in the direction perpendicular to the direction of solidification, the grain boundary spacing is roughly 3000 Å. Therefore, the average number of grain boundaries in the 8μm long channel could be calculated to be zero for the devices parallel to the grain boundaries, 26 for the devices perpendicular to the grain boundaries, and 18 for the devices at a 45° angle.

Figure 2 shows the TFT characteristics according to the relative orientation of the active channels and the directionally solidified grains. The values of channel mobility for the parallel case were 340cm²/Vsec for n-channel devices and 145 cm²/Vsec for p-channel devices, whereas for the perpendicular case, the values were 141 cm²/Vsec and 80 cm²/Vsec, respectively. For the case of 45° misorientation, the values were 227 cm²/Vsec and 105 cm²/Vsec, which are intermediate. These results show clear dependence of device performance on the number of grain boundaries in the channel region, or grain-boundary density, per orientation.

We have also confirmed that other device characteristics such as sub-threshold slope and threshold voltage also show a similar dependence on grain-boundary density. For the parallel case, when the number of grain boundaries in the channel region is minimized, the values of sub-threshold slope and threshold voltage are 0.46 V/dec and 1.1 V respectively for n-channel devices, 0.38 V/dec and -2.1 V for p-channel devices. The values for the perpendicular case, where the number of grain boundaries in the channel is maximized, are 0.56 V/dec and 2.0 V for n-channel, 0.68 V/dec and -3.1 V for p-channel devices. The case of 45° misorientation again produces intermediate values; 0.47 V/dec and 1.4 V for n-channel, and 0.46 V/dec and -2.5 V.

Figure 3 shows the change in the device output characteristics under hot-carrier stress of n-channel TFTs aligned parallel, perpendicular, and 45° to the grain boundaries. It was found that the devices aligned parallel to the grain boundaries - and therefore had the highest mobilities - were affected the most by hot carrier stress.

This result is consistent with the view that hot-carrier stress is a result of high electric fields in the drain region. Carriers are trapped at the grain boundaries in the channel region, resulting

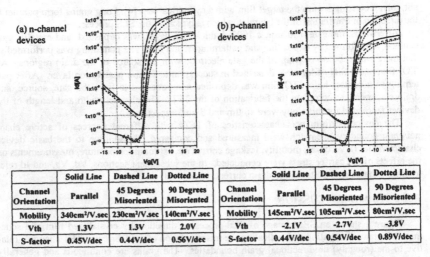

	Solid Line	Dashed Line	Dotted Line
Channel Orientation	Parallel	45 Degrees Misoriented	90 Degrees Misoriented
Mobility	340cm²/V.sec	230cm²/V.sec	140cm²/V.sec
Vth	1.3V	1.3V	2.0V
S-factor	0.45V/dec	0.44V/dec	0.56V/dec

	Solid Line	Dashed Line	Dotted Line
Channel Orientation	Parallel	45 Degrees Misoriented	90 Degrees Misoriented
Mobility	145cm²/V.sec	105cm²/V.sec	80cm²/V.sec
Vth	-2.1V	-2.7V	-3.8V
S-factor	0.44V/dec	0.54V/dec	0.89V/dec

Figure 2. Comparison of n- and p-channel TFT output curves (top) and characteristics (bottom charts) depending on the relative orientation of the channel with the direction of solidification.

in reduction of the electric field in the drain. Reduction in the electric field decreases the mobility on the one hand, but also decreases electron-hole pair production in the drain region arising from impact ionization, resulting a reduced hot carrier stress effect on the TFT performance. In devices parallel to the grain boundaries, the mobility values decreased from 325 cm²/Vsec to 187 cm²/Vsec after stress. In the devices perpendicular to the grain boundaries, the decrease was from 129 cm²/Vsec to 120 cm²/Vsec. Hot carrier stress did not have any significant effects on sub-threshold slope or threshold voltage.

The superior hot-carrier performance of devices fabricated on SLS-processed films with channels parallel to the grain boundaries, as compared to ELA-processed films - despite the greater number of grain boundaries in the latter – that was reported in Reference 6 is attributed to the improved surface roughness of SLS-processed films compared with that of ELA-processed films. Surface roughness is considered to be a source of defect states that contribute to the hot-carrier effect.

SUMMARY

We have confirmed that the number of grain boundaries in the channel region indeed has an effect on the device performance. The microstructure of the SLS-processed directionally solidified material enabled us to control the number of grain boundaries in the channel region, and to observe drastic decreases in the device performance with increasing grain boundary density. This dependence of device performance on relative orientation of the channel with respect to long unidirectional grains and grain boundaries is in agreement with results that have been reported by others [8].

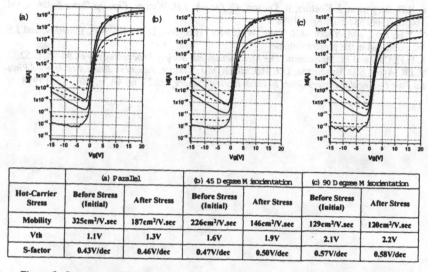

Hot-Carrier Stress	(a) Parallel		(b) 45 Degree M isorientation		(c) 90 Degree M isorientation	
	Before Stress (Initial)	After Stress	Before Stress (Initial)	After Stress	Before Stress (Initial)	After Stress
Mobility	325cm²/V.sec	187cm²/V.sec	226cm²/V.sec	146cm²/V.sec	129cm²/V.sec	120cm²/V.sec
Vth	1.1V	1.3V	1.6V	1.9V	2.1V	2.2V
S-factor	0.43V/dec	0.46V/dec	0.47V/dec	0.50V/dec	0.57V/dec	0.58V/dec

Figure 3. Output curves (top) and device characteristics (bottom chart) relating to hot carrier stress performance of devices according to the relative orientation of the channel with the direction of solidification. Solid lines are the output curves before stress, and dashed lines are the output curves after stress. Stress conditions: Vg = Vth, Vd = 15V, Stress time = 1 hour

However, increasing grain boundary density mitigates the degradation of performance due to hot-carrier stress, resulting in better device reliability. Similar grain-boundary density-dependence of the hot-carrier stress performance has been observed in conventional ELA poly-Si TFTs, and can be remedied by the implementation of a gate overlapped lightly doped drain (GOLDD) device architecture [9].

REFERENCES

1. J. Seto, *Journal of Applied Physics*, Vol. 46, no.12, December 1975
2. J.S. Im and R.S. Sposili, *Crystalline Si Films for Integrated Active Matrix Liquid Crystal Displays (review article), MRS Bulletin*, March 1996
3. R.S. Sposili and J.S. Im, *Applied Physics Letters*, Vol. 69, no.19, (2864) 1996
4. M.A. Crowder, P.G. Carey, P.M. Smith, R.S. Sposili, H.S. Cho, and J.S. Im, *IEEE Electron Device Letters*, Vol. 19, no. 8, (306), August 1998
5. A.B. Limanov, V.M. Borisov, A.Yu.Vinokhodov, A.I. Demin, A.I. El'tsov, Yu.B. Kiirukhin, O.B. Khristoforov, Development of Linear Sequential Lateral Solidification Technique to Fabricate Quasi-Single-Crystal Super-Thin Si Films for High-Performance Thin Film Transistor Devices, *Perspectives, Science, and Technologies for Novel Silicon on Insulator Devices, 55-61, P.L.F. Hemment et al. © 2000 Kluwer Academic Publishers*

6. R.B. Bergmann, J. Koehler, R. Dassow, C. Zaczek, J.H. Werner, *Physica Status Solidi A*, vol. 166, no.2, pp. 587-602, April 1998
7. Y.H. Jung, J.M. Yoon, M.S. Yang, W.K. Park, H.S. Soh, H.S. Cho, A.B. Limanov, and J.S. Im, published in the present volume
8. D.H. Choi and M. Matsumura, *The Electrochemical Society Proceedings*, Vol. 94-35, p. 52
9. J.R. Ayres, S.D. Brotherton, D.J. McCulloch, M.J. Trainor, *Japanese Journal of Applied Physics Part I – Regular Papers Short Notes & Review Papers*, vol.37, no.4A, p.1801, April 1998

AUTHOR INDEX

SUBJECT INDEX

Printed in the United States
By Bookmasters